Springer Series on Atomic, Optical, and Plasma Physics

Volume 95

The Springer Series on Atomic, Optical, and Plasma Physics covers in a comprehensive manner theory and experiment in the entire field of atoms and molecules and their interaction with electromagnetic radiation. Books in the series provide a rich source of new ideas and techniques with wide applications in fields such as chemistry, materials science, astrophysics, surface science, plasma technology, advanced optics, aeronomy, and engineering. Laser physics is a particular connecting theme that has provided much of the continuing impetus for new developments in the field, such as quantum computation and Bose-Einstein condensation. The purpose of the series is to cover the gap between standard undergraduate textbooks and the research literature with emphasis on the fundamental ideas, methods, techniques, and results in the field.

More information about this series at http://www.springer.com/series/411

Shi Nguyen-Kuok

Theory of Low-Temperature Plasma Physics

Springer

Shi Nguyen-Kuok
Laboratory of Plasma Physics
National Research University "MPEI"
Moscow
Russia

ISSN 1615-5653 ISSN 2197-6791 (electronic)
Springer Series on Atomic, Optical, and Plasma Physics
ISBN 978-3-319-43719-4 ISBN 978-3-319-43721-7 (eBook)
DOI 10.1007/978-3-319-43721-7

Library of Congress Control Number: 2016949105

Printed on acid-free paper

This Springer imprint is published by Springer Nature
The registered company is Springer International Publishing AG
The registered company address is: Gewerbestrasse 11, 6330 Cham, Switzerland

Preface

Presently, low-temperature plasma is widely used in science and engineering. Many technological problems such as cutting and welding, melting of metals, welding and spraying of refractory, wear-resistant, corrosion-resistant coatings, obtaining of spherical and dispersion of powders, thermal surface treatment of refractory materials, plasma polishing of optical glass and metals, recycling of heat-resistant plasma chemical waste and others may be successfully solved using low-temperature plasma. Different types of devices for obtaining of plasma flows are developed—plasma torches, where the high-frequency induction (RF), the high-frequency capacitance, ultrahigh-frequency (UHF) and optical discharges are used as the plasma sources.

Effective use of these types of plasma torches requires extensive studies of plasma processes in order to: identify their laws; receive the plasma stream with the desired properties; determine the optimal operating conditions and the development of new structures of plasma torches; direct control of the plasma; and others. These problems can be solved on the basis of experimental studies and mathematical modeling of processes in the plasma.

Field experiments (with the ability to implement them) give detailed information about the plasma. Many phenomena in plasma have been predicted by theoretical calculations, and discovered and confirmed experimentally, in particular, the results of field experiments can be used for comparison with the experimental data in order to analyze the physical processes and the direct control of the plasma.

Unfortunately, many experiments are expensive and experimental methods are still far from perfect. The errors of the experimental methods are often large enough, and often the experimental data of different authors have large discrepancies and contradictory. Some experiments are difficult to carry out in conditions of low-temperature plasma, and their data provide only its integral characteristics.

An important problem of determining the spatial distributions of quantities such as temperature and composition of the plasma, gas flow rate, the electromagnetic field inside the plasma flow and others, their relationships with the external parameters of the plasma torch is a difficult experimental problem. The solution of this problem becomes more difficult when you study the properties of multivariate, nonequilibrium, transient, nonsymmetric plasma flows that is determined solely by a wide variety of physical processes in the plasma and by the complex nature of its components interaction.

Such difficulties in obtaining experimental data for a wide range of plasma flows parameters make actual the development of theoretical modeling techniques.

Theoretical studies of the plasma by mathematical simulation methods allow to eliminate many of the natural experiment limitations, complement and extend the possibilities in the study of plasma, as well as replace complex, expensive, and sometimes impracticable principal experiment. Based on the developed mathematical models you can study the physical processes occurring in the plasma, set optimal modes of plasma sources, to determine changes of the external adjustable parameters, measuring point and the minimum number of measured physical quantities needed for solving the problem of plasma control. In particular, progress in the study of plasma and use cannot be without an in-depth theoretical studies.

One of the main tasks of theoretical studies of plasma processes is to obtain local physical parameters of the plasma and the establishment of their links with external power and the geometric parameters of the plasma torch. Typically, in engineering design it is required to make some form of temperature distribution and plasma flows, corresponding to the special conditions of practical use. This is typically achieved by altering the current, plasma gas flow rate, the operating mode of the generator, and the geometric characteristics of the plasma torch, i.e., only control the external parameters of the plasma torch.

Significant progress in the use of the plasma can be achieved by providing direct control of plasma flows. This requires a structural parametric identification of the plasma. In this case, you first need to examine and define the properties and parameters of the plasma, to find the place of measurement, to build functional relationships, etc. This problem can be solved with the help of mathematical modeling as a real object for study using different mathematical models describing physical processes in the plasma. However, the theoretical description of the plasma is often carried out with the help of highly nonlinear differential equations in partial derivatives. The solution of these equations belongs to the class of rigid and ill-conditioned problems, and is a serious mathematical and information problem.

The obtained results allow us to establish the flow pattern and gas heating, of energy in the plasma, to determine the area of cold channel, thermal and ionic imbalance in the plasma region preference models, create the necessary conditions for solving the problem of structural parametric identification and on the basis of the decision to make the transition to the system automatic control of plasma parameters. You can also use the results as the necessary material for the study and design of plasma torches, optimization of various plasma processes and the creation of simulation models of plasma, which have been free from stability problems of computational procedures.

To learn the basics of theory of low-temperature plasma readers are invited to first familiarize themselves with the contents of the first five chapters. In Chap. 1 the general approach to the description of plasma processes, the basic model of the plasma, their assumptions and applicability are outlined. Models and cross sections for the interaction of particles in the plasma are considered using the classical model of elastic collisions in Chap. 2, and with the help of quantum-mechanical theory of scattering particles in Chap. 3.

Determinations of the composition, thermodynamic functions, transport coefficients of the plasma on the basis of model and the mean free path are given in Chap. 4. In Chap. 5 the solution of the Boltzmann kinetic equation by the Chapman–Enskog approach and calculation of the transport coefficients of the plasma are provided.

In Chap. 6 the main method of control volume for solving the problems of heat transfer and plasma dynamics is presented along with the various numerical schemes (methods) of solutions, their analysis in terms of accuracy, and the ease of implementation. Stability analysis and determining the cause and the remedy of the instability are also discussed in this chapter. We introduce the discrete analogues of various equations of plasma mathematical models. The structural organization of the computational procedures for the simulation of plasma processes is also given.

In Chap. 7 the methodology and the main results of the simulation of the RF plasma torches is set out. As an addition to those obtained in Chap. 6 algorithms, there are the method of calculating the two-dimensional electromagnetic field in the RF plasma torches and simulation of the RF plasma torches based equilibrium and nonequilibrium models.

Similarly, in Chap. 8 the methodology and the main results of the simulation of Arc plasma torches, different design and technological applications (Arc in the channel for cutting and spraying, free Arc welding and melting of metals, etc.) are presented.

In Chap. 9 the modern models and algorithms for the calculation of the near-electrode processes taking into account the working conditions of the majority of electric Arcs are described. The calculations for thermionic cathodes based on pure and thoriated tungsten in a wide range of parameters of the cathode and plasma, as well as algorithms for calculating the heating of the cathode and anode processes are provided.

In Chap. 10 the issues of heat transfer and dynamics of solid particles in plasma, model of heat and motion of particles in the plasma and the main results of the calculations with respect to the jet of the RF plasma torches are described. The results of the interaction model and the dispersed flow with the plasma are given.

Description of the experimental stand, the main methods of diagnosis, and the results of measurements of plasma parameters in the induction and Arc plasma torches are described in Chap. 11.

I would like to specially thank Profs. E. Son and S. Bobashev for carefully reading the manuscript and for helpful comments during the preparation of this book, as well as the staff of the Department of Electrical Engineering and Energy, Saint-Petersburg State Polytechnic University and the General Physics and Nuclear Fusion, National Research University «Moscow Power Engineering Institute» for the help in the preparation of material of the book.

This book was written after my many years of working in the field of low-temperature plasma physics. However, this would not have been possible if it were not for many years of creative collaborations with Prof. S. Dresvin. He was who introduced me to the world of low-temperature plasma physics, when I was still a student of the Electromechanical Faculty of St. Petersburg State Polytechnic University.

I would like to emphasize the special contribution to my creative extension of the work of the Prof. O. Sokolov. During the many sleepless «White nights» in St. Petersburg, we had a conversation with him about a systematic approach to solve a multidimensional system of nonlinear differential equations. It is a great idea that the internal feedback about physical processes found their realization in the simulation model for a sustainable solution of a system of differential equations for plasma.

I have many excellent professors, colleagues, friends, my talent graduate students in colleges and universities of St. Petersburg and Moscow, and I would like to emphasize their help and support in the process of working on the material of this book.

I am infinitely grateful to my parent, my wife, and my sons for their help and support. Their daily smiles are the driving motor of the writing of this book.

Hence, this book was written on behalf of all these wonderful people.

Moscow, Russia Shi Nguyen-Kuok

Reviewers

Dr. Sc., Professor, Corresponding Member of the Russian Academy of Sciences E. Son (Joint Institute of High Temperature of Russian Academy of Sciences); Dr. Sc., Professor, S. Bobashev (Ioffe Physic-Technical Institute)

Shi Nguyen-Kuok
Theory of Low-Temperature Plasma Physics, 2017 - p.
ISBN

The basic approaches to the theoretical description of the plasma, mathematical models, modeling techniques, numerical methods, and modes of ensure the stability of solutions of plasma equations systems are represented in this book. The main methods of calculating the cross sections of plasma particles interaction and solution of the Boltzmann kinetic equation for determine the transport coefficients of the plasma are also represented in the book. The results of calculations of the composition, thermodynamic properties, transport coefficients, the particle interaction cross sections of the equilibrium and two-temperature nonequilibrium plasma. The results of simulation and calculation of plasma parameters in RF and Arc plasma torches are presented. The complex physical processes in the induction plasma and Arc plasma, internal and external parameters of the plasma torches, near-electrode processes, heat transfer and the flow of solid particles in the plasma are considered in the book etc.

The book is intended for professionals involved in the theoretical study of Low-temperature plasma and design of plasma torches, and will be useful for students and postgraduates of corresponding profiles.

Contents

Chapter 1
The Theoretical Basis of the Low-Temperature Plasma

1.1 The Basic Approaches to the Description of Plasma

There are three approaches to the theoretical description of plasma and plasma processes: the microscopic description using the drift approximation; the statistical description based on the solution of the Boltzmann equation, together with Maxwell's equations for the description of the field, and the macroscopic description of the equations of magnet- and gas dynamics (MGD).

The microscopic description reduces to determining the tracks of particles in the plasma fields, produced by the particles and external sources. At the same time completely neglect the mutual collisions of particles. Obviously, this simple approach does not allow getting a complete quantitative description of the properties of plasma, but the results of this method are clear, so it largely enables a qualitative understanding of many phenomena in plasma physics. For example, microscopic description applied to the ensemble of charged particles, allows establishment the presence of macroscopic charges and currents, which create the additional magnetic and electric fields, etc.

The basis of the statistical description are the Boltzmann kinetic equation with respect to the distribution function and the Maxwell equations for the description of the field, and the particle distribution function $f(\vec{r}, \vec{v}, t)$ in the general case is seven-dimensional, and the kinetic equation should be drawn up for each ith type of plasma particles. In the case of elastic interaction of particles in plasma the kinetic Boltzmann equation has the form:

$$\frac{\partial f_i}{\partial t} + \vec{v}_i \cdot \frac{\partial f_i}{\partial \vec{r}} + \vec{F}_i \cdot \frac{\partial f_i}{\partial \vec{v}_i} = \sum_j \int_0^{\infty} \int_0^{2\pi} \int_0^{\infty} \left[f_i(v_i') f_j\left(v_j'\right) - f_i(v_i) f_j(v_j) \right] g_{ij} b \mathrm{d}b \mathrm{d}\varepsilon \mathrm{d}v_j \tag{1.1}$$

S. Nguyen-Kuok, *Theory of Low-Temperature Plasma Physics*,
Springer Series on Atomic, Optical, and Plasma Physics 95,
DOI 10.1007/978-3-319-43721-7_1

Here $f_i(v_i), f_j(v_j), f_i(v_i'), f_j(v_j')$—the particle distribution function of *i*th and *j*th type before the collision and after it, respectively; v_i, v_j, v_i', v_j'—the velocities of particles of *i*th and *j*th type before the collision and after it, respectively; $\vec{F}_l = \frac{e\vec{E}}{m_l}$—the external force acting to the unit of mass of *i*th type particles; $\vec{E}$—the electric field intensity; m_l—the mass of *i*th type of particles; g_{ij}—the initial relative velocity of *i*th and *j*th types of particles; *b*—the impact parameter; ε—the azimuth angle of orientation of the particle scattering plane.

In solving the Boltzmann kinetic equation some mathematical difficulties arise because of the need to determine the collision integral in the right side of (1.1), that, in its turn, depends on the specific conditions of plasma and the distribution function, i.e., from the solution of the kinetic equation.

In many practical problems, the distribution function is close to Maxwell equilibrium distribution function ($\vec{F}_l = \frac{\partial f_i}{\partial t} = \frac{\partial f_i}{\partial \vec{r}} = 0$).

$$f_{\mathrm{M}}(v) = 4\pi v^2 n \left(\frac{m}{2\pi kT}\right)^{3/2} \exp\left[-\frac{mv^2}{2kT}\right] \tag{1.2}$$

Here *n*, *m*, *v*—the concentration, mass, and velocities of particles, respectively; *T*—the temperature of particles; *k*—the Boltzmann constant.

At the same time, the detailed information about the distribution function of particles is not always need. In solving the practical problems not the distribution functions, but such integral quantities as temperature, velocity, pressure, obtained by averaging of these parameters for the whole types of particles, are typically of great interest. In general, for the calculation of these quantities it is necessary to solve the Boltzmann kinetic equation and to find the distribution function of the plasma particles, then to average and to determine the unknown quantity. However, this direct method is rather difficult because of the mathematical complexity of the Boltzmann equation solution, and in practice it is preferable to solve this problem in a different way.

It is well known that the equations of magnet- and gas dynamic can be obtained by the average of the kinetic equation, such as the continuity equation, the momentum transfer equation, the equation of energy conservation, and others. The most of mathematical models describing plasma processes depending on the specific physical conditions is constructed by these equations. These models are called magnet- and gas dynamic (MGD or MHD—magnetic- and hydro- dynamic) models. In some practical applications for study the most important processes and building a simplified model it is necessary to neglect by various phenomena because of the complexity of the processes occurring in the plasma. Such MGD models are often used in the analysis of individual physical processes, or in determining the degree of influence of external parameters on the plasma properties. Some conservation laws with some preset known variables are usually used in the open model. So the number of unknown variables can be kept to a minimum, and many additional variables can be neglected or can be determined by indirect or special way.

For example, to determine the temperature field of the arc in a long cylindrical channel and the heat flux, carried away by convection, one can apply the one-dimensional model of mass, current, and energy balance. The axial component of velocity in the channel is determined from the total gas flow rate, and a one-dimensional electromagnetic field—from a given current and the estimated temperature field. Based on this model, a qualitative picture of gas heating in a long channel of arc can be obtained, the temperature, heat flows, and distribution of the electromagnetic field inside the channel can be found. Of course, such a greatly simplified model does not allow determination the heating degree of the individual components of plasma, as well as consideration the near-electrode processes and analyze their impact for the formation of the physical processes in the Arc column.

Here are the other examples. To estimate the heating degree of the injected gas according to its quantity and the predetermined value of electrons temperature, the one-dimensional model of gas heating in half-space is widely used. To determine the composition and properties of plasma the model of thermodynamic equilibrium is applied. In this model the influence of the electromagnetic field to the distribution of the various components of the velocity and excitation degree are neglected, and the emission is optically thin for the plasma region where the mean free path of a photon is more than the characteristic size of the plasma. Such similar models are often used in practice and are very useful, especially when you want to obtain quickly a qualitative picture of the different physical processes, occurring in plasma, to analyze the influence degree of external parameters of plasma torches on plasma properties, and to align them with the power supply, to determine the optimal area for regulation of plasma torches parameters.

In some cases, the accuracy of the simplified models may not meet to the requirements due to many unaccounted factors, and the plasma parameters can be very different from real ones. Therefore, it is necessary to introduce some limitations of applicability of various models of plasma processes. Historically, the original models for description of plasma are based on the notion of thermodynamic equilibrium of plasma.

Obviously, the plasma in laboratory and in nature can almost never be in a complete thermodynamic equilibrium. However, it is extremely important to know the ratio in case of thermodynamic equilibrium. First, many of the plasma are in a state which is not much different from the complete thermodynamic equilibrium. This condition is called, for example, the local thermodynamic equilibrium (LTE), when the internal and external energy states of individual particles in plasma are the same as for the complete thermodynamic equilibrium, but the blackbody radiation is completely absent or weakened. Second, very often the ratio for full equilibrium, are applied by using the principle of detailed balance, which means that in a thermodynamic system in a state of complete equilibrium, each process has the same velocity as the reverse one. The number of important relations for the plasma in this state should be called.

1.1.1 The Plasma Relations for the Full Thermodynamic Equilibrium

1. *Equality of temperatures* for all groups of particles: electrons T_e, ions T_i, atoms T_a, molecules T_m: $T_e = T_i = T_a = T_m = T$.
2. *The quasi-neutrality of plasma* in a volume of space with radius more than Debye radius r_D:

$$n_e = \sum_{s,z} (z-1) n_{s,z-1} \tag{1.3}$$

The summation in the right side of (1.3) is over all stages of ionization z and on all elements of s that are present in the plasma.
3. *The Boltzmann distribution of particles* in the excited levels. The ratio of the particles of this type at the upper kth level of energy to their concentration on the lower ith level is determined by the Boltzmann's formula:

$$\frac{n_k}{n_i} = \frac{g_k}{g_i} \exp[-\Delta E_{ki}/(kT)], \tag{1.4}$$

Here g_k, g_i—the statistic weights (degeneracy) for kth and ith levels; ΔE_{ki}—the difference of energies between the kth and ith levels.
For atoms and atomic ions the statistic weights g_i that defines a degenerate of levels depend on the quantum numbers of the total orbital momentum L, the total spin S and total angular momentum of the electrons J_i:

$$\begin{aligned} g_i &= (2L+1)(2S+1), \text{ for the } L\text{–}S \text{ term} \\ g_i &= 2L_i + 1, \text{ for } i\text{th level with } J_i \end{aligned} \tag{1.5}$$

In the literature, for example in tables of Moore [1], almost all elements and values of J_i and ionization degrees E_i can be found.
Accordingly, the ratio of the concentration of particles n_i on the ith quantum level with the energy E_i and the total concentration of particles of the same kind will be:

$$\frac{n_i}{n} = \frac{g_i}{U_{in}(T)} \exp[-E_i/(kT)] \tag{1.6}$$

The statistic sum $U_{in}(T)$, internal to the bound states $U(T)$ is defined as:

$$U_{in}(T) = \sum_i g_i \exp[-E_i/(kT)] \tag{1.7}$$

In (1.7) the summation is over all levels. However, the levels that are very close to the ionization are not threshold, may not be realized, because the quantum orbits of highly excited energy states perturb by the surrounding particles. Energy of the last particle level that must be taken into account is $E_i = E_i^+ - \Delta E_i^+$, ΔE_i^+—the reduction of the ionization potential. Practically, the ground state and a large number of levels near the ionization limit make significant contribution to the statistical sum. It is convenient to carry the calculation of statistical sums for three components: the statistical weight of the ground state g_1; the sum that includes the weights of all the lower levels (from 2 to m); the sum of the weights of highly excited hydrogen-reduced levels up to the ionization limit.

$$U_{z-1}(T, n_e) = g_1 + \sum_{i=2}^{m} g_i \exp[-E_i/kT] + \frac{2}{3}\left(\frac{z^2 E_H}{\Delta E_{z-1}^+}\right)^{3/2} \exp\left[-E_{z-1}^+/kT\right] \sum_k g_{z,k} \exp\left[-E_{z,k}/(kT)\right] \quad (1.8)$$

Here E_H is the ionization energy of hydrogen atoms.

The third term in the right side of (1.8) is generalized to the case where the system of atomic levels is based on the different kth quantum levels of the original ion. It should be noted that in many cases the static weight of the ground state g_1 is a good approximation for the statistical sum of the weights. The statistical sum of electrons, atoms, and ions of single, double and triple ionization are given in "Appendix A.1".

The internal statistical sum of the weights of the molecules is determined by the kind of internal degrees of freedom: $U_{in} = U_e U_v U_r$, here U_e, U_v, and U_r—the electronic, vibrational, and rotational statistical sum of the weights of the molecules respectively.

Determination of electronic sum of the weights U_e is carried out in the same way as for atoms. Calculation of the vibrational statistical sum of the weights is increasingly done in an approximate of description of molecular vibrations by the harmonic oscillators. For diatomic molecules $U_v = \frac{1}{1-\exp(-1.44v/T)}$, here v, cm^{-1}—the frequency of oscillator.

In practice the model of rigid rotor is most commonly used during the calculating the rotational static sum of weights. So, for diatomic molecules $U_r = \frac{kT}{\sigma h c B_r}$, here B_r—the rotational constant; $\sigma = 1$ for heteronuclear and homonuclear molecules; h—the Planck's constant; c—the speed of light in vacuum. The calculation of the partition function of polyatomic molecules can be found in the special books.

Note that there is a statistical sum of the weights of the translational motion of particles with mass m that is defined as $U_{tr} = \frac{2\pi mkT}{h^2}$. The total sum of the statistical weights of the internal and translational motions of the particles $U(T) = U_{tr} U_{in}$.

4. *Implementation of the law of acting mass* for the ionization equilibrium in plasma. If you compare the populations of two successive stages of ionization particles of sth type, you will see that the application of Boltzmann formula (1.4)

to the ionization particles leads to the equation of Saha for the ratio of the concentration of particles on the $(z-1)$ stage and the concentration of particles on the z stage $n_{s,z}$.

$$\frac{n_e n_{s,z}}{n_{s,z-1}} = 2\frac{U_{s,z}(T)}{U_{s,z-1}(T)}\left(\frac{2\pi m_e kT}{h^2}\right)^{3/2}\exp\left(-\frac{E^+_{s,z-1} - \Delta E^+_{s,z-1}}{kT}\right) = S_{s,z-1}(T, n_e) \tag{1.9}$$

Here $U_{s,z}(T)$, $U_{s,z-1}(T)$—the statistical sum of the weights for the particles of sth chemical type for two stages of ionization zth and $(z-1)$-st; the value $z = 1$ corresponds to the neutral particle, $z = 2$—to the once ionized particle, etc.; $E^+_{s,z-1}$—the ionization potential of the particles of sth chemical grade of the zth stage.

5. *The reduction of the ionization potential* for the particles of sth chemical grade on the zth stage is $\Delta E^+_{s,z-1} = z\frac{e^2}{4\pi\varepsilon_0 r_D}$, here e—the charge of electron; ε_0—the electrical constant.
6. *The radius of the Debye screening* $r_D = \sqrt{\frac{\varepsilon_0 kT}{e^2\left[n_e + \sum_{s,z}(z-1)^2 n_{s,z-1}\right]}}$. The summation in this formula is carried out for all zth stages of ionization and all the sth chemical elements.
7. *The Maxwell velocity distribution* for each ith grade of plasma particles:

$$f_M(v_i) = 4\pi v_i^2 n_i \left(\frac{m_i}{2\pi kT}\right)^{3/2}\exp\left[\frac{-m_i v_i^2}{2kT}\right] \tag{1.10}$$

8. *The equation of state or the Dalton's law* for partial pressures. For low-density plasma the equation of state of an ideal gas is right. In plasma physics, this equation for a unit volume can be written as:

$$p = \left(n_e + \sum_{s,z} n_{s,z-1}\right)kT \tag{1.11}$$

The summation in the right side of (1.11) is carried out for all zth stages of ionization and all the sth chemical elements.

9. *The law of Kirchhoff.* Photons have integer spin and obey to the statistics of Bose–Einstein. If photons and plasma have the same temperature T, then according to this statistics, the number of photons, distributed in plasma: $n_\nu = \frac{1}{\exp[h\nu/(kT)]-1}$. In this case, the plasma radiation has a continuous spectrum of a blackbody and the equilibrium value of the spectral density of radiation and is given by equation of Planck.

$$\left.\begin{aligned} B_\nu(T) &= \frac{2h\nu^3/c^2}{\exp[h\nu/(kT)]-1} \\ B_\lambda(T) &= \frac{c}{\lambda^2}B_\nu(T) = \frac{2hc^2/\lambda^5}{\exp[hc/(\lambda kT)]-1} \end{aligned}\right\} \qquad (1.12)$$

Here λ—the wavelength.

The volumetric spectral density over all solid angles Ω is described as

$$\left.\begin{aligned} u_\nu &= \frac{1}{c}\int_0^{4\pi} B_\nu d\Omega = \frac{8\pi h\nu^3/c^3}{\exp[h\nu/(\lambda kT)]-1} \\ u_\lambda &= \frac{1}{c}\int_0^{4\pi} B_\lambda d\Omega = \frac{8\pi hc/\lambda^5}{\exp[h\nu/(\lambda kT)]-1} \end{aligned}\right\} \qquad (1.13)$$

The integral bulk density of blackbody radiation is given in accordance with the Stefan–Boltzmann law: $u = \int_0^\infty u_\nu d\nu = aT^4$, here $a = 7.56 \times 10^{-16}$ J/(m^3 K^4). Consequently, the integrated flux of radiation from the surface of blackbody:

$$P = \frac{cu}{4} = \frac{ca}{4}T^4 = \sigma T^4$$

Here the coefficient of the Stefan–Boltzmann $\sigma = 5.67 \times 10^{-8}$ W/(m^2 K^4). The maximum intensity of blackbody radiation exists at $\lambda T = 0.2898$ cm K (Wien's displacement law).

In scientific literature in addition to the considered model of the complete thermodynamic equilibrium the most known models are: the model of LTE, the partial local thermodynamic equilibrium (PLTE), the coronal approximation model (CAM), the collision radiation model (CRM). We restrict ourselves to a qualitative description of the plasma models, details can be found out in [2–4].

1.1.2 The Model of Local Thermodynamic Equilibrium in Plasma

The complete thermodynamic equilibrium plasma is rarely observed, but it is often in a state close to LTE. The term LTE was originally intended to describe the state of the plasma in the inner parts of the stars, where the radiation field was close to the equilibrium. In stellar atmospheres the temperature gradient and, consequently, the power flow always exist. In spite of this fact the local state of the plasma can be quite clear described, assuming that the thermodynamic equilibrium takes place. The LTE condition can be often found in laboratory plasmas with the sufficiently high density of electrons, and where the collision processes involving the electrons play a dominant role.

The LTE model generalizes the case when deviations from the thermodynamic equilibrium are relatively small and the model is based on the following assumptions:

1. All of the distribution functions, except the distribution function of radiant energy, are the Boltzmann functions with the temperature T;
2. The volumetric spectral density of radiation is not equal to this parameter for a blackbody u_ν;
3. The concentration of particles in plasma is sufficient to the approximately satisfaction of the detailed balance of particles in each volume and the Maxwell distribution on velocities for all particles—the functional row of the particles distributions in the phase space coordinates is preserved, but the particle concentration, that are average over the distributions, change with the change of the spatial coordinates;
4. In the real cases plasma is optically transparent for many frequency bands, and a significant fraction of the radiation leave the plasma volume free, thereby violating the conditions of complete thermodynamic equilibrium. In other words, in LTE the detailed balance of processes of emission and absorption in the plasma is broken. The speed of collision processes, leading to the settlement and release of different levels of energy of substances have to exceed the applicable rates of radiation processes—it is necessary for the standing of plasma in a state of LTE.

The probability of collision transition should be in an order of magnitude greater than the probability of spontaneous emission. In line with these arguments Griem [5], McWhirter [6], and Wilson [7] independently of one to another brought a simple criterion for the existence of LTE in optically thin plasma with a predominant role of collisions:

$$n_e \geq 10^{14} T_e^{1/2} (E_k - E_i)^3 \tag{1.14}$$

Here, n_e is measured in cm^{-3}, T_e, E_k, E_i—in eV. The most energy "gap" $E_k - E_i$ usually equals $E_2 - E_1$. For the atoms like hydrogen with $T_e = 1$ eV the condition of LTE can be obtained for $n_e \geq 10^{17}$ cm^{-3}. Note, that radiation-dominated plasma in laboratory conditions is very rare. It can only be in stellar atmospheres and in objects of astrophysics, so we do not give the conditions when we can expect the existence of LTE radiation-dominated plasma.

1.1.3 The Model of Partial Local Thermodynamic Equilibrium in Plasma

The model of PLTE corresponds to the further compared with the model of LTE increase of the role of transport processes and external influences. This trend is increasing with decreasing plasma density, especially of electrons. Because of the smallness of the energy transfer δ in one elastic collision, that is proportional to the ratio of masses of the colliding particles (for electron and atom collisions $\delta = 2m_e/m_a$), the electron temperature T_e can be very different from the temperature of the heavy components T—the gas is in a state of PLTE, both types of particles can have Maxwell distribution on velocities in different temperatures. In this case, the populations of the excited levels are still described by the Boltzmann distribution, but at the temperature T_e.

The presence of a set of different temperatures is a characteristic feature of PLTE approximation. It is based on the difference in relaxation times–times of the establishment of the stationary distributions within the different degrees of freedom for particles of each types and, as a consequence, on the difference between the corresponding average energies and temperatures. Accordingly, temperature of electrons, ions, and neutral particles is used to describe their velocity distributions, rotational, oscillation temperatures, and electronic levels to describe the distribution of the particle populations on the corresponding levels and temperatures of ionization to describe the distribution concentrations of ions of one chemical type and different multiplicities. Convection and diffusion of electrons from one area of plasma to another one can also lead to a violation of the ionization equilibrium in plasma, etc. As a result, the usage of LTE and PLTE models for research most of plasmas in channels of plasma torches is very important and we will consider them in details in subsequent sections of this book.

The model of coronal approach has its name because of its use to describe the conditions in the solar crown. It is the rarefied plasma with concentration $n_e = 10^6 - 10^{10}$ cm^{-3} and the average kinetic energy of electrons $\varepsilon_e = 0.1 - 1$ keV. For such case for very small densities of atoms and electrons the deviation from thermodynamic equilibrium is so much that a detailed balance does not hold for almost all processes. A simple description of the state of such plasma is possible only in very special conditions, namely, when:

1. For all optical transitions plasma is optically thin;
2. The electron density and the density of radiant energy are so small that with a probability close to unity, the excited atom is deactivated due to spontaneous emission;
3. The only mechanism of recombination of ionized atoms is photorecombination;
4. Particles (atoms and ions) are ionized only by direct electron impact from the ground state. Virtually, no collision excitation or de-excitation of the particles from the excited levels and processes are carried out in three-body recombination.

If we denote by $S_{z-1,1}$ and $\alpha_{z,1}$—ionization and recombination coefficients from the ground state (the index "1"), the balance between the processes of ionization and recombination will be $n_{z-1,1}n_e S_{z-1,1} = n_{z,1}n_e\alpha_{z,1}$. Hence we can write well-known equation of "coronal" equilibrium $\frac{n_z}{n_{z-1}} = \frac{S_{z-1,1}(T_e)}{\alpha_{z,1}(T_e)}$. The function in the right side is not universal as in Saha equation and depends on the type of atoms or ions.

Similarly balance for the collision excitation and spontaneous emission is $n_{z,1}n_eK_{l<k} = n_{z,k}\sum_{l<k}A_{kl}$, here $K_{l<k}$—the coefficient of excitation in the transition from the ground state to the lth level; A_{kl}—the probability of an optical transition from kth to the lth level in spontaneous emission.

For estimations and determine the conditions of application the model of the coronal approach, the following expression can be used [8]:

$$\begin{aligned} S_{z-1,1}(T_e) &= 2.34 \times 10^{-7}\frac{(z-1)T_e^{1/4}}{E_{z-1}^{+7/4}}\exp\left(-1.16\times 10^4\frac{E_{z-1}^+}{kT_e}\right) \\ \alpha_{z,1}(T_e) &= 2.8\times 10^{-11}\frac{(z-1)^2}{T_e^{1/2}} \\ K_{l<k}(T_e) &= 1.7\times 10^{-4}\frac{f}{(E_k-E_l)T_e^{1/2}}\exp\left(1.16\times 10^4\frac{E_k-E_l}{kT_e}\right) \end{aligned} \tag{1.15}$$

Here f—the oscillator strength for the transition from the lth to the kth level; energies of the corresponding levels of excitation E_k, E_l and ionization E_{z-1}^+ can be expressed in eV; $S_{z-1,1}$, $\alpha_{z,1}$, $K_{l<k}$—in cm^3/s.

In addition to use in the spectroscopy of the solar crown, this model was analyzed with respect to its application to more dense plasmas like plasma in TOKOMAK reactor and beam discharges.

The collision–radiation model is applied to describe the plasma with a strong disequilibrium when the above models can't be used. This model includes a scheme of full description of particles levels together with the equations for the kinetics of the continuous spectrum of energy and equations of electrodynamics to find the field. The balance of the populations of the quantum levels of particles of each chemical types is described by the general kinetic equations like

$$\frac{dn_k}{dt} = \sum_l (n_l\nu_{lk} - n_k\nu_{kl}) + \sum_\alpha S_{\alpha k} + Q_k \tag{1.16}$$

Here ν_{lk}, ν_{kl}—the frequencies of collision and radiation transitions between the kth and lth levels; $S_{\alpha k}$—the source function that describes the creation and annihilation of particles with the concentration n_k and with creation the particles of the other chemical type; Q_k—the traffic carried by different particles.

Thus, in the considered plasma models the Boltzmann and Saha equations are often applied to plasmas with complex processes of collision and radiation, leading to different degrees of deviation from the thermodynamic equilibrium.

The above-mentioned plasma models, of course, can be regarded as special cases of CRM. To solve the problems of theoretical analysis plasma in laboratory and astrophysics is more often expressed with regard to its deviation from the temperature, ionization, and radiation equilibrium with the other specific conditions of collision–radiation processes.

Most equations in these models are integral–differential with distributed parameters. They can be obtained by converting the kinetic Boltzmann equation to the system of equations for the moments corresponding to different conservation laws; for example, for a system of 13 MHD equations [2, 9] allows description the evolution of plasma on a macroscopic scale. On the basis of kinetic equations enough rigorous theory of transport phenomena for multicomponent plasma is now established that justify the hydrodynamic approximation and represent the appropriate magnet-hydrodynamic equations for plasma.

Among the conditions of application of the hydrodynamic approximation the usual restrictions for gas dynamics includes. The main among them is the requirement that the mean free path and the time between collisions for all plasma components should be much less than the characteristic spatial and temporal scales of the macroscopic changes. The fulfillment of these conditions allows distraction from the discrete structure of the plasma to treat it as a continuous medium, significantly simplify the equations of moments of the kinetic theory and use the linear transport relations. The coefficients of proportionality in these equations, as a rule, are corresponding coefficients to the transfer coefficients, and the equations themselves—to the laws of conservation of energy, momentum, and mass.

Thus, the MHD model for plasma description is the most widely used in the study of plasma processes, and a mathematical description adequately meets the challenges of modeling the physical processes occurring in the channel of plasma torch. Because of the wide practical use of MHD models in the study of plasma their basic positions are given in following.

1.2 Model of Equilibrium Plasma

The model of equilibrium plasma is based on the assumption of thermodynamic equilibrium in plasma. The great number of theoretical and experimental data indicates that the conditions of complete thermodynamic equilibrium in plasma (1.3)–(1.13) are almost not carried out for real plasma. However, the model of thermal equilibrium in plasma is easy to calculate the heating of the gas and plasma in the plasma torches. This fact is explained by the relative simplicity of the equilibrium plasma model and lots of real physical factors specific for the plasma. Here are some of them:

The dense thermal equilibrium plasma creates in channels of most plasma torches that operate at atmospheric pressure or close to atmospheric pressure. The current density in this plasma is $j > 10^2$ A/cm^2 and the concentration of charged particles $n_e > 10^{16}$ cm^{-3}. In these cases, the elastic collisions between electrons and heavy particles (ions and atoms) in plasma occurs with a frequency of $\nu > 10^{11}$ s^{-1} and so effective that the difference in temperature groups of electrons and heavy particles is negligible and the assumption of the equality of the temperature for all groups of particles will be justified [10];

The mean free path of the plasma particles is often much smaller than the spatial dimension of its flow that allows us to consider the plasma as a continuous medium having electrical conductivity, thermal conductivity, viscosity and other characteristics, as well as the thermodynamic functions of plasma depend only on its temperature and pressure. In this case, the elementary processes of interaction between groups of particles and between particles of each group are taken into account in determining the transport coefficients and thermodynamic functions in terms of averaged over the distribution function the interaction cross sections;

The plasma radiation is optically thin for a wide range of frequencies and photons go out of its volume free, without participating in the activation of processes such as "recolonization" of the various excited levels of particles.

The model of equilibrium plasma typically consists of: energy equation that express the law of energy conservation; the momentum equations—the law of momentum conservation; Maxwell's electromagnetic equations—the law of the field conservation; the continuity equation for hole gas—the law of mass conservation; the system of equations defining the dependence of the transport coefficients and thermodynamic functions of the plasma on temperature.

1.2.1 The Energy Balance Equation

The energy balance equation is the fundamental equation in the analytical description and mathematical modeling of plasma and plasma torches. It allows obtaining the distribution of the plasma temperature in the channel of plasma torch and connects this temperature with external electric circuit parameters: current I, voltage U, channel diameter d, flow of the heated gas G. This circumstance is most important to the plasma torch engineering and technology, because it gives the opportunity to see that external parameters of the plasma torch affect to the temperature and velocity of plasma, and to see the connection between them, it is necessary for solving the problems of regulation and control of plasma processes.

The complete change of the plasma temperature in space and time $\frac{\mathrm{D}T}{\mathrm{D}t} = \frac{\partial T}{\partial t} + (\vec{v}\nabla)T$—the substantial derivative is determined by the combined effect of energy sources and its sinks. The power σE^2 in the form of Joule heating is allocated in the unit of plasma volume. In addition, at the high speed of plasma flow and its large gradients heat is added released to the power source because of the forces of the viscous friction Φ_{diss} (dissipation function).

The retraction of energy from the volume unit occurs due to the influence temperature gradients $\mathrm{div}(\lambda \mathrm{grad} T)$ to the thermal conduction and due to convection associated with the gas movement $\mathrm{div}(\rho\vec{v}H)$, here H—the enthalpy of plasma. The radiation is optically thin and photons leave the plasma volume free without arousing the different degrees of freedom of the plasma particles. Thus, the power loss $U_{\mathrm{rad}}(T)$ due to radiation can be represented as a function of temperature and enter it at the energy equation with the sign "minus":

$$\rho c_p \frac{\mathrm{D}T}{\mathrm{D}t} = \mathrm{div}(\lambda \mathrm{grad}T) + \sigma E^2 - U_{\mathrm{rad}} + \mu \Phi_{\mathrm{diss}} \tag{1.17}$$

Here $\rho, c_p, \lambda, \sigma, \mu$—the density, heat capacity, thermal conductivity, electrical conductivity and viscosity of plasma, respectively; E—the electric field intensity.

For ideal gases, liquids, and plasma at the constant pressure: $c_{\mathrm{p}}\mathrm{grad}T = \mathrm{grad}H$. Substituting this expression into Eq. (1.17), we get another entry of the energy equation with the enthalpy H:

$$\frac{\partial(\rho H)}{\partial t} + \mathrm{div}(\rho \vec{v} H) = \mathrm{div}\left(\frac{\lambda}{c_p} \mathrm{grad}H\right) + \sigma E^2 - U_{\mathrm{rad}} + \mu \Phi_{\mathrm{diss}} \tag{1.18}$$

The synopsis of the energy equation with enthalpy (1.18) is rather than with temperature (1.17), because the enthalpy characterizes all kinds of plasma energy that allows account accurately for all channels and influx of heat removal from the unit of plasma volume. However, the solution of the energy equation with enthalpy is inconvenient because of the fact that the properties of the plasma including the thermodynamic transfer function are often represented as a function not of enthalpy H, but the plasma temperature T.

If $c_p = \left.\frac{\mathrm{d}H}{\mathrm{d}T}\right|_{p=\mathrm{const}} = \mathrm{const}$ is taken at the constant pressure p for a small plasma volume, the energy equation can be written with temperature:

$$\frac{\partial(\rho T)}{\partial t} + \mathrm{div}(\rho \vec{v} T) = \mathrm{div}\left(\frac{\lambda}{c_p} \mathrm{grad}T\right) + \frac{1}{c_p}\left[\sigma E^2 - U_{\mathrm{rad}} + \mu \Phi_{\mathrm{diss}}\right] \tag{1.19}$$

The energy balance equation for the equilibrium plasma, written in the form (1.19), is convenient for solving, and all possible types of stored energy are taken into account through the total heat capacity c_p of plasma.

1.2.2 The Momentum Equations of the Plasma Flow

An important property of plasma generated by plasma torch, in addition to its high temperature T, is a great $\vec{v}$. The plasma velocity in modern plasma torches of various designs in different modes of operation may be from a few meters to several kilometers per second. The high temperature and velocity of plasma provide the basic technological advantages of plasma torches for the metal cutting, spray coating, treatment (melting) of fine refractory particles, plasma chemical synthesis and decomposition of substances, and others.

The movement of the plasma flow is described by the momentum conservation law. Complete change of momentum of the system in time and space is the result of the action of various forces applied to the system: $\vec{g}\rho$—gravity, here $\vec{g}$—the acceleration of gravity; grad$p$—the force of static pressure gradient p; $\vec{F}_B = \vec{j} \times \vec{B}$—the electromagnetic force, here $\vec{j}$—vector conduction current density, $\vec{B}$—the magnetic induction vector of the electromagnetic field; $\vec{F}_{\mu}$—the viscous friction forces.

$$\frac{\partial(\rho\vec{v})}{\partial t} + (\rho\vec{v}\nabla)\vec{v} = \vec{g}\rho - \nabla p + \vec{F}_B + \vec{F}_{\mu} \tag{1.20}$$

Gas medium that include low-temperature plasma are compressible. The viscous friction force is determined by the tangent and some related with volume (in compression and expansion of an elementary volume of plasma) and shear deformations with additional requirements:

$$F_{\mu} = \mathrm{div}\vec{\tau} \tag{1.21}$$

Viscous stress tensor is defined as [3]:

$$\tau_{ij} = \mu\left[\left(\frac{\partial v_i}{\partial x_j} + \frac{\partial v_j}{\partial x_i}\right) - \frac{2}{3}\delta_{ij}\mathrm{div}\vec{v}\right] \tag{1.22}$$

Here v_i, v_j—the velocity components on coordinates x_i, x_j; μ—the plasma viscosity; δ_{ij}—the delta function of Cronecker, $\delta_{ij} = 1$ $i \equiv j$ and $\delta_{ij} = 0$ for $i \neq j$.

The first term in (1.22) is the shear viscous friction, and the second—the second volume friction with the viscosity that characterizes the dissipation of the kinetic energy of heavy components in thermodynamic disequilibrium and volumetric strain [3].

1.2.3 The Continuity Equation of the Plasma Flow

To describe the motion of plasma flow another fundamental law—the law mass of conservation is needed. In general, this law is expressed by the equation of continuity of the entire gas:

$$\frac{\partial\rho}{\partial t} + \mathrm{div}(\rho\vec{v}) = 0 \tag{1.23}$$

1.2.4 The Maxwell's Equations

The Maxwell's equations are the basic equations for a wide variety of different types of electromagnetic problems for plasma torches. In the energy equation σE^2—power of Joule heating in the unit of plasma volume is a function of the source expression and in the momentum equations $\vec{F}_B = \vec{j} \times \vec{B}$—electromagnetic force in plasma is the result of the interaction of the plasma current with the magnetic field. Thus, at first it is necessary to define the electromagnetic field in the channel of the plasma torch solutions for the solving of momentum equations and energy equation. Maxwell's equations include four basic equations:

$$\mathrm{rot}\vec{H} = \sum \vec{J}, \quad \mathrm{rot}\vec{E} = -\frac{\partial \vec{B}}{\partial t}, \quad \mathrm{div}\vec{B} = 0, \quad \mathrm{div}\vec{E} = \frac{\rho_c}{\varepsilon\varepsilon_0} \tag{1.24}$$

Here $\sum\vec{j}$—the sum of the current densities of the plasma: current conduction $\sigma\vec{E}$, current transfer $\rho_e\vec{v}$ and bias current $\partial\vec{D}/\partial t$; $\vec{H}$—the magnetic field strength; $\vec{B} = \mu\mu_0\vec{H}$—magnetic induction; $\vec{E}$—the electric field strength; $\vec{D}$—the electric shift; $\rho_c = e(Zn_i - n_e)$—the density of electrostatic charges; (Z-ion charge; for spatial scale greater then the Debye radius ($r > r_D$) and quasi-neutral plasma ($Zn_i = n_e$) and ($\rho_c = 0$); μ_0, ε_0—magnetic and dielectric permittivity in a vacuum; μ, ε—relative magnetic and dielectric permeability of plasma (in many cases, they take the same values as in a vacuum, i.e., $\mu = \varepsilon = 1$).

1.2.5 The Equations of Equilibrium Plasma Model in a Cylindrical Coordinate System

In most cases, the design of channels is a cylindrical symmetric system. Therefore, for convenience of the solutions, it is necessary to write equations of the model of equilibrium plasma in a cylindrical coordinate system. The equation will be symmetric in a cylindrical coordinate system around the axes r and z:

The energy balance equation:

$$\begin{aligned} \frac{\partial}{\partial t}(\rho T) + \frac{\partial}{\partial z}(\rho v_z T) + \frac{1}{r}\frac{\partial}{\partial r}(r\rho v_r T) &= \frac{\partial}{\partial z}\left(\frac{\lambda}{c_p}\frac{\partial T}{\partial z}\right) \\ &+ \frac{1}{r}\frac{\partial}{\partial r}\left(r\frac{\lambda}{c_p}\frac{\partial T}{\partial r}\right) + \frac{1}{c_p}\left[\sigma E^2 - U_{\mathrm{rad}} + \mu\Phi_{\mathrm{diss}}\right] \end{aligned} \tag{1.25}$$

In (1.25) the dissipative function is:

$$\begin{aligned} \Phi_{\mathrm{diss}} = 2\left[\left(\frac{\partial v_z}{\partial z}\right)^2 + \left(\frac{\partial v_r}{\partial r}\right)^2 + \left(\frac{v_r}{r}\right)^2\right] + \left[r\frac{\partial}{\partial r}\left(\frac{v_\varphi}{r}\right)\right]^2 + \left(\frac{\partial v_\varphi}{\partial z}\right)^2 \\ + \left[\frac{\partial v_z}{\partial r} + \frac{\partial v_r}{\partial z}\right]^2 - \frac{2}{3}\left(\frac{\partial v_z}{\partial z} + \frac{1}{r}\frac{\partial (r v_r)}{\partial r}\right)^2 \end{aligned}$$

In the cases when the velocity of the plasma is much less than the speed of sound, the power of thermal sources is much less electrical power $\Phi_{\mathrm{diss}} \ll \sigma E^2$ due to viscous friction and it can be omitted from the energy equation.

The momentum equations of the gas relative to the three components of velocity v_z, v_r, v_φ:

$$\frac{\partial}{\partial t}(\rho v_z) + \frac{\partial}{\partial z}(\rho v_z v_z) + \frac{1}{r}\frac{\partial}{\partial r}(r\rho v_r v_z) = -\frac{\partial p}{\partial z} + \left[\vec{j}\times\vec{B}\right]_z + 2\frac{\partial}{\partial z}\left(\mu\frac{\partial v_z}{\partial z}\right)$$
$$+\frac{1}{r}\frac{\partial}{\partial r}\left[r\mu\left(\frac{\partial v_z}{\partial r} + \frac{\partial v_r}{\partial z}\right)\right] - \frac{2}{3}\frac{\partial}{\partial z}\left[\mu\left(\frac{\partial v_z}{\partial z} + \frac{1}{r}\frac{\partial(rv_r)}{\partial r}\right)\right] + \rho g_z \tag{1.26}$$

$$\frac{\partial}{\partial t}(\rho v_r) + \frac{\partial}{\partial z}(\rho v_z v_r) + \frac{1}{r}\frac{\partial}{\partial r}(r\rho v_r v_r) = -\frac{\partial p}{\partial r} + \left[\vec{j}\times\vec{B}\right]_r + \frac{2}{r}\frac{\partial}{\partial r}\left(r\mu\frac{\partial v_r}{\partial r}\right)$$
$$+\rho\frac{v_\varphi^2}{r} + \frac{\partial}{\partial z}\left[\mu\left(\frac{\partial v_z}{\partial r} + \frac{\partial v_r}{\partial z}\right)\right] - \mu\frac{2v_r}{r^2} - \frac{2}{3}\frac{\partial}{\partial r}\left[\mu\left(\frac{\partial v_z}{\partial z} + \frac{1}{r}\frac{\partial(rv_r)}{\partial r}\right)\right] \tag{1.27}$$

$$\frac{\partial}{\partial t}(\rho v_\varphi) + \frac{\partial}{\partial z}(\rho v_z v_\varphi) + \frac{1}{r}\frac{\partial}{\partial r}(r\rho v_r v_\varphi)$$
$$= \frac{\partial}{\partial z}\left(\mu\frac{\partial v_\varphi}{\partial z}\right) + \frac{1}{r}\frac{\partial}{\partial r}\left(r\mu\frac{\partial v_\varphi}{\partial r}\right) - \frac{v_\varphi}{r^2}\frac{\partial}{\partial r}(r\mu) - \rho\frac{v_r v_\varphi}{r} \tag{1.28}$$

The continuity equation:

$$\frac{\partial \rho}{\partial t} + \frac{\partial}{\partial z}(\rho v_z) + \frac{1}{r}\frac{\partial}{\partial r}(r\rho v_r) = 0 \tag{1.29}$$

If in the task the analyzed time interval is longer than the time of different processes of ionization, recombination and dissociation in the plasma, and the speed of fluctuations for different physical processes is much less than the reciprocal speed of their setting, the processes in channels of plasma torches can be stationary.

In the specific case *Maxwell's electromagnetic equations* (1.24) can be different. For example, for a two-dimensional electromagnetic field of RF plasma torches because of the cylindrical symmetry of the limit inductor current and conduction currents in the plasma (transfer and offset currents equal to zero), we have the following relations for the components of intensities of the electric and magnetic fields: $E_z = E_r = 0, E_\varphi \neq 0$; $H_\varphi = 0, H_z \neq 0, H_r \neq 0$. Then the Maxwell's equations will be:

$$\frac{\partial E_\varphi}{\partial z} = \mu_0\frac{\partial H_r}{\partial t};\quad \frac{\partial H_r}{\partial z} - \frac{\partial H_z}{\partial r} = \sigma E_\varphi;$$
$$\frac{1}{r}\frac{\partial}{\partial r}(rE_\varphi) = -\mu_0\frac{\partial H_z}{\partial t};\quad \frac{\partial H_z}{\partial z} + \frac{1}{r}\frac{\partial}{\partial r}(rH_r) = 0 \tag{1.30}$$

For two-dimensional electromagnetic field of Arc plasma torches $E_\varphi = 0$, $E_z \neq 0, E_r \neq 0$; $H_\varphi \neq 0, H_z = H_r = 0$, Maxwell's equations have the following form:

$$-\frac{\partial H_\varphi}{\partial z} = \sigma E_r; \frac{\partial E_r}{\partial z} - \frac{\partial E_z}{\partial r} = -\mu_0 \frac{\partial H_\varphi}{\partial t}$$
$$\frac{1}{r}\frac{\partial}{\partial r}(rH_\varphi) = \sigma E_z; \frac{\partial E_z}{\partial z} + \frac{1}{r}\frac{\partial}{\partial r}(rE_r) = 0 \quad (1.31)$$

In addition to energy equation, momentum equations, continuity equation, and Maxwell's equations the model of thermal equilibrium plasma include Saha equation and Dalton's law of partial pressures for determining the composition of the plasma, as well as a system of equations for the calculation of transport coefficients and thermodynamic functions depending on the temperature of the plasma: $\sigma(T)$, $\lambda(T)$, $\mu(T)$, $U_{\text{rad}}(T)$, $\rho(T)$, $c_p(T)$, $H(T)$, and others.

Algorithms for the solution of Eqs. (1.25)–(1.31) are given in a number of textbooks and monographs on the computational mathematics and mathematical physics. There are some features in the solution of such a system. The equations describing the thermal equilibrium plasma (1.25)–(1.31) together with a system of equations for determining the composition, the transport coefficients and thermodynamic functions of plasma form a closed system of nonlinear differential equations, i.e., we have a model with distributed parameters that depend on the solution of equations. Thermodynamic functions and transport coefficients depend on the temperature of plasma that should be found from the energy equation. However, the solution can be found if the solutions of Maxwell's equations (for calculation of specific power σE^2 in plasma) and the momentum equations (for determination of the velocity of plasma particles) are known. But they cannot be solved without knowledge about the temperature field, etc.

Thus, to solve the closed system of equations of the model plasma iterative and recursive procedures that allow for the nonlinearity of the equations and the interconnectedness between these equations should be applied. Experience shows that such a sequential iterative procedure for solving the system of equations for the plasma is very unstable, especially in the global iterative loop in the transition from solutions of an equation to solve the other. A small change in the iterative procedure of solving a single equation often causes large changes in the solution of the following equations, which contributes to instability of the solutions of equations in general. Thus, one of the key issues in modeling highly nonlinear systems has been and remains the question of the sustainability of computing and iterative procedures. In computational mathematics, besides the widely used method of under-relaxation that can partially compensate for the lack of information after each decision is still in general, there is no stable algorithm for solving systems of nonlinear differential equations, including those in the simulation of the plasma.

1.3 The Two-Temperature Model of Nonequilibrium Plasma

The state of thermodynamic equilibrium (1.3)–(1.13), that is required as a basis for constructing a model of equilibrium plasma, is very rare. Application of the equilibrium plasma model for description of the physical processes in plasma torches is not always justified. As it is shown by numerous experiments, plasma in its different manifestations in the channels of plasma torches in nature has some spatial regions where the equality of temperatures of electrons and heavy particles (atoms, ions, molecules) is violated. Also it is found that there is no a single Maxwell's velocity distribution and the Boltzmann's distribution over the population of the excited energy levels for all types of particles, there is no the ionization equilibrium. In such cases, plasma is nonequilibrium, and the model of equilibrium plasma is insufficient for its description.

If the thermal plasma will be considered like a continuous medium with a common temperature, we have deliberately refused to consider two main processes: ionization and heating of the plasma particles. Ionization from the collision process turned into a thermal process described by the law of active mass, and the exchange of energy between electrons, atoms, and ions is generally not included in the energy equation (the law of energy conservation). However, plasma in plasma torch becomes an effective means of thermal impact only if it is heated by heavy components: atoms, ions, molecules. In order to obtain such plasma, a mechanism, that provides efficient energy transfer from electrons to atoms, molecules and ions, is needed. Such mechanisms are the elastic collisions of electrons with atoms, ions and molecules, accompanied by the transfer of energy and momentum. The challenge is that it is required to determine not only the mechanisms of energy in the plasma, but also their most favorable conditions for the flow and the degree of plasma heating subject to further its practical use.

In the two-temperature plasma model it is assumes that the temperature of ionization, excitation, and settlement are close to the kinetic electron temperature T_e and the temperature distribution is identified with the temperature T of the heavy components of the atom–ion gas. The validity of these assumptions was confirmed in many researches [11–13]. However, it should be noted that due to the spatial heterogeneity and temporal changes in the relevant portable and kinetic processes of balance, the overall temperature for individual or all degrees of freedom can not be installed. To describe multistage dynamic plasma systems, in general, it is necessary to introduce several temperatures, concentrations, and distributions [14]. Moreover, the nonequilibrium plasma conditions can lead to a more complicated form of the distribution functions that are impossible to characterize any temperature [12]. Therefore, consistent description of nonequilibrium plasma requires the use of sophisticated methods of physical kinetics that if applied to research problems of gas heating, greatly complicates their decision.

In such a situation it is essential to explore the most typical types of nonequilibrium state and identify the conditions when plasma is in the state of local thermal

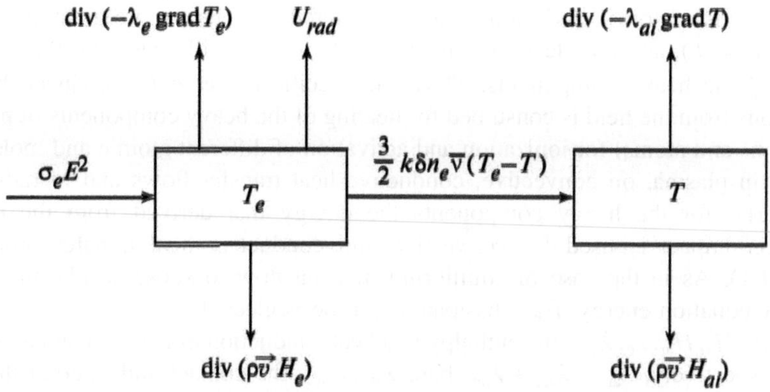

Fig. 1.1 Scheme of energy exchange in the two-temperature plasma model

equilibrium. In this respect, the most interesting is analysis of the thermal and ionization nonequilibrium plasma. Conditions for the creation of such a plasma and its save for a quite long time are determined by the energy balance of the electron gas and heavy components and the processes of ionization and recombination, bipolar diffusion, thermal diffusion, radiation and other processes, occurring simultaneously in plasma.

The scheme of energy exchange in the two-temperature model of nonequilibrium plasma is shown in Fig. 1.1 (for simplicity, the case of the atomic plasma is considered). Specific power of the supplied electric field $\sigma_e E^2$ is directly perceived by the easiest and the most moving electron gas, and atoms and ions are heated primarily as a result of elastic collisions with electrons [4, 13]. Efficiency of energy transfer in one elastic collision is proportional to the ratio of the masses of the colliding particles m_i, m_j:

$$\delta = \frac{2m_i m_j}{(m_i + m_j)^2} \tag{1.32}$$

For the collisions electron–atom: $\delta = 2m_e/m_a$, atom-ion $\delta = 1/2$, etc. The noticeable energy metabolism in a single collision occurs between the particles with similar mass. That is why it is possible to assume that the kinetic temperatures of atoms, ions, and molecules in plasma are the same.

A relatively small fraction of energy transfers in collisions of electron–atom and electron–ion so the local thermal equilibrium establishes separately for the electron gas and for the heavy components (atoms, ions) with temperatures T_e and T, respectively. The presence of a possible low pressure, as well as the convection, diffusion and other transport processes with the continuous energy input from the field to the electron gas can also enhance the thermal nonequilibrium in plasma. This fact makes it possible to introduce into consideration the two-temperature model of thermally nonequilibrium plasma, where the temperature of heavy components (atoms and ions) is considered to be approximately equal ($T_a = T_i = T$), and the electron temperature T_e may be different from the temperature of the heavy components. Thus, the specific power $\sigma_e E^2$ produced by the electrons from the field is consumed for heating of the heavy components of plasma (the ions and atoms) for ionization and activation of different atomic and molecular levels in plasma, on convective, conductive heat transfer flows and to radiation. Similarly, for the heavy components the energy that derived from the elastic electron impact is used for convective and conductive heat transfer processes (Fig. 1.1). As in the case of equilibrium plasma flow to subsonic plasma in the energy equation energy Φ_{diss} dissipation can be neglected.

Here: $H_e, H_{ai}, \lambda_e, \lambda_{ai}$—the enthalpy and heat conduction of electron and atom-ion gas, respectively, $\lambda_{ai} = \lambda_{aic} + \lambda_{int}$, here $\lambda_{aic}, \lambda_{int}$—the contact and internal thermal conductivity of the atom-ion gas; σ_e—the electrical conductivity of electrons; U_{rad}—the power loss due to radiation.

For many frequency bands the radiation can be considered optically thin, and photons can easy go out of the spatial region of the discharge. The kinetic energy of the electrons, transmitted to atoms and ions by elastic collisions per unit of time per unit of volume:

$$\frac{3}{2}k\delta n_e \bar{v}(T_e - T) = b(T_e - T) \tag{1.33}$$

Here: $\bar{v} = \bar{v}_{ea} + \bar{v}_{ei} = \bar{v}_e(n_a Q_{ea} + n_i Q_{ei})$—the sum of the average frequency of electron collisions with atoms and ions with the momentum transfer (the average number of collisions of electrons with atoms and ions per unit of volume and per unit of time); $\bar{v}_e = \sqrt{\frac{8kT_e}{\pi m_e}}$—the average thermal velocity of the electron; n_a, n_i, n_e—the concentration of atoms, ions, and electrons; Q_{ea}, Q_{ei}—the average effective collision cross sections of electron–atom and electron–ion; $b = \frac{3}{2}k\delta n_e \bar{v} = \frac{3}{2}k\frac{2m_e}{m_a}n_e\sqrt{\frac{8kT_e}{\pi m_e}}(n_a Q_{ea} + n_i Q_{ei})$—the coefficient of energy exchange between electrons, atoms and ions.

Equation (1.33) can be used to determine the effectiveness of the heating of heavy components of the plasma and it is a major one in the physics of plasma heating.

The presence of two temperatures in plasma requires the use of two energy equations instead of a single one: for electrons and for heavy components. The relationship between these equations is established through the energy of elastic interaction in collisions between electrons and heavy particles.

Taking into account the convective and conductive transport streams the energy equation of electron gas can be written as follows:

$$\frac{\partial(\rho H_e)}{\partial t} + \mathrm{div}(\rho \vec{v}_e H_e) = \mathrm{div}(\lambda_e \mathrm{grad} T_e) + \sigma_e E^2 - b(T_e - T) - U_{rad} \tag{1.34}$$

Note that the speed of electron gas

$$\vec{v}_e = \vec{v} + \vec{v}_D + \vec{v}_A + \vec{v}_T$$

is the sum of the velocity $\vec{v}$ of total gas flow, the drift of electrons in an electric field,

$$\vec{v}_D = -\frac{\vec{j}}{en_e},$$

the bipolar diffusion

$$\vec{v}_A = -D_{amb} \frac{T_e + T}{2T} \frac{\mathrm{grad} n_e}{n_e},$$

the thermal diffusion [15]

$$\vec{v}_T = -D_T \frac{\mathrm{grad}(T_e + T)}{2T}.$$

Heat transfer due to bipolar diffusion and thermal diffusion of electrons is often expressed through the reactive thermal conductivity and it is taken into account when determining the overall thermal conductivity of the electron gas, $\lambda_e = \lambda_{ec} + \lambda_R$, here λ_{ec}, λ_R—contact and reactive heat conduction of electrons (Sect. 4.3.1). The exit of the drift electron flow energy in the direction of the electric field $\sigma_e E^2$ is determined by the longitudinal component of the electric conductivity ($\sigma_e = \sigma_{e\|}$).

Light and mobile electrons that receive energy from the electric field, play a major role in the processes of ionization, recombination and emission of plasma therefore the ionization energy should be included in the general formula for the electron enthalpy:

$$H_e = \frac{1}{\rho} \left(\frac{5}{2} kT_e + E_i \right) n_e$$

Similarly, in the energy balance of two-temperature plasma model the radiation power takes into account. The power losses on the total radiation of plasma can be logically included in the energy equation for electrons, but not to the energy equation for the heavy components, although the radiation is largely the result of interaction of electrons in the free–free, free–bound and bound–bound transitions. With these assumptions, the energy equation in the steady state for electron gas will be as follows:

$$\mathrm{div}\left[\left(\frac{5}{2} kT_e + E_i \right) n_e \vec{v}_e \right] = \mathrm{div}(\lambda_e \mathrm{grad} T_e) + \sigma_e E^2 - b(T_e - T) - U_{rad} \quad (1.35)$$

In a cylindrical coordinate system and taking into account the cylindrical symmetry of the energy equation of the electron gas (1.35) can be written as:

$$\begin{aligned}&\frac{\partial}{\partial z}\left[\left(\frac{5}{2}kT_e+E_i\right)n_e v_{ez}\right]+\frac{1}{r}\frac{\partial}{\partial r}\left[r\left(\frac{5}{2}kT_e+E_i\right)n_e v_{er}\right]\\&\quad=\frac{\partial}{\partial z}\left(\lambda_e\frac{\partial T_e}{\partial z}\right)+\frac{1}{r}\frac{\partial}{\partial r}\left(r\lambda_e\frac{\partial T_e}{\partial r}\right)+\sigma_e E^2-b(T_e-T)-U_{rad}\end{aligned} \tag{1.36}$$

With the help of (1.33) a source of energy in the energy equation of heavy (atom–ion) components is considered. This equation has the following general form:

$$\frac{\partial(\rho H_{ai})}{\partial t}+\mathrm{div}(\rho\vec{v}H_{ai})=\mathrm{div}(\lambda_{ai}\mathrm{grad}T)+b(T_e-T) \tag{1.37}$$

Here $H_{ai}=\frac{1}{\rho}\frac{5}{2}kT(n_a+n_i)$—the enthalpy of the atom–ion component; λ_{ai}—heat conductivity of atom–ion component. Atom–ion gas moves with the total gas flow velocity. Given the stationary energy equation for atom–ionic component has the form

$$\mathrm{div}\left[\frac{5}{2}kT(n_a+n_i)\vec{v}\right]=\mathrm{div}(\lambda_{ai}\mathrm{grad}T)+b(T_e-T) \tag{1.38}$$

In a cylindrical coordinate system the energy equation for atom–ionic gas is written as

$$\begin{aligned}&\frac{\partial}{\partial z}\left[\frac{5}{2}kT(n_a+n_i)v_z\right]+\frac{1}{r}\frac{\partial}{\partial r}\left[r\frac{5}{2}kT(n_a+n_i)v_r\right]\\&\quad=\frac{\partial}{\partial z}\left(\lambda_{ai}\frac{\partial T}{\partial z}\right)+\frac{1}{r}\frac{\partial}{\partial r}\left(r\lambda_{ai}\frac{\partial T}{\partial r}\right)+b(T_e-T)\end{aligned} \tag{1.39}$$

For plasma in a state of ionization equilibrium, the concentration of particles is determined by the law of action mass (1.9). According to this law, the amount of energy spent on ionization of the unit of plasma volume at any point of space and per unit of time equals to the amount of energy released in the same volume in the reverse recombination reaction. For the plasma flow this model does not correspond to the real picture of the physical processes in plasma. Because of its heterogeneity the region, where the energy release and active field ionization mainly exists in plasma, i.e., the energy of system is mainly spent on ionization, and vice versa, there is an area, where the reverse recombination reaction of particles and release of energy recombination are predominant. Thus, there is a direct connection between the particle and energy balance or between ionization and thermal equilibrium states of plasma. In other words, there is a connection between the laws of conservation of energy and particles. We draw attention to the fact that for low-temperature plasma the ionization energy of particles is usually more than their thermal energy, such as argon single ionization energy $E_i^+ = 15.755$ eV compared with the thermal energy

$kT = 1 - 3$ eV for most plasma in RF, Arc and other types of plasma torches. Consequently, in determination of the energy and particles balance in plasma, it is important to consider its model, taking into account not only a violation of the temperature equilibrium, but also the ionization one.

1.3.1 The Violation of Ionization Equilibrium in Plasma. Equation of Ionization Equilibrium

Ionization equilibrium is violated if the processes of ionization and recombination of particles occur with the different speeds, i.e., the number of ionization acts is not equal to the number of recombination acts per unit of time and per unit of plasma volume. In the steady-state conditions the radiation from plasma volume, the diffusion of charged particles and forced convection of heat contribute to a violation of the ionization equilibrium. To neglect the first of these mechanisms a high concentration of electrons $n_e \geq 10^{16}$ cm^{-3} requires for example at the atmospheric pressure. The role of diffusion (bipolar diffusion and thermal diffusion) clearly reveals for plasma flows in the channel of plasma torch with large gradients of electron density and temperature. In this case, the diffusion of charged particles from the axial zone reduces the electron density on the axis and on the periphery of their increase in the comparison with the equilibrium model. The forced convection can play a major role in particles balance and it increases with increasing of the gas flow, blown through the discharge. The special analysis is carried out in [16], it shows that the convection can lead to a significant deviation of plasma from the ionization equilibrium under a stream of cold gas through the discharge with specific consumption $\rho v > 0.1$ g/(cm^2 s) at temperature $T_e = 10000$ K and pressure $p = 10^5$ Pa.

However, the main mechanisms of the appearance and disappearance of charged particles in low-temperature plasma, especially in dense plasma of high pressure are the ionization of atoms by direct electron impact and the triple recombination [3, 4]. The velocity of charged particles formation as a result of the ionization of atoms by direct electron impact is proportional to the concentrations of electrons n_e, atoms n_a, and some ratio of ionization of this process k_i (it is determined by the ionization cross section) and is equal to $k_i n_e n_a$.

The step ionization is very significant along with the direct ionization of atoms from the ground state in collisions with electrons: the average thermal energy of the electrons in the normal plasma torch is about $1 - 3$ eV, and the ionization energy of atoms from the ground level for most gases $E_i > 5$ eV. Therefore, the number of electrons with such energy in the low-temperature plasma can be determined by the Maxwell distribution function, and this number is very small. In dense plasma of high pressure the ionization from the excited level (the step ionization) is the most probable process of ionization in plasma torches. In this case, the ionization coefficient k_i takes into account the total probability (i.e., the effective cross section) of processes of the direct and stepwise ionizations.

In a dense plasma the recombination of electron–ion pairs often occurs with the participation of the third particle (the Thomson theory)—atom, ion, and electron [2–4]. Analysis of specific conditions of recombination in plasma torches at a high concentration of charged particles is given in [3, 4], it shows that the cross section of three-body recombination is extremely high in conditions of the presence of electron as the third particle. This process is particularly effective with the participation of slow electrons in the field of ions for a long time. First of all the capture of an electron to a distant ion orbit occurs, and the role of the third electron is in the selection of energy released during the deceleration of electron in the field of ion. Next, the recombination occurs in stages, covering all the excited states of an atom with radiation transition, and the quanta of radiation energy is emitted in the spectral lines, i.e., shock radiation or triple recombination occurs. In this case, the recombination velocity of particles is proportional to the concentration of ions n_{i}, and to the square of the electron density n_{e}^2 (two electrons take part in the recombination reaction simultaneously) and is proportional to the coefficients k_r, characterized the probability of triple recombination process, i.e., it is equal to $k_{\mathrm{r}} n_{\mathrm{i}} n_{\mathrm{e}}^2$.

Thus, the velocity of charged particles formation as a result of two opposing processes: the electron impact ionization and triple recombination, involving the third electron $\dot{n}_{\mathrm{e}} = k_{\mathrm{i}} n_{\mathrm{a}} n_{\mathrm{e}} - k_{\mathrm{r}} n_{\mathrm{i}} n_{\mathrm{e}}^2$.

Taking into account the transport of charged particles due to the convection and bipolar diffusion, *the equation of ionization equilibrium of plasma* (the single ionization) is:

$$\frac{\partial n_{\mathrm{e}}}{\partial t} + \mathrm{div}(n_{\mathrm{e}} \vec{v}_{\mathrm{e}}) = \mathrm{div}(D_{\mathrm{amb}} \mathrm{grad} n_{\mathrm{e}}) + k_{\mathrm{i}} n_{\mathrm{a}} n_{\mathrm{e}} - k_{\mathrm{r}} n_{\mathrm{i}} n_{\mathrm{e}}^2 \tag{1.40}$$

For quasi-neutral plasma ($n_{\mathrm{e}} = n_{\mathrm{i}}$), using the principle of the detailed balance [2, 3], the equation of ionization equilibrium can be written as follows:

$$\frac{\partial n_{\mathrm{e}}}{\partial t} + \mathrm{div}(n_{\mathrm{e}} \vec{v}_{\mathrm{e}}) = \mathrm{div}(D_{\mathrm{amb}} \mathrm{grad} n_{\mathrm{e}}) + k_{\mathrm{r}} n_{\mathrm{e}} n_{\mathrm{a}} \left(\frac{n_{\mathrm{e}}^{*2}}{n_{\mathrm{a}}^*} - \frac{n_{\mathrm{e}}^2}{n_{\mathrm{a}}} \right) \tag{1.41}$$

Here $n_{\mathrm{e}}^*, n_{\mathrm{a}}^*$—the concentration of electrons and atoms in ionization equilibrium (it is determined by the Saha equation).

In the case of stationary and cylindrical axe-symmetry the equation of ionization equilibrium can be written as:

$$\begin{aligned} \frac{\partial (n_{\mathrm{e}} v_{\mathrm{ez}})}{\partial z} + \frac{1}{r} \frac{\partial (r n_{\mathrm{e}} v_{\mathrm{er}})}{\partial r} &= \frac{\partial}{\partial z} \left(D_{\mathrm{amb}} \frac{\partial n_{\mathrm{e}}}{\partial z} \right) + \frac{1}{r} \frac{\partial}{\partial r} \left(r D_{\mathrm{amb}} \frac{\partial n_{\mathrm{e}}}{\partial r} \right) \\ &\quad + k_{\mathrm{r}} n_{\mathrm{e}} n_{\mathrm{a}} \left(\frac{n_{\mathrm{e}}^{*2}}{n_{\mathrm{a}}^*} - \frac{n_{\mathrm{e}}^2}{n_{\mathrm{a}}} \right) \end{aligned} \tag{1.42}$$

The sources function in equation of ionization equilibrium is often referred $\dot{n}_{\mathrm{e}} = k_{\mathrm{r}} n_{\mathrm{e}} n_{\mathrm{a}} \left(\frac{n_{\mathrm{e}}^{*2}}{n_{\mathrm{a}}^*} - \frac{n_{\mathrm{e}}^2}{n_{\mathrm{a}}} \right)$.

As in the case of equilibriums plasma, in the model of nonequilibrium plasma it is considered, that the particle velocity distribution is the Maxwell one and the "population" distribution—the Boltzmann one. The two-temperature model must also contain the following equations: the momentum equations [(1.26)–(1.28)]; the equation of continuity for all gas (1.29); the Maxwell equations describing the field [(1.30), (1.31)]; the system of equations for determination of two temperature transfer coefficients and the thermodynamic functions of plasma; the Dalton's law of partial pressures; equations for calculating the coefficients of impact ionization and triple recombination of charged particles and others.

1.4 The Two-Speed Plasma Model

The difference in mass of electrons and other particles not only leads to disruption of the equilibrium temperature between electrons and heavy components in plasma, but also to the difference of velocities of flow. Because of the small mass electrons drift in the field with a directional speed different from the speed of the total gas flow, despite the fact that one effective electron blow can transfer almost all of its directional momentum to heavy particles. Unlike the electrons, ions have greater inertia, their mobility is much smaller than the mobility of electrons, and they move with the same speed as the other neutral particles that are weight comparable in plasma. This fact makes it possible to build a two-speed plasma model with the different mechanisms of acceleration and deceleration of the particle fluxes.

So consider the unit volume of plasma where the electrons move in the field under the influence of the sum of forces: the electric and magnetic forces $-n_e e(\vec{E} + \vec{v}_e \times \vec{B})$, and the electric field only makes a work on the electrons; static pressure gradient $-\nabla p_e$, here $p_e = n_e k T_e$—the partial pressure; the frictional force caused by the scattering of electrons on heavy components $-\rho_e \vec{v}_e v_m$, here v_m—the average effective collision frequency of electrons with heavy particles and the convective forces $-(\rho_e \vec{v}_e \nabla)\vec{v}_e$.

The equation of motion of the electron gas in a two-speed plasma model has the form:

$$\frac{\partial(\rho_e \vec{v}_e)}{\partial t} + (\rho_e \vec{v}_e \nabla)\vec{v}_e = -n_e e(\vec{E} + \vec{v}_e \times \vec{B}) - \nabla p_e - \rho_e \vec{v}_e v_m \tag{1.43}$$

Comparable in mass heavy particles move in the flow at the same speed $\vec{v}$. The flow of heavy components is determined by the action of them the sum of forces: the electric and the magnetic forces $n_i e(\vec{E} + \vec{v} \times \vec{B})$; acceleration forces $\rho_e \vec{v}_e v_m$ caused by the scattering of electrons on the heavy components; static pressure gradient $-\nabla p_{ai}$, here $p_{ai} = (n_a + n_i)kT$—the sum of the partial pressures; viscous friction force $\vec{F}_\mu = \vec{\nabla} \times \vec{\tau}$, consisting of the tangent and the associated bulk and shear deformations (1.22) additional friction and force convection $-(\rho_{ai}\vec{v}\nabla)\vec{v}$. In this case, the equation of motion of heavy components can be written as

$$\frac{\partial(\rho_{ai}\vec{v})}{\partial t} + (\rho_{ai}\vec{v}\nabla)\vec{v} = n_i e(\vec{E} + \vec{v} \times \vec{B}) + \rho_e \vec{v}_e \nu_m - \nabla p_{ai} + \nabla \times \vec{\tau} \quad (1.44)$$

Here $\rho_{ai} = m_a(n_a + n_i)$—the density of atom–ion gas.

In the model it is necessary to include the continuity equation and the energy balance equation for the electron gas and for heavy components. Thus, the two-speed plasma model allows not only to investigate the mechanisms of acceleration and deceleration of its streams, to find the effective conditions for acceleration or deceleration of plasma flows, their gas dynamics, but also to determine the energy transfer in two-temperature (two-fluid) model of a nonequilibrium plasma (Sect. 1.3). In practice, the two-speed movement of the plasma flow is very common in the case where the electrons drift in the field coincides with the total gas stream direction. So in an electric arc with a longitudinal electric field the velocity of electron gas may be significantly different with the total gas flow rate. In the approximate model the velocity of electrons is taken equal to the sum of the total gas flow velocity, the electron drift velocity in the field, the bipolar diffusion and thermal diffusion. In the two-speed model drift in the field is determined by the acceleration of electrons in the field and by their scattering on the heavy particles, and the velocity of bipolar and thermal diffusion is determined by gradient of static pressure. The difference is only the fact that in the two-speed models these mechanisms can be considered simultaneously.

Earlier we have considered the model of laminar plasma flow. The presence of electric and magnetic fields in plasma contributes to regulate movement of the plasma flow. However, the mechanisms of diffusion and thermal diffusion, the presence of the velocity and temperature gradients in the gas mixture, in many cases, can lead to disruption of the plasma stream moving order and occurrence of turbulence, particularly in places where the field is weak effect, i.e., in the plasma jet and in the wall zone of plasma torch. Consider the basic models of turbulence of the computational fluid dynamics based on the terms of a continuous medium taken to the plasma flow.

1.5 Gas Dynamic Models of Plasma Turbulence

One of the important applications of computational fluid mechanics (Computational Fluid Dynamics—CFD) modeling of plasma flows is the modeling of turbulent flow in streams and in the near-wall zone of the plasma torch. Turbulent flows in the channels and in the jets of plasma torches are characterized by different fluctuations of the velocity field. Orderly and calm steam, called as laminar, occurs only at a low flow rate v, small size of the region L, and the high viscosity of the liquid μ. Quantitative measure of the fulfillment of these conditions is the numerical value of the Reynolds rate Re $= \frac{vL}{\mu}$. With increasing Reynolds rate the number the transition from laminar to turbulent flow occurs. It was at first demonstrated in experiments of O. Reynolds, performed in 1883 at Manchester University. The fundamental significance of the results is that the order of Reynolds rates, where a loss of stability and transition to turbulence take place, is the same in the areas of different sizes in fluid of different physical nature.

Currently, the description of turbulent flows methods, based on the solution of the Navier–Stokes equations, are mainly used.

Direct numerical simulation (DNS) [17] assumes the full solution of unsteady Navier–Stokes equations and the continuity equation. The difficulties of DNS model are in the limited computing resources even now. In solving the Navier–Stokes equations, approach based on the use of time-averaged quantities is often used. As a result of the decision of the modified equations requires less computer resources, but there are additional variables. The different turbulence models are used to close the obtained equations. Today, there are a large number of models. But there is not universal model for all classes of engineering problems. Selection of the optimal model of turbulence depends on the type of course, a specific class of the problem, the required accuracy of the solution, available computing resources, etc. From a practical standpoint, the statistics obtained from the DNS, could be used to test and calibrate the models, based on the Reynolds-averaged equations.

1.5.1 The Reynolds-Averaged Navier–Stokes Equations

The turbulence models used in engineering calculations are usually based on the solution of the Reynolds-averaged Navier–Stokes equations (RANS). All the characteristics of the turbulent flow (velocity, pressure, temperature, concentration of mixture components) have the random fluctuations around the average values. In this regard, the value of the velocity components ($u_j, j = x, y, z$) and scalar quantities (Φ) represent the sum of the mean and fluctuating components (the bar atop denotes the averaging over the time, and the bar corresponds to the fluctuating component):

$$u_j = \bar{u}_j + u'_j \tag{1.45}$$

$$\Phi_j = \overline{\Phi}_j + \Phi'_j \tag{1.46}$$

The intensity of the velocity fluctuations is characterized by the turbulent kinetic energy

$$k = \frac{1}{2}\overline{u'_j u'_j} \tag{1.47}$$

The moving of a continuous medium (liquid or gas) is described by the Navier–Stokes equations, which consists of the continuity equation

$$\frac{\partial \rho}{\partial t} + \frac{\partial (\rho u_j)}{\partial x_j} = 0 \tag{1.48}$$

and the momentum equation

$$\frac{\partial (\rho u_i)}{\partial t} + \frac{\partial (\rho u_i u_j)}{\partial x_j} = -\frac{\partial p}{\partial x_j} + \frac{\partial \tau_{ij}}{\partial x_j} + F_i \tag{1.49}$$

The momentum equations in the form (1.49) represent the three transport equations for the velocity components, where $p = \bar{p} + p'$—the pressure, the mean pressure and the pulsation pressure, respectively; τ_{ij}—the components of tensor of the shear stress; F_i—volumetric forces acting on the fluid.

For Newtonian liquids according to the Stokes hypothesis there is a linear relationship between the components and the rate of deformation tensor

$$S_{ij} = \left(\frac{\partial u_i}{\partial x_j} + \frac{\partial u_j}{\partial x_i}\right)/2 = \bar{S}_{ij} + S'_{ij},$$

Here

$$\bar{S}_{ij} = \left(\frac{\partial \bar{u}_i}{\partial x_j} + \frac{\partial \bar{u}_j}{\partial x_i}\right)/2, \quad S'_{ij} = \left(\frac{\partial u'_i}{\partial x_j} + \frac{\partial u'_j}{\partial x_i}\right)/2;$$

i.e.,

$$\tau_{ij} = 2\mu\left(S_{ij} - \frac{1}{3} S_{kk}\delta_{ij}\right) \tag{1.50}$$

After the substituting (1.50) into (1.49) we obtain the momentum equations in the form of:

$$\frac{\partial \rho u_i}{\partial t} + \frac{\partial \rho u_j u_i}{\partial x_j} = -\frac{\partial}{\partial x_j}\left(p + \frac{2}{3}\mu \frac{\partial u_k}{\partial x_k}\right) + \frac{\partial}{\partial x_j}\mu\left(\frac{\partial u_i}{\partial x_j} + \frac{\partial u_j}{\partial x_i}\right) + F_i \tag{1.51}$$

It is easy to show that the average value of the result of multiplication $\overline{u_j u_i}$ is defined as

$$\overline{u_j u_i} = \overline{\left(\bar{u}_j + u'_j\right)\left(\bar{u}_i + u'_i\right)} = \bar{u}_j \bar{u}_i + \overline{u'_j u'_i} \tag{1.52}$$

Substituting (1.45) into (1.51) and averaging, and taking into account (1.52), gives the transport equation for the average velocity components:

$$\frac{\partial \rho \bar{u}_i}{\partial t} + \frac{\partial \rho \bar{u}_j \bar{u}_i}{\partial x_j} = -\frac{\partial}{\partial x_j}\left(p + \frac{2}{3}\mu \frac{\partial \bar{u}_k}{\partial x_k}\right) + \frac{\partial}{\partial x_j}\left(\mu\left(\frac{\partial \bar{u}_i}{\partial x_j} + \frac{\partial \bar{u}_j}{\partial x_i}\right) - \rho \overline{u'_j u'_i}\right) + F_i \tag{1.53}$$

In (1.53) $-\rho\overline{u'_j u'_i}$ correspond to the value of shear stress τ_{ij} in Eq. (1.49), the correlation $-\overline{u'_j u'_i}$ is called the Reynolds (or turbulent) stresses. In the physical interpretation of the Reynolds stress the analogy with viscous stresses, that characterize the transport of momentum due to molecular collisions, is used. The Reynolds stresses also characterize the transport of momentum due to mixing in a pulsating turbulent flow. They form a second-order tensor (3 × 3 matrix), the components of it are symmetric with respect to the diagonal: $\overline{u'_j u'_i} = \overline{u'_i u'_j}$.

Similarly the averaged continuity equation can be written as:

$$\frac{\partial \rho}{\partial t} + \frac{\partial (\rho \bar{u}_j)}{\partial x_j} = 0 \tag{1.54}$$

So, with the help of decomposition of Reynolds and averaging procedure we obtain a system of Eqs. (1.53) and (1.54) for the mean velocity field. These equations are called the Navier–Stokes equations, Reynolds-averaged (Reynolds averaged Navier–Stokes equations or equation RANS). They can not be solved, as yet unknown turbulent stresses. In other words, the system-averaged Reynolds equations are not closed, and to address the circuit $\overline{u'_j u'_i}$ must be determined.

In solving problems of mathematical modeling of plasma processes not only the definition of the vector velocity field, but scalar fields, such as temperature or concentration requires. General transport equation for the generalized scalar value is:

$$\frac{\partial \rho \Phi}{\partial t} + \frac{\partial (\rho u_j \Phi)}{\partial x_j} = \frac{\partial}{\partial x_j}\left(\rho D \frac{\partial \Phi}{\partial x_j}\right) + F_i \tag{1.55}$$

Here D—the diffusion coefficient.

Averaging this equation is similar to the averaged equations for each component of the velocity. As in the equations of motion, the unknown correlation between the velocity fluctuations and a scalar quantity Φ arise during the averaging of nonlinear expression for the convective flow:

$$\overline{u_j \Phi} = \overline{\left(\bar{u}_j + u'_j\right)\left(\overline{\Phi} + \Phi'\right)} = \bar{u}_j \overline{\Phi} + \overline{u'_j \Phi'} \tag{1.56}$$

The averaging transport equation takes the form:

$$\frac{\partial \rho \overline{\Phi}}{\partial t} + \frac{\partial (\rho u_j \overline{\Phi})}{\partial x_j} = \frac{\partial}{\partial x_j}\left(\rho D \frac{\partial \overline{\Phi}}{\partial x_j} - \rho \overline{u_j' \Phi'}\right) + F_i \tag{1.57}$$

Here $\overline{u_j' \Phi'}$ is the turbulent flow of sense scalar quantities Φ that is a result of the turbulent mixing.

As in the case of the equations of motion for $\overline{u_j' \Phi'}$ the transfer equation that contain unknown higher order correlations can be obtained. Thus, in using the Reynolds averaging there is a problem of closing the system of averaged Eqs. (1.53) and (1.57). To solve this problem the physically based mathematical model requires that will allow expression of the desired correlation by the mean values and the correlation of a lower order. The goal is achieved by means of a particular model of turbulence.

Experiments have shown that the turbulence is most intense in areas of high gradients of average speed. In 1877, J. Boussinesque suggested that the relationship between the turbulent stresses $-\overline{u_j' u_i'}$ and the strain rate tensor $\bar{S}_{ij} = \left(\frac{\partial \bar{u}_i}{\partial x_j} + \frac{\partial \bar{u}_j}{\partial x_i}\right)/2$ of the averaged flow is similar to the corresponding communication that takes place between the viscous stress τ_{ij} and components S_{ij} (1.50) for a Newtonian fluid:

$$-\rho\left(\overline{u_j' u_i'} - \frac{2}{3} k \delta_{ij}\right) = 2\mu_t\left(\bar{S}_{ij} - \frac{1}{3}\bar{S}_{kk}\delta_{ij}\right) \tag{1.58}$$

The coefficient of proportionality μ_t is called the turbulent (or eddy) viscosity. Equation (1.58) can be written as:

$$-\rho \overline{u_j' u_i'} = \mu_t\left(\frac{\partial \bar{u}_i}{\partial x_j} + \frac{\partial \bar{u}_j}{\partial x_i}\right) - \frac{2}{3}\left(\rho k + \mu_t \frac{\partial \bar{u}_k}{\partial x_k}\right)\delta_{ij} \tag{1.59}$$

In view of (1.59) the Eq. (1.53) for the mean velocity takes the form:

$$\begin{aligned} \frac{\partial (\rho \bar{u}_i)}{\partial t} + \frac{\partial (\rho \bar{u}_j \bar{u}_i)}{\partial x_j} = &-\frac{\partial}{\partial x_j}\left(\bar{p} + \frac{2}{3}\left(\rho k + (\mu + \mu_t)\frac{\partial \bar{u}_k}{\partial x_k}\right)\right) \\ &+ \frac{\partial}{\partial x_j}\left((\mu + \mu_t)\left(\frac{\partial \bar{u}_i}{\partial x_j} + \frac{\partial \bar{u}_j}{\partial x_i}\right)\right) + F_i \end{aligned} \tag{1.60}$$

By analogy with the concept of turbulent viscosity, the gradient diffusion allows us to express the turbulent flow $\overline{u_j' \Phi'}$ of scalar quantity Φ with gradients of mean values:

$$\overline{u_j' \Phi'} = -D_t \frac{\partial \overline{\Phi}}{\partial x_j} \tag{1.61}$$

Here D_t—the coefficient of turbulent diffusion. In view of (1.61) the averaged Eq. (1.57) takes the form:

$$\frac{\partial(\rho\overline{\Phi})}{\partial t}+\frac{\partial(\rho u_j\overline{\Phi})}{\partial x_j}=\frac{\partial}{\partial x_j}\left(\rho(D+D_t)\frac{\partial\overline{\Phi}}{\partial x_j}\right)+F_i \tag{1.62}$$

Thus, by using the concepts of turbulent viscosity and gradient diffusion for solving the problem of the closure of averaged equations it is necessary to determine the turbulent viscosity and turbulent diffusion coefficient D_t. Note that the concept of turbulent viscosity μ_t and diffusion gradient D_t are extremely useful in the construction of approximate models of turbulence and is still widely used in numerical simulations. However, we must remember that the relevant concepts have no strictly physical basis and in some cases contradict the experimental data. As a consequence, the application of the concepts of turbulent viscosity and gradient diffusion can lead to the large errors in cases of parietal areas, flow separation, and swirling currents.

1.5.2 The Main Gas Dynamic Model of Turbulence

Using the concept of averaged Navier–Stokes equations for turbulent kinetic energy $k=\frac{1}{2}\overline{u_j'u_j'}$ and its dissipation rate $\varepsilon=2\nu\overline{S_{ij}'S_{ji}'}$, here $\nu=\frac{\mu}{\rho}$—the kinematic viscosity and thermal dissipation rate of turbulent kinetic energy $\omega=\frac{\varepsilon}{k}$, we can get different turbulence models based on transport equations of turbulent flow. The object of the application of the model is to determine the turbulence Reynolds stresses. If the concept of turbulent viscosity is adopted, this problem is reduced to the finding of the turbulent viscosity through other flow characteristics (field of average speed for the calculation of it, for example, the turbulent kinetic energy and its dissipation speed are involved).

In general terms, the transport equations of turbulent flow, are used in the models with one or two differential equations, they can be written as follows:

$$\frac{\partial(\rho\Phi)}{\partial t}+\frac{\partial(\rho\bar{u}_j\Phi)}{\partial x_j}=\frac{\partial}{\partial x_j}\left[(\mu+\Gamma_\Phi)\frac{\partial\Phi}{\partial x_j}\right]+P-D+A \tag{1.63}$$

Table 1.1 The parameters of the generalized transport Eq. (1.63)

Determines the amount	Φ	P	D	Γ_Φ
Kinetic energy	k	$\tau_{ij}\frac{\partial\bar{u}_i}{\partial x_j}$	$\rho\varepsilon$ or $\beta^*\rho k\omega$	$\frac{\mu_t}{\sigma_k}$
The dissipation rate of kinetic energy	ε	$c_{\varepsilon 1}\frac{\varepsilon}{k}\tau_{ij}\frac{\partial\bar{u}_i}{\partial x_j}$	$c_{\varepsilon 2}\rho\frac{\varepsilon^2}{k}$	$\frac{\mu_t}{\sigma_\varepsilon}$
The specific rate of dissipation	ω	$\alpha\frac{\omega}{k}\tau_{ij}\frac{\partial\bar{u}_i}{\partial x_j}$	$\beta\rho\omega^2$	$\sigma_\omega\mu_t$

Table 1.2 The coefficients for the models of turbulent flow with two differential equations

Models	Coefficients ($K = 0.41$)				
$k-\varepsilon$ [19]	$c_{\varepsilon 1} = 1.44$	$c_{\varepsilon 2} = 1.92$	$\sigma_k = 1.0$	$\sigma_\varepsilon = 1.3$	$c_\mu = 0.09$
$k-\omega$ [20]	$\beta^* = 0.09$	$\beta = 0.075$	$\alpha \approx 0.556$	$\sigma_k = 0.5$	$\sigma_\omega = 0.5$
$k-\omega$ (SST) [21]	$\beta_1^* = 0.09$	$\beta_1 = 0.075$	$\alpha_1 = \beta_1/\beta_1^* - \sigma_{\omega 1}K^2/\sqrt{\beta_1^*}$	$\sigma_{k1} = 0.85$	$\sigma_{\omega 1} = 0.5$
	$\beta_2^* = 0.09$	$\beta_2 = 0.0828$	$\alpha_2 = \beta_2/\beta_2^* - \sigma_{\omega 2}K^2/\sqrt{\beta_2^*}$	$\sigma_{k2} = 1$	$\sigma_{\omega 2} = 0.856$

Designations of the parameters in (1.63) for each type of equations are shown in Table 1.1 [18].

In the case of the model of Shear Stress Transport (SST) for ω, the last term of (1.63)

$$A = 2(1 - F_1)\rho\sigma_{\omega 2}\frac{1}{\omega}\frac{\partial k}{\partial x_j}\frac{\partial \omega}{\partial x_j} \tag{1.64}$$

In other cases $A = 0$. The mass function F_1 defined as in [18].

Constants for the models with two differential equations are summarized in Table 1.2.

We give a brief overview of the main gas dynamic models.

Laminar model is used to calculate the laminar flows. In this case, a direct solution of the Navier–Stokes equations without any additional any turbulent model occurs. We can assume that the laminar model is required to describe the laminar flow where the Reynolds stresses disappear. The condition of this model is the low value of the Reynolds rate, it would otherwise need to apply turbulence model below.

Zero-order model (the algebraic model) where the turbulent viscosity is determined from a simple algebraic equation. The model of Baldwin–Lomax (BL) is a typical representative of this one. The algebraic models of turbulence differential equations do not contain transport and do not employ the nonlocal characteristics such as boundary layer thickness, wall friction, etc. They are very nonuniversal, difficult to implement in modern codes, and are not used at the moment.

The model of the first order, where the eddy viscosity is determined by solving a differential equation (models with a differential equation). Differential transport equation can be written with respect to different variables: the turbulent viscosity; turbulent kinetic energy or the like. As the linear scale the distance from the wall is commonly used. The most successful of these models is the Spalart–Allmaras (SA) model. This one contains a single differential equation for "high Reynolds" turbulent viscosity associated with the turbulent viscosity with the algebraic relation. This is a one-parameter model was developed for aerospace applications. It contains a number of amendments expanding its scope: the amendment on the curvature and rotation; correction for roughness and others. This model gives good results for the boundary layers, characterized by a positive pressure gradient.

The model of the second order, where the eddy viscosity is determined from two differential equations. These models are models that have become the industry standard and widely used in the practice of computing, for example, the model Shear Stress Transport (SST). Consider the basic model with two differential equations ($k - \varepsilon$ and $k - \omega$).

Models $k - \varepsilon$ can be divided into high- and low Reynolds and by the method of describing the effect of the near-wall effects. In high Reynolds models ("standard" model, Reynolds Normalization Group—RNG) use the parietal function, and in low Reynolds and (Launder–Sharma)—special damping function. The model $k - \varepsilon$ uses two differential transport equations for calculating the turbulent kinetic energy k and the dissipation rate ε. This model is stable, does not require significant computing resources and has long been the industry standard, and has worked well in the calculation of internal flows, but when it is used, there are problems in the calculation of flows with large pressure gradients and in the case of flow separation. Model $k - \varepsilon$ is often used to calculate the free flow. Start of separation is defined too late and the sizes of separated flows are too small. The model $k - \varepsilon$ gives too optimistic results for the flow that is detached. Due to incorrect predictions of the separation point model $k - \varepsilon$ is gradually replaced by more sophisticated models.

Models $k - \omega$. The experience of calculations of separated turbulent flows over bluff bodies showed that the models based on the equation for the dissipation rate ε predict the onset of flow separation greatly downstream than that is observed in the experiments, with settlement size of the recirculation zone and traffic intensity in it are too low. At the same time, it appeared that another class of models, where instead of the transport equation for ε the equation for the specific dissipation rate of turbulent kinetic energy ω is used, resulting in better agreement with the experimental results with this flows, if the separation zone is small. The model $k - \omega$ was developed in the works of Wilcox [22] and was the basis of a combined model of shear stress SST, created later by Menter [23] and combining the models $k - \omega$ and $k - \varepsilon$.

In the standard model $k - \omega$ Low Reynolds effects are accounted, the influence of the compressibility and the shear perturbation, but it is significantly inferior to the breadth of application of the models $k - \varepsilon$. Models $k - \omega$ can be used to calculate the wall turbulence without the introduction of special functions and have extremely high sensitivity to the boundary conditions in the external flow.

Menter's model of Shear Stress Transport (SST) includes the models $k - \omega$ of Wilcox for the wall region and the converted model $k - \varepsilon$ for the region far from the wall [23]. Thus, in calculating the total flux the SST model gives good results, especially in cases of separation of the flow and large pressure gradients therein. In addition, this model proved to be reliable and not demanding the computing power. Therefore, it combines the advantages of models $k - \varepsilon$ and $k - \omega$, but in terms of computation this one yields to the models with only single equation. SST model is a new industrial standard. It gives very good results even in the calculations of phenomena such as flow separation by blowing air into the boundary layer, or in the process of turbulent heat transfer.

In the Reynolds Stress Models—(RSM) the turbulent viscosity is not used (1.59). Instead, this class of models assumes the decision of a single transport equation for each component of the Reynolds stress tensor $\overline{u_j'u_i'}$ and the transport equations for the dissipation rate ε. As a result the possibility of taking into account the anisotropy of turbulent fluctuations opens, it significantly expands the potential range of application of these models, including for flows with large curvature of the streamlines, swirled flow and large deformation of the velocity field.

Reynolds stress model is potentially superior the model of turbulent viscosity in calculation abilities of different currents. They are much more complicated than models of turbulent viscosity and have a number of problems as computational and physical (the principle of feasibility) characters. Taking into account the symmetry of the Reynolds stress tensor in RSM with differential equations six transport equations for different components $\overline{u_j'u_i'}$ and one equation for isotropic dissipation rate in three-dimensional space must be considered. Algebraic Stress Models (ASM) include two transport equations for the scales and six transcendental equations for the stresses and explicit algebraic Reynolds stress model (EARSM)—2 transport equations for the scales and explicit expressions for the stresses.

The main problem of the usage of RSM models is the reliability of approximate simulation of many unclosed correlations arising in the derivation of transport equations for $\overline{u_j'u_i'}$. Additional difficulties arise in the formulation of the boundary conditions near the solid surfaces. Despite the potential benefits, RSM model Reynolds stresses are used less often than the models of turbulent viscosity.

Large Eddy Simulation (LES) is becoming an increasingly popular method of calculation of the turbulent flows not only in research, but also in engineering practice. LES computing strategy differs greatly from the use of the Reynolds averaging (RANS). Instead of the solutions of averaged equations, that are deterministic, the nonstationary solutions, representing a random process are paid here. The aim of LES method is to reproduce numerically the largest vortex structures of the long wavelength part of the inertial range of the energy spectrum. Therefore, the numerical solution arises as a result of the fluctuations in a turbulent flow. In the contrast to the direct numerical simulation the shortwave (High Frequency) part of the spectrum does not play, but its impact is taken into account by using the approximation models of the SubGrid-Scale (SGS) viscosity. In accordance with the hypotheses A. Kolmogorov the small-scale turbulent fluctuations are spatially isotropic and statistically universal. In other words, the dynamics of small (sub-grid) vortices does not depend on the geometric characteristics of the flow and characteristics of the formation of large vortex structures. This fact simplifies the problem of creating a universal model sub-grid. In addition, the increasing of the resolution of the calculation algorithm (primarily at the expense of reducing the size of the grid cells, as well as in the case of increasing the accuracy of approximation of differential equations by algebraic ones) leads to the decrease of the sub-grid turbulence and expand the range of oscillations, explicitly permitted.

Detached Eddy Simulation (DES) [24] The characteristic dimension of turbulent eddies in the boundary layer near the solid surface is much smaller than in the developed turbulent flow far away from it. In this case the explicit permission of even the largest eddies in the boundary layer requires grinding the calculated mesh that may lead to an unacceptable increase in the duration of computation. At the same time, for near the wall surface flows and flows with a small separation quite reliable and cost-effective model based on Reynolds averaging exist. For these currents the RANS models can adequately assess the situation of flow separation, resistance and the lift force without excessive grid refinement near the surface. However for a wide range of practically important problems, accompanied by a large area of flow separation, the existing models RANS lead to large errors in the calculation of these quantities. Moreover, the principle possibility of a foundation of universal model based on the Reynolds averaging is questioned, because this model does not account for the large-scale turbulent flow disturbances, caused by the boundary conditions, the shape and the dimensions of the computational domain. At the same time the LES method has the same aim. In this case of creating the idea of a hybrid approach, where the RANS model is activated for the wall region, appeared and the LES model is activated for the region far from the wall.

The most successful and widespread hybrid approach was the method of Detached Eddy Simulation (DES), proposed in 1997 by P. Spalart and later developed in the works of M. Strelets, M. Shur, A. Travin and others. The ability of the selection of the specified turbulence method generates a whole family of the DES model and some of them are available in the software package Ansys Fluent and Ansys CFX.

DES method gives good results for the flows with sufficiently thin boundary layers, where near-wall grid cells are strongly elongated along the solid surface (where the model is activated RANS), and extensive zones of separation, for that the grid is close to isotropic (in this region model LES is activated). Thus, the method of DES is a hybrid approach where the switching models RANS and LES is used for the different regions of space. Unified structure of averaged equations and switching models and others open other than DES, approaches to the construction of hybrid models. In particular, it is possible to use average values of Reynolds and residual stresses (or turbulent and sub-grid).

These turbulence models give different results in the simulation of separated flows, boundary layers and so on. The accuracy of different models also depends on the width of the computational grid in the wall layer of cells (that are usually characterized by y^+, it can be regarded as a local Reynolds number in the cell). At the same time a large number of engineering hydrodynamic problems can be solved using commercial software packages (Ansys Fluent, Ansys CFD, FlowVision et al.).

In these packages the computational grid is created directly in the package, or imported from another editor. In many cases, the kind of computational grid determines the further calculation algorithm. Currently, two types of grid: tetrahedral and Cartesian are used in software packages. The tetrahedral grid allows you to create cells, similar in form to the boundaries of the computational domain and regions with large gradients of the unknown quantities and this fact enables a good approximation of the boundary layers. At the same time, the construction of a tetrahedral grid is very difficult. Cartesian grid allows only create rectangular cells, that may lead to worse approximation of the boundary layer, but its construction is simple.

There are also a number of ways to approximate the area of the boundary layers with high gradients. Examples of such methods include ALRG technology (adaptive local refinement grid) and the technology of sub-grid approximation of geometric shape, implemented, for example, in a software package FlowVision. ALRG technology allows splitting the selected cells in all directions (adapted to a predetermined level). In this cell, close to the fragmented crushed that the size of two adjacent cells does not differ by more than 2 times. Grinding cells can be set both in the bulk and on the surface. At the same time, the technology can automatically sub-grid approximation more accurately reproduce the shape of the surface.

All of these models and techniques, of course, can be taken in order to simulate the plasma stream flows, especially in the edge zone of the plasma torches and plasma jets.

1.6 Models of Light Emission and Absorption in Plasma

Dense low-temperature plasma is a powerful source of radiation. Emission and absorption in plasma are the result of various physical processes occurring inside it in the interaction of particles: electrons, ions, atoms and molecules. Power emission and absorption directly affect the energy balance and temperature of plasma, so the calculation of emission and absorption of light by plasma is the major problem in plasma physics and gas discharge. Light quanta appear and disappear during the radiation transition of electrons from one energy state to another. The interaction of the electrons with the ions play an important role in the formation of the light flux of radiation, and electron–electron collisions are not contribute to the flow of the emission and absorption of light.

Consider a system that consists of electron and a positive ion. Figure 1.2 shows a diagram of the energy states of electron. The energy is measured from the boundary of the continuous spectrum, separating free and bound states. The lower level corresponds to the ground state of atom. According to the circuit shown in Fig. 1.2, the following transitions of electrons can be implemented: (1) free–free (ff), (2) free–bound (fb) and bound–free (bf), (3) the bound–bound (bb). In the process of transition from top to bottom the quanta $\hbar\omega$ the quanta emitted that is equal to the difference between the energies of the upper and lower levels; during the transition from the bottom to up level the quanta $\hbar\omega$ the quanta absorbed.

Free–free transitions are bremsstrahlung during the Coulomb collisions of electrons with ions and inverse bremsstrahlung absorption processes. Free–bound transitions are the processes of photo– and bound–free transitions (the photo–ionization process). Free–free, free–bound and bound–free transitions provide the continuous spectra (continuum) because either the initial or final, or both electron states lie in the continuous energy spectrum. Bound–bound electron transitions

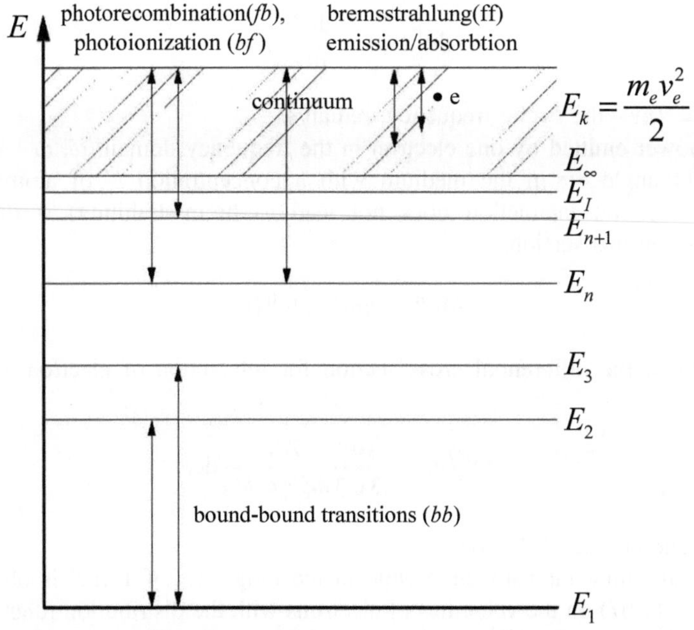

Fig. 1.2 Diagram of energy levels and transitions of electron in the field of ion

between discrete energy levels in atoms lead to the emission and absorption of photons in the spectral lines. The band spectra of molecules are the variation of the line spectra that are also formed during the bound–bound transitions. For each state of electrons in the molecule corresponds a set of oscillation and rotational levels, and there are many possible transitions between them. The same transitions occur in the interaction of the electron–atom in plasma.

1.6.1 Radiation in the Continuous Spectrum

1.6.1.1 Bremsstrahlung (ff) Radiation

In the motion of a free electron in the continuum area a quantum $\hbar\omega$ that is equal to the difference between the kinetic energies of the initial v_e and final v'_e speeds:

$$\hbar\omega = \frac{m_e v_e^2}{2} - \frac{m_e v_e'^2}{2} \tag{1.65}$$

To describe the ff-spectrum of the differential cross section of emissions is introduced, that depends on the speed of the electron v_e:

$$d\sigma_{\omega}^{\rm ff}(v_{\rm e}) = \frac{d\sigma_{\omega}^{\rm ff}}{d\Omega}\omega d\omega \tag{1.66}$$

Here $\omega = 2\pi\nu$—the cyclic frequency, radian/s.

The power emitted by one electron in the frequency domain ω, $\omega + d\omega$ in the total solid angle 4π in the medium with a concentration N of atoms or ions (electron–electron interaction does not lead to bremsstrahlung) is determined depending on this section:

$$\delta W_{\omega}^{\rm ff} = \hbar\omega N v_{\rm e} d\sigma_{\omega}^{\rm ff}(v_{\rm e}) \tag{1.67}$$

By Kramers, the differential cross section for interaction of electron and ion is equal to:

$$d\sigma_{\omega}^{\rm ff}(\omega) = \frac{16\pi}{3\sqrt{3}}\frac{Z^2 e^6}{m_{\rm e}^2 v_{\rm e}^2 c^3 \hbar\omega} d\omega \tag{1.68}$$

Here Z—the charge of the ion.

Bremsstrahlung intensity in plasma in the range $d\omega$ of 1 cm^3 is obtained by integration (1.67) on the velocities of electrons with the distribution function $f(v_{\rm e})$, normalized to the unity. The maximum speed when the electron can emit a photon $\hbar\omega$ is $v_{\rm e}^* = \sqrt{\frac{2\hbar\omega}{m_{\rm e}}}$. Then, the spectral intensity is defined as:

$$I_{\omega}^{\rm ff} = \int_{v_{\rm e}^*}^{\infty} \hbar\omega N n_{\rm e} f(v_{\rm e}) d\sigma(v_{\rm e}) v_{\rm e} dv_{\rm e} \tag{1.69}$$

The results of Maxwell function $f(v_{\rm e})$ averaging and $N = N_i$ is:

$$I_{\omega}^{\rm ff} = C\frac{Z^2 N_{\rm i} n_{\rm e}}{\sqrt{T_{\rm e}}} g^{\rm ff} \exp\left\{-\frac{\hbar\omega}{kT_{\rm e}}\right\} d\omega \tag{1.70}$$

$$C = \frac{16}{3}\left(\frac{2\pi}{3}\right)^{1/2}\frac{e^6}{m_{\rm e}^{3/2} c^3 k^{1/2}} = 1.08 \times 10^{-45}\,{\rm J\,cm^3\,K^{1/2}} \tag{1.71}$$

The (1.70) the factor $g^{\rm ff}$ called Gaunt factor is introduced. It refines the result of the semiclassical consideration and is calculated by quantum mechanics. If you design $\omega^* = mv_{\rm e}^2/2\hbar = E_{\rm k}/\hbar$ the boundary frequency, which can radiate electron with the kinetic energy $E_{\rm k}$, $g^{\rm ff}$ will depend on the dimensionless parameter $x = \omega/\omega^*$. Gaunt factor values depending on the parameters x are listed below [25]:

x	0	0.1	0.2	0.3	0.4	0.5	0.6	0.7	0.8	0.9	1
$g^{\rm ff}$	∞	2.01	1.61	1.34	1.13	0.97	0.81	0.68	0.53	0.36	0

For small in comparison with kT quanta, in particular in relation to the infrared region [26], we can write

$$g^{ff} = \frac{\sqrt{3}}{\pi} \ln \frac{4kT_e}{e^2 n_e^{1/3}} = 1.27\left(3.38 + \lg T_e - \frac{1}{3}\lg n_e\right) \qquad (1.72)$$

Here T_e, K; n_e, cm^{-3}. In the dense low-temperature plasma: $g^{ff} = 1.5 - 2.5$.

The emission spectrum is determined by the exponential factor. According to (1.70) the spectral intensity of the light isotropic emitted (by averaging over a large number of electron–ion pairs) in the angle 4π that corresponds to the total radiation power per unit volume of plasma.

These results can be used to calculate the intensity of transitions due to collisions with neutral atoms, if in the calculation of the cross sections to replace the Coulomb scattering cross section σ_C on the transport σ_{tr} one. The Coulomb cross section and its dependence on the rate determined by the Coulomb radius r_C, when the kinetic energy of the electron is equal to the potential energy of interaction:

$$\frac{m_e v_e^2}{2} = \frac{Ze^2}{r_C}; \quad \sigma_C = \pi r_C^2 = \frac{4\pi Z^2 e^4}{\left(m_e v_e^2\right)^2} \qquad (1.73)$$

The ratio of the cross sections for bremsstrahlung scattering on ions and neutral atoms will be

$$\frac{d\sigma_{\omega,i}^{ff}}{d\sigma_{\omega,n}^{ff}} = \frac{\pi a_0^2 Z^2}{\sigma_{tr}} \left(\frac{2I_H}{E_k}\right)^2 \qquad (1.74)$$

Here $E_k = \frac{m_e v_e^2}{2}$—energy of electron, $I_H = \frac{2\pi^2 e^4 m_e}{h^2} = 13.6$ eV—the ionization potential of a hydrogen atom, $a_0 = h^2/(4\pi^2 m_e e^2)$—the first Bohr radius, $\sigma_{tr} = \sigma_C(1 - \langle\cos\theta\rangle)$, $\langle\cos\theta\rangle$—the averaged cosine of the scattering angle.

In a typical relation $\sigma_{tr}/(\pi a_0^2) \approx 1 - 10$: electrons with energies of a few electron volts per heavy particle emitting in 10–100 times more energy on ions than on neutral particles, i.e., the intensity of bremsstrahlung (and absorption) for electron scattering on ions and neutral particles is comparable with the degree of ionization $\delta \approx (1 - 10)\,\%$. In a weakly ionized plasma $\delta \leq 0.1\,\%$, in addition it is typical for it the lack of multiply charged ions. As a result electron bremsstrahlung is predominant in collisions with neutral particles.

1.6.1.2 Recombination (fb) Radiation

In the radiative electron capture the rate v_e with the transition to a bound state of negative energy E_n the quanta is emitted:

$$\hbar\omega = |E_n| + \frac{m_e v_e^2}{2} \tag{1.75}$$

The cross section for this process σ_n^{ff} can be found using the following method. The validity of (1.68), if we consider it for the differential bremsstrahlung cross section is limited to a maximum photon $\hbar\omega_{max} = m_e v_e^2/2$ but electron stay still free. The formula (1.68) can be applied for large emission of photons. Electron in such cases will still have negative energy, i.e., remains bound. If the excess $\hbar\omega$ over $\hbar\omega_{max}$ is small, an electron is transferred to the upper levels of the excited atom.

But at a fixed initial energy of electron $\varepsilon = m_e v_e^2/2$ quanta with discrete energy values are emitted (1.75), but $d\sigma_\omega^{fb}$—this section of the photons emission in the continuous spectral range $d\omega$. To fix, the fact that the upper levels of the atom are very close to each other is used, and the total cross section for the emission of photons in a small, but finite interval $\Delta\omega$ around ω will be important. Electron falls on one level at a distance of Δn belonging to the energy interval $\Delta E = \hbar\Delta\omega$ around $E_n = \varepsilon - \hbar\omega$. If $\Delta E \ll |E_n|$, then the locks on all these levels are almost equally. Consequently,

$$\frac{d\sigma_\omega}{d\omega}\Delta\omega = \sigma_n^{fb}\Delta n, \quad \sigma_n^{fb} = \frac{d\sigma_\omega}{d\omega}\frac{\Delta E}{\hbar\Delta n} \tag{1.76}$$

We will consider the hydrogen-like system. Electron in such a system is moving in the field of a point charge Z, and for it $E_n = -I_H Z^2/n^2$, here n—the main quantum number. But $\Delta E/\Delta n \approx |dE_n/dn| = 2I_H Z^2/n^3$, and $I_H = me^4/2\hbar^2$ and using (1.68), we obtain the equation for the capture cross section in cm^2:

$$\sigma_n^{fb} = \frac{16\pi}{3\sqrt{3}}\frac{e^{10}Z^4}{c^3\hbar^4 m_e v_e^2 \omega}\frac{1}{n^3} = \frac{2.1\times 10^{-22}(I_H Z^2)^2}{\varepsilon\hbar\omega n^3} \tag{1.77}$$

Equation (1.68) for calculation of σ_n^{fb} have to be applied in the conditions associated, in particular, with the quasi-classical motion of the electron in orbits with large quantum numbers n, and is consistent with the principle of spectroscopic stability.

The spectral intensity of radiation by electrons, moving in a range of speeds v_e, $v_e + dv_e$, in their capture to a level with n per unit of plasma volume will be determined:

$$I_{\omega n}^{fb} = \hbar\omega N_i n_e \sigma_n^{fb} f(v_e) v_e dv_e \tag{1.78}$$

Here N_i—the ion concentration.

Substituting σ_n^{fb} (1.77) in Eq. (1.78) and replacing v_e in accordance with (1.75), to ω, i.e. $m_e v_e dv_e = \hbar d\omega$, taking into account the Maxwell function $f(v_e)$ we can get:

$$I^{\text{fb}}_{\omega n} = C\frac{Z^2 N_{\text{i}} n_{\text{e}}}{\sqrt{T_{\text{e}}}}\left(\frac{2x_n}{n}\right)\exp\left(-\frac{\hbar\omega}{kT_{\text{e}}}\right)\exp(x_n)\text{d}\omega \tag{1.79}$$

Here $x_n = (m_e e^4 Z^2)/(2\hbar^2 n^2 kT_{\text{e}}) = (I_{\text{H}} Z^2)/(kT_{\text{e}} n^2) = E_{\text{n}}/(kT_{\text{e}})$, C—the constant (1.71).

Quanta of the certain energy $\hbar\omega$ can be emitted in the capture of electron on the different levels, and the law of conservation of energy is respecting (1.75).

Thus, for the same reasons (the decrease of electron density is because of reduced Maxwell function), and for the bremsstrahlung recombination radiation intensity decreases exponentially with increasing frequency. However, this applies to the photoelectron capture with its transition to a bound state with a fixed n. Radiation with the same frequency ω can also occur in the capture of electron to other levels with different E_{n} in the conditions of (1.75). Therefore, the intensity of the spectrum is determined with the addition of (1.79) on the possible quantum numbers n:

$$I^{\text{fb}}_{\omega} = \sum_{n^*}^{\infty} I^{\text{fb}}_{\omega n} = C\frac{Z^2 N_{\text{i}} n_{\text{e}}}{\sqrt{T_e}}\exp\left(-\frac{\hbar\omega}{kT_{\text{e}}}\right)\sum_{n^*}^{\infty}\left(\frac{2x_n}{n}\right)\exp(x_n)\text{d}\omega \tag{1.80}$$

The lower limit of the addition n^* is determined by the condition $|E_{n^*}| \le \hbar\omega$. At the different frequencies of recombination spectrum a different number of levels of grip gives contribute to the intensity of spectrum. With the increasing frequency the number of such levels increases, and apparently it is manifested in the "sawtooth" shape of the spectrum (Fig. 1.3). Typically, at not so high temperatures of electrons the neighboring quantum numbers n are summed up.

The total intensity of radiation, J/(s cm^3) is the sum of the brake (1.70) and recombination (1.80) components and it is determined by the following approximate equation:

$$I_{\omega} = C\frac{Z^2 N_{\text{i}} n_{\text{e}}}{T_{\text{e}}^{1/2}}\Psi\left(\frac{\hbar\omega}{kT_{\text{e}}}\right)\text{d}\omega; \tag{1.81}$$

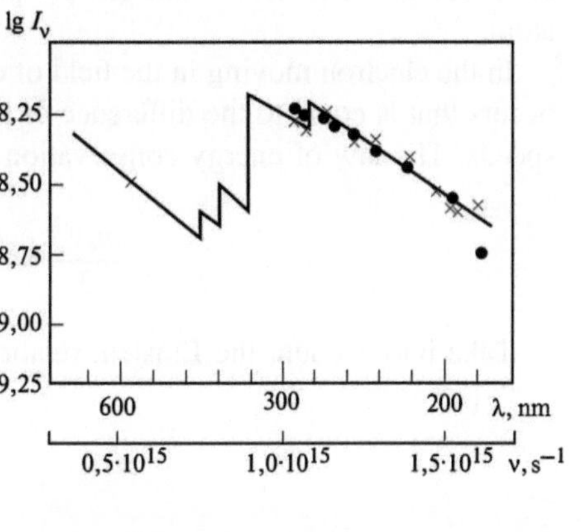

Fig. 1.3 Continuum of helium plasma with $n_{\text{e}} = 5 \times 10^{16}\ \text{cm}^{-3}$: *points* the experimental data, *lines* the calculated results

$$\Psi(x) = \begin{cases} 1, & x = \hbar\omega/kT_e < x_g = |E_g|/(kT_e); \\ e^{-(x-x_g)}, & x_g < x < x_1 = I/(kT_e); \\ e^{-(x-x_g)} + 2x_1 e^{-(x-x_1)}, & x > x_1. \end{cases} \tag{1.82}$$

Here $|E_g|$—the energy of the lowest excited state, I—the ionization potential. In the small quanta region ($\hbar\omega < |E_g|$) the contribution of free–bound and free–free transitions in emissivity are characterized by the relation

$$I_\omega^{fb} : I_\omega^{ff} \approx \{\exp[\hbar\omega/(kT_e)] - 1\} : 1 \tag{1.83}$$

Consequently, the photons $\hbar\omega < 0.7kT_e$ are emitted mainly in the brake mechanism, and $\hbar\omega > 0.7kT_e$—at recombination one. At $T_e = 10^4$ K photons with $\hbar\omega = 0.7kT_e = 0.6$ eV ($\lambda = 2000$ nm) corresponds to the infrared region of the spectrum. These values are significantly less than $|E_g|$ that for hydrogen, for example, is $I/2^2 = 3.37$ eV ($n = 2$) ($\lambda = 370$ nm—in the near ultraviolet region). According to (1.83), nearly all the spectrum emission, except for infrared region, occurs by the recombination mechanism. This situation is typical of the low-temperature plasma. Brake processes play a small role there and come to the fore only in the emission in a rather far infrared and microwave bands.

1.6.2 Absorption in the Continuum

1.6.2.1 The Bremsstrahlung (ff) Absorption

The above information about bremsstrahlung is valid for the description of absorption with regard to the principle of detailed balance. Full balance of ff process consists of spontaneous emission, absorption and stimulated emission. We can introduce the coefficients of the bremsstrahlung absorption a_ω^{ff} and stimulated emission b_ω^{ff}, corresponding to absorption and by stimulated emission coming under the influence of external radiation per pair of interacting electron–ion or electron–atom.

In the electron moving in the field of continuum the absorption of a quantum $\hbar\omega$ occurs that is equal to the difference of kinetic energies in the final v_e' and initial v_e speeds. The law of energy conservation in bremsstrahlung (ff) absorption is:

$$\frac{m_e v_e'^2}{2} = \frac{m_e v_e^2}{2} + \hbar\omega \tag{1.84}$$

Take into account the Einstein relation

$$\begin{aligned} v'_e b^{ff}_\omega(v'_e) &= v_e a^{ff}_\omega \\ b^{ff}_\omega(v'_e) &= \frac{\pi^2 c^2 v'_e \, d\sigma^{ff}_\omega(v'_e)}{\omega^2 \quad d\omega} \end{aligned} \tag{1.85}$$

The absorption coefficient calculated per electron with speed and one ion can be found with the Kramers Eq. (1.68):

$$a^{ff}_\omega(v_e) = \frac{16\pi^3}{3\sqrt{3}} \frac{Z^2 e^6}{m_e^2 c \hbar \omega^3 v_e} \tag{1.86}$$

Multiplying $a^{ff}_\omega(v_e)$ and Nn_e by averaging this relation on velocity v_e with the Maxwell distribution, we obtain the equation for the coefficient of the brake light absorption in plasma, cm^{-1}:

$$\begin{aligned} \chi^{ff}_\omega &= C_1 \frac{Z^2 N n_e g^{ff}}{T_e^{1/2} \omega^3}; \\ C_1 &= \frac{16\pi^2}{3} \left(\frac{2\pi}{3}\right)^{1/2} \frac{e^6}{m_e^{3/2} c k^{1/2} \hbar} = 1.45 \times 10^{10}\, \text{cm}^5/\left(\text{s}^3\, \text{K}^{1/2}\right) \end{aligned} \tag{1.87}$$

where N—the concentration of ions or atoms; here is similar to (1.70), we introduce a correction Gaunt factor g^{ff}.

The photoionization cross section at the bf absorption σ^{bf}_n (an inverse process for the process of photorecombination) can be obtained from (1.77) taking into account the conditions of detailed balance (Milne equation) [27]:

$$\sigma^{bf}_n = \frac{g_Z}{g_{Z-1}} \left(\frac{\hbar\omega}{m_e v_e c}\right)^2 \sigma^{fb}_n \tag{1.88}$$

Here g_Z, g_{Z-1}—the statistical weights of ions corresponding to the degree of ionization. For a neutral atom: $Z = 1$, for a hydrogen atom statistical weights of levels with the main quantum number n will be: $g_{Z-1,n} = 2n^2$, $g_Z = 1$:

$$\sigma^{bf}_n = \frac{8\pi}{3\sqrt{3}} \frac{e^{10} m_e Z^4}{c \hbar^6 \omega^3 n^5} \tag{1.89}$$

If the threshold frequency of photoionization of atoms from the level with $\omega_n = |E_n|/\hbar$, $|E_n| = m_e e^4 Z^2/(2\hbar^2 n^2)$, the photoionization cross section, cm^2:

$$\sigma^{bf}_n = 7.9 \times 10^{-18} \frac{n}{Z^2} \left(\frac{\omega_n}{\omega}\right)^3 \tag{1.90}$$

With increasing frequency, from the threshold value, photoionization cross section is proportional to ω^{-3}. Equations (1.88) and (1.77) are obtained on the basis of the semiclassical approximation by the generalizing of formula (1.68) for the bremsstrahlung the results should be as closer to the exact quantum solution as n is higher.

As in the case of bremsstrahlung, in the quasi-classical equation for the photorecombination and photoionization introduce appropriate corrective factors of Gaunt g^{fb} and g^{bf}. The total absorption coefficient χ_ω is the sum of the coefficients of the brake χ_ω^{ff} and photoionization χ_ω^{bf} absorption: $\chi_\omega = \chi_\omega^{ff} + \chi_\omega^{bf}$.

The coefficient χ_ω^{bf} is found by the sum of the photoionization cross sections for all levels of atom: $\chi_\omega^{bf} = \sum_{n^*}^{\infty} N_n \sigma_n^{bf}$, where the weighting factors are the populations of levels N_n. Substituting equation (1.89), the Saha equation (1.9) and the Boltzmann equation (1.4), using (1.87) in the last equation, we obtain the absorption coefficient without stimulated emission in the photoionization:

$$\begin{aligned}\chi_\omega^{bf} &= \sum_{n^*}^{\infty} N_n \sigma_n^{bf} = C_1 \frac{z^2 N n_e g^{bf}}{T_e^{1/2} \omega^3} \exp\left(-\frac{I}{kT}\right)\left[\exp\left(\frac{\hbar\omega}{kT}\right) - 1\right] \cdot \xi^{bf} \\ &= \frac{2\alpha^{-4} kT \lambda^3}{3^{3/2} \pi^2 e^2} n_a \frac{g_i}{g_a} \exp\left(-\frac{I}{kT}\right)\left[\exp\left(\frac{\hbar\omega}{kT}\right) - 1\right] \cdot \xi^{bf}\end{aligned} \tag{1.91}$$

Here for many-electron atoms in the calculation of photoionization absorption coefficients a factor ξ^{bf}, that takes values from 0.1 to 3 s with increasing wavelength λ from 200 to 1000 nm, is introduced.

Accordingly, the total absorption coefficient χ_ω, cm^{-1} is written as [26]

$$\chi_\omega = C_1 \frac{Z^2 N_i n_e}{T_e^{1/2} \omega^3} \exp(x) \Psi(x) = 4.05 \times 10^{-23} \frac{Z^2 N_i n_e}{T_e^{7/2}} \frac{\exp(x) \Psi(x)}{x^3} \tag{1.92}$$

Here $x = \hbar\omega / kT_e$; C_1 is determined by (1.87), and $\Psi(x)$—by (1.82).

The contribution of photoionization and braking mechanisms to the absorbance at $\hbar\omega < |E_g|$ is described by a ratio $\chi_\omega^{bf} : \chi_\omega^{ff} \approx [\exp(\hbar\omega/kT_e) - 1] : 1$.

At the not high temperatures, when the degree of ionization is small and the quanta are not large $(\hbar\omega < |E_g|)$, we can obtain for χ_ω, cm^{-1} the equation of Unsold–Kramers [26]:

$$\chi_\omega = \frac{16\pi}{3\sqrt{3}} \frac{Z^2 e^6 kT_e N}{\hbar^4 c \omega^3} \frac{g_i}{g_a} \exp\left(-\frac{I - \hbar\omega}{kT_e}\right) = 1.95 \times 10^{-7} \frac{Z^2 N}{T_e^2} \frac{g_i}{g_a} \frac{\exp\left(-\frac{I - \hbar\omega}{kT_e}\right)}{(\hbar\omega/kT_e)^3} \tag{1.93}$$

Here g_i, g_a—the statistical weights of ion and atom and N—the atom concentration.

1.6.3 Radiation in the Spectral Lines

Radiation in the spectral lines has a significant energy contribution to the radiant loss of plasma. In bound–bound transitions of electron in atom or the ion between different energy levels photon in the spectral lines is emitted (or absorbed). Thus, in spontaneous transition of atom or ion from the upper *k*th energy level to lower *i*th the photon $\hbar\omega_{ki} = E_k - E_i$ is emitted. The intensity (W/cm^3 sr) in the spectral line at such a transition of atom or ion from the upper level *k* of to the lower level *i* is expressed by the relation:

$$I_{ki} = \frac{1}{4\pi} h\nu_{ki} A_{ki} n_k = \frac{1}{4\pi} \hbar\omega_{ki} A_{ki} n_k = \int_0^\infty i_{ki}(\lambda) \mathrm{d}\lambda, \tag{1.94}$$

Here ν_{ki}—the frequency of the line center ($\nu_{ki} = \nu_0 = c/\lambda_0$); A_{ki}—the probability of optical (spontaneous) transition from the *k*th to *i*th level (the first Einstein coefficient); n_k—concentration of the emitting particles of the upper *k*th level; $i_{ki}(\lambda)$—distribution of the spectral intensity at a wavelength that is often fixed by spectrometer. Figure 1.4 shows the distribution of the spectral intensity of argon plasma (free burning arc) in the wavelength range λ = 400–450 nm.

The relation (1.94) is exact at low optical thickness of the plasma layer with the depth L:

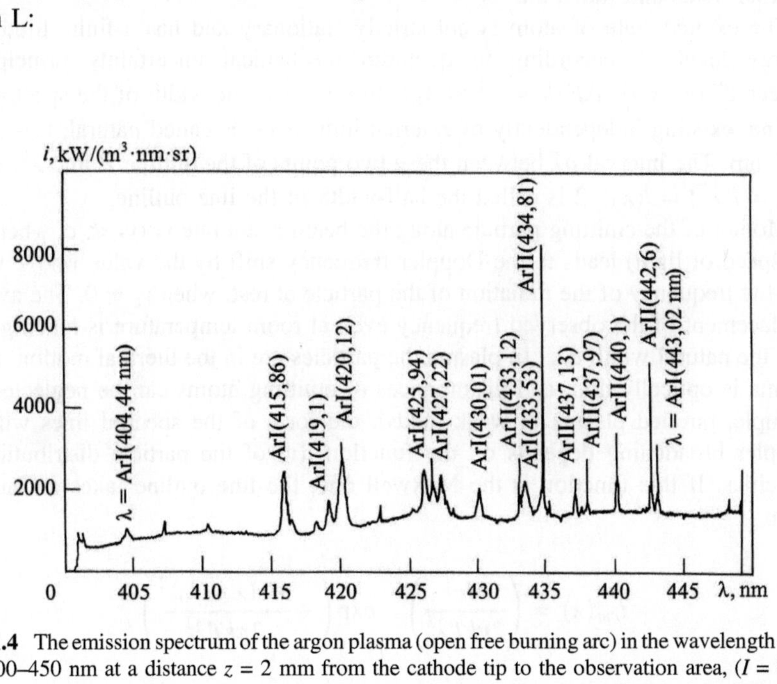

Fig. 1.4 The emission spectrum of the argon plasma (open free burning arc) in the wavelength range λ = 400–450 nm at a distance z = 2 mm from the cathode tip to the observation area, (I = 80 A, a diffraction grating with 600 lines per 1 mm): ArI—the atom of argon, ArII—the ion of argon

$$\tau_o = \int_0^L k_0 \mathrm{d}l_0 < 1$$

Here k_0—the absorption coefficient at the center of the line.

The probability of optical transitions $A_{ki} = 10^7 - 10^8\ \mathrm{s}^{-1}$ and corresponds to the lifetime of atom (or ion) in the kth excited state with $\tau_{ki} = 1/A_{ki} = 10^{-7} - 10^{-8}$ s. The spectral measurements for the extended objects of finite optical thickness, however, require the consideration of reabsorption and reemission of light. Spontaneously re-radiated photon can have a different frequency, polarization, and the direction of propagation, dissipate in the quenching process. The totality of this kind of nonlocal processes in the phase space is considered in the framework of the general problem of radiation transfer.

1.6.3.1 The Spectral Lines Outline

The part of the spectrum corresponding to the transition between the bound states of particles is understood as the spectral line. Mechanisms of line broadening and shaping their contours can be associated with the interactions between the particles, external fields and radiation.

The excited state of atom is not strictly stationary and has a finite lifetime τ. Energy level E_n according to quantum mechanical uncertainty principle is "smeared" on $\Delta\omega \sim \Delta E/\hbar \sim \tau^{-1} \approx A_{ki}^{-1}$. In this case, the width of the spectral line outline, existing independently of external influences, is called natural, it is $\delta\lambda_n \approx 10^{-5}$ nm. The interval $\delta\lambda$ between these two points of the outline λ' and λ'', where $i(\lambda') = i(\lambda'') = i(\lambda_0)/2$ is called the half-width of the line outline.

Motion of the emitting particle along the beam z at a rate v_z ($v_z \ll c$, where c—the speed of light) leads to the Doppler frequency shift by the value $\nu_0 v_z/c$ where ν_0—the frequency of the radiation of the particle at rest, when $v_z = 0$. The average displacement of the observed frequency even at room temperature is much greater than the natural width $\delta\lambda_n$. In plasma the particles are in the thermal motion, and if plasma is optically thin, and disturbances of emitting atoms can be neglected (for example, rarefied plasma in weak fields), the form of the spectral lines with the Doppler broadening depends on the function $f(v)$ of the particle distribution on velocities. If this function is the Maxwell one, the line outline takes a Gaussian form:

$$i_{mn}(\lambda) = \left(\frac{Mc^2}{2\pi kT\lambda_0^2}\right)^{1/2} \exp\left(-\frac{Mc(\lambda - \lambda_0)^2}{2\pi kT\lambda_0^2}\right) \tag{1.95}$$

Table 1.3 Values of the Doppler width at $T = 10^4$ K

Atom	M	λ_0 (nm)	$\delta\lambda_D$ (nm)
H	1	656.3	0.047
He	4	587.6	0.021
Ar	40	426.6	0.0049
Hg	200	546.1	0.0028

Here M—the relative atomic (molecular) mass of radiating particles. We have $\int_0^\infty i_{mn}(\lambda)\mathrm{d}\lambda = 1$.

The width of such outline is called the Doppler width and it is defined as:

$$\delta\lambda_D = 2\frac{\lambda_0}{c} \cdot \sqrt{\frac{2kT}{M}\ln 2} \tag{1.96}$$

As an example, the values of the Doppler width of the lines outline at temperature $T = 10^4$ K are represented at Table 1.3.

However, the effect of Doppler is not always the main mechanism of the spectral lines broadening in plasma with a high density, where the indignations of radiating atoms due to collisions with other particles arise very often. These collisions lead to the Lorentz broadening. The shape of lines is the dispersion one (the term $(\lambda - \lambda_0)^2$ is present in denominator):

$$i_{mn}(\lambda) = \frac{i_{mn}(\lambda_0)\delta\lambda_L^2}{4(\lambda - \lambda_0)^2 + \delta\lambda_L^2} \tag{1.97}$$

When the Lorentz width $\delta\lambda_L$ becomes more than natural, it can be expressed in the frequency of the "broadening" collisions:

$$\delta\lambda_L = \frac{\lambda_0^2}{2\pi c} = \frac{\sigma N}{\pi c} = \sqrt{\frac{2kT}{\pi M}} \tag{1.98}$$

Here σ—the cross section of the "broadening" collisions; N—the concentration of the perturbing particles.

The cross sections of "broadening" collisions are usually higher than the gas-kinetic ones. According to (1.98) at atmospheric pressure and $T \sim 10^4$ K the Lorentz width, roughly estimated, is in times less than the Doppler one. From (1.98) it also follows that in a weakly ionized plasma ($n_e/N \ll 1$) the Lorentz width $\delta\lambda_L \sim p/\sqrt{T}$.

The dispersion (Lorentz) and Doppler (Gaussian) shapes of lines with the same width are compared on Fig. 1.5.

It is essential that the decrease in intensity in the wing of the dispersion (Lorentz) line is much slower than in the wing of Doppler (Gauss).

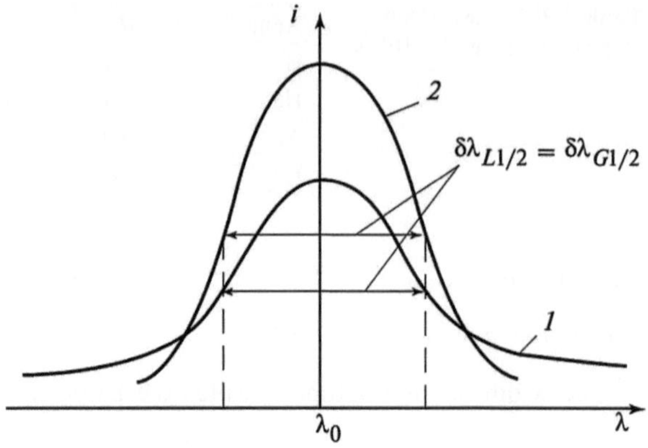

Fig. 1.5 Comparison of Lorentz (*lower curve 1*) and Gaussian (*upper curve 2*) outlines

The collisions with electrons, as a rule, play the main role in the mechanism of the spectral lines broadening. It is defined as large cross sections of these collisions and high-speed electrons. Cross section depends essentially on the nature and type of interaction between the emitting atom and electrons. The discovery of the phenomenon of splitting and shift of atomic levels under the influence of external electric field with the intensity E by R. Stark in 1913 has been called the Stark effect. The splitting of the levels in high electric fields is proportional to E and it corresponds to the linear Stark effect, in weak fields it is proportional to E^2 and leads to the quadratic Stark effect. Both circuits are described in the first approximation in the form (1.97). More details of these interactions are presented in [28]. The most typical values of half-width of the lines outlines are broadened as a result of the quadratic Stark effect, for the low-temperature plasma are equal to 0.0005–0.5 nm.

In practice, profiles of the spectral lines that are registered with the spectral equipment present a convolution of the Gaussian and dispersion outlines. The result of the simultaneous broadening of the Lorentz line shape and the Gaussian one is the convolution of the last lines:

$$I_V(\Delta\lambda) = I_{0V}\int\limits_{-\infty}^{+\infty} I_G(\Delta\lambda - \Delta\lambda')I_L(\Delta\lambda)\mathrm{d}(\Delta\lambda') \tag{1.99}$$

It is known as the Voigt profile. After the transformation (1.99), we obtain the convenient for numerical calculations equation

$$I_V(\Delta\lambda) = I_{0V}\frac{a}{\pi}\int\limits_{-\infty}^{+\infty}\frac{\exp(-y^2)\mathrm{d}y}{(u-y)^2 + a^2} \tag{1.100}$$

Here $u = \frac{\Delta\lambda}{\delta\lambda_\Sigma^G}$; $a = \frac{\delta\lambda_L}{2\delta\lambda_\Sigma^G}$; $\delta\lambda_\Sigma^G = [(\delta\lambda_D^G)^2 + (\delta\lambda_A^G)^2]^{1/2}$; $\delta\lambda_A^G$—the width of the hardware outline.

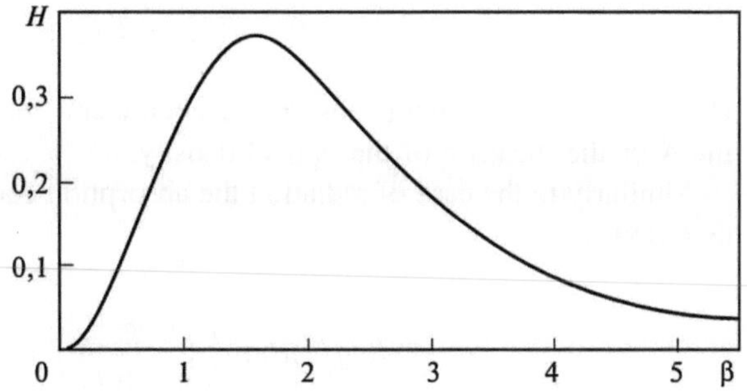

Fig. 1.6 The Holtsmark function

Lorentz, Gaussian and, as a consequence, Voigt outlines have the symmetrical appearance. The presence of ions in plasma leads to a further displacement of the result outline and to the asymmetry of the last one. This mechanism of broadening by ions is described by statistical theory [28]. The influence of it to the lines outlines is represented in experimental proceedings. However in plasma with a high concentration of ions, simultaneous impact to atoms of a large number of ions, effects to the line broadening, and it should be taken into account.

The basis of the mechanism of broadening by ions is the Holtsmark function [28] that is the probability of the presence of the field with the relative value β (caused by the presence of ions with a concentration n_{i}):

$$H(\beta) = \frac{2}{\pi\beta}\int_0^{\infty} x\,\sin x \exp\left[-\left(\frac{x}{\beta}\right)^{3/2}\right]\mathrm{d}x \qquad (1.101)$$

Here $\beta = E/E_0$, $E_0 = Ze/r_0^2$—the average electric field intensity created by ions, Z—the ion charge, $r_0 = (4\pi n_{\mathrm{i}}/3)^{-1/3}$—the average distance between the ions.

The function $H(\beta)$, calculated by the numerical methods is shown at Fig. 1.6.

In practical terms, the statistical broadening by ions (manifested in the statistical wing of the spectral line) should be expected for lines with large values of the constants of the quadratic Stark effect. This asymmetry of the spectral lines has been frequently observed in the Arc discharge.

1.6.4 The Absorption in Spectral Lines

The decrease in the light intensity with a spectrum in a narrow range of wavelengths $(\lambda, \lambda + \mathrm{d}\lambda)$ at the passing of a homogeneous medium with the layer thickness l and the atomic transition line from the kth to ith level (k—the upper level) occurs exponentially in Bouguer–Lambert–Berr law:

$$I(\lambda, z = l) = I(\lambda, z = 0)\exp[-\chi_{ik}(\lambda)l] \tag{1.102}$$

Here $\chi_{ik}(\lambda)$—the spectral absorption coefficient; $\chi_{ik}(\lambda)l$—the spectral absorption index or the measure of the optical density.

Similarly to the case of radiation the absorption coefficient in the lines is defined as follows:

$$\chi_{ik} = (h\nu_{ki}/c)B_{ik}n_i\left(1 - \frac{g_i}{g_k}\frac{n_k}{n_i}\right) = \int\limits_0^\infty x_{ik}(\lambda)\mathrm{d}\lambda \tag{1.103}$$

Here ν_{ki}—the frequency at the center of the radiation line outline; B_{ik}—the Einstein coefficient for absorption; n_i and n_k—the concentrations of particles in the lower and upper levels; $x_{ik}(\lambda)$—the spectral intensity of absorption depending on the wavelength λ.

The Einstein coefficients for spontaneous A_{ki} and induced B_{ki} radiation can also be used:

$$B_{ki} = \frac{g_i}{g_k}B_{ik}, \quad A_{ki} = \frac{g_i}{g_k}\frac{8\pi h\nu_{ik}^3}{c^3}B_{ik} \tag{1.104}$$

Note that according to (1.103), using the measured relative changes in the intensity of light passing through the object, the absolute value of the absorption coefficient and respectively, the density of absorbing states (neglecting the induced transitions) can be obtained.

References

1. C. Moore. Atomic energy levels. As derived from the analyses of optical spectra. — Washington, U.S.G., 1958. – 309 p.
2. V. Golant, A. Gilinskiy, S. Sakharov. Basics of plasma physics. – M.: Atomizdat, 1977.
3. M. Mitchner, C. Kruger. The partially ionized gases — M.: Mir, 1976 — 496p.
4. Y. Raiser. Physics of gas discharge. — M.: Science, 1987 – 592 p.
5. H. Griem. Phys. Rev., V. 131, p. 1170, 1963.
6. R. McWhirter, A. Hearn // Proc. Roy. Soc. 1963, V.82, p. 641.
7. Wilson R., J. Quant. Spectr. Radiative transfer – V.2, 1962, p. 477.
8. R. Huddlestone and S. Leonard. The diagnostics of plasma: Mir, 1967.
9. M. Alievsky, V. Zhdanov. The transport equations for non-isothermal multicomponent plasma. – 1963, №5 – p. 11 – 17.
10. Shi Nguyen-Kuok. The modeling of equilibrium plasma in RF- and Arc plasma torches // Proceedings of the Int. scientific. conf. "Electro-physical and electrochemical techniques" — St. Petersburg, 1997, pp. 63—66.
11. V. Kolesnikov The arc discharge in inert gases // Physical Optics: Proceedings of Phys. Inst named by P. Lebedev, the USSR Academy of Sciences. – M.: 1964. - T.30. — pp. 66 — 157.
12. L. Biberman, V. Vorobev, I. Yakubov. Low-temperature plasma with nonequilibrium ionization. // Successes of physical sciences. — 1979. — T. 128, №2. — pp. 233 — 271.

13. E. Velikhov, A. Kovalev, A. Rakhimov. Physical phenomena in gas-discharge plasma. – M.: Science, 1987 – p. 160.
14. V. Goldfarb, E. Ilyina. The nonequilibrium low-temperature plasma sources. The populations of the levels and spectral diagnostics. // Proc. of the Siberian department of the USSR Academy of Sciences — №3, V.1, 1975, pp. 28 — 38.
15. J. Beulens, D. Milojevic, D. Schram, P. Vallinga. A two dimensional nonequilibrium model of cascaded arc plasma flows. // J. Phys. Fluids, 3, 2548, 1991.
16. S. Dresvin, J. Amouroux, Nguen Quoc Shi. Analisis of deviation from thermal and ionization equilibrium in an Argon plasma flow. // J. High temperature Material Processes. — V. 1, №3, 1997 — pp. 369—381.
17. I. Belov, S. Isaev. Modeling of turbulent flow — St. Petersburg: Publishing House of the Baltic State Technical University, 2001 — p. 108.
18. A. Yun. Development and Analysis of Advanced Explicit Algebraic, Turbulence and Scalar Flux models for Complex Engineering configuration — Doctor thesis. Darmstadt. 2005.
19. W. Jones, B. Launder. The prediction of laminarization with a two-equation model of turbulence. // J. Heat Mass Transfer, 15: pp. 301—314, 1972.
20. D. Wilcox. Turbulence modeling for CFD — California, 1994.
21. F. Menter Zonal two Equation k-ω Turbulence Models for Aerodynamic Flows // AIAA paper 1993. P. 2906.
22. D. Wilcox. Turbulence Modeling for CFD. La Canada, California: DCW Industries Inc., 1998.
23. F. Menter. Two-equation eddy-viscosity turbulence models for engineering applications // AIAA Journal. 1994. Vol. 32, No 8. pp. 1598–1605.
24. A. Snegirev. High performance computing in technical physics. Numerical modeling of turbulent flows: Proc. allowance. — St. Petersburg: Polytechnic University Press, 2009.
25. V. Ochkin. Spectroscopy of low-temperature plasma. — M.: FIZMATLIT, 2006.
26. Y. Raiser // Success of Phys. Sciences — 1969 — T. 99, № 4 — p. 545.
27. Optical pyrometer of plasma. // Ed. N. Sobolev. — M.: Publishing House of Foreign. lit., 1960.
28. B. Smirnov. Ions and excited atoms in plasma. — M.: Atomizdat 1974.

Chapter 2
Classical Theory of the Particle Scattering

2.1 Classical Consideration of the Particle Scattering

For the determination of the plasma transport coefficients (electric conductivity, thermal conductivity, viscosity, the ambipolar and thermal diffusion coefficient) it is necessary to know the cross sections of particles interaction in plasma. The calculation of the interaction cross sections should be carried out using the quantum mechanical theory, but in the case of low-temperature plasma, when you need to take into account the large number of types of interactions between the particles, the bulky size and the complex method of calculation significantly limit its practical application. In practice, for the calculation of the interaction cross sections in plasma classical theory of binary collisions is often used according to the different functions of the potential energy (potential functions) $\phi(r)$ that can be chosen using the experimental data about the transfer coefficients. The paired elastic collisions of neutral–neutral, ion–neutral, electron–neutral, and the Coulomb interaction of charged particles are determined for transport phenomena in low-temperature plasma.

The Classical Problem of Particles Paired Collisions The classical problem of particles with the mass m_i scattering on the particle with the mass m_j in a spherically symmetric potential field $\phi(r)$ is often reduced to an equivalent problem of the scattering of a particle with the reduced mass $m = \frac{m_i m_j}{m_i + m_j}$ and the relative initial velocity $\vec{g} = \vec{v}_j - \vec{v}_i$ in the same potential field. Schematically, the formulation of this problem is shown in Fig. 2.1.

To analyze the dynamics of movement in equivalent case we write the equations of conservation of angular momentum and energy in the form:

$$mgb = mr^2\dot{\alpha} \tag{2.1}$$

S. Nguyen-Kuok, *Theory of Low-Temperature Plasma Physics*,
Springer Series on Atomic, Optical, and Plasma Physics 95,
DOI 10.1007/978-3-319-43721-7_2

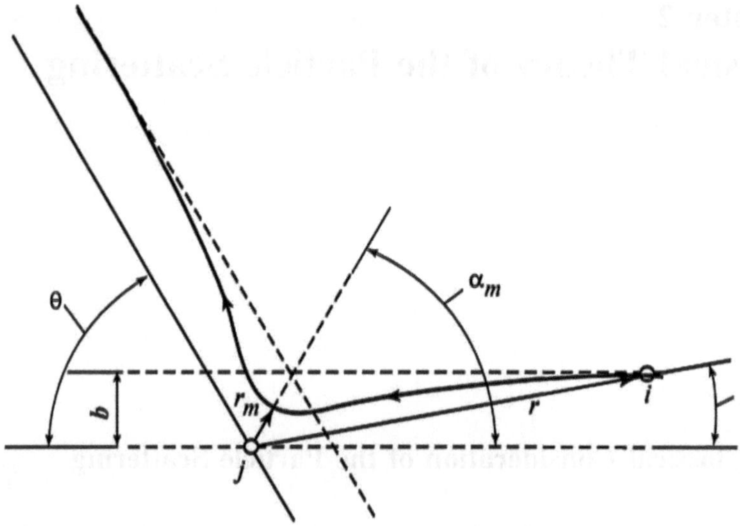

Fig. 2.1 Driving motion of the particle with mass m_i in a spherically symmetric potential field $\phi(r)$ in a collision with a particle of mass m_j. α the angle that determines the direction of a particle motion (polar angle); α_m the angle, corresponding to the maximum convergence of particles; b the impact parameter; θ the scattering angle; r the current distance between the particles; r_m the distance of the maximum particle convergence

$$\frac{1}{2}mg^2 = \frac{1}{2}m\dot{r}^2 + \frac{1}{2}mr^2\dot{\alpha}^2 + \phi(r), \tag{2.2}$$

where $\dot{\alpha} = \frac{d\alpha}{dt}$; $\dot{r} = \frac{dr}{dt}$.

If we put a value $\dot{\alpha}$ from (2.1) to (2.2), we get:

$$\frac{1}{2}mg^2 = \frac{1}{2}m\dot{r}^2 + \frac{1}{2}mg^2\frac{b^2}{r^2} + \phi(r) \tag{2.3}$$

The Eq. (2.3) does not contain a variable α and it can be used to describe the one-dimensional motion of a particle with mass m with and a total energy $\frac{1}{2}mg^2$ moving in an effective potential field $\phi_{\text{eff}}(r) = \phi(r) + \frac{1}{2}mg^2\frac{b^2}{r^2}$, here $\frac{1}{2}mg^2\frac{b^2}{r^2}$—the centrifugal potential.

The scattering angle $\theta(g, b)$ is defined as:

$$\theta(g, b) = \pi - 2\alpha_m \tag{2.4}$$

here α_m—the angle, corresponding to the maximum convergence of particles.

Taking into account (2.1) and (2.3), we have the following relation:

$$d\alpha = \frac{d\alpha/dt}{dr/dt}dr = \frac{b/r^2}{\sqrt{1 - \frac{\phi(r)}{1/(2mg^2)} - \frac{b^2}{r^2}}}dr \tag{2.5}$$

Integrating (2.5), we obtain an expression for α_m

$$\alpha_m = \int_{r_m}^{\infty} \frac{b/r^2}{\sqrt{1 - \frac{\phi(r)}{1/(2mg^2)} - \frac{b^2}{r^2}}}dr \tag{2.6}$$

Substituting (2.6) into (2.4), we have

$$\theta(g, b) = \pi - 2b\int_{r_m}^{\infty} \frac{dr/r^2}{\sqrt{1 - \frac{\phi(r)}{1/(2mg^2)} - \frac{b^2}{r^2}}}, \tag{2.7}$$

where the distance of the maximum convergence of particles r_m is determined from the condition $\left.\frac{dr}{dt}\right|_{r=r_m} = 0$, i.e., by solving the equation $1 - \frac{\phi(r_m)}{1/(2mg^2)} - \frac{b^2}{r_m^2} = 0$.

If we take into consideration the binary collisions only, some of plasma transport coefficients can be expressed using the deflection angle. The Eq. (2.7) is true for any spherically symmetric interaction potentials $\phi(r)$, so the knowledge about the potential function $\phi(r)$ is important for determination of the binary collision dynamics.

Differential, Efficient, and Averaged Effective Interaction Cross Sections The differential cross section is the ratio of the differential probability of particles scattering to the flow of incident particles in the elementary solid angle $d\Omega = 2\pi \sin\theta d\theta$. The differential cross section is $d\sigma = 2\pi b db$ and $\sigma(E, \theta) = \frac{d\sigma}{d\Omega} = \frac{2\pi b db}{2\pi \sin\theta d\theta}$. The effective cross section $Q^{(l)}(E)$ for interaction of order l, depending on the given initial relative power $E = mg^2/2$ is defined as:

$$Q^{(l)}(E) = 2\pi \int_0^{\infty} (1 - \cos^l\theta) b db = 2\pi \int_0^{\pi} \sigma(E, \theta)(1 - \cos^l\theta) \sin\theta d\theta \tag{2.8}$$

In the particular case $l = 1$ we have the cross section of the momentum transfer or the diffusion cross section, for $l = 2$—the viscosity section, etc.

The average cross section depending on the temperature, using the Maxwellian velocity distribution, can be determined as:

$$Q^{(l,s)}(T) = \frac{2(l+1)}{(s+1)!\left[2l+1-(-1)^l\right]}\int\limits_0^\infty e^{-\gamma^2}\gamma^{2(s+1)}Q^{(l)}(E)\mathrm{d}\left(\frac{E}{kT}\right)$$
$$= \frac{2(l+1)}{(s+1)!\left[2l+1-(-1)^l\right]}\int\limits_0^\infty e^{-x}x^{s+1}Q^{(l)}(kTx)\mathrm{d}x \tag{2.9}$$
$$= \frac{4(l+1)}{(s+1)!\left[2l+1-(-1)^l\right]}\sqrt{\frac{2\pi m}{kT}}\Omega^{(l,s)}(T),$$

Here $x = \gamma^2 = \frac{mg^2}{2kT} = E/(kT)$—the initial relative velocity of the colliding particles.

Accordingly, the collision integral $\Omega^{(l,s)}(T)$ of the order (l, s), depending on the temperature, is determined as follows:

$$\Omega^{(l,s)}(T) = \sqrt{\frac{\pi kT}{2m}}\int\limits_0^\infty e^{-x}x^{s+1}\int\limits_0^\pi \sigma(E,\theta)\left(1-\cos^l\theta\right)\sin\theta\mathrm{d}\theta\mathrm{d}x$$
$$= \sqrt{\frac{kT}{8\pi m}}\int\limits_0^\infty e^{-x}x^{s+1}Q^l(kTx)\mathrm{d}x = \frac{(s+1)!\left[2l+1-(-1)^l\right]}{4(l+1)}\sqrt{\frac{kT}{2\pi m}}Q^{(l,s)}(T) \tag{2.10}$$

In the future, data about the averaged effective cross sections $Q^{(l,s)}(T)$ are used to calculate coefficients of plasma transport. The procedure of calculating the transport coefficients usually begins with a definition of the function of the interaction potential of the particles $\phi(r)$ and the differential cross section $\sigma(E,\theta)$. These data dependent on the chemical composition and density of plasma, difficult collisional and radiative processes in it as well as on the specific conditions of plasma existence. There are a huge number of literatures on these issues.

For a gas that consists of the neutral particles, the interaction potential is short-range at the long distance ($r > R$) that decreases so rapidly that the integral $\int\limits_R^\infty \phi(r)\mathrm{d}r$ tends to zero. As an example, we note that at large distances the two charged particles interact by the Coulomb law $\phi(r) \sim 1/r$, charged and neutral particles are attracted because of the interaction of charged particles and the induced dipoles $\phi(r) \sim 1/r^4$ and two neutral particles are attracted due to the interaction of instantaneous dipoles. In all cases, the particles are attracted to large distances to each other, i.e., the potential energy is negative, and repel at short distances due to the overlap of the wave functions of the particles. Here are some potential functions depending on the intermolecular distance r that are often used in calculation of the interaction cross sections of particles and plasma transport coefficients [1]:

(1) $\phi(r) = \frac{Z_i Z_j e^2}{r}\exp(r/r_D)$—the potential of the Coulomb interaction. Here Z_i, Z_j—the charges of the particles, r_D—the Debye radius;
(2) $\phi(r) = \phi_0 \exp(-r/\rho_0)$—the interaction potential between atoms and molecules. Here ϕ_0, ρ_0—the some constants;
(3) $\phi(r) = 4\varepsilon\left[\left(\frac{\sigma}{r}\right)^{12} - \left(\frac{\sigma}{r}\right)^{6}\right]$—the Lennard–Jones potential with polynomials of 6th and 12th degrees, that is often used to describe the induced dipole–dipole interaction of molecules. The value σ—distance r, where $\phi(r) = 0$, ε—the maximum energy of attraction or the depth of the potential well that is reached at $r = \sqrt[6]{2}\sigma$;
(4) $\phi(r) = \infty$, $r \le \sigma$; $\phi(r) = 0$, $r \ge \sigma$—model of hard spheres with a diameter σ;
(5) $\phi(r) = \mathrm{d}r^{-\delta}$—a model of the point center of repulsion. The value $\delta = 9 - 15$ is called as the repulsive index, d—is a constant;
(6) $\phi(r) = \frac{\varepsilon}{1-6/\alpha}\left[\frac{6}{\alpha}\exp\left[\alpha\left(1 - \frac{r}{r_m}\right)\right] - \left(\frac{r_m}{r}\right)^{6}\right]$; where $r \ge r_{\max}$; $\phi(r) = \infty$, $r \le r_{\max}$—the modified potential of Buckingham that is used to describe the induced dipole–dipole interaction of molecules. Here r_m—the equilibrium distance between the nucleus of the molecule that is formed with the repulsive particles; $r_{\max}$—the distance r, that corresponds to the relative maximum $\phi(r)$. The ratio $r_{\max}/r_m$ is determined by the smallest root from the transcendental equation

$$\left(\frac{r_{\max}}{r_m}\right)^{7} \exp\left[\alpha\left(1 - \frac{r_{\max}}{r_m}\right)\right] - 1 = 0;$$

(7) $\phi(r) = \varepsilon[\exp(-2\beta(r - r_m)/r_m) - 2\exp(\beta(r - r_m)/r_m)]$—the Morse potential where $\beta = 1.6739$ Å^{-1}.
(8) $\phi(r) = A/r^n$—the inverse-repellent potential. Where A—a constant.

The Trajectories of the Particle in a Pair Collision The scattering angle is π for the small values of the impact parameter, because the direct reflection of the particles occurs (Fig. 2.2, 3).

For small impact parameters the scattering angle is small and positive (Fig. 2.2, 1). At the large values of the impact parameter, the scattering angle of impact is small and negative due to the attraction of particles (Fig. 2.2, 2).

Between these two cases the function $\theta(g, b)$ if it is continuous, it has at least one zero value (Fig. 2.2, 4).

As a result, the dynamics of interaction between colliding particles can be treated several times around the scattered particles (Fig. 2.2, 5).

In the experimental determination of the angular dependence of the scattered particles the Ramseur device is used, that consists of multiple detectors positioned

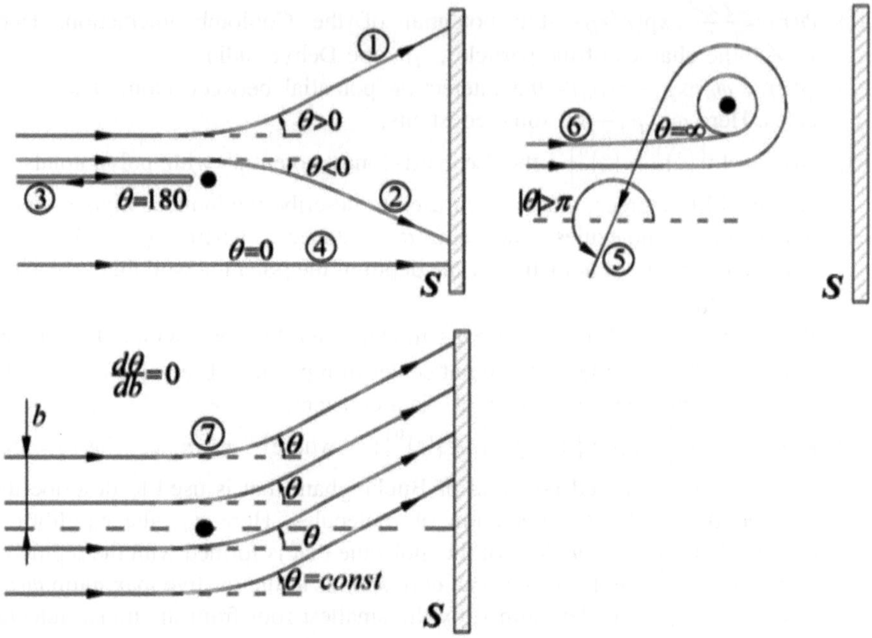

Fig. 2.2 Possible tracks of particle motion in a pair collision: *S* the screen

at different angles and recording the scattered particles. These detectors cannot differ the particles that made several turns.

The function $\theta(g, b)$ for some impact parameter can tend to infinity. This one corresponds to the fact that the attraction is compensated with the repulsion, i.e., the particles are captured or "orbiting" (Fig. 2.2, 6).

In practice, there are two cases of dispersion that are known as "radiance" when $\theta(g, b) = 0$ (Fig. 2.2, 4) and "rainbow scattering" when $\frac{\partial\theta(g,b)}{\partial b} = 0$ (Fig. 2.2, 7).

For the case of "radiance" particle does not change its direction of movement, despite the fact that part of the track in the field of the potential action, for example, when the condition $\phi(r) = -\frac{1}{2}mg^2\frac{b^2}{r^2}$ is performed.

The problem of light propagation in the atmosphere drops leads to the optical effect "radiance".

The Conditions for Application of the Theory of Classical Mechanics for the Scattering of Particles In classical mechanics particles are considered infinitely small. In quantum mechanics a wave packet with the de Broglie wavelength corresponds to a particle. Provisions of classical mechanics, especially violated in the scattering at small angles because of occurrence of diffraction. The small angle at the classical scattering can be defined as the ratio of the transverse and longitudinal pulses:

$$\theta_{\text{class}} \approx \frac{\Delta p_{\perp}}{p} \approx \frac{b}{pv}\frac{\partial \phi}{\partial r}.$$

In the scattering of particles in the power center the cross change in momentum is defined as $\Delta p_{\perp} \approx \hbar/a$, where $a \approx b$—the size of the gap. Quantum scattering angle or the diffraction angle is estimated as

$$\theta_{\text{dif}} \approx \frac{\Delta p_{\perp}}{p} \approx \frac{\hbar}{pa} \approx \frac{\lambda\!\!\!^{-}}{a} \approx \frac{\lambda\!\!\!^{-}}{b}.$$

The condition of the classical mechanics application is as follows:

$$\frac{b}{pv}\frac{\partial \phi}{\partial r} = b\frac{\partial}{\partial r}\left(\frac{\phi}{2E}\right) \gg \frac{\lambda\!\!\!^{-}}{b}. \tag{2.11}$$

As a rough approximation (2.11) has the form: $\frac{\phi(b)}{2E} \gg \frac{\lambda\!\!\!^{-}}{b}$. In this condition, we can restrict with the classical consideration. $\lambda\!\!\!^{-} \sim 1/m$ and the classical description is mainly affects for electrons. In this case, quantum effects play a significant role, and the scattering of electrons in plasma should be considered by the theory of quantum mechanics.

2.2 Determination of Cross Sections for Plasma Particles Interaction

2.2.1 *Coulomb Scattering: The Cross Section of the Interaction of Charged Particles*

Among all the forces of interaction between atomic particles the Coulomb forces has the slowest decrease with distance ($\sim r^{-2}$) and potential energy of interaction $\phi(r) \sim r^{-1}$. Coulomb forces have the greatest long-range, i.e., collision occurring at large impact parameters, make a major contribution to the loss of a particle momentum. The low-temperature plasma with the degree of ionization of more than 0.1–1 % a collision of charged particles with each other play the decisive role for the drift and diffusion of charges and energy exchanges. The small-angle scattering in the short-range potential can be considered, that is valid for almost all impact parameters. The scattering angle θ in this case can be determined by the relation of the longitudinal cross-pulse $\Delta p_{\perp}$ pulse to the $p_{||}$ of dissipating particles (Fig. 2.3). Given the fact that $r^2 = b^2 + v^2t^2$ we have:

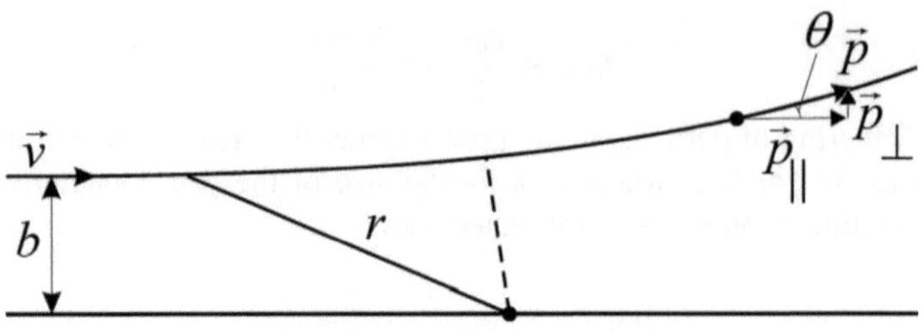

Fig. 2.3 To the derivation of Rutherford's equation

$$\theta \approx \frac{\Delta p_{\perp}}{p_{||}} = \frac{1}{mv}\int_0^{\infty} F_{\perp}\mathrm{d}t = \frac{2}{mv}\int_0^{\infty} F\frac{b}{r}\mathrm{d}t = \frac{2b}{mv}\int_b^{\infty}\left(-\frac{\partial\phi}{\partial r}\right)\frac{1}{v\sqrt{r^2-b^2}}\mathrm{d}r$$
$$= \frac{Z_iZ_je^2b}{mv^2/2}\int_b^{\infty}\frac{\mathrm{d}r}{r^3\sqrt{1-(b/r)^2}} \tag{2.12}$$

where Z_i, Z_j—the charges of the particles. If we change the variable $b/r = \sin\theta$, we have $r = b/\sin\theta$; $\mathrm{d}r = -\frac{b\cos\theta\mathrm{d}\theta}{\sin^2\theta}$. Substituting the last relation in (2.12) and integrating, we obtain

$$\theta(E, b) = \frac{Z_iZ_je^2}{bmv^2/2} = \frac{Z_iZ_je^2}{Eb} \tag{2.13}$$

As a result,

$$b(E, \theta) = \frac{Z_iZ_je^2}{E\theta} \tag{2.14}$$

The differential cross section for the Coulomb potential is

$$\sigma(E, \theta) = \frac{d\sigma}{d\Omega} = \frac{2\pi b}{2\pi\sin\theta}\frac{db}{d\theta} = \left(\frac{Z_iZ_je^2}{E}\right)^2\frac{1}{\theta^4} \tag{2.15}$$

Equation (2.15) is called the Rutherford's equation. The rapid growth of the differential cross section with decreasing scattering angle ($\sigma \sim 1/\theta^4$) and inversely proportionality of θ to the square of kinetic energy of the relative motion is a characteristic of Coulomb scattering.

Rutherford's equation in up to a factor coincides with the equation obtained for the differential cross section in accordance with the quantum mechanical theory.

This is due to the condition of classical mechanics application $\hbar v/e^2 \ll 1$, that expresses the smallness of the de Broglie wavelength $\lambdabar$ compared to the characteristic size of the force field r_0, i.e., the distance between the particles, when the potential energy of the Coulomb interaction $\phi(r) = e^2/r$ is comparable to the kinetic energy of the relative motion $E = mv^2/2$. The value $r_0 = \frac{Z_i Z_j e^2}{mv^2}$ is called as the Coulomb radius. The inequality $\hbar v/e^2 \ll 1$ is equivalent to $\lambdabar \ll r_0$. It imposes a restriction on the kinetic energy: $E \ll me^4/(2\hbar^2)$. For electrons in the ion field the condition $E \ll I_H = 13.6$ eV must be satisfied. In most cases that are interested in the physics of low-temperature plasma: $E \ll I_H$. Thus, the calculation of the electrons scattering in the field of the other charges in up to a factor can be considered in accordance with the theory of classical mechanics. For ions it can be fair to energies of tens of keV.

For small scattering angles $(1 - \cos\theta \approx \theta^2/2)$ scattering cross section of the vehicle in the Coulomb potential

$$
\begin{aligned}
Q^{(1)}(E) &= 2\pi \int\limits_{\theta_{\min}}^{\theta_{\max}} \left(\frac{Z_i Z_j e^2}{E}\right)^2 \frac{1}{\theta^4} \frac{\theta^2}{2} \theta \mathrm{d}\theta = \pi \left(\frac{Z_i Z_j e^2}{E}\right)^2 \ln \frac{\theta_{\max}}{\theta_{\min}} \\
&= \pi \left(\frac{Z_i Z_j e^2}{E}\right)^2 \ln \Lambda,
\end{aligned}
\tag{2.16}
$$

where $\theta_{\max}$, $\theta_{\min}$—the maximum and minimum scattering angles; $\ln \Lambda = \ln \frac{\theta_{\max}}{\theta_{\min}} = \ln \frac{r_{\max}}{r_0}$—the Coulomb logarithm; $r_{\max}$—the maximum radius of the scattering.

Using the Coulomb potential of a charged particle $\phi(r) = Z_i Z_j e^2/r$ without shielding by surrounding charged particles would lead to a divergence of the integral (2.7). In order to get the final value of the cross section, many authors consider it possible to limit the upper limit of integration in (2.7) with the screening parameter, taking it equal to the radius of the Debye r_D, or even the average interatomic distance $(r = n^{-1/3})$.The different values of the effective interaction cross sections follow from here.

As for the potential function, interaction of charged particles with small scattering angles are widely described by a screened Coulomb potential: $\phi(r) = \frac{Z_i Z_j e^2}{r} \exp(-r/r_D)$. Such a potential function shows that the averaged effective cross sections of the electron–ion interaction in the fourth order of approximation have the form [2]

$$\begin{aligned}
Q^{(1,s)} &= \frac{4\pi}{s(s+1)} b_0^2 \left[\ln \Lambda - \frac{1}{2} - 2\bar{\gamma} + \psi(s) \right]; \\
Q^{(2,s)} &= \frac{12\pi}{s(s+1)} b_0^2 [\ln \Lambda - 1 - 2\bar{\gamma} + \psi(s)]; \\
Q^{(3,s)} &= \frac{12\pi}{s(s+1)} b_0^2 \left[\ln \Lambda - \frac{7}{6} - 2\bar{\gamma} + \psi(s) \right]; \\
Q^{(4,s)} &= \frac{16\pi}{s(s+1)} b_0^2 \left[\ln \Lambda - \frac{4}{3} - 2\bar{\gamma} + \psi(s) \right],
\end{aligned} \tag{2.17}$$

Here $\Lambda = \frac{2r_D}{b_0}$; $b_0 = \frac{Z_i Z_j e^2}{2kT}$—the distance of maximum convergence of colliding particles; $\bar{\gamma} = 0.5772$—Euler's constant; $\psi(s) = \sum_1^{s-1} 1/n$, ($\psi(1) = 0$).

The averaged cross section of the electron–ion interaction, obtained from (2.17) for $l = 1$, $s = 1$ [2] is shown in Fig. 2.4.

In the literature, there are other equations for calculation of the effective cross section of interaction of charged particles. One of them [3, 4] has the form

$$Q_{\mathrm{ei}} = \frac{a \times 10^{-10}}{T_{\mathrm{e}}^2} \ln \left(\frac{b \times 10^4 T_{\mathrm{e}}}{n_{\mathrm{e}}^{1/3}} \right), \tag{2.18}$$

Here a, b—are numerical coefficients. These factors vary among different authors. Thus, in [3] $a = 5.85$, $b = 5.35$; in [4] $a = 2.79$, $b = 5.98$. These

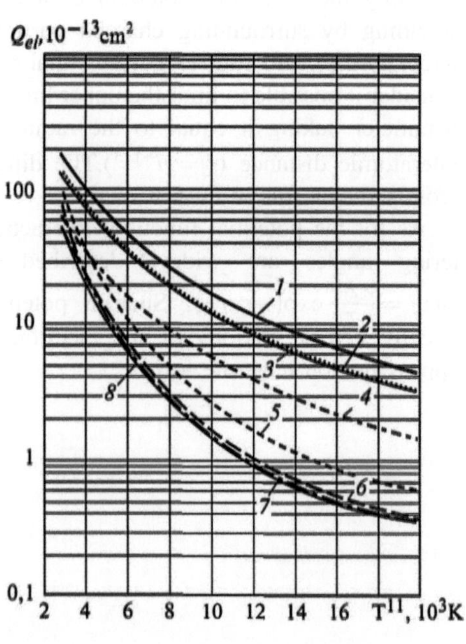

Fig. 2.4 The effective cross section of the electron–ion $Q_{\mathrm{ei}}(T_{\mathrm{e}})$ at the thermal equilibrium. *1–6, 8* calculation of equations, respectively (2.20), (2.21), (2.18) with $a = 5.85$, $b = 5.35$, (2.18) when $a = 2.79$, $b = 5.98$, (2.19), (2.22), (2.23); *7* data [2]

differences are generally related to the choice of impact parameters in the Coulomb logarithm. Most often, the quantity on the logarithm, is the ratio of Debye radius to sustainable medium parameter of the collision, resulting to the dispersal of particles to the angle of 90°.

Michner and Kruger [5] found a relatively simple equation to calculate the effective cross section of the electron–ion:

$$Q_{\mathrm{el}} = 6\pi \bar{b}_0^2 \ln \Lambda, \quad (2.19)$$

Here $\Lambda = \frac{r_D}{\bar{b}_0}$. In this case, the average sighting distance $\bar{b}_0$ and the Debye radius is calculated as:

$$\bar{b}_0 = \frac{e^2}{12\pi\varepsilon_0 kT}, \quad r_{\mathrm{D}} = \sqrt{\frac{\varepsilon_0 kT}{n_e e^2}}.$$

In [6] the cross section Q_{el} for two-temperature plasma is calculated by the equation

$$Q_{\mathrm{el}} = \frac{8.78 \times 10^{-10}}{T_{\mathrm{e}}^2} \ln\left[\frac{2.79 \times 10^4 T_{\mathrm{e}}}{n_{\mathrm{e}}^{1/3}(1 + T_{\mathrm{e}}/T_{\mathrm{a}})^{1/3}}\right]. \quad (2.20)$$

In [7] for the cross section Q_{el} the equation is:

$$Q_{\mathrm{el}} = \frac{6.58 \times 10^{-10}}{T_{\mathrm{e}}^2} \ln\left(\frac{2.57 \times 10^4 T_{\mathrm{e}}}{n_{\mathrm{e}}^{1/3}(1 + T_{\mathrm{e}}/T_{\mathrm{a}})}\right). \quad (2.21)$$

In [8] using the methods of quantum field theory to derive the expressions are given using the renormalized scattering probability that automatically limit the interaction at the large impact parameters and eliminate the need to artificially introduce a lower limit of integration. Thus obtained expression for the cross section of electron scattering on the ions is given by [9]:

$$Q_{\mathrm{el}}(T) = \frac{7 \times 10^{-10}}{T^2}\left[\lg\left(2.8 \times 10^6 \frac{T}{\sqrt{n_{\mathrm{e}}}}\right) + 3\right]. \quad (2.22)$$

Results of the effective cross section calculation of the electron–ion of different expressions (Fig. 2.4) shows that the latter expression gives the most consistent with the data Devoto [2]. For more matches with the data [2] we suggest the following calculation equation obtained by approximation (2.22):

$$Q_{\mathrm{el}} = \frac{6.4 \times 10^{-10}}{T_{\mathrm{e}}^2}\left[\lg\left(2.8 \times 10^6 \frac{T_{\mathrm{e}}}{\sqrt{n_{\mathrm{e}}}}\right) + 3\right]. \quad (2.23)$$

These calculations [10] shows that in a dense low-temperature plasma the mass difference between electrons and ions do not significantly affect the angle of dispersion for interactions of electron–ion, electron–electron, and ion–ion and average effective cross sections of the highest order $Q_{\rm ei}^{(l,s)}(T)$, $Q_{\rm ee}^{(l,s)}(T)$, $Q_{\rm ii}^{(l,s)}(T)$ are distinguished from each other by not more than 3–7 %, in the future, you can take them equal: $Q_{\rm ei}^{(l,s)}(T) = Q_{\rm ee}^{(l,s)}(T) = Q_{\rm ii}^{(l,s)}(T)$. By (2.17) we also obtain that the average effective cross sections are proportional to the square of the charge of the ions.

Electrons play the dominant role in the dynamics of the process of screening, the temperature in the equations for the cross section of the interaction of charged particles is often taken equal to $T_{\rm e}$ (ions are considered to be fixed in comparison with electrons).

The results obtained by author in the calculation of two-temperature section $Q_{\rm ei}(T_{\rm e}, T_{\rm a})$ also show that it depends mainly on the electron temperature $T_{\rm e}$.

2.2.2 *Ramseur Effect and the Resonance Scattering of the Electrons by Atoms*

Elastic scattering of electrons on neutral particles can be described only by means of quantum mechanical theory. Currently, it is possible to calculate analytically the effective scattering cross sections for simple atoms only. For more complex particles, there is a considerable discrepancy between the experimental data and the results of theoretical calculations, especially for low-energy electrons (0.1–2 eV). In this range, the Ramseur minimum effect significantly affects the determination of the effective interaction cross sections of electron–atom, such as the scattering of electrons in argon. The wave properties of the electron appear in the existence of minimums. In some of the wave energy that describe the motion of the electron (de Broglie wavelength) and dispersed potential field, interfere and cancel each other out. The probability and the scattering cross section depend on the intensity of the scattered wave and it is very small. Such a situation occurs when a low-energy, different for different gases, when the de Broglie wavelength of an electron is comparable with the size of the atom. At the same time as a result of the diffraction of electrons per atom there is an amplitude maximum of the electron wave and the electron as “skirts” atom without scattering on it.

Results of calculation of the effective cross section of the electron–atom interaction, depending on the power using the so-called phase shift [11] η_l are in good agreement with the experimental data. Consequently, experimental data remain a reliable source for determining the effective interaction cross section of electron–atom in low-temperature plasma. Results of calculation of the average cross section of the interaction of the electron–atom [2] for $l = 1$, $s = 1$ are shown in Fig. 2.5.

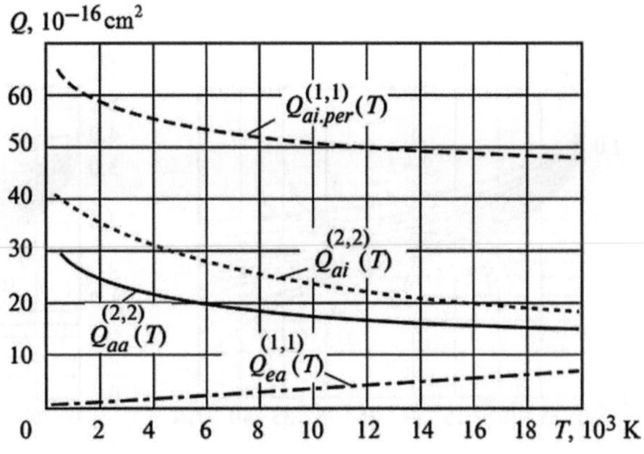

Fig. 2.5 The average effective interaction cross sections for argon depending on the temperature. $Q^{(2,2)}_{\mathrm{aa}}(T)$, $Q^{(2,2)}_{\mathrm{ai}}(T)$, $Q^{(1,1)}_{\mathrm{ea}}(T)$ the interaction cross sections atom–atom, atom–ion, electron–atom, respectively; $Q^{(1,1)}_{\mathrm{ai.per}}(T)$ the exchange cross section of atom–ion

2.2.3 The Cross Sections for the Interaction of Molecule–Molecule, Atom–Atom

The potentials $\phi(r)$ of the heavy particles are divided into long-range and short-range. Long-range potentials between the heavy particles determine the values of the effective cross sections only at low energies of particles in the conditions of a temperature not more than 3000 K. Below this limit as a function approximating the interaction potential, the Lennard–Jones potential with polynomials 6th and 12th degree $\phi(r) = 4\varepsilon\left[\left(\frac{\sigma}{r}\right)^{12} - \left(\frac{\sigma}{r}\right)^{6}\right]$ [1] or a modified capacity of Buckingham [10] well established itself. At $T > 3000$ K the contribution of the short-range potential in the effective cross section becomes significant, and it is necessary to use the exponential type of function for approximation:

$$\phi(r) = \phi_0 \exp(-r/\rho_0), \tag{2.24}$$

Here ϕ_0, ρ_0—some constants. For example, the potential of atom–atom interaction in argon at $\phi_0 = 7100$ eV, $\rho_0 = 0.258$ Å [2] (or $\phi_0 = 32300$ eV, $\rho_0 = 0.224$ Å [10]) $\phi(r) = 0.19 - 14.5$ eV, ($r = 2.716 - 1.598$ Å).

Another approximation is the Morse potential:

$$\phi(r) = \varepsilon\{\exp[-2\beta(r - r_m)/r_m] - 2\exp[\beta(r - r_m)/r_m]\} \tag{2.25}$$

here r_m—the equilibrium distance between the kernels of molecules formed of repulsive particles. Detailed tables of collision integrals for the Morse potentials can be found, for example, in [12, 13].

In selection of appropriate potential functions the polarity of the particles, especially with unfilled outer electron shells and others, should be also considered.

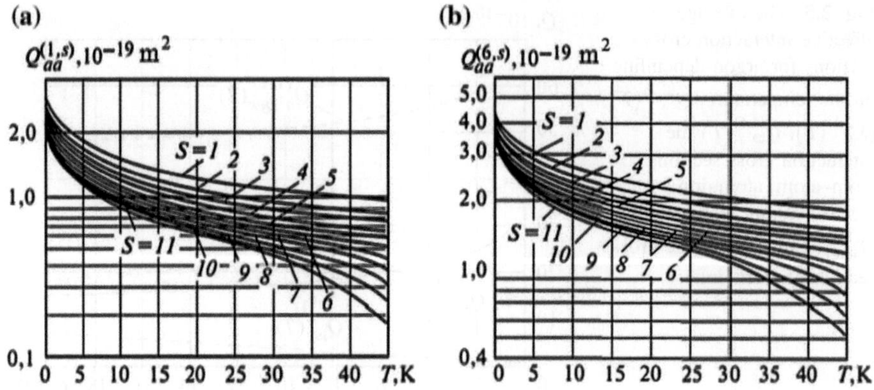

Fig. 2.6 The results of calculations of collision integrals with the order l = 1 (**a**) and l = 6 (**b**) for the atom–atom interaction cross sections

The Average Effective Interaction Cross Section of Atom–Atom Theoretical differential scattering cross sections were calculated by A. Phelps and others [14–18] using the scattering phase shifts estimated by approximation of G. Ventzela, H. Kramers, and L. Brillouin. The theoretical results of asymmetric scattering of argon atoms in an Argon plasma are approximated by the following function:

$$\frac{d\sigma}{d\Omega}(\theta) = \frac{1}{2}a[1 + 2\eta - \cos\,\theta]^{-2} + \frac{1}{2}b \tag{2.26}$$

Here

$$\begin{cases} a(\varepsilon) = 0.27\varepsilon^{0.2}\left[1 + \left(\frac{\varepsilon}{9}\right)^{0.8} + \left(\frac{\varepsilon}{350}\right)^{1.3} + \left(\frac{\varepsilon}{800}\right)^{2.2}\right]^{-1}; \\ b(\varepsilon) = 3.0\varepsilon^{-0.2}\left[1 + \left(\frac{\varepsilon}{15}\right)^{0.5} + \left(\frac{\varepsilon}{800}\right)^{1.1} + \left(\frac{\varepsilon}{10\,000}\right)^{2}\right]^{-1}; \\ \eta(\varepsilon) = 0.002\varepsilon^{0.75}\left[1 + \left(\frac{\varepsilon}{6}\right)^{1.15} + \left(\frac{\varepsilon}{80}\right)^{1.95} + \left(\frac{\varepsilon}{500}\right)^{2.5}\right]^{-1}. \end{cases} \tag{2.27}$$

$d\Omega = \sin(\theta)d\theta d\varphi$—differential solid angle of scattering.

The collision integrals to l = 6, s = 11 were obtained by direct integration of the Eqs. (2.8) and (2.9) for the quantum elastic scattering differential cross sections (Fig. 2.6).

2.2.4 The Cross Sections for the Interaction of the Atom–Ion

When the ion closes to molecule or atom, it polarizes them by their Coulomb field, and this ion experiences the induced dipole moment in the molecule. The interaction potential of the atom–ion is a more long-range than the potential of the atom–atom, as, in addition to these components, there are electrostatic attraction of charge-induced dipole and pull-type charge-induced dipole.

Located at a distance r from the molecule, the singly charged ion induces a dipole moment $d = \alpha e/r^2$ in the molecule by its electric field ($E = e/r^2$), here α—the polarizability of the molecule. Positive ions attract electrons that are molecular closer to him, and the negative pushes them away. Therefore, the polarizing nature of the interaction is attractive. The electric field intensity of the induced dipole in the point, where the ion is located, $E = 2d/r^3 = 2\alpha e/r^5$, its potential $\varphi(r) = -\alpha e/(2r^4)$. The force that acts to ion is $F = 2\alpha e^2/r^5$, the potential energy of interaction between the ion and the molecule $\phi(r) = -\alpha e^2/(2r^4)$, and the energy is negative regardless of the sign of the ion charge.

Approximation of the ion–atom interaction potential using (2.24) or (2.25) can be written as

$$\phi(r) = \phi_0\{\exp[-2C(r - r_e)/\sigma] - 2\exp[-C(r - r_m)/\sigma]\} \tag{2.28}$$

Here $C = \sigma \ln 2/(r_e - \sigma)$; σ—the distance of the maximum convergence of two particles. For example, to approximate the interaction potential of the atom–ion in argon $\phi_0 = 1.25$ eV, $r_e = 2.434$ Å, $C = 3.382$, $r = 2 - 2.9$ Å [2]. Results of calculation of the average effective interaction cross section of atom–ion for argon plasma $Q_{ai}^{(2,2)}(T)$ using the function (2.28) are shown in Fig. 2.5.

Determination of the average effective atom–ion cross section is extremely important for the transport coefficients. The low-temperature plasma with ion concentration $n_i > 10^{14}$ cm^{-3} ($p = 10^5$ Pa = 1 bar) [19] a significant process of atom–ion recharge, when the ions move in their own gas, is observed. The positive ion is accelerated by electric field between collisions or just wandering in the gas passing by a neutral atom, the latter selects the outer electron is neutralized and continues the inertia path in the form of a neutral atom. Recharging takes place even in their flying at large impact parameters that occurs very often. For electron transfer the contact time it is sufficient, as it goes from an atom to ion with a speed about its rate in atom (10^6 m/s), that is much larger than the velocity of the ions. If the process is in an electrostatic field, the new formed ion begins to accelerate in the direction of the field with an average zero directed velocity. So, in an act of recharging occurs total loss of directional momentum of the ion. Recovery is possible in the case of heterogeneous particles, but it most often occurs one time and as a result of not always feasible conditions on the energy state of the particles with different ionization potentials.

Recharge in a gas is interesting because of their result of energy transfer is equivalent to elastic collision, although it accompanied by a transformation of particles as an inelastic collision. The recharging processes are quantum mechanical in nature, so their cross sections are more than elastic scattering cross section of atom–ion interaction. It is extremely difficult to investigate the overcharging slow ions experimentally and the spread of the experimental data is more than 100 %. Typically, the atom–ion charge sections are measured at ion energy of more than

1 eV or more. In this case, the theoretical equation is considerable interesting but its calculation error is roughly 50 % [19]:

$$Q_{\text{ai.per}} = \pi a_0^2 \frac{E_H}{E_i} \ln^2\left(\frac{100 v_0}{\bar{v}_i}\sqrt{\frac{E_i}{E_H}}\right) K \tag{2.29}$$

Here $a_0 = 0.529 \times 10^{-10}$ M—the Bohr radius of the hydrogen atom; E_i—the ionization potential of atoms (molecules) when the ion is formed; $E_H = 13.6$ eV—the ionization potential of the hydrogen atom; $v_0 = 2.19 \times 10^6$ m/c—the speed of the electron in the first Bohr orbit in the hydrogen atom; $v_i = \sqrt{\frac{8kT_i}{\pi m_i}}$—the average thermal velocity of the ion; K—a correction coefficient.

In practical use of the Eq. (2.29) to calculate the cross section the coefficient K should be selected and the follow calculated effective viscosity and thermal conductivity, where the charge-section are included, must coincide with the experimental ones. For the argon plasma $K = 0.5$. These calculations of recharge section with $K = 0.5$ are shown in Fig. 2.5 and they have a good agreement with the experimental data [20].

The Average Effective Interaction Cross Section of Atom–Ion in Argon For the interaction of the atom–ion in argon (Ar–Ar^+) the processes of elastic collisions should be taken into account, as well as the collision processes with the charge transfer. They correspond to the collision integrals $\Omega_{ij}^{(l,s)}$ of odd l order. Another approximation was proposed by *E.* A. Mason and others [2, 14, 21–23], it does not take into account the contribution of elastic collisions there:

$$Q^{(l)} = 2\sigma_{ex}, \tag{2.30}$$

where σ_{ex}—the cross section of the charge transfer.

The collision integrals $Q^{(l,s)}$ of odd l—order have been obtained using the Devoto equation [2, 11, 24]:

$$\begin{aligned} Q^{(l,s)} = {} & A^2 - ABx + \left(\frac{Bx}{2}\right)^2 + \frac{B\zeta}{2}[Bx - 2A] + \frac{B^2}{4}\left(\frac{\pi^2}{6} - \sum_{n=1}^{s+1}\frac{1}{n^2} + \zeta^2\right) \\ & + \frac{B}{2}[B(x+\zeta) - 2A]\ln\left(\frac{T}{M}\right) + \left[\frac{B}{2}\ln\left(\frac{T}{M}\right)\right]^2, \end{aligned} \tag{2.31}$$

Here $x = \ln(4R)$, $\zeta = \sum_{n=1}^{s+1}\frac{1}{n} - \gamma$, and $\gamma \approx 0.57721$—the Euler's constant. The parameters A and B are the approximating coefficients of the charge transfer section (its dimension $Å^2$) as a function of the collision velocity v, cm/s:

$$\sigma_{ex} = \frac{1}{2}[A - B \ln v]^2 \tag{2.32}$$

Estimated values for these coefficients were determined by Cressault [25] for the even and odd divisions ${}^2\Sigma^+_{g/u}$ and ${}^2\Pi^+_{g/u}$ states of the molecular ion Ar_2^+, that correlates with the ground state of the atom $Ar(^1s)$ and ion $Ar^+(^2p)$:

$$\begin{cases} {}^2\Sigma^+_{g/u} \Rightarrow 26.39;\ B = 1.12; \\ {}^2\Pi^+_{g/u} \Rightarrow 18.96;\ B = 0.83. \end{cases} \tag{2.33}$$

Results of the collision integrals calculation for the interaction cross sections of atom–ion argon at l = 1; 5 and l = 2; 6 are shown in Fig. 2.7 and 2.8.

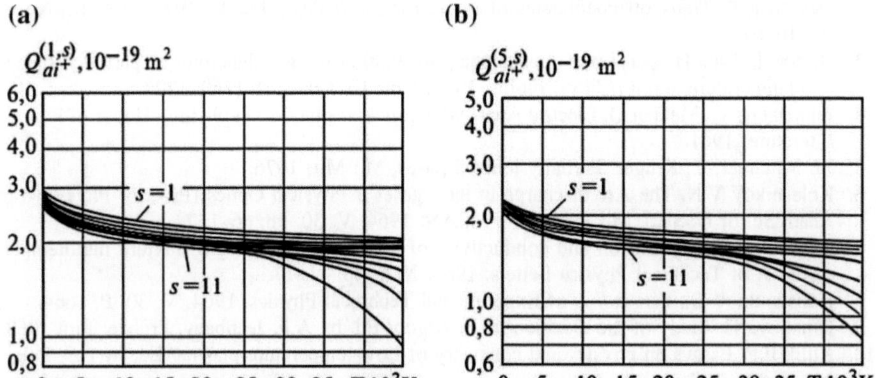

Fig. 2.7 Results of the collision integrals calculation for the interaction cross sections of the recharge atom–ion Ar–Ar$^+$, l = 1 (**a**) and l = 5 (**b**)

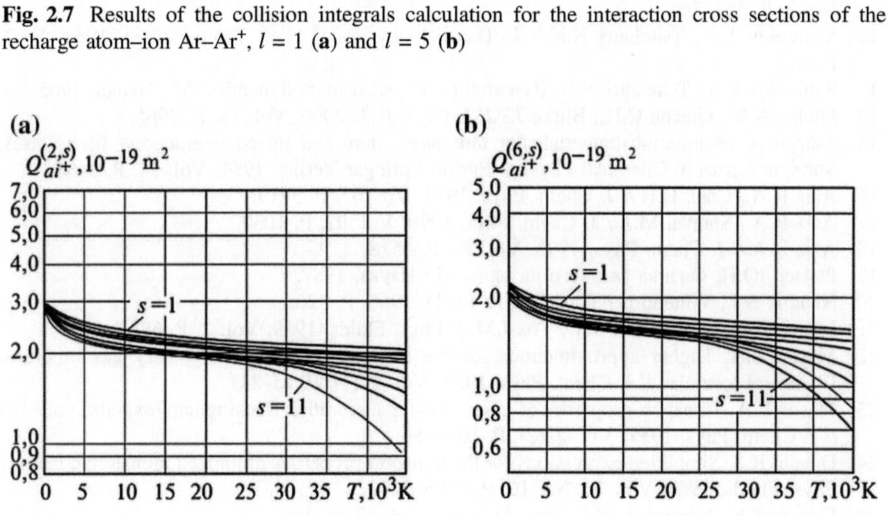

Fig. 2.8 Results of the collision integrals calculation for the interaction cross sections of the atom–ion Ar–Ar$^+$, l = 2 (**a**) and l = 6 (**b**)

Full collision integrals were obtained by averaging of the pair over all states using statistical weights equal to the multiplicity of states:

$$Q^{(l,s)} = \frac{\sum_n \omega_n Q_n^{(l,s)}}{\sum_n \omega_n} = \frac{Q_\Sigma^{(l,s)} + 2Q_\Pi^{(l,s)}}{3}. \quad (2.34)$$

References

1. J. O. Hirschfelder, Ch. F. Curtiss, R. B. Bird. Molecular theory of gases and liquids / John Wiley and Sons, New York. 2nd Edition: 1964.
2. Devoto R.S. Transport coefficients of ionized argon // Phys. Fluids. 1973. Vol. 16. No. 5. P. 616–623.
3. Hoffert I., Lien H. Quasi one-dimensional, nonequilibrium gas dynamics of partially ionized two-temperature argon // Phys. Fluids. 1977. Vol. 10. No. 8. P. 1769–1777.
4. Finkelberg V., Mekker G. Electric Arcs and thermal plasma. M.: Publishing House of Foreign Literature, 1961.
5. M. Mitchner, C. Kruger. Partially ionized gases. M.: Mir, 1976.
6. Kolesnikov V.N. The Arc discharge in inert gases // Physical Optics (Proc. of Phys. Inst of Acad. Sc. of USSR.). M.: Publishing PhIAN, 1964. V. 30. pp. 66–157.
7. Polyansky V.A. Diffusion and conductivity of a partially ionized gas mixture multitemperature// J. of Technical Physics Letters. 1964. № 5. pp. 11–17.
8. Konstantinov O., Perl B.// J. of Experimental Technical Physics. 1964. V. 30. P. 1664.
9. Ismailova D. Study of the electric Arc in Argon/ Ed. by A.F. Jeenbaev. Frunze: Ilim 1963.
10. Kulik P.P. Essays on physics and chemistry of Low-temperature plasma/ Ed. by L.S. Polak. M.: Nauka, 1971.
11. Devoto R.S. Transport coefficients of partially ionized argon// Phys. Fluids. 1967. Vol. 10. No. 2. P. 354–364.
12. Samuylov E.V. Tsitelauri N.N.// J. Thermal physics of High Temperatures. 1964. V. 2. P. 565.
13. Samuylov E.V. Tsitelauri N.N. Research on Physical hydrodynamics. M.: Nauka, 1965.
14. Phelps A.V., Greene C.H., Burke J.P.// J. Phys. B 3. 2000. Vol. 33. P. 2965.
15. Aziz R.A. Interatomic potentials for rare-gases: pure and mixed interaction// Inert Gases, Springer Series in Chemical Physics. Berlin: Springer-Verlag. 1984. Vol. 34. P. 5–86.
16. Aziz R.A., Chen H.H.// J. Chem. Phys. 1977. Vol. 67. P. 5719.
17. Aziz R.A., Slaman M.J.// J. Chem. Phys. 1990. Vol. 92. P. 1030.
18. Aziz R.A.// J. Chem. Phys. 1993. Vol. 99. P. 4518.
19. Райзер Ю.П. Физика газового разряда. М.: Наука, 1987.
20. Nichols B.J., Witteborn F.C. // NASA TN D. 1966. P. 3265.
21. Mason E.A., Vanderslice J.T., Yos J.M. // Phys. Fluids. 1959. Vol. 2. P. 688.
22. Mason E.A. Higher approximations for the transport properties of binary gas mixtures. I. General formulas // J. Chem. Phys. 1957. Vol. 27(1). P. 75–84.
23. Mason E.A. Transport properties of gases obeying a modified Buckingham (exp-six) potential // J. Chem. Phys. 1957. Vol. 22(2). P. 169–186.
24. Devoto R.S. Simplified expressions for the transport properties of ionized monatomic gases // Phys. Fluids. 1967. Vol. 10. No. 10. P. 2105–2112.
25. Cressault Y., Gleizer A. // J. Phys. D. 2004. Vol. 37. P. 560.

Chapter 3
Quantum Mechanical Theory of the Particle Scattering

3.1 The Schrödinger Equation

The fundamental equation of quantum mechanics is the Schrödinger equation. It is intended to describe the dynamics of particles motion without spin at speeds much less than the speed of light. In the case of fast particles and particles with spin its generalizations are applied (the Klein–Gordon, Pauli, Dirac equations, and others).

In quantum physics, the wave function ψ (complex-valued) is used to describe the state of a moving particle. The most common Copenhagen interpretation of the wave function is related to the probability of finding a particle in a pure state (the square of the wave function is a probability density). Let the wave function to be set in an n-dimensional space and $\vec{r}(x_1, x_2, \ldots, x_n)$ at each point at the time t, it will look like $\psi(\vec{r}, t)$. For the three-dimensional space in spherical we have $\vec{r}(r, \theta, \varphi)$. In general, the Schrödinger equation can be written as:

$$-\frac{\hbar^2}{2\mu}\Delta\psi(\vec{r}, t) + \phi(\vec{r})\psi(\vec{r}, t) = i\hbar\frac{\partial}{\partial t}\psi(\vec{r}, t), \tag{3.1}$$

Here $\hbar = \frac{h}{2\pi}$, h—the Planck's constant; $\mu = m$—the particle mass. In the center of mass system for the particles scattering μ—the reduced mass of the particles; $\phi(\vec{r})$—the external to the particle potential energy in the point $\vec{r}(x_1, x_2, \ldots, x_n)$.

The solution of Schrödinger equation with respect to the time should be simple, because the time is included in this equation using the first derivative in its right-hand side. Indeed, a particular solution for the case $\phi(\vec{r})$ is not a function of time and it can be written as

$$\psi(\vec{r}, t) = \psi(\vec{r})\exp(-iEt/\hbar), \tag{3.2}$$

Here the function $\psi(\vec{r})$ should satisfy the equation

S. Nguyen-Kuok, *Theory of Low-Temperature Plasma Physics*,
Springer Series on Atomic, Optical, and Plasma Physics 95,
DOI 10.1007/978-3-319-43721-7_3

$$-\frac{\hbar^2}{2\mu}\Delta\psi(\vec{r}) + \phi(\vec{r})\psi(\vec{r}) = E\psi(\vec{r}), \tag{3.3}$$

This expression is obtained from Eq. (3.1) by substituting of (3.2) in it. The resulting Eq. (3.3) does not contain the time and called stationary Schrödinger equation.

Expression (3.2) is a particular solution of the time-dependent Schrödinger Eq. (3.1), the general solution is a linear combination of particular solutions of the form (3.2). The dependence on time for $\psi(\vec{r}, t)$ is the elementary function, but its dependence on the coordinates has not always an elementary form, since the Eq. (3.3) with one form of the potential function $\phi(\vec{r})$ is quite different from the same equation for its other. In fact, the Eq. (3.3) can be solved analytically only for a small number of special types of functions $\phi(\vec{r})$. The interpretation of the value E in the Eq. (3.2) is very important. The time dependence of the function $\psi(\vec{r}, t)$ in (3.2) is exponential, and the coefficient before t in the upper index of exponent function is selected so that the right side of (3.3) contains only a constant factor E. In the left part of (3.3) the function $\psi(\vec{r})$ is multiplied by a function of potential energy $\phi(\vec{r})$. Consequently, the value E should have the dimension of energy. The one value with a dimension of energy that is constant in mechanics is the total energy of the system. According to the physical interpretation of the Schrödinger equation E is the total energy of the particle motion, described by the function $\psi(\vec{r}, t)$.

The function of a plane wave can be represented as $A\exp[i(kz - \omega t)]$, where A—the amplitude; k—the wave number; $\omega = 2\pi\nu = \frac{E}{\hbar}$—the angular frequency; ν—the frequency.

Let us call the sum of the incoming and reflected waves as the incident wave. Then, the total wave function ψ is a sum of functions of incident ψ_{inc} and ψ_{scatt} waves (Fig. 3.1):

$$\psi = \psi_{\text{inc}} + \psi_{\text{scatt}}. \tag{3.4}$$

During an elastic collision, the wave energy does not change in time, so you can only consider only that part of the incident wave function that depends on the coordinates in the space.

$$\psi_{\text{inc}}(r, \theta) = A\exp(ikz). \tag{3.5}$$

The scattered wave is a spherical wave that has the asymptotic form far from the center of scattering:

$$\psi_{\text{scatt}}(r, \theta) \approx \frac{A}{r} f(\theta)\exp(ikr), \tag{3.6}$$

Here $f(\theta)$—the scattering amplitude that is related to the differential cross section by the equation

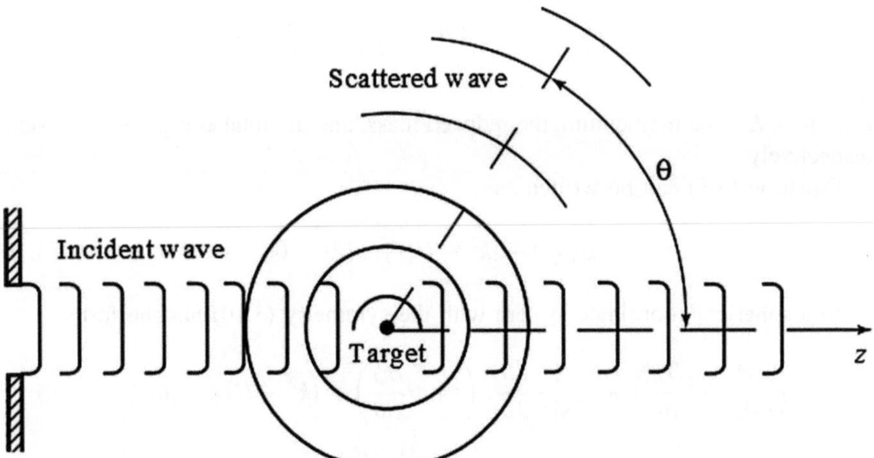

Fig. 3.1 To the description of the wave–particle scattering

$$\sigma(\theta) = |f(\theta)|^2. \tag{3.7}$$

Thus, the total wave function can be written as

$$\psi = A\left[\exp(ikz) + \frac{\exp(ikr)}{r}f(\theta)\right]. \tag{3.8}$$

The amplitude A is often assumed to be equal to unity.

The quantum mechanical theory of particles scattering based on the solution of the Schrödinger wave Eq. (3.1) or (3.3) in view of the potential energy $\phi(\vec{r})$ of interaction between the particles. In practice, quantum mechanical calculation of the particles scattering is determined as follows:

(1) Functions of incident and scattered waves;
(2) Total wave function;
(3) Phase shift;
(4) Amplitude of the scattered wave;
(5) Differential cross sections for the particles interaction;
(6) The total cross section of collisions between particles.

According to the Heisenberg uncertainty principle $\Delta x \Delta p \geq \frac{\hbar}{2}$, here Δx and Δp—the standard deviations of the measured values of the position and momentum, in quantum theory both the position and velocity of the particles, i.e., the angle of the particles scattering after the collision, cannot be accurately determined. Quantum mechanical description allows getting only the probability of scattering (i.e., the differential cross sections) of the particles at a certain angle θ. This probability is usually expressed using the phase shift η_l of the radial wave function $\psi(\vec{r})$. The phase shift η_l depends on the wave number k and function $U(r)$, that are defined as

$$k = \frac{2\pi}{\lambda} = \frac{p}{\hbar} = \frac{\sqrt{2\mu E}}{\hbar}; \quad U(r) = \frac{2\mu\phi(r)}{\hbar^2}, \tag{3.9}$$

Here p, μ, E—the momentum, the reduced mass, and the total energy of the system, respectively.

Equation (3.3) can be written as

$$\Delta\psi(\vec{r}) + \left[k^2 - U(\vec{r})\right]\psi(\vec{r}) = 0. \tag{3.10}$$

In a spherical coordinate system with the symmetry (3.10) has the form

$$\frac{1}{r^2}\frac{\partial}{\partial r}\left(r^2\frac{\partial\psi}{\partial r}\right) + \frac{1}{r^2 \sin\theta}\frac{\partial}{\partial\theta}\left(\sin\theta\frac{\partial\psi}{\partial\theta}\right) + (k^2 - U)\psi = 0. \tag{3.11}$$

3.2 Solution of the Schrödinger Equation for the Elastic Interactions

The solution of the wave Eq. (3.11) is carried out using the method of variables separation. For this the wave function $\psi(r,\theta)$ is represented as a result of multiplication of two independent one from the other functions, one of them depends only on the coordinates r, and the other—from θ

$$\psi(r,\theta) = L(r)Y(\theta), \tag{3.12}$$

Here $L(r)$—the radial function that depends on the coordinates $Y(\theta)$—the spherical harmonic function.

Substituting (3.12) into (3.11), we obtain

$$\frac{1}{L}\frac{\mathrm{d}}{\mathrm{d}r}\left(r^2\frac{\mathrm{d}L}{\mathrm{d}r}\right) + r^2\left[k^2 - U(r)\right] = -\frac{1}{Y}\left[\frac{1}{\sin\theta}\frac{\mathrm{d}}{\mathrm{d}\theta}\left(\sin\theta\frac{\mathrm{d}y}{\mathrm{d}\theta}\right)\right]. \tag{3.13}$$

The left side of (3.13) depends only on r, and the right—on θ. Both parts of the equation must be equal to each other r и θ, each of them must be equal to a constant, for example $l(l+1)$, where l—the orbital quantum number.

$$\frac{\mathrm{d}}{\mathrm{d}r}\left(r^2\frac{\mathrm{d}L}{\mathrm{d}r}\right) + r^2\left[k^2 - U(r) - \frac{l(l+1)}{r^2}\right]L = 0; \tag{3.14}$$

$$\frac{1}{\sin\theta}\frac{\mathrm{d}}{\mathrm{d}\theta}\left(\sin\theta\frac{\mathrm{d}y}{\mathrm{d}\theta}\right) + l(l+1)Y = 0. \tag{3.15}$$

Equation (3.15) is the Legendre equation of the second order and has two linearly independent solutions, each of them can be written as the sum of the power series for $\cos\theta$. Both solutions are equal to infinity (i.e., unacceptable) for $\theta = 0$, if l is not zero or a positive integer number. Taking into account (3.12), the general solution of Eq. (3.11) in accordance with the cylindrical symmetry of the problem and the physical condition that the wave functions are finite at $r = 0$ can be written in the form [1]:

$$\psi(r,\theta) = \sum_{l=0}^{\infty} A_l L_l(r) P_l(\cos\theta). \tag{3.16}$$

Here A_l—is a constant; $L_l(r)$—the solution of (3.14); functions $P_l(\cos\theta)$—the Legendre polynomial, is a solution of Legendre Eq. (3.15).

The solution of Legendre equation is as usual

$$Y = a_0 F(\cos\theta) + a_1 G(\cos\theta), \tag{3.17}$$

Here the function F contains only even degrees of $\cos\theta$, and G—only odd degrees, a_0 and a_1—some constants. To simplify (3.17) instead of F and G we write the function, given that for l = 0, 2, 4, … constant a_0 corresponds to it, and for l = 1, 3, 5, …—constant a_1.

The expression for $L_l(r)$ is represented as

$$L_l(r) = \frac{G_l(r)}{kr}. \tag{3.18}$$

Substituting (3.18) into (3.14), we obtain the equation

$$\frac{d^2G_l(r)}{dr^2} + \left[k^2 - U(r) - \frac{l(l+1)}{r^2}\right] G_l(r) = 0. \tag{3.19}$$

At large values of r the function $U(r)$ tends to zero as the term containing $1/r^2$. Consequently, the asymptotic solution has the form:

$$G_l(r) \approx \sin(kr + C_l), \tag{3.20}$$

Here C_l—is a constant, that is determined by the boundary condition that the wave function of the relative motion must vanish when the distance between the particles r, is greater then r_{max}, that subsequently assumed to be infinity.

At large values of r the expression (3.14) has an asymptotic solution

$$L_l(r) \approx \frac{1}{kr} \sin\left(kr - \frac{\pi l}{2} + \eta_l\right), \tag{3.21}$$

Here η_l—the phase shift of the lth partial wave due to the influence of the potential energy $\phi(r)$.

In the ideal case of noninteracting particles $\phi(r) = 0$, $\eta_l = 0$, i.e., there is no phase shift if there is no scattering. The phase shift depends on the orbital quantum numbers l (or angular momentum $\hbar\sqrt{l(l+1)}$) and k (or momentum of the relative motion $\hbar k$). You can draw an analogy with the classical case when the scattering angle θ is calculated as a function of the impact parameter b, and the relative collision velocity g.

Substituting Eq. (3.21) to (3.16), we obtain an asymptotic expression for the total wave

$$\psi(r,\theta) \simeq \sum_{l=0}^{\infty} A_l \frac{1}{kr} \sin\left(kr - \frac{\pi l}{2} + \eta_l\right) P_l(\cos\theta). \tag{3.22}$$

To determine the constant A_l the full wave function should be represented as a sum of functions of incident ψ_{inc} and scattered ψ_{scatt} waves. For the incident wave propagating along the axis z, we have

$$\psi_{\text{inc}}(r,\theta) = \exp(ikz) = \exp(ikr\cos\theta). \tag{3.23}$$

Expression (3.23) can be expanded in a series of Legendre polynomials; expansion coefficients contain Bessel functions

$$\begin{aligned}\psi_{\text{inc}}(r,\theta) &= \sum_{l=0}^{\infty} (2l+1) i^l P_l(\cos\theta) \sqrt{\frac{\pi}{2kr}} J_{l+1/2}(kr) \\ &= \sum_{l=0}^{\infty} (2l+1) i^l P_l(\cos\theta) j_l(kr),\end{aligned} \tag{3.24}$$

Here $j_l(kr)$—the spherical Bessel function that has the asymptotic solution for large values of r the following:

$$j_l(kr) \approx \frac{1}{kr} \sin\left(kr - \frac{\pi l}{2}\right). \tag{3.25}$$

Thus, the asymptotic form of the incident wave at large r is expressed as

$$\psi_{\text{inc}}(r,\theta) \approx \sum_{l=0}^{\infty} (2l+1) i^l P_l(\cos\theta) \frac{1}{kr} \sin\left(kr - \frac{\pi l}{2}\right). \tag{3.26}$$

Substituting (3.26) into (3.22), we find the expression for the asymptotic form of the scattered wave:

$$\psi_{\text{scatt}}(r,\theta) = \psi(r,\theta) - \psi_{\text{inc}}(r,\theta)$$
$$\approx \sum_{l=0}^{\infty} \frac{1}{kr} P_l(\cos\theta)\left[A_l \sin\left(kr - \frac{\pi l}{2} + \eta_l\right) - (2l+1)i^l \sin\left(kr - \frac{\pi l}{2}\right)\right]. \tag{3.27}$$

To determine the coefficient A_l, given that $\sin(x) = \frac{\exp(ix)-\exp(-ix)}{2i}$ the latter expression can be transformed as follows:

$$\begin{aligned}\psi_{\text{scatt}}(r,\theta) \approx \sum_{l=0}^{\infty} \frac{1}{kr} P_l(\cos\theta)\frac{1}{2i} \\ \times \left\{\exp\left[i\left(kr - \frac{\pi l}{2}\right)\right]\left[A_l \exp(i\eta_l) - (2l+1)i^l\right]\right. \\ \left. - \exp\left[-i\left(kr - \frac{\pi l}{2}\right)\right]\left[A_l \exp(-i\eta_l) - (2l+1)i^l\right]\right\}.\end{aligned} \tag{3.28}$$

Here, according to the physical meaning, the term with the exponential function corresponds to a negative power of the incident wave and the exponential term with a positive power—reflected. Since there is no term in the expression for the function of the incident wave, the second term (in square brackets) must be equal to zero, i.e.,

$$\begin{aligned}A_l \exp(-i\eta_l) - (2l+1)i^l = 0, \\ A_l = (2l+1)i^l \exp(i\eta_l).\end{aligned} \tag{3.29}$$

Thus, we have obtained the expression for the coefficient A_l. Substituting it into (3.22), we obtain an expression for the total wave function

$$\psi(r,\theta) = \sum_{l=0}^{\infty} (2l+1)i^l \exp(i\eta_l)\frac{1}{kr}\sin\left(kr - \frac{\pi l}{2} + \eta_l\right)P_l(\cos\theta). \tag{3.30}$$

Equation (3.27) after the substitution (3.29) takes the form

$$\begin{aligned}\psi_{\text{scatt}}(r,\theta) \approx \sum_{l=0}^{\infty} \frac{1}{kr} P_l(\cos\theta)(2l+1)i^l \\ \times \left[\exp(i\eta_l)\sin\left(kr - \frac{\pi l}{2} + \eta_l\right) - \sin\left(kr - \frac{\pi l}{2}\right)\right] \\ = \frac{\exp(ikr)}{kr}\sum_{l=0}^{\infty} \frac{1}{2i}(2l+1)[\exp(2i\eta_l) - 1]P_l(\cos\theta)\end{aligned} \tag{3.31}$$

The expression for the scattering cross section can also be obtained by using (3.6) and (3.31). It will look like

$$\frac{\exp(ikr)}{r}f(\theta) = \frac{\exp(ikr)}{kr}\sum_{l=0}^{\infty}\frac{1}{2i}(2l+1)[\exp(2i\eta_l) - 1]P_l(\cos\theta). \quad (3.32)$$

Consequently,

$$\begin{aligned} f(\theta) &= \frac{1}{k}\sum_{l=0}^{\infty}\frac{1}{2i}(2l+1)[\exp(2i\eta_l) - 1]P_l(\cos\theta) \\ &= \frac{1}{k}\sum_{l=0}^{\infty}(2l+1)\exp(i\eta_l)\sin\eta_l P_l(\cos\theta). \end{aligned} \quad (3.33)$$

3.3 Determination of the Phase Shift

The phase shift can be used to determine the differential cross section in the following form:

$$\frac{d\sigma}{d\Omega} = \frac{1}{k^2}\left|\sum_{l=0}^{\infty}(2l+1)\exp(i\eta_l)\sin\eta_l P_l(\cos\theta)\right|^2. \quad (3.34)$$

According to the classical theory of angular momentum, the particle in the collision depends on the *pa*, where *p*—momentum of the center of mass of the system; *a*—the range of the radius, or "black disc," where the wave is completely absorbed. In this way

$$l \leq \frac{pa}{\hbar} = ka = L. \quad (3.35)$$

The phase shift can be found by using the equation for the calculation of the scattered waves and full waves [2, 3]:

$$\mathrm{tg}\eta_l(k) = \frac{kj_l'(ka)j_l(k_1a) - k_1j_l(ka)j_l'(k_1a)}{kn_l'(ka)j_l(k_1a) - kn_l(ka)j_l'(k_1a)}, \quad (3.36)$$

where j_l—Bessel function; n_l—Newman function j_l', n_l'—derivatives of the argument of the Bessel and Newman functions; $k_1 = \frac{\sqrt{2m(E+\phi_0)}}{\hbar}$—the wave number based on potential field; ϕ_0—the depth of the potential barrier ($\phi_0 > 0$ for attractive potential).

For the scattering cross section of particles using the modified effective range theory (MERT) phase shifts for the first time were used in the study of neutron–neutron and neutron–proton scattering [1, 4]. Processing scattering of particles is usually made by comparing the phase shifts that were determined from the experimental interaction cross sections and theoretical calculations using nuclear potentials. It was necessary to do a lot of calculations for it and the semiempirical equations for a given nuclear capability were introduced.

The experimental data are used to find some characteristics of the nuclear potential, semiempirical methods allow relatively easy to determine which characteristics of nuclear capability can be obtained from experimental data. The values of the phase shifts change with the energy of the incident particle for each nuclear capability. The most famous potentials are square, Gauss, exponential, and Yukawa. Energies of nuclear potentials are expressed as a function of the depth of the potential well and the inner distance, defining the boundaries of existence potential. The internal distance cannot be found by direct measurement, it must be obtained from the experimental data by entering some values of the active length r_0 and scattering length A. The active length also depends on the depth of the potential pit and decreases with its increase. The relative values of the current domestic and distances depend on the accepted form of the potential well. At low energies in the scattering process the s-waves are dominated.

The basis of the approach using the existing distance lies in the formation of the assumptions of the phase shifts functions in addiction of the energy in the form of simple functions, using the most common assumptions about nuclear potentials. These functions include some uncertain parameters. This allows the comparison of theoretical and experimental results that are not obtained by phase shifts in the parameters and functions that reduces the amount of computation required. Thus, for a given potential energy it is necessary to find three parameters instead of a large number of phase shifts. The nature of the scattering is determined by the parameters of the functions found. To obtain the functional form of the phase shifts and errors the method of least squares is used. Designating the phase shift η and the wave number k, we can write the following expression for $k\eta$:

$$k \operatorname{ctg} \eta = -\frac{1}{A} + \frac{1}{2} r_0 k^2, \tag{3.37}$$

Here r_0—the active length; A—the scattering length obtained for zero energy; factor 1/2 is introduced because the function value begins to decrease near the edge of the potential pit.

Equation (3.37) contains only two parameters—the length of the scattering and the given length. It is known as independent of the form of the potential approximation relations, since in any case you can choose these parameters so that the form of the function fit for any shape, depth, and size of the potential pit.

The scattering cross section at zero energy is defined as $Q_0 = 4\pi A^2$. For potentials acting on a short distance, marking the angular momentum L, the Eq. (3.37) can be written as

$$k^{2L+1} \operatorname{ctg} \eta(L) = -\frac{1}{A(L)} + \frac{1}{2} r_0(L) k^2. \tag{3.38}$$

Note that (3.37) is independent of the potential type and gives similar results for the scattering of any type. The utility of the MERT theory lies in the fact that the scattering amplitude for a range of energy with a certain accuracy, expressed in terms of a small number of parameters. To obtain such experimental parameters points that lie in the area can be used. For a scattering neutron–proton of the order of 10^{-15} m at a certain value r_0, and we can assume that the approximation that does not depend on the type of potential, it can be used at energies up to 10 meV.

There are strict limits on the use of MERT theory to the nuclear field to the field of atomic scattering. The modified effective range theory that is briefly presented above is ill-suited to describe of the electron–atom scattering, since the forces are proportional to $1/r^4$ at the long distances. Nevertheless, it managed to change the way that it gave acceptable results for the scattering of electrons by neutral particles that can be polarized. Coefficients used in it are chosen that the resultant values agree with the experimental sections in the lower energy range (several electron volts), including the results of Ramseur–Collat experiments. Next there is the extrapolation to zero energy.

For the s-wave can apply Eq. (3.38) with $l = 0$, at sufficiently low energy the follow approximation is used [4–6]:

$$tg\,\eta_0 \approx \eta_0 \approx -Ak - \frac{\pi\alpha}{3a_0}k^2; \tag{3.39}$$

$$tg\,\eta_1 = \frac{\pi\alpha}{15a_0}k^2 - A_1k^3 \quad (l = 1); \tag{3.40}$$

$$tg\,\eta_l = \frac{\pi\alpha k^2}{(2l+3)(2l+1)(2l-1)a_0} \quad (l > 1). \tag{3.41}$$

If the value A is positive, such as helium and neon, the phase shift will be in the form of a monotonically decreasing function. If the value of A is negative, the phase shift will initially increase by a positive factor π at zero energy to a maximum value, and then decrease, passing a value of zero when $A \approx -(\pi/3a_0)\alpha k$. The scattering length in Ramseur minimum can be found from the last equation. To obtain A it is necessary to have data points that lie in the region of the minimum sections. For the momentum transfer cross sections of some inert gas the Ramseur–Townsends effect is also observed, and the minimum of the cross section of the energy can be determined in this way. In this case it is necessary to take into account an additional factor η_1 even for the s-waves as opposed to total scattering cross section, where this ratio is only present in the case of p-waves. Buckman and Lohmamn [5] have analyzed the scattering cross sections by using MERT theory data, but in their approach, they used two additional members with k of a higher degree

$$tg\,\eta_0 = -Ak\left(1 + \frac{4\alpha}{3\alpha_0}k^2\ln(ka_0)\right) - \frac{\pi\alpha}{3\alpha_0}k^2 + Dk^3 + Fk^4 \quad (l = 0). \tag{3.42}$$

Table 3.1 The coefficients for the approximation MERT4

Parameter	References				
	[10]	[5]	[6]	[7]	Author
A	$-1.609\ a_0$	$-1.442\ a_0$	$-1.436\ a_0$	$-1.492\ a_0$	$-1.55\ a_0$
A1	$8.094\ a_0^3$	$9.229\ a_0^3$	$8.57\ a_0^3$	$9.104\ a_0^3$	$7.8\ a_0^3$
D	$60.52\ a_0^3$	$70.784\ a_0^3$	$75.2\ a_0^3$	$64.3\ a_0^3$	$61.5\ a_0^3$
F	$-59.47\ a_0^4$	$-97.188\ a_0^4$	$-103.9\ a_0^4$	$-77.3\ a_0^4$	$-71.4\ a_0^4$

For the p-waves ($l = 1$) and waves of higher degree, the equations have the same form as (3.40), (3.41). In the analysis of Backman and Lohman [5] it was noted that the limit of the application of MERT theory at low energies has been set in the determination of the scattering length for comparison of the relative values of Q_T and Q_M. The phase shifts for MERT theory ($l > 0$) in the electron–atom scattering [4–7] are

$$tg\,\eta_1 = \frac{\pi\alpha}{15a_0}k^2 - A_1k^3 \quad (l = 1); \tag{3.43}$$

$$tg\,\eta_l = \frac{\pi\alpha k^2}{(2l+3)(2l+1)(2l-1)a_0} \quad (l > 1), \tag{3.44}$$

Here a_0—Bohr radius, α—the dipole polarizability of atom, $\alpha = 11,08a_0^3$ for Argon; k—number of the wave associated with the energy E, eV by the equation $E = 13,6057(ka_0)^2$; A, A_1, D, and F—the parameters, that should be determined for a correspondence between theoretical and experimental data. On the basis of the physical meaning, the parameter A is of greatest interest, which is the expression for the cross section at zero energy. Data for A, A_1, D, and F are shown in Table 3.1. Cross sections depending on the phase shifts are determined by summing of an infinite series of higher order partial waves [8, 9].

3.4 Born Approximation for Calculating the Amplitudes of the Scattered Waves

A finite number of partial waves make the contribution to the scattering amplitude in the communication of the scattering and incident particles over a short distance. Their number increases with increasing collision energy. If the particle interaction is essential at a great distance, theoretically to obtain accurate values of the scattering amplitude it is necessary to take into account the large number of partial waves. In practice, the phase shifts of the first few partial waves l can be calculated accurately, using the asymptotic radial part of the wave function and phase shifts of partial waves of higher orders (from $l+1$ to infinity), they are estimated using the approximation of Born using only interaction at large distances.

$$f^{(1)}(\theta) = -\frac{\mu}{\hbar^2}\frac{1}{k\,\sin\frac{\theta}{2}}\int\limits_0^\infty r\phi(r)\sin\left(2rk\,\sin\frac{\theta}{2}\right)\mathrm{d}r. \tag{3.45}$$

As a test of the Born approximation, the calculation of differential scattering cross section of charged particles in the plasma will be considered using the function of potential energy based screening

$$\phi(r) = \frac{Z_iZ_je^2}{4\pi\varepsilon_0}\frac{\exp(-r/r_D)}{r}, \tag{3.46}$$

Here r_D—the Debye radius.

Note that $\phi(r)$ reduces to zero very quickly when $r \gg r_D$. We have

$$f^{(1)}(\theta) = -\frac{2\mu\frac{Z_iZ_je^2}{4\pi\varepsilon_0}}{\hbar^2}\frac{1}{2k^2(1-\cos\,\theta)+\left(\frac{1}{r_D}\right)^2}. \tag{3.47}$$

Thus, for the first ($l = 1$) Born approximation the differential cross section with Yukawa potential function is

$$\begin{aligned}\left(\frac{\mathrm{d}\sigma}{\mathrm{d}\Omega}\right) &\simeq \left(\frac{2\mu\frac{Z_iZ_je^2}{4\pi\varepsilon_0}}{\hbar^2}\right)^2\frac{1}{\left[2k^2(1-\cos\,\theta)+\left(\frac{1}{r_D}\right)^2\right]^2}\\ &= \left(\frac{Z_iZ_je^2/(4\pi\varepsilon_0)}{4E\sin^2(\theta/2)+(\hbar^2/2mr_D^2)}\right)^2,\end{aligned} \tag{3.48}$$

where the total energy of the system $E = \frac{k^2\hbar^2}{2\mu}$ (3.9).

In the absence of screening (i.e., $r_D \to \infty$), this expression reduces to the well-known formula of Rutherford.

Consider the elastic scattering of a particle of mass μ with energy $E = \hbar^2k^2/(2\mu)$ using a function $\phi(r)$. The Thompson expression (for the partial waves) for the elastic scattering amplitude has the form of

$$f(\theta) = \frac{1}{k}\sum_{l=0}^{\infty}(2l+1)T_lP_l(\cos\,\theta) = \frac{1}{k}\sum_{l=0}^{L}(2l+1)T_lP_l(\cos\,\theta)+\Delta f_L(\theta), \tag{3.49}$$

Here the elements of the transition matrix T_l are associated with a phase shift η_l by ratio

$$T_l = \exp(i\eta_l)\sin(\eta_l). \tag{3.50}$$

If the values T_l for $l \gg L$ are approximated by T_{lB}, that can be found by using of the Born approximation, we can write

$$\Delta f_L(\theta) = \frac{1}{k}\sum_{l=L+1}^{\infty}(2l+1)T_{lB}P_l(\cos\,\theta) = f_B(\theta) - \frac{1}{k}\sum_{l=0}^{L}(2l+1)T_{lB}P_l(\cos\,\theta). \tag{3.51}$$

Here $f_B(\theta)$—the total scattering amplitude in the Born approximation,

$$f_B(\theta) = -\frac{2\mu}{\hbar^2}\int_0^{\infty}\phi(r)r^2\sum_{l=0}^{\infty}(2l+1)j_l^2(kr)P_l(\cos\,\theta)dr = \frac{1}{k}\sum_{l=0}^{\infty}(2l+1)T_{lB}P_l(\cos\,\theta), \tag{3.52}$$

Consequently,

$$T_{lB} = \frac{2\mu k}{\hbar^2}\int_0^{\infty}\phi(r)r^2 j_l^2(kr)dr. \tag{3.53}$$

The long distance interaction only contributes to the phase shifts of higher partial waves, so it is reasonable to take for $\phi(r)$ the function of the form $\phi(r) = C/r^4$, here C—is a constant. Then we get

$$\Delta f_L(\theta) = \frac{\pi\alpha k}{a_0}\left(\frac{1}{3} - \frac{1}{2}\sin\frac{\theta}{2} - \sum_{l=1}^{L}\frac{P_l(\cos\,\theta)}{(2l+3)(2l-1)}\right). \tag{3.54}$$

Note that the upper limit of summation in (3.51) is L but not infinity, then we have

$$f(\theta) = \frac{1}{k}\sum_{l=0}^{L}(2l+1)\exp(i\eta_l)\sin\eta_l P_l(\cos\,\theta) + \frac{\pi\alpha k}{a_0}\left[\frac{1}{3} - \frac{1}{2}\sin\frac{\theta}{2} - \sum_{l=1}^{L}\frac{P_l(\cos\,\theta)}{(2l+3)(2l-1)}\right]. \tag{3.55}$$

The resulting expression for $f(\theta)$ allows to find the averaged interaction cross section $Q_{el}^{(L,S)}$ (Fig. 3.2).

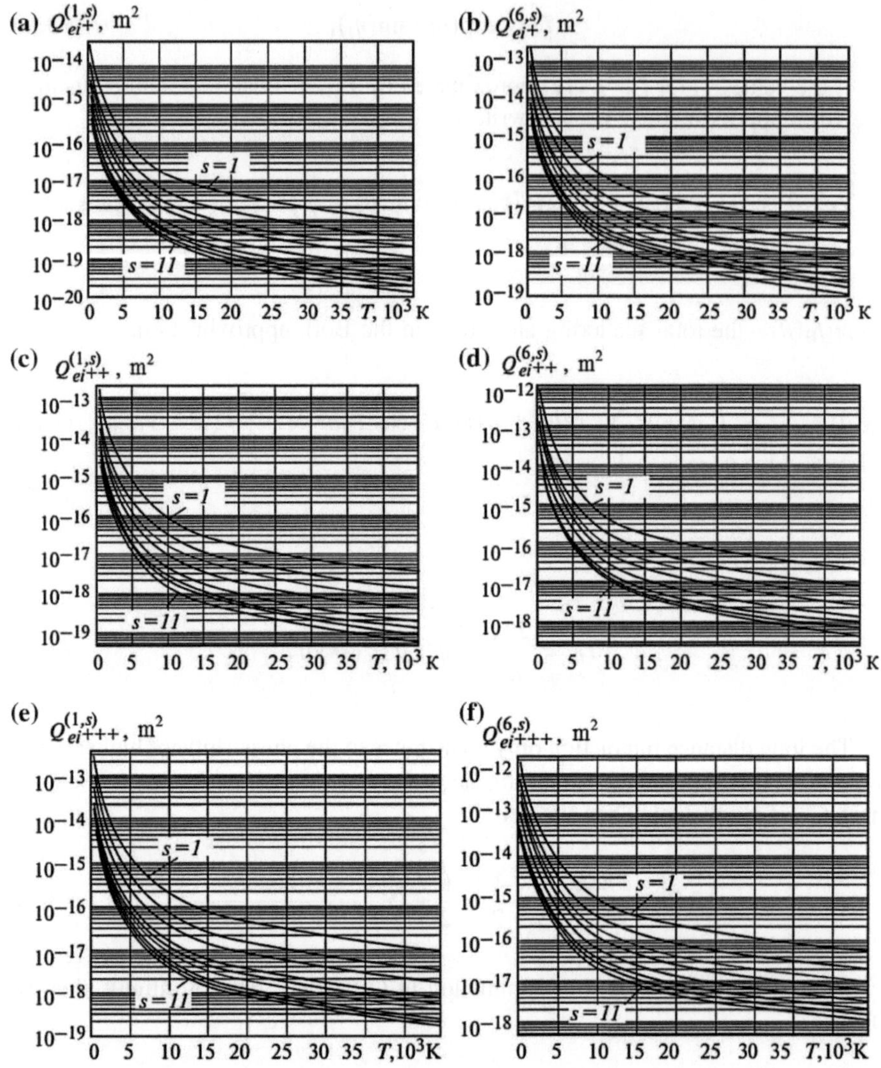

Fig. 3.2 The average effective cross sections of charged particles interaction in an argon plasma, depending on the temperature: **a**, **b** the interaction of electrons with single ions at $l = 1$ and 6; **c**, **d** the same with a double ions; **e**, **f** the same with the three time ions

3.5 Determination of Differential and Total Cross Sections of Elastic Interactions

In general, the differential cross section can be defined as

$$\sigma = \frac{d\sigma}{d\Omega} = |f(\theta, \varphi)|^2. \tag{3.56}$$

At the same time, if the potential depends on the spin, then there may be some azimuthal function φ. The physical meaning of the differential cross section can be shown as follows:

$$\frac{d\sigma}{d\Omega} d\Omega = \frac{A}{B}, \tag{3.57}$$

Here *A*—the amount of particles dispersed in a solid angle $d\Omega$ per unit time; *B*—amount of particles passing through a unit area per unit time.

The total cross section is defined as:

$$\sigma_{tot}(k) = \int \frac{d\sigma}{d\Omega} d\Omega. \tag{3.58}$$

Substituting the expression for $\frac{d\sigma}{d\Omega}$ into (3.58), we have

$$\begin{aligned} \sigma_{tot}(k) = 2\pi \int_0^{\pi} \frac{1}{k^2} \left\{ \sum_l (2l+1) \exp[i\eta_l(k)] \sin \eta_l(k) P_l(\cos \theta) \right\} \\ \times \left\{ \sum_{l'} (2l'+1) \exp[i\eta_{l'}(k)] \sin \eta_{l'}(k) P_{l'}(\cos \theta) \right\} \sin \theta d\theta, \end{aligned} \tag{3.59}$$

Using the equation $\int_0^{\pi} P_l(\cos \theta) P_{l'}(\cos \theta) \sin(\theta) d\theta = \frac{2}{2l+1} \delta_{ll'}$, where $\delta_{ll'}$—the Kronecker function, we obtain

$$\sigma_{tot} = \frac{4\pi}{k^2} \sum_{l=0}^{\infty} (2l+1) \sin^2 \eta_l(k). \tag{3.60}$$

The amplitude at zero scattering angle *f*(0) is associated with a full cross section

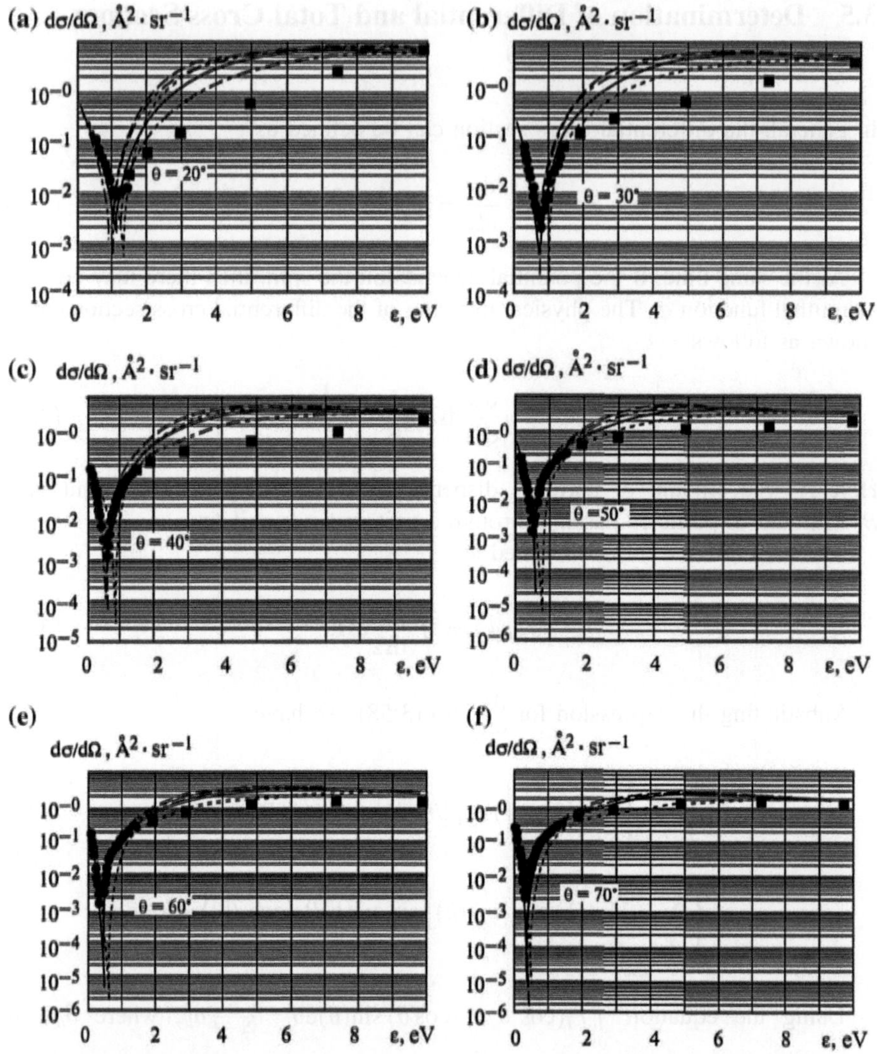

Fig. 3.3 The differential elastic scattering cross section for the scattering of electrons by atoms in argon for different scattering angles: points—the experimental data, *curves*—the designed data: *filled circle*—[12]; *filled triangle*—[16]; *filled square*—[17]; *space with dash lines*—[5]; *doted lines*—[6]; *dash with doted line*—[7]; *solid line*—results of the author

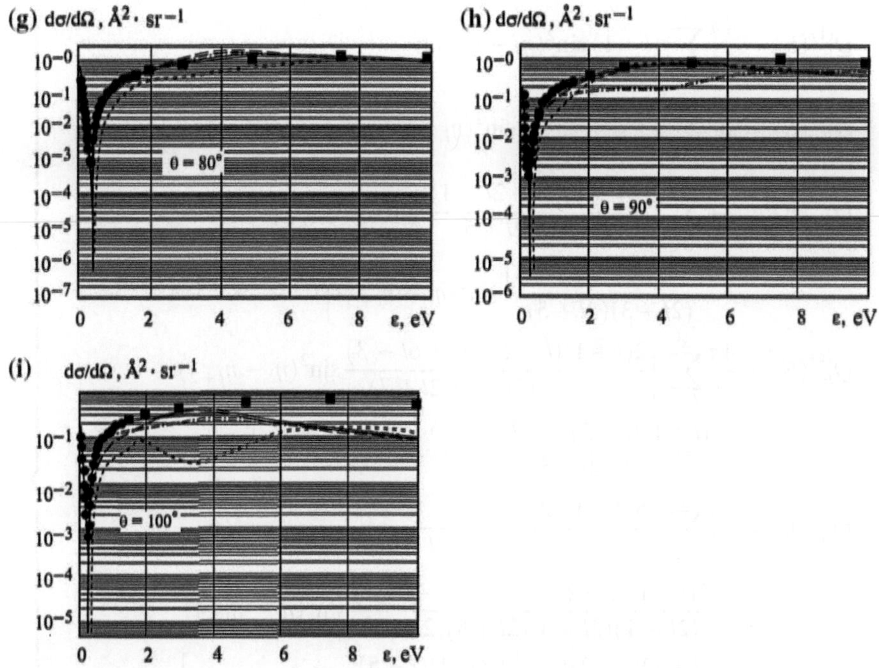

Fig. 3.3 (continued)

$$\begin{aligned}\mathrm{Im}f(0) &= \frac{1}{k}\sum_{l=0}^{\infty}(2l+1)\mathrm{Im}[\exp(i\eta_l(k))\sin\eta_l(k)]P_l(1)\\ &= \frac{1}{k}\sum_{l=0}^{\infty}(2l+1)\sin\eta_l(k) = \frac{k}{4\pi}\sigma_{\mathrm{tot}}.\end{aligned} \tag{3.61}$$

This ratio (the optical theorem) holds even for inelastic processes such as scattering processes in nuclear physics and particle physics. The form of this relationship is consistent with the following fact: the total scattering cross section reflects the decrease in the flow of the incident beam. This phenomenon can be only as a result of destructive interference that, in turn, can only occur between the incoming wave and the reflected elastic wave propagating further. This explains the linear form of $f(0)$ [2, 9].

In our case, the spin effects are not taken into account, these effects may be relevant for the heavy inert gases. The results of the differential cross sections calculations for scattering of electrons by atoms in argon using the fitting procedure MERT4 are generally in good agreement with the experimental data [7, 11–14] (Fig. 3.3). The author has calculated the scattering angles (θ = 20–100°). In the calculations for higher energies it is necessary to use a large number of members (partial waves). We give expression for $Q_B^{(1)}(E)$—$Q_B^{(6)}(E)$ [8, 15]:

$$\left.\begin{aligned}
Q_{\mathrm{B}}^{(1)}(E) &= \frac{4\pi}{k^2}\sum_{l=0}^{\infty}(l+1)\sin^2(\eta_l-\eta_{l+1});\\
Q_{\mathrm{B}}^{(2)}(E) &= \frac{4\pi}{k^2}\sum_{l=0}^{\infty}\frac{(l+1)(l+2)}{(2l+3)}\sin^2(\eta_l-\eta_{l+2});\\
Q_{\mathrm{B}}^{(3)}(E) &= \frac{4\pi}{k^2}\sum_{l=0}^{\infty}\left[\frac{3(l+1)(l^2+2l-1)}{(2l-1)(2l+5)}\sin^2(\eta_l-\eta_{l+1})\right.\\
&\quad\left.+\frac{(l+1)(l+2)(l+3)}{(2l+3)(2l+5)}\sin^2(\eta_l-\eta_{l+3})\right];\\
Q_{\mathrm{B}}^{(4)}(E) &= \frac{4\pi}{k^2}\sum_{l=0}^{\infty}\left[\frac{2(l+1)(l+2)(2l^2+6l-3)}{(2l-1)(2l+3)(2l+7)}\sin^2(\eta_l-\eta_{l+2})\right.\\
&\quad\left.+\frac{(l+1)(l+2)(l+3)(l+4)}{(2l+3)(2l+5)(2l+7)}\sin^2(\eta_l-\eta_{l+4})\right];\\
Q_{\mathrm{B}}^{(5)}(E) &= \frac{4\pi}{k^2}\sum_{l=0}^{\infty}\left[\frac{5(l+1)(2l^4+8l^3-4l^2-24l+9)}{(2l-3)(2l-1)(2l+5)(2l+7)}\sin^2(\eta_l-\eta_{l+1})\right.\\
&\quad+\frac{5(l+1)(l+2)(l+3)(l^2+4l-2)}{(2l-1)(2l+3)(2l+5)(2l+9)}\sin^2(\eta_l-\eta_{l+3})\\
&\quad\left.+\frac{(l+1)(l+2)(l+3)(l+4)(l+5)}{(2l+3)(2l+5)(2l+7)(2l+9)}\sin^2(\eta_l-\eta_{l+5})\right];\\
Q_{\mathrm{B}}^{(6)}(E) &= \frac{4\pi}{k^2}\sum_{l=0}^{\infty}\left[\frac{15(l+1)(l+2)(l^4+6l^3+l^2-24l+9)}{(2l-3)(2l-1)(2l+3)(2l+7)(2l+9)}\sin^2(\eta_l-\eta_{l+2})\right.\\
&\quad+\frac{3(l+1)(l+2)(l+3)(l+4)(2l^2+10l-5)}{(2l-1)(2l+3)(2l+5)(2l+7)(2l+11)}\sin^2(\eta_l-\eta_{l+4})\\
&\quad\left.+\frac{(l+1)(l+2)(l+3)(l+4)(l+5)(l+6)}{(2l+3)(2l+5)(2l+7)(2l+9)(2l+11)}\sin^2(\eta_l-\eta_{l+6})\right].
\end{aligned}\right\}\tag{3.62}$$

Total scattering cross section is determined by the equation

$$Q_{\mathrm{T}}(E) = \frac{4\pi}{k^2}\sum_{l=0}^{\infty}(2l+1)\sin^2\eta_l.\tag{3.63}$$

The cross section of the pulse transmission $Q_{\mathrm{M}}(E) = Q_{\mathrm{B}}^{(1)}(E)$.

At very low energies ($\varepsilon < 0.1$ eV) the partial waves of lower orders make the main contribution to the scattering amplitude. Next (at $0.1 < \varepsilon < 1$ eV) it should take into account the contribution of higher order partial waves. They give a dip in the scattering cross section of particles that is a manifestation of the Ramseur–Townsend observed experimentally (Table 3.2). At the same time, measurements of the differential cross section of elastic scattering in crossed beams can in principle

Table 3.2 The differential cross sections for elastic scattering of electrons by atoms in Argon, $Å^2$ sr^{-1} [16]

Scattering angle θ (°)	The electron energy E (eV)								
	0.1	0.5	1	3	5	7.5	10	15	20
0	5.4888×10^{-1}	3.304×10^{-1}	2.3103×10^{-1}	4.761×10^{-1}	1.826	6.0209	1.201×10	1.593×10	1.327×10
5	5.0127×10^{-1}	2.529×10^{-1}	1.4338×10^{-1}	2.968×10^{-1}	1.372	5.0407	1.064×10	1.481×10	1.227×10
10	4.5367×10^{-1}	1.859×10^{-1}	7.9252×10^{-2}	1.815×10^{-1}	1.017	4.1446	9.241	1.353×10	1.120×10
15	4.0886×10^{-1}	1.313×10^{-1}	3.6685×10^{-2}	1.269×10^{-1}	7.617×10^{-1}	3.3325	7.841	1.210×10	1.011×10
20	3.6405×10^{-1}	8.821×10^{-2}	1.2294×10^{-2}	1.224×10^{-1}	5.993×10^{-1}	2.6604	6.497	1.053×10	8.961
25	3.1925×10^{-1}	5.461×10^{-2}	3.0244×10^{-3}	1.582×10^{-1}	5.209×10^{-1}	2.1087	5.237	8.877	7.729
30	2.7640×10^{-1}	2.996×10^{-2}	6.0769×10^{-3}	2.246×10^{-1}	5.181×10^{-1}	1.6915	4.145	7.253	6.441
35	2.3467×10^{-1}	1.330×10^{-2}	1.8819×10^{-2}	3.136×10^{-1}	5.713×10^{-1}	1.4086	3.192	5.713	5.181
40	1.9687×10^{-1}	3.837×10^{-3}	3.8646×10^{-2}	4.117×10^{-1}	6.721×10^{-1}	1.2546	2.476	4.285	3.977
45	1.6466×10^{-1}	2.487×10^{-4}	6.2449×10^{-2}	5.125×10^{-1}	8.009×10^{-1}	1.207	1.946	3.080	2.884
50	1.3890×10^{-1}	9.997×10^{-4}	8.7373×10^{-2}	6.049×10^{-1}	9.381×10^{-1}	1.2406	1.602	2.086	1.915
55	1.1930×10^{-1}	4.537×10^{-3}	1.1118×10^{-1}	6.861×10^{-1}	1.070	1.3218	1.411	1.333	1.143
60	1.0446×10^{-1}	9.577×10^{-3}	1.3162×10^{-1}	7.421×10^{-1}	1.173	1.4226	1.333	8.177×10^{-1}	5.769×10^{-1}
65	9.2974×10^{-2}	1.515×10^{-2}	1.4842×10^{-1}	7.785×10^{-1}	1.243	1.5094	1.325	5.125×10^{-1}	2.249×10^{-1}
70	8.3732×10^{-2}	2.084×10^{-2}	1.6074×10^{-1}	7.869×10^{-1}	1.269	1.557	1.341	3.809×10^{-1}	6.441×10^{-2}
75	7.5051×10^{-2}	2.638×10^{-2}	1.6915×10^{-1}	7.701×10^{-1}	1.243	1.5458	1.344	3.725×10^{-1}	6.105×10^{-2}
80	6.6370×10^{-2}	3.192×10^{-2}	1.7363×10^{-1}	7.337×10^{-1}	1.176	1.473	1.305	4.341×10^{-1}	1.669×10^{-1}
85	5.7969×10^{-2}	3.697×10^{-2}	1.7503×10^{-1}	6.749×10^{-1}	1.067	1.3358	1.207	5.153×10^{-1}	3.276×10^{-1}
90	4.9567×10^{-2}	4.173×10^{-2}	1.7279×10^{-1}	6.049×10^{-1}	9.269×10^{-1}	1.1398	1.045	5.769×10^{-1}	4.957×10^{-1}
95	4.2286×10^{-2}	4.593×10^{-2}	1.6774×10^{-1}	5.209×10^{-1}	7.645×10^{-1}	9.0733×10^{-1}	8.317×10^{-1}	5.937×10^{-1}	6.273×10^{-1}
100	3.6405×10^{-2}	4.873×10^{-2}	1.5934×10^{-1}	4.313×10^{-1}	5.909×10^{-1}	6.553×10^{-1}	5.853×10^{-1}	5.545×10^{-1}	7.001×10^{-1}
105	3.1645×10^{-2}	5.041×10^{-2}	1.4842×10^{-1}	3.417×10^{-1}	4.257×10^{-1}	4.1726×10^{-1}	3.389×10^{-1}	4.705×10^{-1}	7.029×10^{-1}
110	2.8284×10^{-2}	5.097×10^{-2}	1.3526×10^{-1}	2.588×10^{-1}	2.800×10^{-1}	2.1619×10^{-1}	1.375×10^{-1}	3.613×10^{-1}	6.441×10^{-1}

(continued)

Table 3.2 (continued)

Scattering angle θ (°)	The electron energy E (eV)								
	0.1	0.5	1	3	5	7.5	10	15	20
115	2.5932×10^{-2}	5.041×10^{-2}	1.2126×10^{-1}	1.851×10^{-1}	1.686×10^{-1}	8.4853×10^{-2}	1.728×10^{-2}	2.677×10^{-1}	5.433×10^{-1}
120	2.3804×10^{-2}	4.929×10^{-2}	1.0698×10^{-1}	1.257×10^{-1}	9.969×10^{-2}	4.5367×10^{-2}	1.683×10^{-2}	2.252×10^{-1}	4.285×10^{-1}
125	2.1563×10^{-2}	4.817×10^{-2}	9.3534×10^{-2}	8.289×10^{-2}	8.037×10^{-2}	1.1314×10^{-1}	1.641×10^{-1}	2.722×10^{-1}	3.332×10^{-1}
130	1.9043×10^{-2}	4.733×10^{-2}	8.1492×10^{-2}	5.797×10^{-2}	1.129×10^{-1}	2.9404×10^{-1}	4.733×10^{-1}	4.397×10^{-1}	2.940×10^{-1}
135	1.6354×10^{-2}	4.649×10^{-2}	7.057×10^{-2}	5.013×10^{-2}	1.946×10^{-1}	5.8529×10^{-1}	9.465×10^{-1}	7.449×10^{-1}	3.304×10^{-1}
140	1.3806×10^{-2}	4.593×10^{-2}	6.1329×10^{-2}	5.741×10^{-2}	3.192×10^{-1}	9.7454×10^{-1}	1.568	1.187	4.565×10^{-1}
145	1.1762×10^{-2}	4.509×10^{-2}	5.2928×10^{-2}	7.757×10^{-2}	4.789×10^{-1}	1.4366	2.305	1.750	6.777×10^{-1}
150	1.0474×10^{-2}	4.369×10^{-2}	4.5647×10^{-2}	1.07×10^{-1}	6.581×10^{-1}	1.9407	3.108	2.406	9.717×10^{-1}
155	1.0025×10^{-2}	4.201×10^{-2}	3.9206×10^{-2}	1.423×10^{-1}	8.457×10^{-1}	2.4588	3.949	3.108	1.319
160	1.0362×10^{-2}	3.977×10^{-2}	3.3885×10^{-2}	1.792×10^{-1}	1.025	2.9404	4.733	3.781	1.677
165	1.1314×10^{-2}	3.725×10^{-2}	2.9404×10^{-2}	2.134×10^{-1}	1.187	3.3885	5.405	4.397	2.013
170	1.2490×10^{-2}	3.529×10^{-2}	2.6184×10^{-2}	2.414×10^{-1}	1.311	3.7246	5.965	4.873	2.285
175	1.3498×10^{-2}	3.389×10^{-2}	2.4196×10^{-2}	2.599×10^{-1}	1.392	3.9206	6.301	5.181	2.462
180	1.3862×10^{-2}	3.332×10^{-2}	2.3523×10^{-2}	2.660×10^{-1}	1.420	4.0046	6.413	5.293	2.520

be used for the integral scattering cross sections that can also be compared with the results of measurements of attenuation methods. Although this method is quite attractive, there may be problems in extrapolating the cross sections for large and small angles, where the experimental data cannot be obtained. This eventually leads to the limited accuracy of finding the integral scattering cross sections.

In some cases, such as for light inert gases, methods for determining the phase shift can be used for differential scattering cross sections by fitting the measured angular distributions. These techniques can be used to extrapolate the differential scattering cross sections and for immediate calculation of integral cross sections for elastic scattering. This technique is mainly used for low energy, i.e., energies below the first excitation level. Thus it is possible to limit the number of phases required for approximation and avoid the complicated phase shifts.

The transport cross section $Q^{(L)}$ is calculated based on the relative kinetic energy in a collision ε by the numerical integration of the differential cross sections for elastic scattering of electrons by atoms in argon, using the method adopted by Rath [18–20]:

At $0.02 \leq \varepsilon < 1.0$ eV the theoretical values of the collision cross sections are taken from [21].
At $1.0 \leq \varepsilon \leq 10.0$ eV the experimental values given in [11, 22] are used.
At $10.0 \leq \varepsilon \leq 100.0$ eV the experimental values of [23] are applied for small and theoretical values of [24–26]—for the large scattering angles.

The reliable source in the determination of the effective cross section of the electron–atom interaction in low-temperature plasmas are experimental data. The results of averaged cross section calculations of electron–atom in argon [22] for l = 1 and l = 6, s = 1–11 are shown in Fig. 3.4.

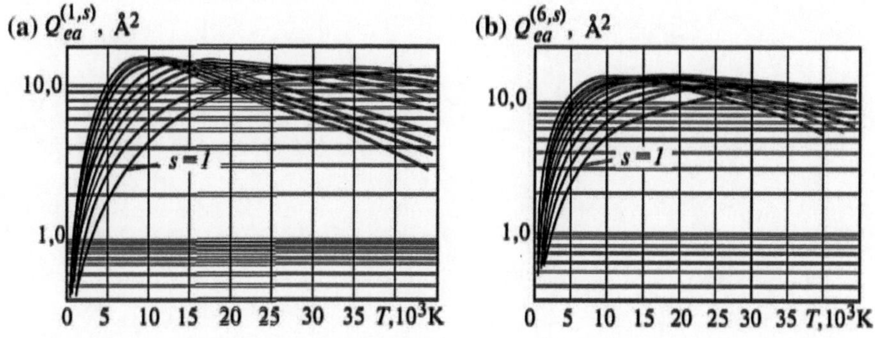

Fig. 3.4 Results of the effective cross sections calculation of interaction of electron–atom in Argon for orders l = 1 (**a**) and l = 6 (**b**)

References

1. Thurston M.O. Gaseous Electronics Theory and Practice. Canada: CRC Press, 2006.
2. Sakurai J.J., Napolitano J.J. Modern Quantum Mechanics. USA, 2010.
3. Gasiorowicz S. Quantum Physics. USA, 2003.
4. Buckman S.J., Mitroy J. // J. Phys. B. 1989. Vol. 22. P. 1365.
5. Buckman S.J., Lohmamn B. // J. Phys. B. 1987. Vol. 20. P. 5807.
6. Ferch J. et al. // J. Phys. B. 1985. Vol. 18. P. 967.
7. Haddad G.N., O'Malley T.F. // J. Phys. 1982. Vol. 35. P. 35.
8. Devoto R.S. Transport coefficients of partially ionized argon // Phys. Fluids. 1967. Vol. 10. No. 2. PP. 354–364.
9. Devoto R.S. // Phys. Fluids. 1966. Vol. 9. P. 1230.
10. Weyhreter M. et al. // J. Phys. D Atoms, Molecules and Clusters. 1988. Vol. 7. P. 333.
11. Gibson C., Gulley R.J., Sullivan J.P. et al. // J. Phys. B. 1996. Vol. 29. P. 3177.
12. Weyhreter M. et al. // J. Phys. D Atoms, Molecules and Clusters. 1988. Vol. 7. P. 333.
13. Plenkiewicz B. et al. // J. Phys. Rev. A. 1988. Vol. 38. P. 4460.
14. Buckman S.J. // J. Phys. 1997. Vol. 50. P. 483.
15. Meeks F.R. et al. // J. Chem. phys. 1994. Vol. 100 (5).
16. Plenkiewicz B., Plenkiewicz P., Jay Gerin J.P. Pseudopotential calculations for elastic scattering of slow electrons (0–20 eV) from noble gases. I. Argon.
17. Gibson J.C. et al. // J. Phys. B. 1996. Vol. 29. P. 3177.
18. Transport properties in a two temperature plasma: theory and application / Rat V. et al. // Phys. Rev. E. 2001. Vol. 64.
19. Rat V., Andre P., Aubreton J. et al. // Phys. D: Applied Phys. 2001. Vol. 34. P. 2191–2204.
20. Transport coefficients including diffusion in a two-temperature argon plasma / V. Rat, et al. // J. Phys. D: Appl. Phys. 2002. Vol. 35. P. 981–991.
21. Bell K.L., Scott N.S., Lennon M.A. // J. Phys. B. 1984. Vol. 17. P. 4757.
22. Bose T.K., Kannappan D. and Seeniraj R.V. // Warme Sioffubertragung. 1985. Vol. 19. P. 3.
23. Panajotovic R., Filipovic D., Marinkovic B. et al. // J. Phys. B. 1997. Vol. 30. P. 5877.
24. Nahar S.N. and Wadehra J.M. // Phys. Rev. A. 1987. Vol. 35. P. 2051.
25. Wadehra J.M., Nahara S.N. // Phys. Rev. A. 1987. Vol. 36. P. 1458.
26. Itikawa Y. At. Data Nucl. Data Tables. 1974. Vol. 14. No. 1.

Chapter 4
Determination of the Composition, Thermodynamic Properties, and Transport Coefficients on the Basis of the Mean Free Path

Determination of composition and properties of plasma, including the calculation of the thermodynamic properties and transport coefficients is always a key issue in plasma physics, in calculation of plasma processes and plasma torches. There are a large number of sources dedicated to this issue. Relatively simple equations can be obtained based on the model of the mean free path of the particles.

The phenomena of diffusion, viscosity, and thermal conductivity are physically similar and they are the transferring of some flows generated by the gradients of concentration, temperature and velocity in gas. The diffusion is a mass transfer from one region to another due to the presence of a concentration gradient. It can also be the result of a temperature gradient (thermal diffusion), and energy transfer can be the result of a concentration gradient (diffusion heat conductivity). Viscosity—the momentum transfer through the gas due to the presence of a velocity gradient; thermal conductivity—the transfer of thermal energy due to the presence of a temperature gradient in gas. These processes are called as transport phenomena.

This chapter provides an analysis of theoretical methods and the basic method of calculating the structure, thermodynamic properties, and transport coefficients of equilibrium and nonequilibrium two-temperature plasma on the basis of the submission of the mean free path of the particles.

4.1 The Plasma Composition

The calculation of the plasma composition in the equilibrium ionization on the basis of the action mass law is performed using the Saha equations. In the equations the speed of ionization and recombination are the same, the source of birth and death of charged particles are equal to zero, all the flows of transport and dissipation of particles are also equal to zero. The right side of the Saha equations the electron temperature is usually taken instead of the temperature of ionization. For a 5-component plasma with atoms, one-, two-, and three-ions and electrons the Saha equation are

S. Nguyen-Kuok, *Theory of Low-Temperature Plasma Physics*,
Springer Series on Atomic, Optical, and Plasma Physics 95,
DOI 10.1007/978-3-319-43721-7_4

$$\frac{n_e n_i^+}{n_a} = 2\frac{U_i^+}{U_a}\left(\frac{2\pi m_e kT_e}{h^2}\right)^{3/2}\exp\left(-\frac{E_i^+ - \Delta E_i^+}{kT_e}\right);$$
$$\frac{n_e n_i^{++}}{n_i^+} = 2\frac{U_i^{++}}{U_i^+}\left(\frac{2\pi m_e kT_e}{h^2}\right)^{3/2}\exp\left(-\frac{E_i^{++} - \Delta E_i^{++}}{kT_e}\right); \tag{4.1}$$
$$\frac{n_e n_i^{+++}}{n_i^{++}} = 2\frac{U_i^{+++}}{U_i^{++}}\left(\frac{2\pi m_e kT_e}{h^2}\right)^{3/2}\exp\left(-\frac{E_i^{+++} - \Delta E_i^{+++}}{kT_e}\right),$$

Here n_e, n_i^+, n_i^{++}, n_i^{+++}, n_a—the concentration of electrons, respectively, single, double, and triple ions and atoms in the plasma; U_i^+, U_i^{++}, U_i^{+++}, U_a—the statistical sums of single, double, and triple ions and atoms in plasma E_i^+, E_i^{++}, E_i^{+++}—the potential of the single, double and triple ionization of atoms; ΔE_i^+, ΔE_i^{++}, ΔE_i^{+++}—the reduction of potentials, respectively, for single, double, and triple ionization of atoms due to the screening of the charged particles in plasma

$$\Delta E_i^+ = \frac{2e^2}{4\pi\varepsilon_0 r_D}, \quad \Delta E_i^{++} = \frac{3e^2}{4\pi\varepsilon_0 r_D}, \quad \Delta E_i^{+++} = \frac{4e^2}{4\pi\varepsilon_0 r_D}. \tag{4.2}$$

Taking into account the thermal non-equilibrium of plasma the Debye radius is defined as

$$r_D = \sqrt{\frac{2\varepsilon_0 kT_e T}{e^2\left(n_e + n_i^+ + 4n_i^{++} + 9n_i^{+++}\right)(T_e + T)}}. \tag{4.3}$$

Hereafter, we will use the following notation: $T_e \neq T_{ai} = T_a = T_i = T$.

The Saha equations can be used for the calculation of the nonequilibrium plasma composition, when $T_e \neq T_{ai}$ and the experimental evidence is required. For this calculation the other equation can be used. The Potapov equations [1] obtained in the analysis of the chemical composition of the equilibrium system are the most famous

$$n_e\left(\frac{n_i^+}{n_a}\right)^{T/T_e} = 2\frac{U_i^+}{U_a}\left(\frac{2\pi m_e kT}{h^2}\right)^{3/2}\exp\left(-\frac{E_i^+ - \Delta E_i^+}{kT}\right);$$
$$n_e\left(\frac{n_i^{++}}{n_i^+}\right)^{T/T_e} = 2\frac{U_i^{++}}{U_i^+}\left(\frac{2\pi m_e kT}{h^2}\right)^{3/2}\exp\left(-\frac{E_i^{++} - \Delta E_i^{++}}{kT}\right); \tag{4.4}$$
$$n_e\left(\frac{n_i^{+++}}{n_i^{++}}\right)^{T/T_e} = 2\frac{U_i^{+++}}{U_i^{++}}\left(\frac{2\pi m_e kT}{h^2}\right)^{3/2}\exp\left(-\frac{E_i^{+++} - \Delta E_i^{+++}}{kT}\right).$$

These equations are valid in cases where the plasma can be regarded as an ideal gas, consisting of quasi-independent subsystems with different temperatures. When thermal equilibrium ($T_e = T$) Potapov equation turns into Saha equation (4.1).

To determine the composition of plasma it is necessary to use two additional equations

The plasma quasi-neutrality condition

$$n_e = n_i^+ + 2n_i^{++} + 3n_i^{+++}; \quad (4.5)$$

The Dalton's law of partial pressures

$$p = n_e kT_e + \left(n_i^+ + n_i^{++} + n_i^{+++} + n_a\right)kT. \quad (4.6)$$

Thus, to calculate the composition of the 5-component plasma we have a system of nine Eqs. (4.1)–(4.6) with unknown variables n_e, n_a, n_i^+, n_i^{++}, n_i^{+++}, ΔE_i^+, ΔE_i^{++}, ΔE_i^{+++}, r_D. The solution of such a system can be performed iteratively with the bottom of relaxation to achieve the convergence. Calculation of the composition of two-temperature plasma is carried out for a given temperature T_e, T and pressure p. The values U_a, U_i^+, U_i^{++}, U_i^{+++}, E_i^+, E_i^{++}, E_i^{+++} are given as input variables. The results of the composition calculation for the main gases of equilibrium plasma at atmospheric pressure are shown in Figs. 4.1 and 4.2.

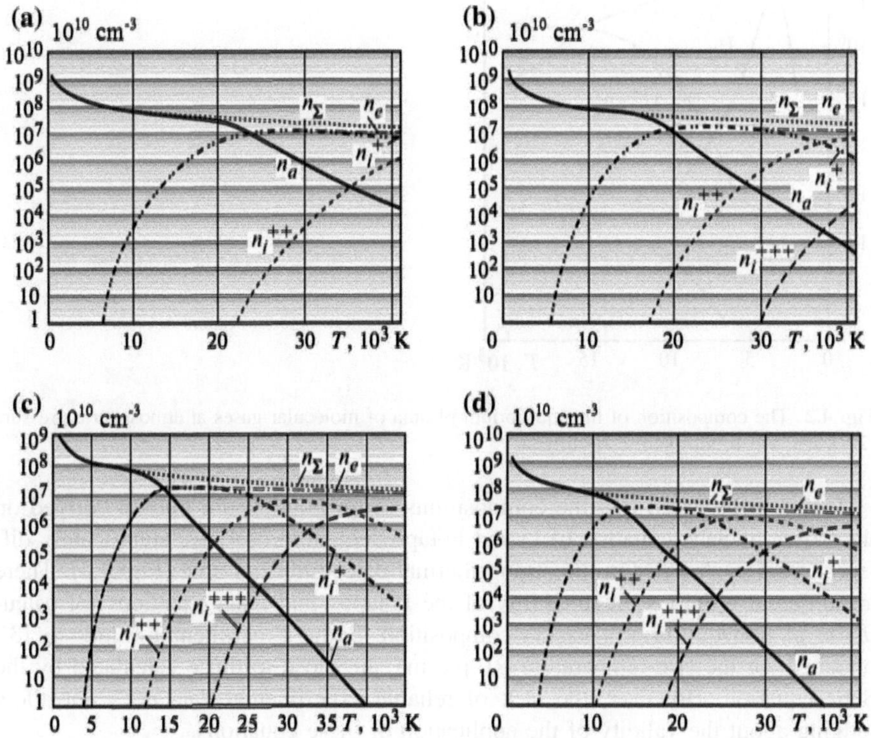

Fig. 4.1 The compositions of equilibrium plasmas: **a** Helium, **b** Neon, **c** Argon, **d** Krypton, $n_\Sigma = n_a + n_e + n_i^+ + n_i^{++} + n_i^{+++}$—the total concentration of particles

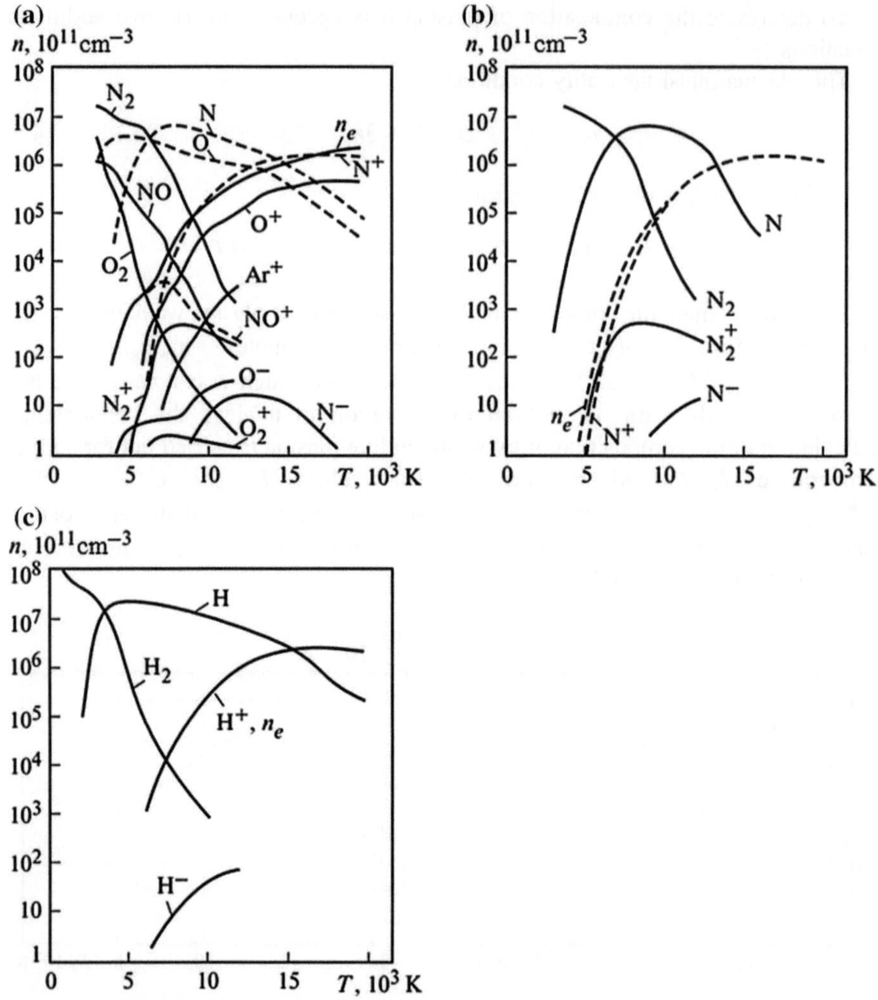

Fig. 4.2 The composition of the equilibrium plasma of molecular gases at atmospheric pressure [2]: **a** the Air, **b** Nitrogen, **c** Hydrogen

It should be noted that the compositions of nonequilibrium plasma derived on the basics of Saha equation (4.1) and Potapov equation (4.4) are significantly different from each other in the strong thermal disequilibrium area (Fig. 4.3). There are different opinions on suitability of the Potapov and Saha equations for calculation of nonequilibrium plasma composition. Some experimental values n_e [3], obtained for the two-temperature RF-plasma, are close to those calculated by the Saha equation. However, the lack of reliable experimental data does not allow judging about the validity of the application of these equations.

In the case where the ionization equilibrium is disturbed, composition of the plasma should be determined by the equation of ionization equilibrium (1.40).

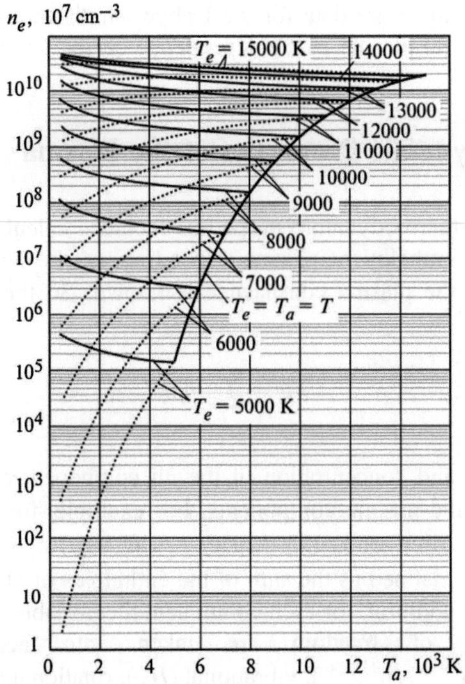

Fig. 4.3 The concentration of electrons in the two-temperature argon plasma: *straight line* the data obtained by the Saha equation; *dotted line* by the same Potapov equation

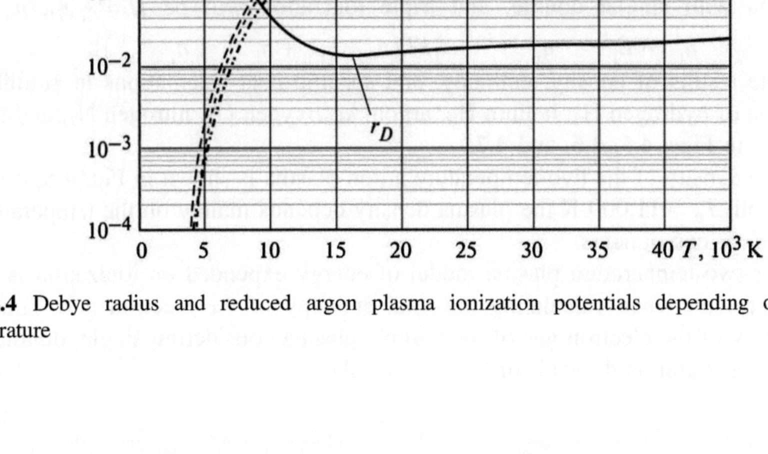

Fig. 4.4 Debye radius and reduced argon plasma ionization potentials depending on the temperature

On the Fig. 4.4, there are data for the Debye length and decreasing ionization potentials of the argon plasma in the dependence on the temperature.

4.2 Thermodynamic Properties of the Plasma

Calculation of the thermodynamic properties includes calculation of the density, heat capacity, and enthalpy of plasma depending on the temperature. They are easily identified if the plasma consistence is known. So the density of a multi-component plasma

$$\rho = \sum_j m_j n_j \tag{4.7}$$

Here m_j, n_j—mass and concentration of the jth component of plasma. The summation is over all the plasma components. For example, for atomic plasma with single, double, and triple ionization density $\rho = m_a\left(n_a + n_i^+ + n_i^{++} + n_i^{+++}\right)$.

Plasma energy is defined as the sum of the enthalpies of all its components and of stored energy. In addition, for each component the possible reserves of energy of different degrees of freedom are taken into account [progressive ($H_a, H_m, H_e, H_i^+, H_i^{++}, H_i^{+++}$), vibrational ($H_V$), rotational ($H_R$) motion], stocks of energy dissociation (H_D) and ionization (H_I), the excitation of different levels (H^*), to perform work in compression advanced (H_p), etc.

$$\begin{aligned} H_\Sigma = \sum_j H_j = H_a + H_m + H_e + H_i^+ + H_i^{++} + H_i^{+++} + H_V \\ + H_R + H_I + H_D + H^* + H_p \end{aligned} \tag{4.8}$$

In determination of the plasma enthalpy it is generally considered that the law of distribution of energy over the degrees of freedom is valid: the same kinetic energy $kT/2$ corresponds to each degree of the freedom in plasma. So, the enthalpy of atomic plasma with single, double, and triple ionization will be $H = \frac{1}{\rho}\left[\frac{5}{2}kTn_e + E_i^+ n_i^+ + E_i^{++} n_i^{++} + E_i^{+++} n_i^{+++} + \frac{5}{2}kT\left(n_a + n_i^+ + n_i^{++} + n_i^{+++}\right)\right]$.

The results of density, enthalpy, and specific heat calculations in equilibrium plasma of hydrogen H_2, helium He, argon Ar, oxygen O_2, nitrogen N_2, and air are shown in Figs. 4.5, 4.6, and 4.7.

The density of the two-temperature argon plasma is shown in Fig. 4.8, it shows that until T_e = 11,000 K the plasma density depends mainly on the temperature of the heavy components.

For two-temperature plasma model of energy expended on ionization is taken into account when calculating the total enthalpy of the electron gas. Thus, the enthalpy of the electron gas of monatomic plasma considering single, double, and triple ionization is determined by the formula

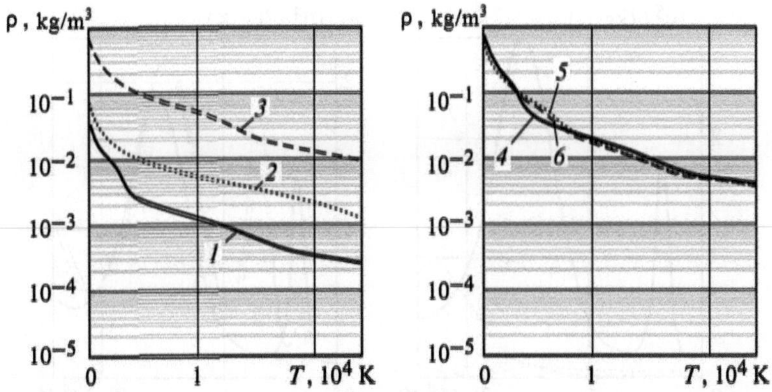

Fig. 4.5 Density of equilibrium plasma at atmospheric pressure depending on the temperature: *1* Hydrogen (H_2), *2* Helium (He), *3* Argon (Ar), *4* Oxygen (O_2), *5* Nitrogen (N_2), *6* Air

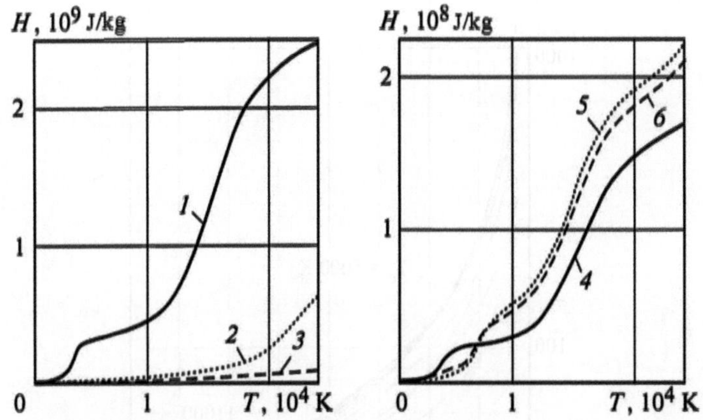

Fig. 4.6 The enthalpy of the equilibrium plasma at atmospheric pressure for gases depending on the temperature: the notations as in Fig. 4.5

$$H_e = \frac{1}{\rho}\left[\frac{5}{2}kT_e n_e + E_i^{+} n_i^{+} + E_i^{++} n_i^{++} + E_i^{+++} n_i^{+++}\right] \quad (4.9)$$

In two-temperature plasma it should also distinguish the specific heat of electrons and heavy components. The heat capacity of the *j*th component of the plasma at a constant pressure can be defined as: $c_{pj} = \frac{dH_j}{dT_j}$. The results of calculations of enthalpy and heat capacity of a two-temperature argon plasma are shown in Figs. 4.9 and 4.10.

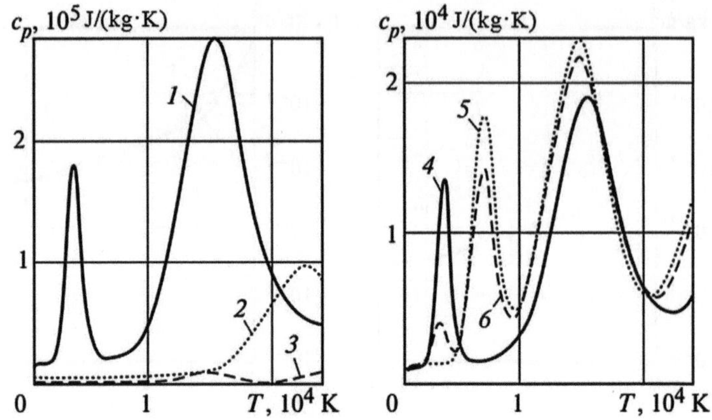

Fig. 4.7 Heat capacity of the equilibrium plasma at atmospheric pressure for gases depending on the temperature: the notations as in Fig. 4.5

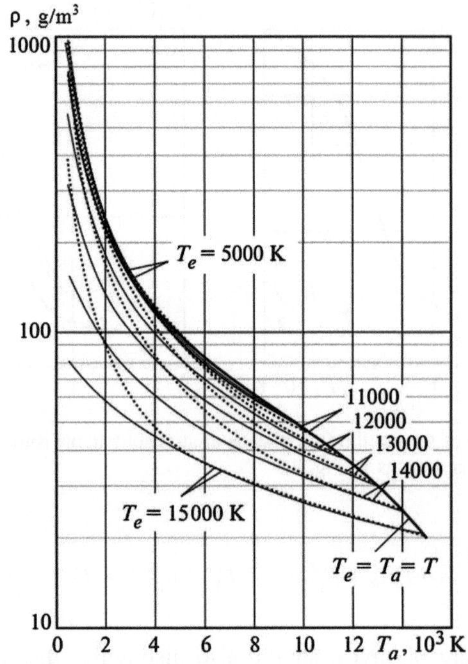

Fig. 4.8 The density of the two-temperature argon plasma at atmospheric pressure: *straight line* calculation of Saha equation (4.1), *dotted line* calculation of Potapov equations (4.4)

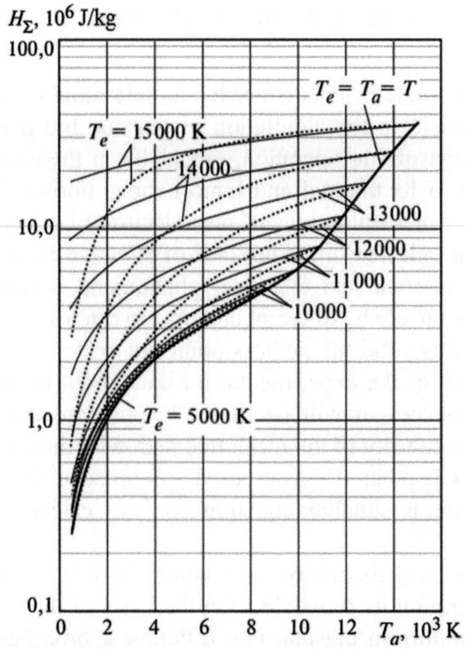

Fig. 4.9 The total enthalpy of a two-temperature argon plasma, depending on the temperature: All the notations are the same as in Fig. 4.8

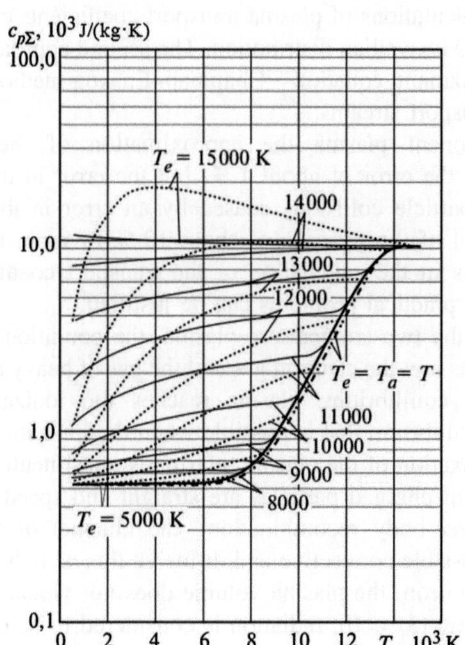

Fig. 4.10 The total heat capacity of a two-temperature argon plasma, depending on the temperature: the notations are the same as in Fig. 4.8 (*dashed–dotted* and *solid lines* are different because of the error of numerical differentiation enthalpy temperature)

4.3 Transport Coefficients of the Plasma

Calculation of transfer coefficients is needed to determine the flow of heat transfer, momentum and particle in the simulation of plasma, but it is often a very cumbersome and time-consuming operation, especially in the case of multicomponent plasma with regard to its thermal and ionic nonequilibrium. In the course of the calculation a number of assumptions of the selection of parameters for calculating is used, for example, selected different kinds of effective cross sections, taking into account of certain thermal and electrical phenomena occurring in plasma. The results of calculations made by different authors are often significantly different, and as a rule they go far beyond the authors point out errors. Unfortunately, the same situation is observed for the experimental for data about the transport coefficients.

In these circumstances, it will be useful to compare the results of a simpler method based on the concept of the mean free path with the data of accurate method for calculating the Chapman–Enskog-based solutions of the Boltzmann transport equation and choose a suitable equation for the calculation of the transport coefficients.

The condition of local thermodynamic equilibrium (LTE) plasma greatly facilitates the task of defining its properties. On the basis of the theory, with respect to thermodynamic equilibrium plasma, two different approaches in determining the interaction of the particles in the plasma and the plasma transfer coefficients are used. The first approach is based on the concept of the mean free path of the plasma particles (atoms, ions, electrons, molecules, excited atoms and molecules, and others). And the calculations of plasma transport coefficients can be obtained from averaging over the Maxwellian distribution. The second approach is based on direct solution of the Boltzmann equation—Chapman–Enskog method with the definition of the relevant transport streams.

For multicomponent plasma the approximation of the Chapman–Enskog although may give the error of about 1 %, but the error in the calculation of the cross sections of particle collisions caused by an error in the knowledge of the interaction potential of the order is of about 10 % or more [4]. Therefore, large computational costs in the calculation of the transport coefficients of the above procedure in many practical problems can be justified.

With regard to the two-temperature plasma, the conditions for calculating the transport coefficients are: the electron gas and the gas of heavy components are both in thermodynamic equilibrium; plasma reaches the ionization equilibrium to achieve thermal equilibrium that is possible when the ionization rate exceeding the rate of thermal relaxation of the plasma; plasma is quasi-neutral, i.e., $n_e = \sum Z_i n_i$; basic mechanisms of charged particles are straight and speed two-particle ionization, while the three-body recombination—the channel of their disappearance; plasma lacks all possible convective and diffusive fluxes; radiation and free access of charged particles from the plasma volume does not violate the ionization equilibrium of the system ($\dot{n}_e = 0$); radiation is considered to be optically thin.

In accordance with such restrictions, the plasma composition, its thermodynamic properties, and the coefficients are uniquely determined by the electron temperature T_e, the temperature of the heavy components T, and pressure p.

To calculate the transport coefficients of a nonequilibrium plasma approach based on the use of simple, yet experimentally proven methods for determining the properties of the plasma are more appropriate, as well as the results of accurate calculations obtained by solving the kinetic equation. For example, in [5] the two-temperature plasma transport coefficients are calculated based on the concept of the mean free path of particles and taking into account the kinetic corrections for accurate calculations in thermal equilibrium ($T = T_e$).

Definition of the argon plasma properties is relevant in connection with the widespread use of argon in the plasma torch technology. The most accurate transfer coefficients for equilibrium in argon plasma are obtained in [6–11]. The coefficients of two-temperature plasma are both theoretically and experimentally in many studies [12–19]. The data of different authors about argon on thermodynamic properties such as density, heat capacity, and enthalpy, have the satisfactory agreement. However, the data of different authors on the dissipative transport coefficients—electrical conductivity, thermal conductivity, viscosity, radiation—vary greatly.

4.3.1 Determination of the Electrical Conductivity and Thermal Conductivity

The mobility of electrons is much larger than mobility of ions, so the conductivity of overall gas is often equated with its electronic conductivity $\sigma \approx \sigma_e$.

Methods for the calculation of electrical conductivity for partially ionized gas of arbitrary composition that are based on the exact theory, are quite complex. Relatively simple and accurate equation for calculation of the electrical conductivity of the ionized gas can only be obtained in two limiting cases. For a weakly ionized gas we have

$$\sigma_e = \frac{4\pi}{3} \frac{n_e e^2}{kT_e} \int_0^\infty \frac{v^2 f_{eM}(v)}{\sum_m \nu_{em}(v)} dv,$$

Here $f_{eM}(v)$—the Maxwell distribution function of the electron velocity; $\sum_m \nu_{em}(v)$—the sum of the frequency of collisions for electrons with all neutral particles.

In the case for the fully ionized plasma, consisting of singly charged ions there is no field, expressions for σ_e, λ_e take the following form:

$$\sigma_e = 1{,}975\frac{n_e e^2}{m_e \nu_{ei}}; \quad \lambda_e = 3{,}203\frac{n_e k^2 T_e}{m_e \nu_{ei}},$$

Here $\nu_{ei} = n_i \frac{4\sqrt{2\pi}}{3}\left(\frac{m_e}{kT_e}\right)^{3/2}\left(\frac{e^2}{4\pi\varepsilon_0 m_e}\right)^2 \ln \Lambda$—the frequency of collisions.

These results were obtained by L. Spitzer, taking into account the Debye screening in determination of the Coulomb potential and, as shown in [5], they are valid for only large values of $\ln \Lambda$.

For the calculation of the electrical conductivity and the electron conductivity the expression of elementary kinetic theory can be used in the form

$$\sigma_e = \frac{k_\sigma e^2 n_e}{m_e \bar{v}_e \sum_j n_j Q_{ej}} = \frac{k_\sigma e^2 n_e}{m_e \sqrt{\frac{8kT_e}{\pi m_e}}(n_a Q_{ea} + n_i Q_{ei})}; \tag{4.10}$$

$$\lambda_e = \frac{k_{\lambda e} n_e k^2 T_e}{m_e \bar{v}_e \sum\limits_j n_j Q_{ej}} = \frac{k_{\lambda e} n_e k^2 T_e}{m_e \sqrt{\frac{8kT_e}{\pi m_e}}(n_a Q_{ea} + n_i Q_{ei})}, \tag{4.11}$$

Here $\bar{v}_e = \sqrt{\frac{8kT_e}{\pi m_e}}$—the average thermal velocity of the electrons; k_σ, k_{λ_e}—the amendments to the kinetic model of equilibrium plasma ($T_a = T_e$) that are the functions of the collisions frequency. The coefficients k_σ, k_{λ_e} indicate how the values of σ_e, λ_e are calculated by the equations of the kinetic theory, and are different from the corresponding values obtained by the elementary equations. The change limits are $k_\sigma = 1 - 2.8$, $k_{\lambda_e} = 0.3 - 1.6$. The calculation results of σ_e, λ_e for the equilibrium argon plasma (4.10) and (4.11) with $k_\sigma = 1.05$, $k_{\lambda_e} = 1.6$ have a good agreement with data obtained by the Chapman–Enskog [6, 7]. Data for the calculation of σ (4.10) for the equilibrium argon plasma at atmospheric pressure are shown in Fig. 4.11. It also shows their comparison with experimental data and the calculated data of other authors. The calculation results for the electrical conductivity of various gases in equilibrium plasma at atmospheric pressure are shown in Fig. 4.12.

The electronic conductivity in [5] with the influence of an alternating electromagnetic field is determined as follows:

$$\sigma_e = \frac{k_\sigma e^2 n_e \nu_m}{m_e(\omega^2 + \nu_m^2)}.$$

Here $\omega = eB/m_e$—the gyromagnetic or cyclotron frequency; $\nu_m = \sum_j \nu_{ej}$—the sum of the collisions frequency of electrons with all the other particles. Electronic thermal conductivity [5] is calculated by the expression

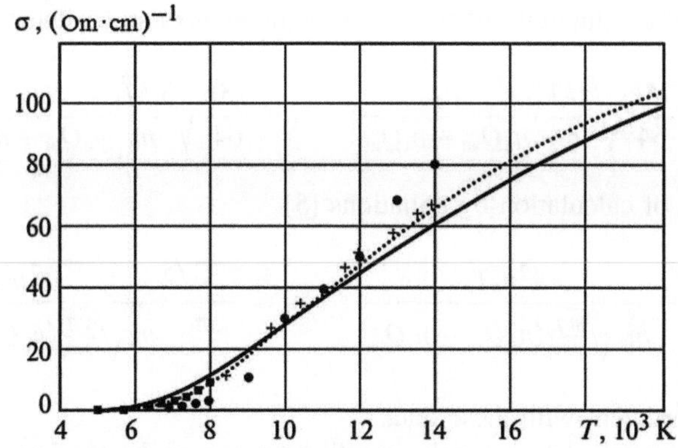

Fig. 4.11 The electrical conductivity of the equilibrium argon plasma depending on the temperature: *straight line* the author's estimates; *dotted line* the results of the calculation of Devoto [6]; *filled square, filled circle, plus sign* experimental data, respectively [21–23]

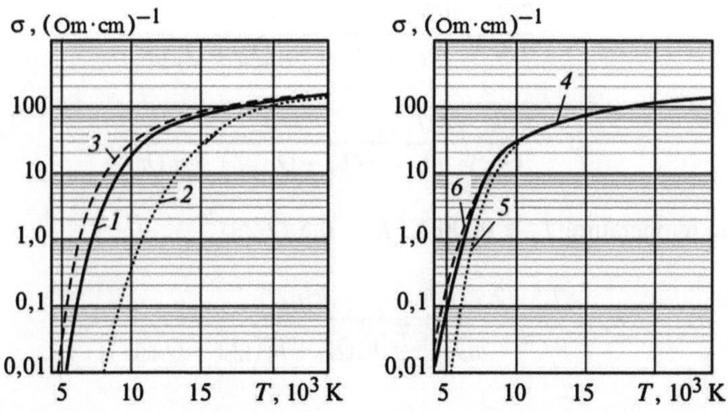

Fig. 4.12 The electrical conductivity of the equilibrium plasma at atmospheric pressure for gas: the notations are the same as in Fig. 4.5

$$\lambda_e = \frac{15/(2\pi)}{1 + \frac{\nu_{ei}}{\sqrt{2}\sum_j \nu_{ej}}} \frac{k_{\lambda_e} n_e k^2 T_e}{m_e \sum_j \nu_{ej}}.$$

The calculation using this formula with the amendment $k_{\lambda_e} = 1.2$ gives the same result as the calculation by (4.11) with the amendment $k_{\lambda_e} = 1.6$.

In some works, the electronic electrical conductivity was calculated from the ratio for the mixture [5], when the resistivity $\frac{1}{\sigma_e} = \frac{1}{\sigma_1} + \frac{1}{\sigma_2}$ is the sum of the resistivity for weakly ionized and fully ionized gases. Similarly $\frac{1}{\lambda_e} = \frac{1}{\lambda_1} + \frac{1}{\lambda_2}$. However, in this approach it should be borne in mind that the additive $\frac{1}{\sigma_e} = \frac{1}{\sigma_1} + \frac{1}{\sigma_2}$ is broken at comparable values of the collisions frequency ν_{ei} and ν_{ea}.

The thermal conductivity of the heavy components can be determined as [20]

$$\lambda_a = \frac{75k}{64}\sqrt{\frac{\pi k T_a}{m_a}}\frac{n_a}{n_a Q_{aa} + n_i Q_{ai}}; \quad \lambda_i = \frac{75k}{64}\sqrt{\frac{\pi k T_i}{m_i}}\frac{n_i}{n_a Q_{ai} + n_i Q_{ii}}.$$

The results of calculation by equations [5]

$$\lambda_a = \frac{7.5\sqrt{2}}{\pi}\frac{k^2 n_a T_a}{m_a\sqrt{\frac{8kT_a}{\pi m_a}}(n_a Q_{aa} + n_i Q_{ai})}; \quad \lambda_i = \frac{7.5\sqrt{2}}{\pi}\frac{k^2 n_i T_i}{m_i\sqrt{\frac{8kT_i}{\pi m_i}}(n_a Q_{ai} + n_i Q_{ii})}$$

have good agreement with these data.

However, the calculation for the equilibrium plasma by the last equation gives some higher results at T > 9000 K compared with the data of [6]. This is mainly due to the neglect of the atom-ion recharge.

For the temperature $T_a < 9000$ K, $K = 0.4$ (2.29)

$$\lambda_a = \frac{75k}{64}\sqrt{\frac{\pi k T_a}{m_a}}\frac{n_a}{n_a Q_{aa} + n_i (Q_{ai} + Q_{ai\cdot per})}; \tag{4.12}$$

$$\lambda_i = \frac{75k}{64}\sqrt{\frac{\pi k T_i}{m_i}}\frac{n_i}{n_a (Q_{ai} + Q_{ai\cdot per}) + n_i Q_{ii}}; \tag{4.13}$$

For the temperature $T_a \geq 9000$ K, $K = 0.5$ (2.29)

$$\lambda_a = \frac{7.5\sqrt{2}}{\pi}\frac{k^2 n_a T_a}{m_a\sqrt{\frac{8kT_a}{\pi m_a}}\left[n_a Q_{aa} + n_i (Q_{ai} + Q_{ai\cdot per})\right]}; \tag{4.14}$$

$$\lambda_i = \frac{7.5\sqrt{2}}{\pi}\frac{k^2 n_i T_i}{m_i\sqrt{\frac{8kT_i}{\pi m_i}}\left[n_a (Q_{ai} + Q_{ai\cdot per}) + n_i Q_{ii}\right]}. \tag{4.15}$$

The existence of spatial gradients of temperature and concentration of the particles in plasma causes the different diffusion fluxes. Thus, the electrons and ions diffuse from the hot spots in the cold, where they recombine and emit energy spent before their ionization. There is a special type of heat transfer—reactive thermal conductivity. Reactive heat conductivity is determined by the corresponding flow of heat transfer like $\lambda_R = E_i D_{amb}\left(\frac{\partial n_e}{\partial T}\right)_p$. In [28], the thermal conductivity is determined as

$$\lambda_R = n_\Sigma D_{amb}\frac{E_i^2}{kT_a}\frac{n_a n_e}{(n_a + n_e)^2}. \tag{4.16}$$

Here $n_{\sum} = n_a + n_e + n_i$—the total concentration of particles.

The total thermal conductivity for the equilibrium argon plasma is shown in Fig. 4.13. The results show that the total thermal conductivity consists essentially of reactive and electronic conductivity. In the temperature range of $T = 10{,}000$–$15{,}000$ K the reactive thermal conductivity can be greater than the contact (conduction atoms, ions and electrons) and provides a maximum total conductivity. The values of the total thermal conductivity of the equilibrium plasma at atmospheric pressure in various gases are shown in Fig. 4.14. The molecular gas first maximum thermal conductivity is achieved mainly at $T < 10{,}000$ K due to the dissociation of molecules (Fig. 4.14).

Electrical and thermal conductivity of two-temperature argon plasma are designed using the above equations, the results of these calculations are shown in Figs. 4.15 and 4.16.

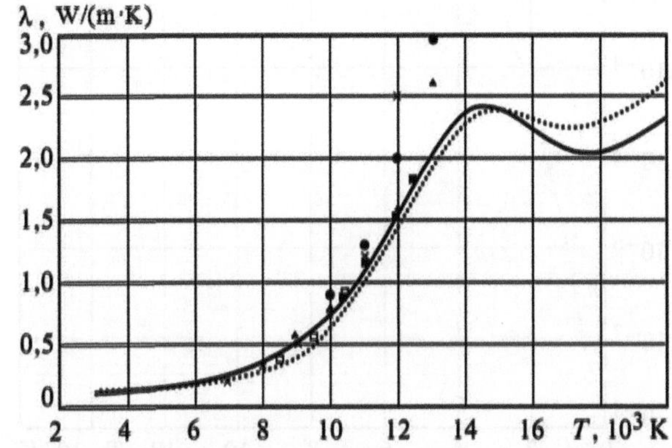

Fig. 4.13 The total thermal conductivity of the argon plasma depends on the temperature: experimental data: *filled square* [24]; *filled triangle* [25]; *cross line* [23]; *filled circle* [26]; *open square* [27]; the *other symbols* are the same as in Fig. 4.11

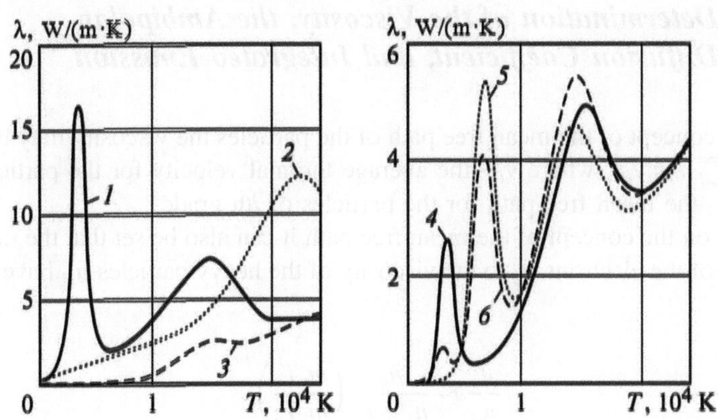

Fig. 4.14 The thermal conductivity of the equilibrium plasma at atmospheric pressure: the notations as in Fig. 4.5

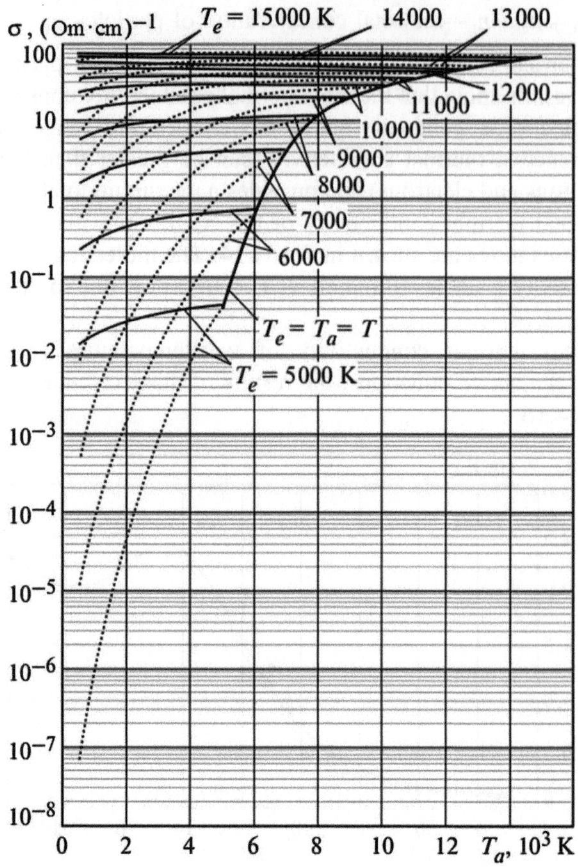

Fig. 4.15 The electrical conductivity of two-temperature argon plasma: the notations are the same as in Fig. 4.8

4.3.2 Determination of the Viscosity, the Ambipolar Diffusion Coefficient, and Integrated Emission

With the concept of the mean free path of the particles the viscosity may be defined as $\mu = \sum_i m_i n_i \bar{v}_i l_i$, where $\bar{v}_i$—the average thermal velocity for the particles of *i*th grade; l_i—the mean free path for the particles of *i*th grade.

Based on the concept of the mean free path it can also be set that the ratio of the viscosity of the electrons μ_e to the viscosity of the heavy particles μ_{ai} have the order of [5]

$$\frac{\mu_e}{\mu_{ai}} \sim \frac{n_e}{n_a + n_i} \left(\frac{m_e T_e}{m_a T_a} \right)^{1/2}.$$

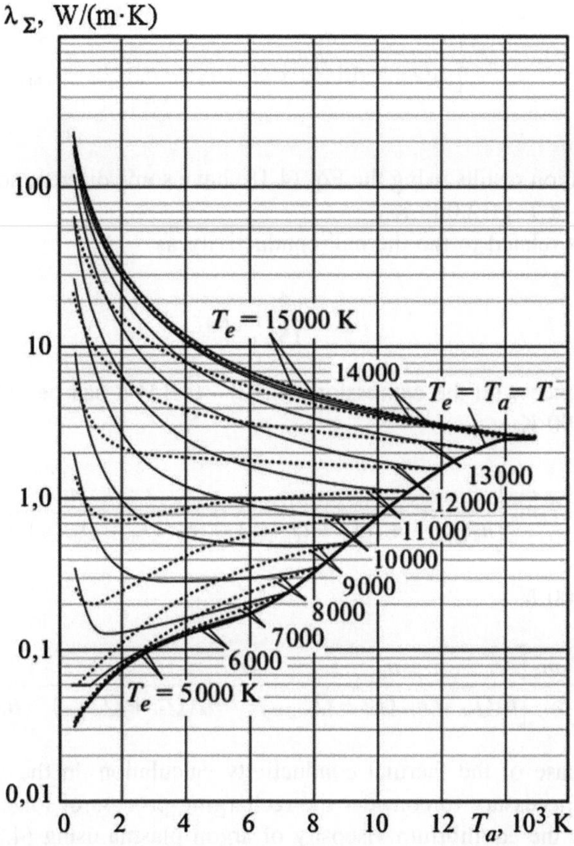

Fig. 4.16 The total thermal conductivity of two-temperature plasma depending on temperature: the notations as in Fig. 4.8

It follows that the plasma viscosity is mainly determined by the momentum transfer at the contact movement of the heavy particles (atoms, ions, molecules), therefore, the influence on the viscosity of the electrons can be neglected.

In particular, the viscosity of plasma is caused mainly by ion pulse transfer, so it is determined by the relation [5]

$$\mu_{\mathrm{i}} = \frac{0.406(4\pi\varepsilon_0)^2\sqrt{m_{\mathrm{i}}}(kT_{\mathrm{i}})^{5/2}}{Z^4 e^4 \ln \Lambda} \tag{4.17}$$

Here Z—ion charge.

There are other equations for the viscosity calculation in the references, e.g.,

$$\mu_{ai} = 6 \times 10^{-25} T_a^{0.5} \left[\frac{n_a}{n_a Q_{aa} + n_i (Q_{ai} + Q_{ai \cdot per})} + \frac{n_i}{n_a (Q_{ai} + Q_{ai \cdot per}) + n_i Q_{ii}} \right]. \tag{4.18}$$

The calculation results using the Eq. (4.18) have some discrepancy with the data of Devoto [6] at $T < 12{,}000$ K.

Viscosity is related to the thermal conductivity as

$$\mu_{ai} = \frac{4}{15k} m_a \lambda_{ai}, \tag{4.19}$$

and in accordance with the expressions (4.12)—(4.15) it can be written that

For $T < 9000$ K

$$\mu_{ai} = \frac{5}{16} \sqrt{\pi k T_a m_a} \left[\frac{n_a}{n_a Q_{aa} + n_i (Q_{ai} + Q_{ai \cdot per})} + \frac{n_i}{n_a (Q_{ai} + Q_{ai \cdot per}) + n_i Q_{ii}} \right]; \tag{4.20}$$

For $T > 9000$ K

$$\mu_{ai} = \sqrt{\frac{k T_a m_a}{\pi}} \left[\frac{n_a}{n_a Q_{aa} + n_i (Q_{ai} + Q_{ai \cdot per})} + \frac{n_i}{n_a (Q_{ai} + Q_{ai \cdot per}) + n_i Q_{ii}} \right]. \tag{4.21}$$

As in the case of the thermal conductivity calculation, in the viscosity determination it is necessary to consider the recharging process of ions. The results of calculation for the equilibrium viscosity of argon plasma using (4.20), (4.21) and taking into account exchange cross section [(2.29) with $K = 0.5$] at a temperature

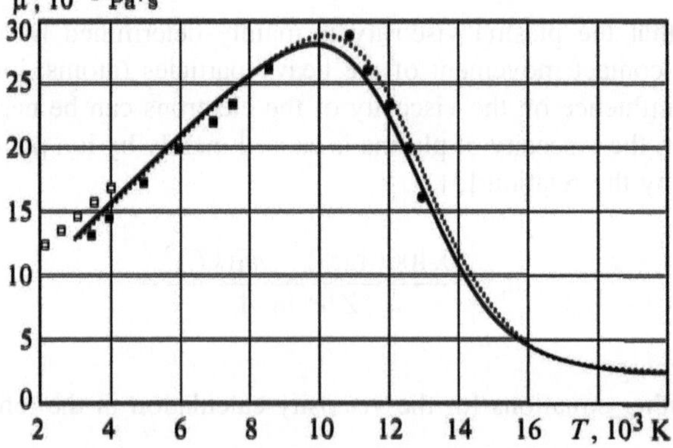

Fig. 4.17 Viscosity of the equilibrium argon plasma at atmospheric pressure, depending on the temperatures: Experimental data: *filled square* [29]; *filled circle* [30]; *open square* [31]; the other symbols are the same as in Fig. 4.11

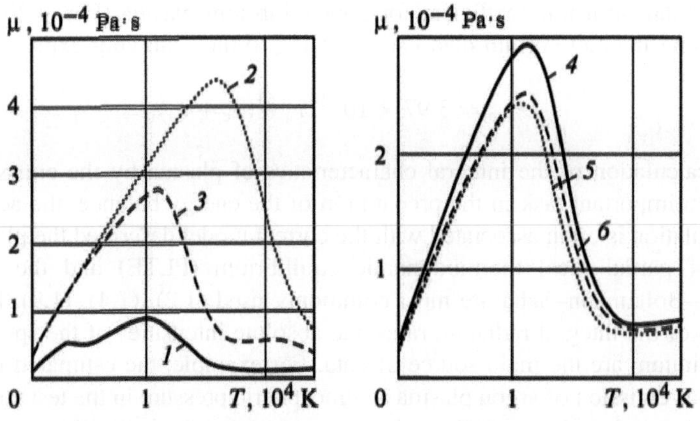

Fig. 4.18 Viscosity of the equilibrium plasma at atmospheric pressure: the notations as in Fig. 4.5

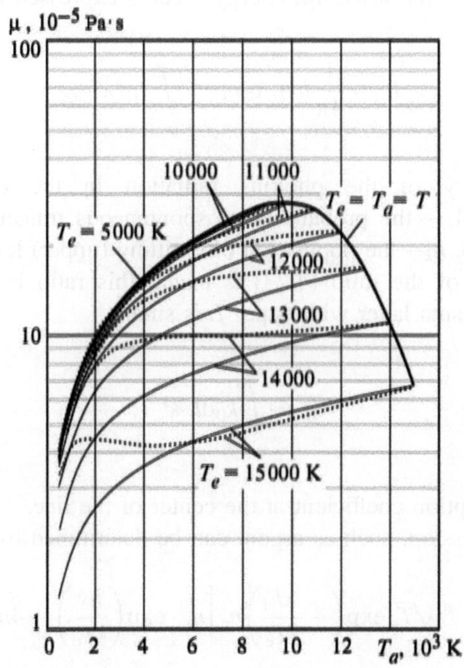

Fig. 4.19 Viscosity of the two-temperature argon plasma, depending on the temperatures: the notations as in Fig. 4.8

$T < 20{,}000$ K are in good agreement with the data of [6] (4.17). There are some experimental data for comparison in Fig. 4.17. The results of viscosity calculation for various gases of equilibrium plasma at atmospheric pressure are shown in Fig. 4.18. By (4.20) and (4.21), we can calculate the viscosity of the two-temperature argon plasma (Fig. 4.19).

Ambipolar diffusion coefficient for nonequilibrium plasma (Fig. 4.20) is determined using the equilibrium data [7] according to the following expression

$$D_{\mathrm{amb}} = 3.97 \times 10^{-10} T_{\mathrm{a}}^{0.63} (T_{\mathrm{e}} + T_{\mathrm{a}}). \tag{4.22}$$

The calculation of the integral characteristics of plasma by the emission spectrum is an important task in the preparation of the energy balance, the accuracy of this calculation is often associated with the correct model described the plasma. The model of partial local thermodynamic equilibrium (PLTE) and the equations Maxwell–Boltzmann–Saha are most commonly used (1.2), (1.4), (1.9). For determination of the integral radiation ratio, the absolute intensities of the spectral lines and continuum are the main source of data. For example, the estimated linear and continuous emission of argon plasma at atmospheric pressure in the test temperature range and wavelength $\lambda > 200$ nm in many cases, are optically thin.

The intensity of the spectral lines, the transition of an atom, or ion to the level of the upper kth level to the lower ith energy level is expressed by

$$I_{ki} = \frac{1}{4\pi} h\nu_{ki} A_{ki} n_k = \int_0^{\infty} i_{ki}(\lambda)\mathrm{d}\lambda, \tag{4.23}$$

Here ν_{ki}—frequency of the photons radiation in the center of the line ($\nu_{ki} = \nu_0 = c/\lambda_0$), A_{ki}—the probability of spontaneous transition $k \to i$ (the first Einstein coefficient), n_k—the population of emitting (upper) level of the transition.

The calculation of the ratio (4.23) is exact, this ratio is used if the optical thickness of the plasma layer with depth L is small

$$\tau_0 = \int_0^{L} k_0 \mathrm{d}l \ll 1,$$

Here k_0—the absorption coefficient at the center of the line.

Full volume emission, such as argon, can be determined by [37]

$$U_{\mathrm{rad}} = 1.14 \times 10^{-40} \sqrt{T_{\mathrm{e}}} \exp\left(-\frac{\Delta E_{\mathrm{i}}}{kT_{\mathrm{e}}}\right) n_{\mathrm{e}} \left[n_{\mathrm{i}}^{+} \exp\left(\frac{h\nu}{kT_{\mathrm{e}}}\right) + 4n_{\mathrm{i}}^{++} \exp\left(\frac{h\nu^{+}}{kT_{\mathrm{e}}}\right) \right], \tag{4.24}$$

Here $h\nu = 2.85$ eV; $h\nu^{+} = 8.2$ eV; $\Delta E_{\mathrm{i}} = \frac{e^2}{4\pi\varepsilon_0 r_{\mathrm{D}}}$—the decrease in the ionization potential; $r_{\mathrm{D}} = \sqrt{\frac{\varepsilon_0 k T_{\mathrm{e}} T_{\mathrm{a}}}{e^2 n_{\mathrm{e}} (T_{\mathrm{e}} + T_{\mathrm{a}})}}$—the Debye radius.

The Eq. (4.24) describes the radiative losses due to brake and recombination radiation, as well as discrete transitions between excited states of atoms and ions. The results of calculation by this equation for the equilibrium argon plasma and

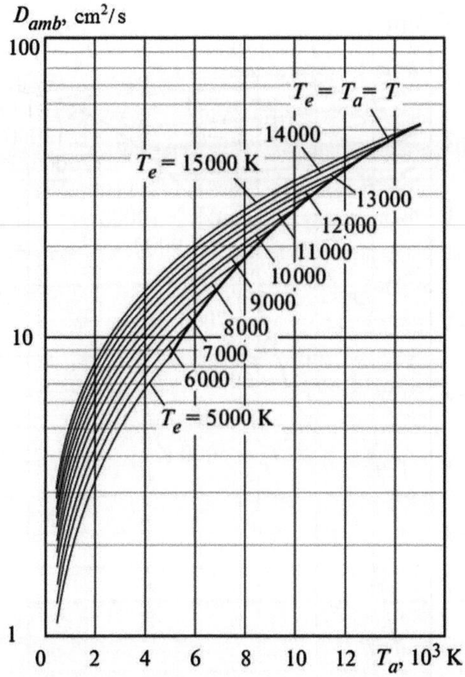

Fig. 4.20 Ambipolar diffusion coefficient of a two-temperature argon plasma, depending on the temperature

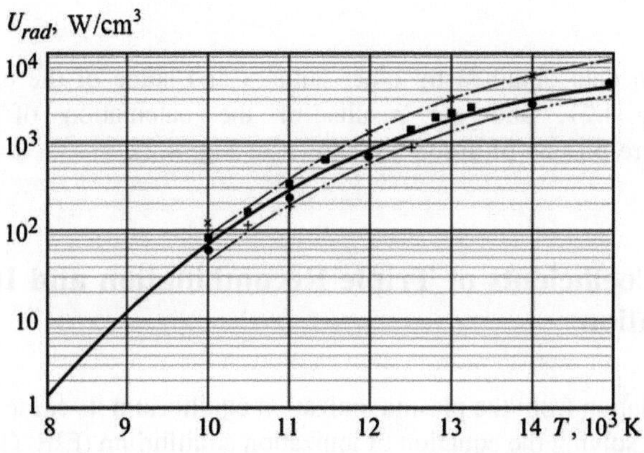

Fig. 4.21 The total radiation of equilibrium argon plasma: Experimental data: *filled square* [32]; *filled circle* [10]; *plus sign* [33]; *cross sign* [34]; *curves* estimates: *single dot* [35], *double dot* [36], *straight line* author

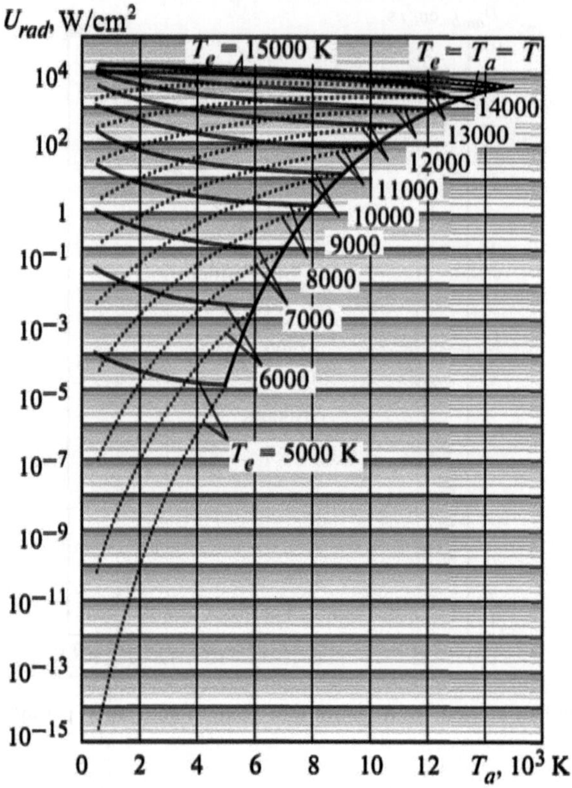

Fig. 4.22 Full two-temperature argon plasma emission in depending on temperature: the notations as in Fig. 4.8

compared with data obtained by other authors and some of the experiments are given in Fig. 4.21, and the results of the calculation of the radiation two-temperature plasma of argon are shown in Fig. 4.22.

4.4 The Coefficients of Triple Recombination and Impact Ionization

In case of deviation from the plasma ionization equilibrium its composition must be determined by solving the equation of ionization equilibrium (EIE) (1.40). The main mechanisms of birth and disappearance of charged particles in a dense low-temperature plasma are stepwise ionization and triple recombination. For solving the EIE, it is necessary to find the coefficients of recombination k_r and ionization k_i that are connected with each other by the principle of detailed balance [5, 38]. To calculate the rates of these processes it is sufficient to know one of these factors.

It should be emphasized that both the experimental and theoretical data on these factors are very scarce, and the main method of their determination is a theoretical calculation.

The best-known analytical expression for the calculation of k_r obtained by Hinnov and Hirchberg [39], is

$$k_r = 3.4 \times 10^{-34} (T_e/1000)^{-9/2}. \tag{4.25}$$

The result of the k_r calculation by (4.25) is in good agreement with the experimental data (Fig. 4.23), but this expression is applicable only in a very narrow range of temperatures ($T_e < 3000$ K).

Some estimates of k_r for hydrogen (Fig. 4.23) were represented in the works of Bates [44, 45]. In the work of A. Gurevich and L. Pitaevskii the k_r factor is defined as

$$k_r = \frac{4\sqrt{2}\pi^{3/2}}{9(4\pi\varepsilon_0)^5} \frac{e^{10} Z^3 \ln\sqrt{Z^2+1}}{m_e^{1/2}(kT_e)^{9/2}} = 0.614 \times 10^{-20} T_e^{-9/2} \tag{4.26}$$

In the work of Biberman [46] the following formula for the calculation of k_r is given:

$$k_r = 4.3 \times 10^{-44} (E_H/T_e)^{9/2} \times \left[1 + \frac{(g_i/g_a)(\Delta E_1/T_e)^{1+b}(E_H/T_e)^{3/2}\exp(-E_2/T_e)}{13.3a}\right]^{-2}, \tag{4.27}$$

Here g_i, g_a—the static weight of the ions and atoms; E_H—the ionization potential of the hydrogen atom; $\Delta E_1 = E_1 - E_2$—the potential difference of two energy levels of corresponding atom ($E_1 = E_i$). When $T_e \leq 0.07\Delta E_1$ the coefficients $a = 0.032$, $b = 4/9$; $T_e \geq 0.07\Delta E_1$—$a = 0.25$, $b = 5/6$.

Note that the calculation Eq. (4.27) gives low results (three decimal orders) compared with the experimental data in the temperature range $T_e = 7000$–20,000 K. Further, in the calculation according to (4.27) there is sharp rise in the values of k_r at $T_e \sim 0.07\Delta E_1$ (Fig. 4.23). Ismailova [47] gives a different expression for determining the rate of recombination

$$k_r = 1.29 \times 10^{-44}\left(\frac{E^*}{kT_e} + 2\right)\exp\left(\frac{E_i - E^*}{kT_e}\right), \tag{4.28}$$

Here $E^* = 11.61$ eV.

However, the value k_r, obtained using (4.28) does not coincide with the results of calculations by (4.27), and all such equations give lower results compared with the experimental data (Fig. 4.23). Good agreement between the calculated and experimental data, however, provides a relatively simple equation of Raiser [48]

$$k_r = 5.2 \times 10^{-35}(T_e/1000)^{-9/2} = 1.644 \times 10^{-21} T_e^{-9/2}. \qquad (4.29)$$

From the comparison of the calculated and experimental data (Fig. 4.23) it follows that at $T_e \leq 3200$ K the formula of E. Hinnov and J. Hirchberg (4.25) is recommended, at $3200\,\mathrm{K} \leq T_e \leq 3800$ K—the L. Bibermans formula (4.27) and at $T_e \geq 3800$ K—the formula of Y. Raiser (4.29).

Impact ionization coefficient k_i is determined by k_r based on the principle of detailed balance

$$k_i = k_r \frac{n_e^{*2}}{n_a^*}, \qquad (4.30)$$

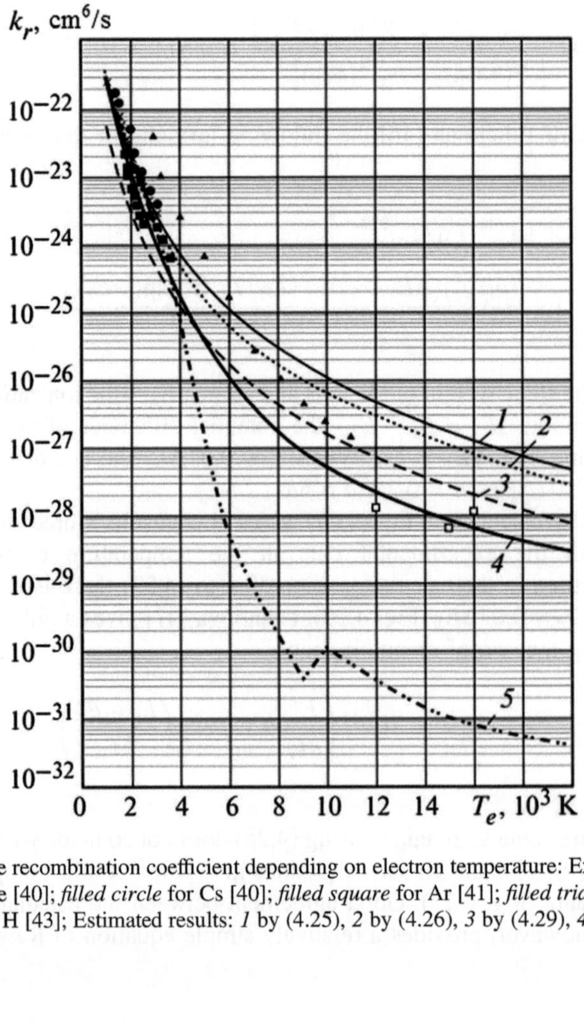

Fig. 4.23 Triple recombination coefficient depending on electron temperature: Experimental data: *cross sign* for He [40]; *filled circle* for Cs [40]; *filled square* for Ar [41]; *filled triangle* for Ar [42]; *open square* for H [43]; Estimated results: *1* by (4.25), *2* by (4.26), *3* by (4.29), *4* for H [44], *5* by (4.27)

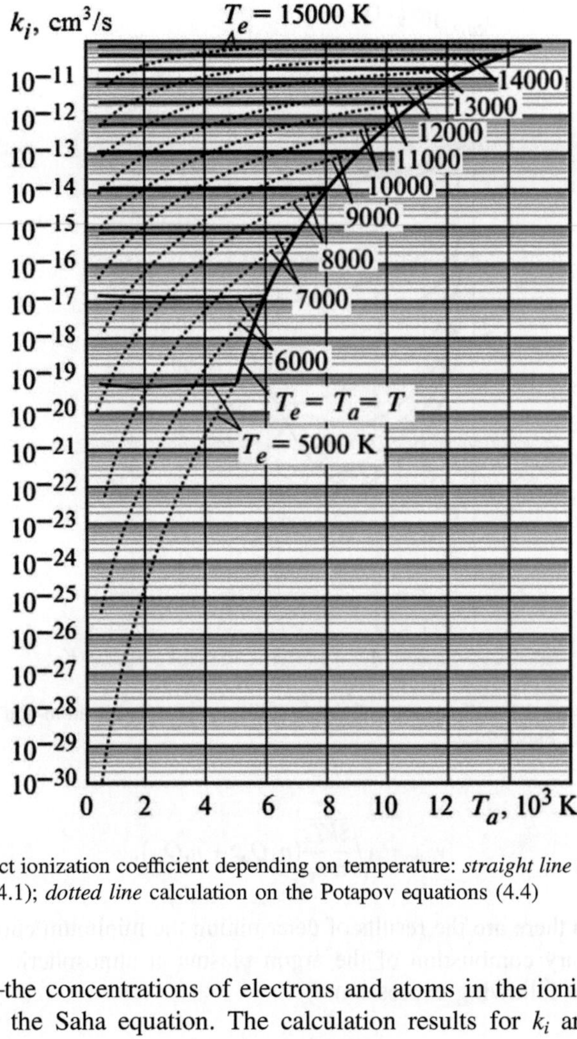

Fig. 4.24 Impact ionization coefficient depending on temperature: *straight line* calculation on the Saha equation (4.1); *dotted line* calculation on the Potapov equations (4.4)

Here n_e^*, n_a^*—the concentrations of electrons and atoms in the ionization balance, calculated by the Saha equation. The calculation results for k_i are presented in Fig. 4.24.

4.5 The Frequency of Particle Collisions, the Current Density, and Heating Capacity of Plasma

For improving the plasma heating and to design the plasma torches or analyze of the plasma flow properties in the channel of plasma torches, it is important to know the frequency of the particles collisions, the current density and the plasma power under heating. The results of calculation of the average total frequency of collisions are shown in Fig. 4.25 and prepared according to the following equation:

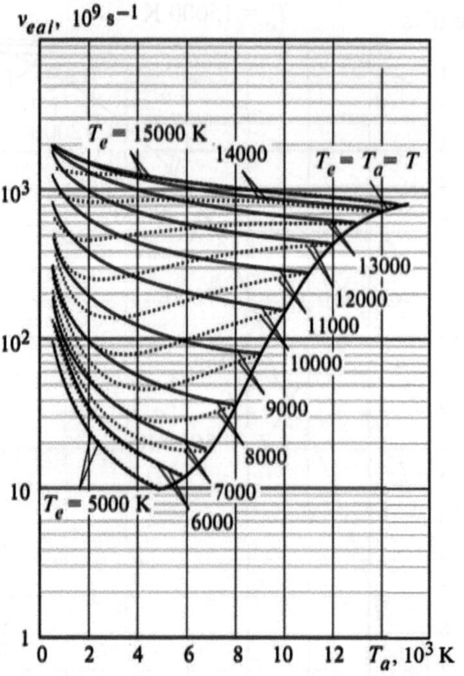

Fig. 4.25 The average total frequency of elastic collisions of electron–atom–ion in argon plasma: the notations as in Fig. 4.24

$$\nu_{\mathrm{eai}} = \sqrt{\frac{8kT_{\mathrm{e}}}{\pi m_{\mathrm{e}}}}(n_{\mathrm{a}}Q_{\mathrm{ea}} + n_{\mathrm{i}}Q_{\mathrm{ei}}). \tag{4.31}$$

In Fig. 4.26 there are the results of determining the minimum current density to for the stationary combustion of the argon plasma at atmospheric pressure, estimated from the following expression:

$$j^2 = \frac{3}{2}k\frac{2m_{\mathrm{e}}}{m_{\mathrm{a}}}n_{\mathrm{e}}\sqrt{\frac{8kT_{\mathrm{e}}}{\pi m_{\mathrm{e}}}}(n_{\mathrm{a}}Q_{\mathrm{ea}} + n_{\mathrm{i}}Q_{\mathrm{ei}})(T_{\mathrm{e}} - T_{\mathrm{a}})\sigma. \tag{4.32}$$

Figure 4.27 shows the calculation results of the power transmitted through the elastic collisions between electrons and heavy particles of argon plasma and is defined as

$$S_{\mathrm{eai}} = \frac{3}{2}k\frac{2m_{\mathrm{e}}}{m_{\mathrm{a}}}n_{\mathrm{e}}\sqrt{\frac{8kT_{\mathrm{e}}}{\pi m_{\mathrm{e}}}}(n_{\mathrm{a}}Q_{\mathrm{ea}} + n_{\mathrm{i}}Q_{\mathrm{ei}})(T_{\mathrm{e}} - T_{\mathrm{a}}). \tag{4.33}$$

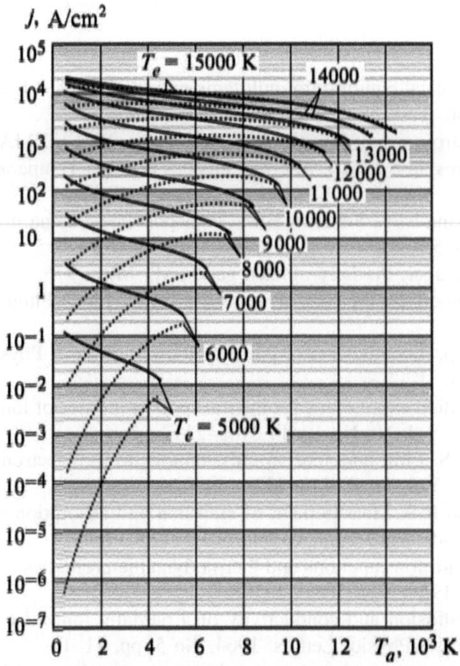

Fig. 4.26 The minimum current density for steady burning of argon plasma: the notations as in Fig. 4.24

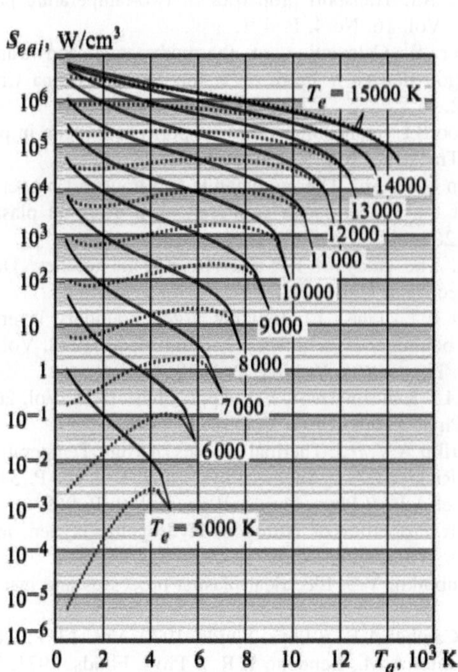

Fig. 4.27 The specific power transmitted in argon plasma at atmospheric pressure by the heavy particles (atoms and ions) through elastic collisions of electrons: the notations as in Fig. 4.24

References

1. A.V. Potapov. Chemical equilibrium multitemperature systems // J. Thermal physics of High Temperatures. 1966. T. 4. No 1. pp 55–58.
2. Ochkin V.N. Spectroscopy of low-temperature plasma. M .: FIZMATLIT 2006.
3. Goldfarb V.M., Dresvin S.V. // J. Thermal physics of High Temperatures. 1965. V. 3. No 3. PP. 420–424.
4. Eletskii A.V., Palkina L.A., Smirnov B.M. Transport phenomena in weakly ionized plasma. M .: Atomizdat 1975.
5. M. Mitchner, C. Kruger. Partially ionized gases. M.: Mir, 1976.
6. Devoto R.S. Transport coefficients of ionized argon // Phys. Fluids. 1973. Vol. 16. No 5. P. 616–623.
7. Devoto R.S. Transport coefficients of partially ionized argon // Phys. Fluids. 1967. Vol. 10. No 2. PP. 354–364.
8. Devoto R.S. Simplified expressions for the transport properties of ionized monatomic gases // Phys. Fluids. 1967. Vol. 10. No 10. P. 2105–2112.
9. Li C.P., Devoto R.S. Fifth and sixth approximations to the electron transport coefficients // Phys. Fluids. 1968. Vol. 11. P. 448–450.
10. Evans D.L., Tankin R.S. Measurement of emission and absorption of radiation by an argon plasma // Fhys. Fluids. 1967. Vol. 10. No 6. P. 1137–1144.
11. Drellischak K.S. Partition functions and thermodynamic properties of high temperature gases // AEDS TRD-64. 1964. Vol. 10. P. 1–148.
12. Polyansky V.A. Diffusion and conductivity of a partially ionized gas mixture multitemperature // J. of Technical Physics Letters. 1964. No 5. pp. 11–17.
13. Zhdanov V.M. Transport phenomena in multicomponent plasma. M.: Energoizdat 1982.
14. Kannappan D., Bose T.K. Transport properties of a two-temperature argon plasma // Phys. Fluids. 1977. Vol. 20. No 1. P. 1668–1673.
15. Miller E.J., Sandler S.I. Transport properties of two-temperature partially ionized argon // Phys. Fluids. 1973. Vol. 16. No 4. P. 491.
16. Hsu K.C., Pfender E. Calculation of thermodynamic and transport properties of a twotemperature argon plasma // Proc. of V Int. Symp. Plasma Chem., Edinburgh. 1981. Vol. 1. P. 144–152.
17. Bohadore M.N., Soo S.L. Nonequilibrium transport phenomena in partially ionized argon // J. Heat and Mass Transfer. 1966. Vol. 9. P. 17–34.
18. Dresvin S., Nguyen Quoc Shi. Thermodynamic and transport properties two temperature Ar plasma // Abst. of the IV European Conference on Thermal plasma Processes—Athens. Greece, 1996. P. 120.
19. Shi Nguyen-Kuok. The study of RF and Arc plasma torches: Dissertation of D.Sc. St. Petersburg State Technical University, 2002.
20. Lui W.S., Whitten B.T., Glass I.I. Ionizing argon boundary layers. Part 1. Quasi-steady flat-plate laminary boundary layer flows // J. Fluid Mech. 1978. Vol. 87. No 4. P. 609–640.
21. Lau J., Canad. // J. Phys. 1964. Vol. 42. P. 1548.
22. Lin S.C., Resler E.L., Kantrowitz. A. // J. Appl. Phys. 1955. Vol. 26. P. 95.
23. Emmons H.W. // Phys. Fluids. 1967. Vol. 10. P. 1125.
24. Asinovsky E.I., Kirilin A.V. // J. Thermal physics of High Temperatures. 1965. V. 5. P. 677.
25. Bues I., Patt H.J., Richter J.Z. // Angew. Phys. 1967. Vol. 22. P. 345.
26. Knopp C.F., Cambel A.B. // Phys. Fluids. 1966. Vol. 9. P. 989.
27. Yos J.M. Transport Properties of Nitrogen, Hydrogen, Oxygen, and Air. RDS-TM-63-7. 1963. AVCO/RAD. Wilmington, Mass.
28. Donskoy A.V., Klubnikin V.S. Electrical plasma processes and installation engineering. L.: Engineering, 1979.
29. Aeschliman D.P., Cambel A.B. // Phys. Fluids. 1970. Vol. 13. P. 2466.
30. Schreiber P.W., Hunter A.M., Bendetto K.R. // Phys. Fluids. 1971. Vol. 14. P. 2696.

31. Soo S.L., Bahadori M. // AGAR Dogaph. 1964. No 84. P. 1013.
32. Krey R.U., Moris J.C. // Phys. Fluids. 1970. Vol. 13. P. 1483–1489.
33. Abu-Romia M., Blair D. // AIAA Paper. 1969. P. 601–602.
34. Barzelau M. // AIAA Paper. 1966. Vol. 4. P. 815–819.
35. Biberman L.M., Norman G.E. // Successes of physical sciences. 1967. No 91. PP. 193–218.
36. I.T. Yakubov // Optics and Spectroscopy. 1965. No 19. PP. 193–199.
37. Kozlov G.I., Kuznetsov V.A., Masukov V.A. Radiant loss of Argon plasma and radiative model of a continuous optical discharge // J. Experimental technical physics. 1974. V. 66. PP. 954–964.
38. Golant V.E., Zhilinski A.P., Sakharov S.A. Basics of plasma physics. M.: Atomizdat 1977.
39. Hinnov E., Hirchberg J.G. Electron-Ion recombination in dense plasmas // Phys. Rev. 1962. Vol. 126. P. 795.
40. Aleskovsky Y.M. Study volume recombination in a Cesium plasma // J. Experimental technical physics. 1963. V. 17. P. 570.
41. Cool T.A., Zukoski E.E. Recombination, ionization and nonequilibrium electrical conductivity in seeded plasmas // Phys. Fluids. 1966. Vol. 9. P. 780.
42. Desai S.V. Corcoran W.H. Recombination of electrons and ions in an atmospheric argon plasma // J. Quant. Spectrosc. Radiat. Transfer. 1969. Vol. 9. PP. 1371–1386.
43. Craggs J.D., Hopwood W. // Proc. Roy. Soc. 1946. Vol. 186a. P. 257.
44. Atomic and Molecular Processes / Ed. by D.R. Bates. M.: Mir, 1964.
45. Bates D.R., Kingston A.E. Collisional-radiative recombination at low temperatures and densities // Proc. Phys. Soc. 1963. Vol. 83. P. 43.
46. Biberman L.M., Vorobiev V.M., Yakubov N.T. Nonequilibrium Low-temperature plasma IV ionization and recombination functions // J. Thermal physics of High Temperatures. 1969. V. 7. No 4. PP. 593–603.
47. Ismailova D. Study of the electric Arc in Argon / Ed. by A.F. Jeenbaev. Frunze: Ilim 1963.
48. Y.P. Raiser. Physics of the gas discharge. M.: Nauka, 1987.

Chapter 5
The Boltzmann Kinetic Equation and Calculation of the Transport Coefficients

5.1 Introduction of the Boltzmann Kinetic Equation

The kinetic theory of gases is based on the knowledge of the distribution function, defined in the phase space (μ-space) of a molecule. This function is the probable number of gas molecules $f_i(\mathbf{r}, \mathbf{v}_i, t)d\mathbf{r}d\mathbf{v}_i$ of ith grade in the elementary volume $d\mathbf{r}^*$ (*—For convenience, in this chapter, vectors in bold) around the point $\mathbf{r}$, with the phase velocities in the element $d\mathbf{v}_i$ near $\mathbf{v}_i$.

Consider a mixture of monatomic gases in nonequilibrium state. Assume that the mixture is sufficiently sparse, so for the description of the properties of such mixture the pair, rather than the triple collision of particles should be considered. For generality, assume that the external force $\mathbf{F}_i$ acts as the mass of particles per unit of ith type. We believe the strength of $\mathbf{F}_i$—the function of coordinates and time, independent of the velocity. It is believed that the external force is much smaller than the forces acting on the molecules during the collision. Intermolecular forces are far superior than gravitation forces and are valid only for the moment of collision.

The basis of the statistical description of plasma is the Boltzmann Kinetic Equation (BKE) with respect to the distribution function $f_i(\mathbf{r}_i, \mathbf{v}_i, t)$ for each ith grade of the plasma particles that is generally seven-dimensional. In the case of elastic interaction of particles, the BKE is

$$\begin{aligned}
&\frac{\partial f_i}{\partial t} + \mathbf{v}_i \cdot \frac{\partial f_i}{\partial \mathbf{r}} + \mathbf{F}_i \cdot \frac{\partial f_i}{\partial \mathbf{v}_i} \\
&= \sum_j \int_0^{\infty} \int_0^{2\pi} \int_0^{\infty} \left(f_i(v_i') f_j\left(v_j'\right) - f_i(v_i) f_j(v_j) \right) g_{ij} b\, db\, d\varepsilon\, dv_j \qquad (5.1) \\
&= \sum_j J(f_i f_j),
\end{aligned}$$

S. Nguyen-Kuok, *Theory of Low-Temperature Plasma Physics*,
Springer Series on Atomic, Optical, and Plasma Physics 95,
DOI 10.1007/978-3-319-43721-7_5

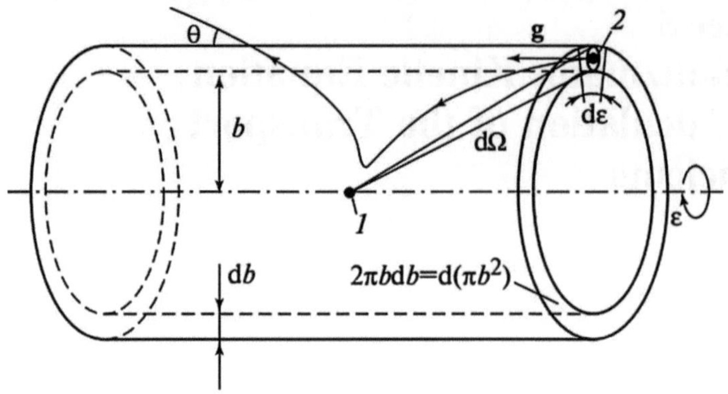

Fig. 5.1 The particle scattering in a pair collision: 1, 2—particles of *i*th and *j*th grades

where $f_i(v_i), f_j(v_j), f_i(v_i'), f_j(v_j')$—the particle distribution function of *i*th and *j*th type before and after collision, respectively; v_i, v_j, v_i', v_j'—the molecular velocities of particles of *i*th and *j*th grades before and after collision, respectively; $\mathbf{F}_i = \frac{e_i \mathbf{E}}{m_i}$—the strength of the electric field acting on a unit mass of the particle; **E**—the strength of electric field; e_i—the charge of *i*th component; $\mathbf{g}_{ij} = -\mathbf{g}_{ji} = \mathbf{v}_i - \mathbf{v}_j = \mathbf{g}$—the initial relative velocity of the particles of *i*th and *j*th grades; *b*—the impact parameter; ε—the angle of azimuthal orientation of the plane where the scattering of particles takes place (Fig. 5.1).

We accept the following notations:

$$\sum_j \int_0^\infty \int_0^{2\pi} \int_0^\infty \left(f_i(v_i') f_j\left(v_j'\right) - f_i(v_i) f_j(v_j) \right) g_{ij} b \mathrm{d}b \mathrm{d}\varepsilon \mathrm{d}v_j = \sum_j \iiint \left(f_i' f_j' - f_i f_j \right) g b \mathrm{d}b \mathrm{d}\varepsilon \mathrm{d}v_j = \sum_j j(f_i f_j) \tag{5.2}$$

The Boltzmann kinetic equation is also often written in the form of a differential operator:

$$\left(\frac{\partial}{\partial t} + \mathbf{v}_i \cdot \nabla_r + \mathbf{F}_i \cdot \nabla_{v_i} \right) f_i(r, v_i, t) = \sum_j \iiint \left(f_i' f_j' - f_i f_j \right) g b \mathrm{d}b \mathrm{d}\varepsilon \mathrm{d}v_j = \sum_j j(f_i f_j) \tag{5.3}$$

The left side of BKE—the total change of the particle distribution function in time, space, and the phase space velocity, respectively, the right side—the sum of the collision integrals $j(f_i f_j)$, it is called the operator of collision. The right side of (5.3) depends on the change of the distribution function due to the interaction of the particles with each other and with the walls. From elementary volume eliminated the particles, have experienced a collision, their number is equal to the number of particles in the phase space, multiplied by the probability of the particle output process from a given volume. Into the same volume the particles are brought after inverse collisions. The right-hand side of Eq. (5.3) for a particle in an unbounded space has a form determined by the process of direct and inverse collisions.

For finding the sum of the collision integrals for all components, interaction between identical components (when $i = j$) is considered as well as between the different, which speeds up to and after the collision, they are considered as independent variables in the process of integration. In the references during the decision of BKE, often the index «i» is omitted (i.e., $\mathbf{v}_i \to \mathbf{v}$ and $f_i \to f$, $j = 1$). The Boltzmann kinetic equation for the one-component gas can be written as

$$\left(\frac{\partial}{\partial t} + \mathbf{v} \cdot \nabla_r + \mathbf{F} \cdot \nabla_v\right) f(r, v, t) = \iiint \left(f' f'_1 - f f_1\right) g b \mathrm{d}b \mathrm{d}\varepsilon \mathrm{d}v_1 = j(f f_1). \quad (5.4)$$

Equation (5.1) is valid only for the low-density gas. If the mean free path of the particles in plasma is small compared with the dimensions of the plasma torch channel, the gas behaves as a continuous medium and the plasma is represented as a complex multicomponent chemical system, where interaction of the components described various functions of potential $\phi(r)$.

Gas in any initial state that has not external perturbations for a long time reaches eventually a steady state. If the system is adiabatically isolated and is not affected by external forces, such stationary state is homogeneous state where all of the distribution functions $f_i(r, v_i, t)$ are reduced to the Maxwell distribution:

$$f_i = 4\pi v_i^2 n_i \left(\frac{m_i}{2\pi kT}\right)^{3/2} \exp\left(-\frac{m_i v_i^2}{2kT}\right). \quad (5.5)$$

Using the H-theorem it can also be proved that the integrand of the right side of each integral in the BKE (5.1) equals to zero. This is the principle of detailed balance, when in equilibrium state, the number of molecules of the ith grade, with the rate of nonspecified interval velocities, due to collisions with molecules of jth grade, exactly compensated by inverse collisions by the number of molecules with velocities corresponding to the specified interval.

When the system is not in equilibrium, the distribution function can be determined from Eq. (5.1).

Usually we are interested in the properties of gases in the environment, only a little different from the equilibrium. Indeed, only in those circumstances where the flows are linear with respect to some derivatives, usual calculation of the transfer coefficients can be used. This limitation of the distribution function is close to Maxwell, and the Boltzmann kinetic equation can be solved by perturbation theory developed by Chapman and Enskog [1]. The solutions are then used to get of the expressions for the fluxes and transport coefficients. In these expressions, it is considered that mass transfer occurs in the presence not only of the concentration gradient (diffusion), but the temperature gradient (thermal diffusion), and vice versa.

These and other phenomena of the second order that cannot be described in terms of a simplified kinetic theory, of course found out only when a more rigorous approach is used. For example, the diffusion coefficient is numerically equal to the particle flux density in this class of particles; the viscosity is numerically equal to the momentum flux density produced by a unit of the velocity gradient, and thermal conductivity—to the energy flux density caused by a unit temperature gradient.

5.2 The Transport Equations

The kinetic theory of rarefied and dense gases provides a transport equation. These equations are differential equations describing the concentration, flow rate, and temperature as a function of coordinate and time.

Hydrodynamic transfer equation—the equation of mass conservation, momentum conservation, and energy conservation—can be obtained directly from the BKE. In deriving these equations, certain values expressed in terms of the distribution function appear. We can show that they are the flows of mass, momentum, energy, and directly related to the speed of the diffusion, tensor of pressure, and heat flux.

The linear velocity of the particle of ith grade relative to the fixed coordinate system is denoted by $\mathbf{v}_i$, and the components of the velocity v_{ix}—through v_{iy}, v_{iz}. The absolute value of the linear velocity is denoted as v_i. The average speed $\bar{\mathbf{v}}_i$ of the particle of ith kind, with concentration of particles n_i is determined as follows:

$$\bar{\mathbf{v}}_i(\mathbf{r},t) = \frac{1}{n_i}\int \mathbf{v}_i f_i(\mathbf{r},\mathbf{v}_i,t)\mathrm{d}\mathbf{v}_i. \tag{5.6}$$

The average speed is a function of coordinate and time, and it is a macroscopic velocity for the particles flow of ith grade.

The average mass velocity is determined by the equation

$$\mathbf{v}_0(\mathbf{r},t) = \frac{1}{\rho}\sum_i \int m_i \mathbf{v}_i f_i(\mathbf{r},\mathbf{v}_i,t)\mathrm{d}\mathbf{v}_i, \tag{5.7}$$

Here $\rho(\mathbf{r},t) = \sum_i n_i m_i$—the total density of the gas.

The thermal velocity for the particle of ith grade is defined as the velocity of the particle in the coordinate system moving with the mean mass velocity $\mathbf{v}_0$:

$$\mathbf{V}_i(\mathbf{v}_i,\mathbf{r},t) = \mathbf{v}_i - \mathbf{v}_0. \tag{5.8}$$

The diffusion rate of the *i*th component of the chemical gas is the velocity of the flow of particles with *i*th grade in the coordinate system moving with the mean mass velocity of the gas

$$\bar{\mathbf{V}}_i(\mathbf{r}, t) = \bar{\mathbf{v}}_i - \mathbf{v}_0. \tag{5.9}$$

To simplify some expressions a quantity called the reduced speed $\mathbf{W}_i$ is used, it is defined as

$$\mathbf{W}_i = \sqrt{\frac{m_i}{2kT}}\mathbf{V}_i. \tag{5.10}$$

The reduced initial relative velocity γ is also used:

$$\gamma = \sqrt{\frac{\mu}{2kT}}g, \tag{5.11}$$

where μ—the reduced mass of the two colliding particles.

The temperature in the kinetic theory is defined in terms of the average kinetic energy of all the particles of the thermal motion:

$$\frac{3}{2}kT = \frac{1}{n}\sum_i \int \frac{1}{2} m_i v_i^2 f(\mathbf{v}_i) d\mathbf{v}_i, \tag{5.12}$$

where $n = \sum_i n_i$—the total concentration of particles.

The mass transfer is expressed by the vector of mass flux density:

$$\mathbf{\Psi}_m = \sum_i \psi_{mi} = \sum_i m_i \int \mathbf{V}_i f_i d\mathbf{V}_i = \sum_i n_i m_i \overline{\mathbf{V}_i}. \tag{5.13}$$

The momentum transfer can be described as

$$\mathbf{\Psi}_{P_x} = \sum_i \mathbf{\Psi}_{P_{xi}} = \sum_i m_i \int V_{ix} \mathbf{V}_i f_i d\mathbf{V}_i = \sum_i n_i m_i \overline{V_{ix} \mathbf{V}_i}. \tag{5.14}$$

The vector $\mathbf{\Psi}_{P_{xi}}$ has components proportional to $\overline{V_{ix}V_{ix}}$, $\overline{V_{ix}V_{iy}}$, $\overline{V_{ix}V_{iz}}$. Similar vectors of flows can be obtained for pulses of y- and z-components. These three flux vectors completely characterize the transfer of momentum. Nine components of these vectors form a symmetric tensor $\mathbf{p}_i$ of second range:

$$\begin{aligned} (p_i)_{xx} &= m_i \int V_{ix} V_{ix} f_i d\mathbf{V}_i = n_i m_i \overline{V_{ix} V_{ix}}; \\ (p_i)_{xy} &= (p_i)_{yx} = n_i m_i \overline{V_{ix} V_{iy}}; \\ &\ldots \end{aligned} \tag{5.15}$$

Symbolically, the tensor $\mathbf{p}_i$, characterizing the partial pressure of the *j*th component of the chemical gas, can be written as

$$\mathbf{p}_i = n_i m_i \overline{\mathbf{V}_i \mathbf{V}_i}. \tag{5.16}$$

Sum of partial pressures tensors for all sorts of particles can be described by gas pressure tensor of mixture:

$$\mathbf{p} = \sum_i \mathbf{p}_i = \sum_i n_i m_i \overline{\mathbf{V}_i \mathbf{V}_i}. \tag{5.17}$$

The physical meaning of the pressure tensor is that it represents the momentum flux of the gas. The individual components of the pressure tensor are expressed in the following: the diagonal elements p_{xx}, p_{yy}, p_{zz} are equal to the normal stresses; e.g., p_{xx} is equal to the force acting on a unit area normal to *x-axis*, etc.

The voltage or pressure is received as the pressure tensor components that can be measured by devices moving in the flow with velocity $\mathbf{v}_0$. The pressure measured in the stream using the fixed gauge depends on $\mathbf{v}_0$ and on the location of manometer.

In a system in equilibrium, the shear stresses are equal to zero, and the normal stresses are equal to each other. In this case, the force acting on any selected surface element in the gas is constant and independent of the orientation of the surface perpendicular to it, i.e.,

$$\begin{aligned} p_{xx} &= p_{yy} = p_{zz} = p; \\ p_{xy} &= p_{yz} = \cdots = 0, \end{aligned} \tag{5.18}$$

where *p*—the hydrostatic pressure for the equilibrium of the system.

The transfer of kinetic energy for the molecules of *i*th grade is characterized by the vector

$$\mathbf{\Psi}_{q_i} = \frac{1}{2} m_i \int V_i^2 \mathbf{V}_i f_i \mathrm{d}\mathbf{V}_i = \frac{1}{2} n_i m_i \overline{V_i^2 \mathbf{V}_i} = \mathbf{q}_i. \tag{5.19}$$

The sum of such vectors for all the components of the gas mixture is equal to the heat flow density vector:

$$\mathbf{q} = \sum_i \mathbf{q}_i = \sum_i \frac{1}{2} n_i m_i \overline{V_i^2 V_i}. \tag{5.20}$$

The physical meaning of the components q_x, q_y, q_z of the vector of heat flux density $\mathbf{q}$ is that they represent the flux density of the kinetic energy in the direction of the axis *x*, *y*, and *z*, respectively.

The transfer of energy and momentum by collisions of particles (internal sources of transfer) is not taken into account in the above equations' transfer streams.

The general transfer equation is

$$\int \psi_i \left[\frac{\partial f_i}{\partial t} + \mathbf{v}_i \cdot \frac{\partial f_i}{\partial \mathbf{r}} + \mathbf{F}_i \cdot \frac{\partial f_i}{\partial \mathbf{v}_i} \right] \mathrm{d}\mathbf{v}_i \\ = \sum_j \iiiint \psi_i \left(f_i' f_j' - f_i f_j \right) g_{ij} b \mathrm{d}b \mathrm{d}\varepsilon \mathrm{d}\mathbf{v}_i \mathrm{d}\mathbf{v}_j, \tag{5.21}$$

therefore,

$$\int \left[\frac{\partial f_i \psi_i}{\partial t} + \frac{\partial}{\partial \mathbf{r}} \cdot (\mathbf{v}_i f_i \psi_i) + \frac{\partial}{\partial \mathbf{v}_i} \cdot (\mathbf{F}_i f_i \psi_i) \right] \mathrm{d}\mathbf{v}_i \\ - \int f_i \left[\frac{\partial \psi_i}{\partial t} + \mathbf{v}_i \cdot \frac{\partial \psi_i}{\partial \mathbf{r}} + \mathbf{F}_i \cdot \frac{\partial \psi_i}{\partial \mathbf{v}_i} \right] \mathrm{d}\mathbf{v}_i \\ = \frac{\partial (n_i \overline{\psi_i})}{\partial t} + \frac{\partial}{\partial \mathbf{r}} \cdot (n_i \overline{\psi_i \mathbf{v}_i}) - n_i \left(\overline{\frac{\partial \psi_i}{\partial t}} + \overline{\mathbf{v}_i \cdot \frac{\partial \psi_i}{\partial \mathbf{r}}} + \mathbf{F}_i \cdot \overline{\frac{\partial \psi_i}{\partial \mathbf{v}_i}} \right) \\ = \sum_j \iiiint \psi_i \left(f_i' f_j' - f_i f_j \right) g_{ij} b \mathrm{d}b \mathrm{d}\varepsilon \mathrm{d}\mathbf{v}_i \mathrm{d}\mathbf{v}_j. \tag{5.22}$$

Equation (5.22) is known as a general Enskog transport equation for ψ_i, corresponding to the particles of the ith grade.

The equations like (5.22) for any ψ_i are little used because of the presence of very difficult integrals in their right parts, but if ψ_i—the molecular mass (for particles) of ith type, given that the mass of individual particles is not changed by the collision, these equations can be written as

$$\sum_j \iiiint m_i \left(f_i' f_j' - f_i f_j \right) g_{ij} b \mathrm{d}b \mathrm{d}\varepsilon \mathrm{d}\mathbf{v}_i \mathrm{d}\mathbf{v}_j = 0. \tag{5.23}$$

The mass, momentum, and energy of the particles are conserved in collisions, we can show that when $\psi_i = m_i$, $m_i \mathbf{V}_i$, and $m_i V_i^2/2$

$$\sum_j \iiiint \psi_i \left(f_i' f_j' - f_i f_j \right) g_{ij} b \mathrm{d}b \mathrm{d}\varepsilon \mathrm{d}\mathbf{v}_i \mathrm{d}\mathbf{v}_j = 0. \tag{5.24}$$

Relations (5.23) and (5.24) make it possible to simplify the general transport equation. The validity of them can be set as follows. Integral

$$\sum_j \iiiint \psi_i \left(f_i' f_j' - f_i f_j \right) g_{ij} b \mathrm{d}b \mathrm{d}\varepsilon \mathrm{d}\mathbf{v}_i \mathrm{d}\mathbf{v}_j \tag{5.25}$$

is equal to the integral

$$\sum_j \iiiint \psi_i'\left(f_if_j - f_i'f_j'\right)g_{ij}'b'\mathrm{d}b'\mathrm{d}\varepsilon\mathrm{d}\mathbf{v}_i'\mathrm{d}\mathbf{v}_j', \tag{5.26}$$

that is record to reverse collision.

Since

$$g_{ij} = g_{ij}', b = b', \mathrm{d}\mathbf{v}_i\mathrm{d}\mathbf{v}_j = \mathrm{d}\mathbf{v}_i'\mathrm{d}\mathbf{v}_j', \tag{5.27}$$

the integral (5.25) can be represented as

$$-\iiiint \psi_i'\left(f_i'f_j' - f_if_j\right)g_{ij}b\mathrm{d}b\mathrm{d}\varepsilon\mathrm{d}\mathbf{v}_i\mathrm{d}\mathbf{v}_j. \tag{5.28}$$

Consequently, we can write

$$\begin{aligned} &\sum_j \iiiint \psi_i\left(f_i'f_j' - f_if_j\right)g_{ij}b\mathrm{d}b\mathrm{d}\varepsilon\mathrm{d}\mathbf{v}_i\mathrm{d}\mathbf{v}_j \\ &= \frac{1}{2}\iiiint\left(\psi_i - \psi_i'\right)\left(f_i'f_j' - f_if_j\right)g_{ij}b\mathrm{d}b\mathrm{d}\varepsilon\mathrm{d}\mathbf{v}_i\mathrm{d}\mathbf{v}_j \end{aligned} \tag{5.29}$$

In the case of $\psi_i = m_i$ (5.29) Eq. (5.23) follows. Equality $\psi_i - \psi_i' = 0$ expresses the fact that the masses of the individual particles are not changed during the collisions.

5.2.1 Principle of the Equilibrium Systems

If we consider only elastic collisions—the system without changing the internal energy and conversion of the external energy into internal one and back, we can distinguish a hypothetical model in a conservative enclosure system, where the particle of the ith grade is scattered by the particle of the jth grade, the momentum ΔP and the kinetic energy ΔE are losing and this particle of the ith grade is deflected by an angle θ in the area of solid angle dΩ. If the particle of ith kind is moving back from the final state and is scattered by the particle of jth grade and obtain the momentum ΔP and the kinetic energy ΔE, it returns to the original state:

$$\begin{aligned} &\frac{1}{2}\sum_{i,j}\iiiint\left(\psi_i - \psi_i'\right)\left(f_i'f_j' - f_if_j\right)g_{ij}b\mathrm{d}b\mathrm{d}\varepsilon\mathrm{d}\mathbf{v}_i\mathrm{d}\mathbf{v}_j \\ &= \frac{1}{2}\sum_{i,j}\iiiint\left(\psi_j - \psi_j'\right)\left(f_i'f_j' - f_jf_i\right)g_{ji}b\mathrm{d}b\mathrm{d}\varepsilon\mathrm{d}\mathbf{v}_j\mathrm{d}\mathbf{v}_i. \end{aligned} \tag{5.30}$$

Replacing the particle of the ith grade to the particle of the jth grade, it can be written as

$$\begin{aligned}&\frac{1}{2}\sum_{i,j}\iiiint \psi_i\left(f_i'f_j'-f_if_j\right)g_{ij}b\mathrm{d}b\mathrm{d}\varepsilon\mathrm{d}\mathbf{v}_i\mathrm{d}\mathbf{v}_j\\&=\frac{1}{4}\sum_{i,j}\iiiint\left(\psi_i+\psi_j-\psi_i'-\psi_j'\right)\left(f_i'f_j'-f_if_j\right)g_{ij}b\mathrm{d}b\mathrm{d}\varepsilon\mathrm{d}\mathbf{v}_i\mathrm{d}\mathbf{v}_j.\end{aligned} \tag{5.31}$$

The sum of the collision invariants $\left(\psi_i+\psi_j-\psi_i'-\psi_j'\right)=0$. Therefore, Eq. (5.24) is observed at $\psi_i=m_i$, $m_i\mathbf{V}_i$, and $m_iV_i^2/2$.

The different transport equations can be obtained from the total Enskog equation (5.22), assuming $\psi_i=m_i$, $m_i\mathbf{V}_i$, and $m_iV_i^2/2$. In particular, $\psi_i=1$, we have

$$\frac{\partial n_i}{\partial i}+\frac{\partial}{\partial\mathbf{r}}\cdot n_i\overline{\mathbf{v}_i}=0. \tag{5.32}$$

Multiplying (5.32) on m_i and summed over all types of particles, we obtain the equation of continuity for the entire gas.

$$\frac{\partial\rho}{\partial t}+\frac{\partial}{\partial\mathbf{r}}\cdot\rho\overline{\mathbf{v}_0}=0. \tag{5.33}$$

If we substitute $\psi_i=m_iV_i$ in (5.22) and if we sum over all types of particles, the collision integral in the right-hand side of the equation disappears. After some transformations, we have the following equation of the gas motion:

$$\frac{\partial\mathbf{v}_0}{\partial t}+\mathbf{v}_0\cdot\frac{\partial}{\partial\mathbf{r}}\mathbf{v}_0=-\frac{1}{\rho}\frac{\partial}{\partial\mathbf{r}}\cdot\mathbf{p}+\frac{1}{\rho}\sum_i n_i\mathbf{F}_i. \tag{5.34}$$

Substituting $\psi_i=m_iV_i^2/2$ in the general transport equation with a conversion that are similar to that performed in the derivation of the previous equation, we get

$$\sum_i\frac{1}{2}m_i\left[\frac{\partial n_i\overline{V_i^2}}{\partial t}+\frac{\partial}{\partial\mathbf{r}}\cdot n_i\overline{V_i^2\mathbf{v}_i}-n_i\frac{\overline{\partial V_i^2}}{\partial t}-n_i\overline{\mathbf{v}_i\cdot\frac{\partial V_i^2}{\partial\mathbf{r}}}-n_i\mathbf{F}_i\cdot\overline{\frac{\partial V_i^2}{\partial\mathbf{v}_i}}\right]=0. \tag{5.35}$$

Introducing the pressure tensor (5.16) and the vector of heat flux density (5.20), the energy balance equation can be written as

$$\frac{\partial}{\partial t}\rho\hat{U}_{\mathrm{trans}}+\frac{\partial}{\partial\mathbf{r}}\cdot\rho\hat{U}_{\mathrm{trans}}\mathbf{v}_0+\frac{\partial}{\partial\mathbf{r}}\cdot\mathbf{q}+\mathbf{p}:\frac{\partial}{\partial\mathbf{r}}\mathbf{v}_0-\sum_i n_im_i\mathbf{F}_i\cdot\overline{\mathbf{V}_i}=0, \tag{5.36}$$

where $\hat{U}_{\mathrm{trans}}$—the internal energy of the gas per unit of mass that is equal to the energy of translational motion of the particles,

$$\hat{U}_{\mathrm{trans}} = \frac{1}{\rho}\sum_i \frac{1}{2} n_i m_i V_i^2. \tag{5.37}$$

In (5.36) and then mark « : » means the dual tensor multiplication.

The internal energy is equal to the total energy per unit of the gas mass in the coordinate system moving with the mean mass velocity $\mathbf{v}_0$ (kinetic energy of the flow is not included in this energy), without taking into account the potential energy of the gas in an external field. Using ratios for $\hat{U}_{\mathrm{trans}}$ and T, Eq. (5.36) can be transformed to

$$\begin{aligned}\frac{3}{2}nk\left(\frac{\partial T}{\partial t} + \mathbf{v}_0\cdot\frac{\partial T}{\partial \mathbf{r}}\right) &= -\frac{\partial}{\partial \mathbf{r}}\cdot\mathbf{q} - \mathbf{p} : \frac{\partial}{\partial \mathbf{r}}\mathbf{v}_0 \\ &+ \sum_i n_i m_i \mathbf{F}_i\cdot\overline{\mathbf{V}}_i + \frac{3}{2}kT\left(\frac{\partial}{\partial \mathbf{r}}\cdot\sum_i n_i\overline{\mathbf{V}}_i\right).\end{aligned} \tag{5.38}$$

The resulting equation is valid for a gas mixture consisting of particles that have no internal degrees of freedom.

5.3 Solution of the Boltzmann Kinetic Equation by Chapman–Enskog Approach

The Boltzmann kinetic equation (5.1) is a nonlinear integral–differential equation, and its exact analytical solution is very difficult. Chapman–Enskog approach received the widespread among the existing methods for solving the Boltzmann kinetic equation to determine the transfer coefficients and the essence of its method is a logical expansion of the distribution function on the small parameter.

According to the method of Chapman–Enskog the solution of (5.1) for the distribution function f_i of ith grade of particles is sought in the form $f_i(v_i) = f_i^0(v_i)[1 - v_i\Phi_i(v_i)]$, where $\Phi_i(v_i)$—a vector of the perturbation that is proportional to the external electric field E. Then the unknown vector $\Phi_i(v_i)$ is decomposed into a series in Sonine polynomials and this expansion of the distribution function is inserted into the Boltzmann kinetic equation (5.1). Executing a sequence of approximate solutions with different initial and boundary conditions, such as conditions for normalization of each component of the gas, we get various approximations of the distribution function of particles. Knowing the distribution function, we can identify the different streams of particles and the corresponding transport coefficients that are often expressed through the collision integrals, taking into account the interaction of groups of particles:

$$\Omega^{(l,s)} = \sqrt{\frac{kT}{2\pi\mu}} \int_0^\infty \exp(-\gamma^2)\gamma^{2s+3} Q^{(l)}(g)\mathrm{d}\gamma,$$

where g—the relative velocity of the particles in the system of the mass center; $\gamma^2 = \frac{\mu}{2kT} g^2$—square of the reduced relative velocity for the particles of ith and jth grades with the reduced mass μ; $Q^{(l)}(g)$—the effective cross section for collisions of particles of the ith and jth grades, calculated by Eq. (2.8).

The final result of the method is that we can express all the transport coefficients through the system of integrals $\Omega^{(l,s)}$. With the help of these integrals the dynamics of molecular collisions and, therefore, the law of intermolecular forces are taken into account.

Solutions in accordance with Chapman–Enskog approach, for example, show that in a first approximation, to calculate the diffusion coefficient it should be taken that $l = 1$, $s = 1$, and for calculation the viscosity and thermal conductivity—$l = 2$, $s = 2$. The difference between the values of integrals is not very large. To calculate the transport coefficients of higher approximations it is necessary to know the collision integrals for other values of l and s.

The above-mentioned Chapman–Enskog approach is based on various assumptions that to some extent limit the application of the results. Consider each constraint separately and discuss the conditions when these limitations become significant.

Chapman–Enskog approach consider only the binary collisions, it does not apply at high densities, when the triple collisions begin to play a significant role. This method makes it possible to introduce the distribution function in a series. The first approach that is considered here is rightly in the case where all the gradients of physical quantities are small. Higher approximations give correction for large gradients. In a first approximation, the flows are proportional to the first derivative of the density, velocity, and temperature; the resulting equations that describe the change in density, velocity, and temperature in time (transfer equation) are called the Navier–Stokes equations. Higher approximations give the terms proportional to higher powers of derivatives and derivatives of less high physical quantities. The transport equations, corresponding to the second approximation are known as the equations of Barnett. Quantitatively, the first approximation is valid when the relative change in density, velocity, and temperature in the region of the mean free path is small compared to unity.

Using Chapman–Enskog approach it is assumed that the volume of the vessel containing the gas is large compared to the mean free path of the particles, so the surface layers take the negligible proportion of the total volume. At very low densities the particles more often collide with the walls of volume containing the gas than with each other; therefore, in this case, the mechanism of establishment of the local equilibrium in the gas is ineffective. In such a gas the concept of local density, velocity, and temperature lose their original meaning, and the gas even approximately is not a continuous medium, but rather exhibits the properties of the discrete environment. Gas in highly diluted conditions is known as Knudsen, and the behavior of such a gas is well known.

Strictly speaking, Chapman–Enskog approach is only applicable to monatomic gases (molecules without internal degrees of freedom, for them the interaction potential is spherically symmetric). Inelastic collisions are possible between molecules with internal degrees of freedom. In these collisions the kinetic energy changes, and the mass and momentum are clearly preserved. Therefore, the viscosity and diffusion insignificantly depend on the internal degrees of freedom, and the theory of atomic gases can be successfully applied to polyatomic molecules, if they are not too different from the spherical ones.

Consider the case of a gas mixture of two components 1 and 2 with the distribution functions f_1 and f_2, respectively. In Chapman–Enskog approach, the distribution functions are presented in the form of series

$$f_1 = f_1^{(0)} + f_1^{(1)} + f_1^{(2)} + \cdots; \tag{5.39}$$

$$f_2 = f_2^{(0)} + f_2^{(1)} + f_2^{(2)} + \cdots, \tag{5.40}$$

where the zero-order approximation corresponds to the local equilibrium of Maxwellian distribution:

$$f_1^{(0)} = 4\pi V_1^2 n_1 \left(\frac{m_1}{2\pi kT}\right)^{3/2} \exp\left(-\frac{m_1}{2kT}(\mathbf{v}_1 - \mathbf{v}_0)^2\right); \tag{5.41}$$

$$f_2^{(0)} = 4\pi V_2^2 n_2 \left(\frac{m_2}{2\pi kT}\right)^{3/2} \exp\left(-\frac{m_2}{2kT}(\mathbf{v}_2 - \mathbf{v}_0)^2\right), \tag{5.42}$$

where m_1, m_2—particles masses of the gas components: k—the Boltzmann constant; n_1, n_2—the concentrations of the gas particles: T—temperature of the mixture; $\mathbf{v}_0$—the speed of the gas mixture; $\mathbf{v}_0 = M_1 x_1 \mathbf{v}_1 + M_2 x_2 \mathbf{v}_2$; $\mathbf{v}_1$, $\mathbf{v}_2$—the velocity of gas components; $M_1 = m_1/m_0$, $M_2 = m_2/m_0$; $m_0 = m_1 + m_2$; $x_1 = n_1/n$, $x_2 = n_2/n$—ratio of the concentrations of the components (molar fractions); $n = n_1 + n_2$—the concentration of the mixture.

Chapman–Enskog approach for determination of the transport coefficients of the Boltzmann kinetic equation is described in detail in the paper of Chapman and Cowling [1].

The boundary condition of BKE has the form

$$\int f_i \mathrm{d}\mathbf{v}_i = n_i; \tag{5.43}$$

$$\sum_i m_i \int \mathbf{v}_i f_i d\mathbf{v}_i = \rho \mathbf{v}_0; \tag{5.44}$$

$$\frac{1}{2}\sum_i m_i \int (\mathbf{v}_i - \mathbf{v}_0)^2 f_i d\mathbf{v}_i = \frac{3}{2} nkT. \tag{5.45}$$

With the use of the following additional conditions:

$$\int f_i^{(r)} d\mathbf{v}_i = 0, \quad r = 1, 2, 3, \ldots; \tag{5.46}$$

$$\sum_i m_i \int \mathbf{v}_i f_i^{(r)} d\mathbf{v}_i = 0, \quad r = 1, 2, 3, \ldots; \tag{5.47}$$

$$\frac{1}{2}\sum_i m_i \int (\mathbf{v}_i - \mathbf{v}_0)^2 f_i^{(r)} d\mathbf{v}_i = 0, \quad r = 1, 2, 3, \ldots \tag{5.48}$$

all the functions $f_i^{(r)}$ can be clearly defined. If the obtained distribution functions satisfy the conditions (5.43)–(5.45), they may be considered as solutions of the BKE.

If limit the further consideration by the first approximation (up to $f^{(1)}$), the distribution function can be written as the sum of the equilibrium distribution functions and small deviations from it, that is $f_i = f_i^{(0)}\left(1 + \Phi_i^{(1)}\right)$. Functions $f_1^{(1)}$ and $f_2^{(1)}$ take the form

$$f_1^{(1)} = f_1^{(0)} \Phi_1; \tag{5.49}$$

$$f_2^{(1)} = f_2^{(0)} \Phi_2, \tag{5.50}$$

where Φ_1 and Φ_2—the perturbation functions.

For any vectors **F** and **K**, we have

$$\begin{aligned} n_i^2 I_i(F) &= \iiint (f_i' f' - f_i f) g b db d\varepsilon dv \\ &= \iiint f_i^{(0)} f^{(0)} (F_i + F - F_i' - F') g b db d\varepsilon dv; \end{aligned} \tag{5.51}$$

$$\begin{aligned} n_i n_j I_{ij}(K) &= \iiint \left(f_i' f_j' - f_i f_j\right) g b db d\varepsilon dv_j \\ &= \iiint f_i^{(0)} f_j^{(0)} (K - K') g b db d\varepsilon dv_j. \end{aligned} \tag{5.52}$$

Expressions (5.51), (5.52) are the linear functions, so $I(\phi+\psi)=I(\phi)+I(\psi)$ and $I(a\phi)=aI(\phi)$ regardless of the values of the constants (a—the arbitrary constant). The right side of the BKE can be linearized as follows:

$$\begin{aligned} D_1^{(1)} &= -n_1^2 I_1(\Phi_1) - n_1 n_2 I_{12}(\Phi_1+\Phi_2); \\ D_2^{(1)} &= -n_2^2 I_2(\Phi_2) - n_1 n_2 I_{21}(\Phi_1+\Phi_2), \end{aligned} \tag{5.53}$$

wherein

$$D_i^{(r)} = \frac{\partial_{r-1} f_i^{(0)}}{\partial t} + \frac{\partial_{r-2} f_i^{(1)}}{\partial t} + \cdots + \frac{\partial_0 f_i^{(r-1)}}{\partial t} + \left(\mathbf{v}_i \cdot \frac{\partial}{\partial \mathbf{r}} + \mathbf{F}_i \cdot \frac{\partial}{\partial \mathbf{v}_i}\right) f_i^{(r-1)}. \tag{5.54}$$

The left part of the BKE can be written as

$$D_1^{(1)} = f_1^{(0)} \left[\left(W_1^2 - \frac{5}{2} \right) \mathbf{V}_1 \cdot \nabla \ln(T) + x_1^{-1} \mathbf{d}_{12} \cdot \mathbf{V}_1 + 2\, \mathbf{W}_1 \overset{o}{\mathbf{W}}_1 : \nabla \mathbf{v}_0 \right]; \tag{5.55}$$

$$D_2^{(1)} = f_2^{(0)} \left[\left(W_2^2 - \frac{5}{2} \right) \mathbf{V}_2 \cdot \nabla \ln(T) + x_2^{-1} \mathbf{d}_{21} \cdot \mathbf{V}_2 + 2\, \mathbf{W}_2 \overset{o}{\mathbf{W}}_2 : \nabla \mathbf{v}_0 \right], \tag{5.56}$$

where $\mathbf{W}_i \equiv \sqrt{m_i/2kT}\mathbf{V}_i$—the reduced speed, $\mathbf{V}_i \equiv \mathbf{v}_i - \mathbf{v}_0$, $\mathbf{W}_1 \overset{o}{\mathbf{W}}_1$ and $\mathbf{W}_2 \overset{o}{\mathbf{W}}_2$—the dual tensors.

The value $\mathbf{d}_{12}$ can be determined by any of the following expressions [1–3]:

$$\mathbf{d}_{12} = \frac{\rho_1 \rho_2}{\rho p} \left[\mathbf{F}_2 - \frac{1}{\rho_2} \nabla p_2 - \left(\mathbf{F}_1 - \frac{1}{\rho_1} \nabla p_1 \right) \right]; \tag{5.57}$$

$$\mathbf{d}_{12} = \nabla x_1 + \frac{n_1 n_2 (m_2 - m_1)}{n \rho} \nabla \ln(p) - \frac{\rho_1 \rho_2}{\rho p} (\mathbf{F}_1 - \mathbf{F}_2), \tag{5.58}$$

where ρ_1 and ρ_2—the mass density of the components of mixture; $\rho = \rho_1 + \rho_2$—the mass density of mixture; p_1 and p_2—the partial pressures of the components; $p = p_1 + p_2$—the pressure of mixture.

Since $\nabla x_2 = -\nabla x_1$, $x_i = \frac{n_i}{n}$, it is possible to show that $\mathbf{d}_{21} = -\mathbf{d}_{12}$. Functions Φ_1 and Φ_2 then can be rewritten as [1, 3, 4]

$$\begin{aligned} \Phi_1 &= -\mathbf{A}_1 \cdot \frac{\partial \ln(T)}{\partial \mathbf{r}} - \mathbf{D}_1 \cdot \mathbf{d}_{12} - 2\mathbf{B}_1 : \frac{\partial}{\partial \mathbf{r}} \mathbf{v}_0; \\ \Phi_2 &= -\mathbf{A}_2 \cdot \frac{\partial \ln(T)}{\partial \mathbf{r}} - \mathbf{D}_2 \cdot \mathbf{d}_{12} - 2\mathbf{B}_2 : \frac{\partial}{\partial \mathbf{r}} \mathbf{v}_0, \end{aligned} \tag{5.59}$$

where $\mathbf{A}_i$ and $\mathbf{D}_i$—vectors of velocity, $\mathbf{B}_i$—tensors without sources such that

$$\mathbf{A}_i(\mathbf{V}_i) = \mathbf{V}_i A_i(V_i), \mathbf{D}_i(\mathbf{V}_i) = \mathbf{V}_i D_i(V_i), \mathbf{B}_i(\mathbf{V}_i) = \mathbf{V}_i \overset{\circ}{\mathbf{V}}_i B_i(V_i), \tag{5.60}$$

where the double tensor is defined as $\mathbf{V}_i\overset{\circ}{\mathbf{V}}_i = \mathbf{V}_i\mathbf{V}_i - \frac{1}{3}V_i^2\mathbf{U}$; $\mathbf{U}$—the unit tensor.

Making the appropriate transformations, we obtain the following pairs of equations for the coefficients $\mathbf{A}_i$,$\mathbf{D}_i$, and $\mathbf{B}_i$ of the disturbance function Φ:

$$f_1^{(0)}\left(W_1^2 - \frac{5}{2}\right)\mathbf{V}_1 = n_1^2 I_1(\mathbf{A}_1) + n_1 n_2 I_{12}(\mathbf{A}_1 + \mathbf{A}_2); \tag{5.61}$$

$$f_2^{(0)}\left(W_2^2 - \frac{5}{2}\right)\mathbf{V}_2 = n_2^2 I_2(\mathbf{A}_2) + n_1 n_2 I_{21}(\mathbf{A}_1 + \mathbf{A}_2); \tag{5.62}$$

$$x_1^{-1} f_1^{(0)}\mathbf{V}_1 = n_1^2 I_1(\mathbf{D}_1) + n_1 n_2 I_{12}(\mathbf{D}_1 + \mathbf{D}_2); \tag{5.63}$$

$$-x_2^{-1} f_2^{(0)}\mathbf{V}_2 = n_2^2 I_2(\mathbf{D}_2) + n_1 n_2 I_{21}(\mathbf{D}_1 + \mathbf{D}_2); \tag{5.64}$$

$$f_1^{(0)}\,\mathbf{W}_1\overset{\circ}{\mathbf{W}}_1 = n_1^2 I_1(\mathbf{B}_1) + n_1 n_2 I_{12}(\mathbf{B}_1 + \mathbf{B}_2); \tag{5.65}$$

$$f_2^{(0)}\,\mathbf{W}_2\overset{\circ}{\mathbf{W}}_2 = n_2^2 I_2(\mathbf{B}_2) + n_1 n_2 I_{21}(\mathbf{B}_1 + \mathbf{B}_2). \tag{5.66}$$

Coefficients $\mathbf{A}_i$ and $\mathbf{D}_i$ have to satisfy the following boundary conditions:

$$\int f_1^{(0)} m_1 \mathbf{V}_1 \cdot \mathbf{A}_1 d\mathbf{v}_1 + \int f_2^{(0)} m_2 \mathbf{V}_2 \cdot \mathbf{A}_2 d\mathbf{v}_2 = 0; \tag{5.67}$$

$$\int f_1^{(0)} m_1 \mathbf{V}_1 \cdot \mathbf{D}_1 d\mathbf{v}_1 + \int f_2^{(0)} m_2 \mathbf{V}_2 \cdot \mathbf{D}_2 d\mathbf{v}_2 = 0. \tag{5.68}$$

Thus, the distribution function can be written as

$$f_1 = f_1^{(0)}\left[1 - A_1(V_1)\mathbf{V}_1 \cdot \nabla \ln(T) - D_1(V_1)\mathbf{V}_1 \cdot \mathbf{d}_{12} - 2B_1(V_1)\,\mathbf{V}_1\overset{\circ}{\mathbf{V}}_1 : \nabla \mathbf{v}_0\right]; \tag{5.69}$$

$$f_2 = f_2^{(0)}\left[1 - A_2(V_2)\mathbf{V}_2 \cdot \nabla \ln(T) - D_2(V_2)\mathbf{V}_2 \cdot \mathbf{d}_{12} - 2B_2(V_2)\,\mathbf{V}_2\overset{\circ}{\mathbf{V}}_2 : \nabla \mathbf{v}_0\right]. \tag{5.70}$$

Taking into account (5.61)–(5.66) denote

$$n^2\{\mathbf{A},\mathbf{a}\} = \int f_1^{(0)}\left(W_1^2 - \frac{5}{2}\right)\mathbf{V}_1\cdot\mathbf{a}_1 d\mathbf{v}_1 + \int f_2^{(0)}\left(W_2^2 - \frac{5}{2}\right)\mathbf{V}_2\cdot\mathbf{a}_2 d\mathbf{v}_2; \quad (5.71)$$

$$n^2\{\mathbf{D},\mathbf{a}\} = x_1^{-1}\int f_1^{(0)}\mathbf{V}_1\cdot\mathbf{a}_1 d\mathbf{v}_1 - x_2^{-1}\int f_2^{(0)}\mathbf{V}_2\cdot\mathbf{a}_2 d\mathbf{v}_2; \quad (5.72)$$

$$n^2\{\mathbf{B},\mathbf{b}\} = \int f_1^{(0)}\,\mathbf{W}_1\overset{o}{\mathbf{W}}_1 : \mathbf{b}_1 d\mathbf{v}_1 + \int f_2^{(0)}\,\mathbf{W}_2\overset{o}{\mathbf{W}}_2 : \mathbf{b}_2 d\mathbf{v}_2, \quad (5.73)$$

where **a**—the vector having two vector components—vectors $\mathbf{a}_1$ and $\mathbf{a}_2$, that are defined in the phase velocity space; **b**—the double speed tensor; $\{F,G\}$—the parenthesis integrals that are defined as follows:

$$n^2\{F,G\} \equiv n_1^2[F,G]_1 + n_1 n_2[F_1+F_2, G_1+G_2]_{12} + n_2^2[F,G]_2, \quad (5.74)$$

where

$$[F,G]_1 \equiv \int G_1 I_1(F)d\mathbf{v}_1, [F,G]_2 \equiv \int G_2 I_2(F)d\mathbf{v}_2; \quad (5.75)$$

$$[F_1+G_2, H_1+K_2]_{12} \equiv \int F_1 I_{12}(H_1+K_2)d\mathbf{v}_1 + \int G_2 I_{21}(H_1+K_2)d\mathbf{v}_2. \quad (5.76)$$

Because the symmetry and linearity have collision integrals, $[F,G]_1 = [G,F]_1$, $[F,G]_2 = [G,F]_2$, and $[F_1+G_2, H_1+K_2]_{12} = [H_1+K_2, F_1+G_2]_{12}$, and so $\{F,G\} = \{G,F\}$, $\{F,G+H\} = \{F,G\} + \{F,H\}$, and $\{F,aG\} = a\{F,G\}$ (a—the arbitrary constant).

These relations need to find the transport coefficients. The coefficients **A**, **B**, and **D** in the future can be calculated using the expansion by Sonine polynomials:

$$S_m^{(n)}(x) = \sum_{p=0}^{n}\frac{(m+n)_{n-p}}{(p)!(n-p)!}(-x)^p = \sum_{p=0}^{n}\frac{(m+n)!}{(p)!(n-p)!(m+p)!}(-x)^p$$
$$= \sum_{p=0}^{n}\frac{\Gamma(m+n+1)}{\Gamma(p+1)\Gamma(n-p+1)\Gamma(m+p+1)}(-x)^p, \quad (5.77)$$

wherein $S_m^{(0)}(x) = 1$ and $S_m^{(1)}(x) = m+1-x$; $\Gamma(x)$—the gamma function. Thus, factors **A**, **B**, and **D** are unnormalized functions of the Sonine polynomials, that were originally used in the gases kinetic theory of D. Barnett in 1935. Note also the property of orthogonal Sonine polynomials:

$$\int_0^\infty \exp(-x) S_m^{(p)}(x) S_m^{(q)}(x) x^m \mathrm{d}x = \frac{\Gamma(m+p+1)}{p!} \delta_{p,q}, \tag{5.78}$$

where $\delta_{p,q}$—the delta function of Kronecker.

5.4 Determination of Viscosity

During the determination of viscosity it can be noted that the pressure of the two-component mixture at least order of approximation is reduced to a simple hydrostatic pressure of the mixture. For a first-order approximation to this system, it is necessary to add the pressure tensor $\mathsf{p}^{(1)}$, defined as [1, 2]

$$\mathsf{p}^{(1)} = n_1 m_1 \left(\overline{\mathbf{V}_1 \mathbf{V}_1}\right)^{(1)} + n_2 m_2 \left(\overline{\mathbf{V}_2 \mathbf{V}_2}\right)^{(1)} \tag{5.79}$$

This tensor characterizes the deviation of the system pressure from the hydrostatic one. In the notation of approximation functions of the first order (5.69) and (5.70) it has the form

$$\begin{aligned} \mathbf{p}^{(1)} &= -2m_1 \int f_1^{(0)} \mathbf{V}_1 \mathbf{V}_1 (\mathbf{B}_1 : \nabla \mathbf{v}_0) \mathrm{d}\mathbf{v}_1 - 2m_2 \int f_2^{(0)} \mathbf{V}_2 \mathbf{V}_2 (\mathbf{B}_2 : \nabla \mathbf{v}_0) \mathrm{d}\mathbf{v}_2 \\ &= -\frac{2}{5} \left\{ m_1 \int f_1^{(0)} \overset{\circ}{\mathbf{V}_1 \mathbf{V}_1} : \mathbf{B}_1 \mathrm{d}\mathbf{v}_1 + m_2 \int f_2^{(0)} \overset{\circ}{\mathbf{V}_2 \mathbf{V}_2} : \mathbf{B}_2 \mathrm{d}\mathbf{v}_2 \right\} S \\ &= -\frac{4}{5} k n^2 T \{\mathbf{B}, \mathbf{B}\} S, \end{aligned} \tag{5.80}$$

where S—the rate-of-shear tensor, element of this tensor is determined by equation

$$S_{\alpha\beta} = \frac{1}{2} \left(\frac{\partial v_{0\beta}}{\partial \alpha} + \frac{\partial v_{0\alpha}}{\partial \beta} \right) - \frac{1}{3} \delta_{\alpha\beta} (\nabla \cdot \mathrm{v}_0).$$

If we write an expression for the viscosity

$$\mu = \frac{2}{5} k n^2 T \{\mathbf{B}, \mathbf{B}\}, \tag{5.81}$$

then by substituting it into (5.80), the it takes the form

$$\mathsf{p}^{(1)} = -2\mu S. \tag{5.82}$$

Then (5.82) will be the expression for the tensor of viscous pressure in any environment.

The pressure tensor p is defined as

$$\mathsf{p} = m \int f \overset{\circ}{\mathbf{V}\mathbf{V}} \, \mathrm{d}\mathbf{v}. \tag{5.83}$$

Using the conditions

$$\int f^{(0)} A(W) W^2 \mathrm{d}\mathbf{v} = 0 \tag{5.84}$$

the pressure tensor can be rewritten as

$$\mathsf{p} = p\mathbf{U} - 2\mu \boldsymbol{S}, \tag{5.85}$$

where p—hydrostatic pressure $\mathbf{U}$—the unit tensor.

For a monatomic gas the viscosity μ is defined as

$$\mu = \frac{2kT}{5n} \int f^{(0)} \overset{\circ}{\mathbf{W}\mathbf{W}} : \overset{\circ}{\mathbf{W}\mathbf{W}} \, B(W) \mathrm{d}\mathbf{v}, \tag{5.86}$$

or in another form

$$\mu = \frac{2kT}{5} [\mathbf{B}(\mathbf{W}), \mathbf{B}(\mathbf{W})]. \tag{5.87}$$

In Eq. (5.87), the term in brackets is called the square bracket integrals, defined as

$$[\mathbf{B}(\mathbf{W}), \mathbf{B}(\mathbf{W})] \equiv \int \mathbf{B}(\mathbf{W}) : I(\mathbf{B}(\mathbf{W})) \mathrm{d}\mathbf{v}. \tag{5.88}$$

Consequently, the viscosity can be calculated as

$$\mu = \frac{4kT}{15n} \int f^{(0)} W^4 B(W) \mathrm{d}\mathbf{v} = \frac{16kT}{15\pi^{1/2}} \int_0^\infty \exp(-W^2) W^6 B(W) \mathrm{d}W. \tag{5.89}$$

Using the Sonine polynomial expansion, the bracket integral and viscosity of the gas can be determined.

5.4.1 Determination of the Elements in the Bracket Integral {B, B} by Sonine Polynomials

Chapman–Enskog coefficients $\mathbf{B}_1$ and $\mathbf{B}_2$ depending on the viscosity of the two-component mixture can be expressed as a sum of

$$\mathbf{B}_1 = \sum_{\substack{p=-\infty \\ p\neq 0}}^{+\infty} b_p \mathbf{b}_1^{(p)}; \mathbf{B}_2 = \sum_{\substack{p=-\infty \\ p\neq 0}}^{+\infty} b_p \mathbf{b}_2^{(p)}, \tag{5.90}$$

where b_p is the coefficient that is used for both expansions, the member $p = 0$ is not included in the sum and functions $\mathsf{b}^{(p)}$ are provided in two phase spaces of the velocity:

$$\left.\begin{array}{l} \mathsf{b}_1^{(p)} = \mathsf{b}_1^{(0)} = 0,\ \mathsf{b}_2^{(-p)} = \mathsf{b}_2^{(0)} = 0,\ (p = 0); \\ \mathsf{b}_1^{(p)} = 0,\ \mathsf{b}_2^{(-p)} = 0,\ (p<0); \\ \mathsf{b}_1^{(p)} = S_{5/2}^{(p-1)}(W)\,\mathbf{W}_1 \overset{\circ}{\mathbf{W}}_1,\ \mathsf{b}_2^{(-p)} = S_{5/2}^{(p-1)}(W)\,\mathbf{W}_2 \overset{\circ}{\mathbf{W}}_2,\ (p > 0). \end{array}\right\} \tag{5.91}$$

From Eq. (5.73) the following expression can be obtained:

$$\left\{\mathbf{B}, \mathbf{b}^{(q)}\right\} = \beta_q, \tag{5.92}$$

wherein

$$n^2\beta_q = \int f_1^{(0)}\, \mathbf{W}_1 \overset{\circ}{\mathbf{W}}_1 : \mathbf{b}_1 d\mathbf{v}_1 + \int f_2^{(0)}\, \mathbf{W}_2 \overset{\circ}{\mathbf{W}}_2 : \mathbf{b}_2 d\mathbf{v}_2. \tag{5.93}$$

Integrating (5.93), we have

$$\beta_1 = \frac{5n_1}{2n^2}; \quad \beta_{-1} = \frac{5n_2}{2n^2}; \quad \beta_q = 0(q \neq \pm 1) \tag{5.94}$$

From (5.90)–(5.93) we obtain the expression

$$\sum_{\substack{p=-\infty \\ p\neq 0}}^{+\infty} b_p b_{pq} = \beta_q, \tag{5.95}$$

where

$$b_{pq} = \left\{\mathbf{b}^{(p)}, \mathbf{b}^{(q)}\right\}. \tag{5.96}$$

If the values b_{pq} are known, all the values b_p can defined by solving the system of algebraic equations (5.95).

Because $\beta_q = 0(q \neq \pm 1)$ (5.94), the expressions for the bracket integrals are

$$\{\mathbf{B}, \mathbf{B}\} = \sum_p b_p \left\{\mathbf{B}, \mathbf{b}^{(p)}\right\} = \beta_1 b_1 + \beta_{-1} b_{-1}. \tag{5.97}$$

In accordance with the work of Chapman and Cowling [1] and with the designations adopted by them, we note that for two-component gas mixture the viscosity approximation of the order m can be expressed by the Sonine polynomials:

$$[\mu]_m = p\left(x_1 b_1^{(m)} + x_2 b_{-1}^{(m)}\right), \tag{5.98}$$

where $b_{-1}^{(m)}$ and $b_1^{(m)}$—the expansion coefficients are determined by solving a system of algebraic equations (5.94) and (5.95).

Using the matrix notation this system of equations can be written as

$$\mathbf{B}\mathbf{b} = \beta, \tag{5.99}$$

where

$$\mathbf{B} = \begin{bmatrix} b_{-m-m} & \cdots & b_{-m-1} & b_{-m1} & \cdots & b_{-mm} \\ \vdots & \ddots & \vdots & \vdots & \ddots & \ddots \\ b_{-1-m} & \cdots & b_{-1-1} & b_{-11} & \cdots & b_{-1m} \\ b_{1-m} & \cdots & b_{1-1} & b_{11} & \cdots & b_{1m} \\ \vdots & \ddots & \vdots & \vdots & \ddots & \vdots \\ b_{m-m} & \cdots & b_{m-1} & b_{m1} & \cdots & b_{mm} \end{bmatrix}, \quad \mathbf{b} = \begin{bmatrix} b_{-m} \\ \vdots \\ b_{-1} \\ b_1 \\ \vdots \\ b_m \end{bmatrix} \quad \text{and}$$

$$\boldsymbol{\beta} = \begin{bmatrix} 0 \\ \vdots \\ \beta_{-1} \\ \beta_1 \\ \vdots \\ 0 \end{bmatrix} \tag{5.100}$$

By increasing the order of approximation (expansion) m in the matrixes there are additional rows and columns.

Thus, for obtaining the expansion coefficients $b_{-1}^{(m)}$ and $b_1^{(m)}$ or viscosities with this order of decomposition it is necessary only to form the matrix $2m \times 2m$ (5.100), and invert it. The elements of the matrix **B** in (5.100) are obtained from the combinations of bracket integrals containing relevant Sonine polynomials. For all (p, q) it is easy to show that $b_{pq} = b_{qp}$ and it is possible to imagine that

$$\begin{aligned} b_{pq} = b_{qp} = {} & x_1^2\left[S_{5/2}^{(p-1)}\left(W_1^2\right)\mathbf{W}_1\overset{o}{\mathbf{W}}_1,\, S_{5/2}^{(q-1)}\left(W_1^2\right)\mathbf{W}_1\overset{o}{\mathbf{W}}_1\right]_1 \\ & + x_1x_2\left[S_{5/2}^{(p-1)}\left(W_1^2\right)\mathbf{W}_1\overset{o}{\mathbf{W}}_1,\, S_{5/2}^{(q-1)}\left(W_1^2\right)\mathbf{W}_1\overset{o}{\mathbf{W}}_1\right]_{12}; \end{aligned} \tag{5.101}$$

$$b_{p-q} = b_{-qp} = x_1x_2\left[S_{5/2}^{(p-1)}\left(W_1^2\right)\mathbf{W}_1\overset{o}{\mathbf{W}}_1,\, S_{5/2}^{(q-1)}\left(W_2^2\right)\mathbf{W}_2\overset{o}{\mathbf{W}}_2\right]_{12}; \tag{5.102}$$

$$b_{-pq} = b_{q-p} = x_1x_2\left[S_{5/2}^{(p-1)}\left(W_2^2\right)\mathbf{W}_2\overset{o}{\mathbf{W}}_2,\, S_{5/2}^{(q-1)}\left(W_1^2\right)\mathbf{W}_1\overset{o}{\mathbf{W}}_1\right]_{21}; \tag{5.103}$$

$$\begin{aligned} b_{-p-q} = b_{-q-p} = {} & x_2^2\left[S_{5/2}^{(p-1)}\left(W_2^2\right)\mathbf{W}_2\overset{o}{\mathbf{W}}_2,\, S_{5/2}^{(q-1)}\left(W_2^2\right)\mathbf{W}_2\overset{o}{\mathbf{W}}_2\right]_2 \\ & + x_1x_2\left[S_{5/2}^{(p-1)}\left(W_2^2\right)\mathbf{W}_2\overset{o}{\mathbf{W}}_2,\, S_{5/2}^{(q-1)}\left(W_2^2\right)\mathbf{W}_2\overset{o}{\mathbf{W}}_2\right]_{21}. \end{aligned} \tag{5.104}$$

Analyzing the expression for the bracket integrals, we can see that Eqs. (5.103) and (5.104) are identical to Eqs. (5.102) and (5.101), and have the difference only in the indexes that indicate the components of the mixture. Thus, in general, a complete solution of the BKE using Chapman–Enskog approach for the viscosity of two-component gas mixture requires the determination of only three types of bracket integrals:

$$\left[S_{5/2}^{(p)}\left(W_1^2\right)\mathbf{W}_1\overset{o}{\mathbf{W}}_1,\, S_{5/2}^{(q)}\left(W_1^2\right)\mathbf{W}_1\overset{o}{\mathbf{W}}_1\right]_1; \tag{5.105}$$

$$\left[S_{5/2}^{(p)}\left(W_1^2\right)\mathbf{W}_1\overset{o}{\mathbf{W}}_1,\, S_{5/2}^{(q)}\left(W_2^2\right)\mathbf{W}_2\overset{o}{\mathbf{W}}_2\right]_{12}; \tag{5.106}$$

$$\left[S_{5/2}^{(p)}\left(W_1^2\right)\mathbf{W}_1\overset{o}{\mathbf{W}}_1,\, S_{5/2}^{(q)}\left(W_1^2\right)\mathbf{W}_1\overset{o}{\mathbf{W}}_1\right]_{12}. \tag{5.107}$$

5.4.2 *The Bracket Integrals*

In order to complete the calculation of viscosity it is necessary to calculate bracket integrals' parentheses used in (5.74) for $\left\{\mathsf{b}^{(p)}, \mathsf{b}^{(q)}\right\}$. First of all, it is necessary to find the square bracket integrals in (5.75), (5.76): $\left[\mathrm{b}_1^{(p)}, \mathrm{b}_1^{(q)}\right]_1$, $\left[\mathrm{b}_1^{(p)}, \mathrm{b}_1^{(q)}\right]_{12}$ and $\left[\mathrm{b}_1^{(p)}, \mathrm{b}_2^{(q)}\right]_{12}$. Integration is performed for all variable parameters of the collision, information about the intermolecular interaction potential is required for it. However, the potential affects only the results of the integration of using variables g and b, they determine the scattering angle θ. The other six integrals can be used without data on the potential. For obtaining these two integrals, we use Eqs. (5.65) and (5.66). Then we substitute the polynomial approximation of Sonine solutions $\mathbf{B}_1$ and $\mathbf{B}_2$, doing multiplication with additional Sonine polynomials and then integrating the equation. Use the property of orthogonal Sonine polynomials (5.78), we obtain a set of equations including matrix coefficients needed for decisions. In the right-hand sides of equations a constant β_q is included, and the left side is a combination of bracket integrals corresponding to the desired ratio b_{pq} of the matrix used for the determination of values b_p. These values are in turn used to find the viscosity according to Eq. (5.98). Thus, expressions for the integrals include the bracket average effective interaction cross sections, the ratio for them were obtained earlier (Chap. 2).

To find the viscosity the following quantities are used:

$$\overline{\overline{W_i}} = \overset{\mathrm{o}}{W_i} W_i . \tag{5.108}$$

The value Q_{ij}^{pq} is defined as [5–7]

$$\begin{aligned} Q_{ij}^{pq} = \sum_{l=1}^{\nu} n_i n_l \Big\{ \delta_{ij} \left[S_n^{(p)}(W_i^2)\overline{\overline{W_i}};\ S_n^{(q)}(W_i^2)\overline{\overline{W_i}} \right]_{il} \\ + \delta_{jl} \left[S_n^{(p)}(W_i^2)\overline{\overline{W_i}};\ S_n^{(q)}(W_l^2)\overline{\overline{W_l}} \right]_{il}, \end{aligned} \tag{5.109}$$

where l and ν—conditioned sort of particles of the gas mixture.

Formulas for the bracket integrals in determining viscosity of plasma are given in Appendices B and C [4, 8].

The elements of matrix that are necessary for the determination of the viscosity can be found using Q_{ij}^{pq}:

$$p_{ij}^{pq} = \sqrt{\frac{2\pi m_i}{kT}} Q_{ij}^{pq}, \tag{5.110}$$

Equation for the calculation of the viscosity of a sixth-order approximation is

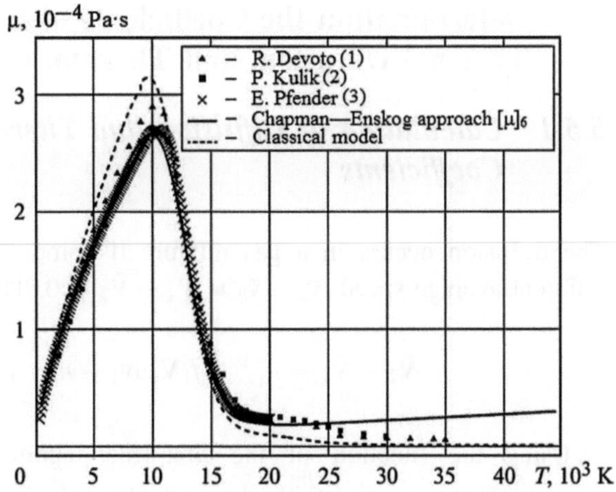

Fig. 5.2 The calculation results of the argon viscosity using Chapman–Enskog approach (6-order), as well as their comparison with the data of other authors and the results of calculation by the classical model: (1)—[9, 10]; (2)—[11]; (3)—[12]

$$[\mu_{ij}]_6 = -\frac{5(2\pi kT_i)^{1/2}}{2|p|}\begin{vmatrix} p_{ij}^{00} & p_{ij}^{01} & p_{ij}^{02} & p_{ij}^{03} & p_{ij}^{04} & p_{ij}^{05} & n_i\sqrt{m_i} \\ p_{ij}^{10} & p_{ij}^{11} & p_{ij}^{12} & p_{ij}^{13} & p_{ij}^{14} & p_{ij}^{15} & 0 \\ p_{ij}^{20} & p_{ij}^{21} & p_{ij}^{22} & p_{ij}^{23} & p_{ij}^{24} & p_{ij}^{25} & 0 \\ p_{ij}^{30} & p_{ij}^{31} & p_{ij}^{32} & p_{ij}^{33} & p_{ij}^{34} & p_{ij}^{35} & 0 \\ p_{ij}^{40} & p_{ij}^{41} & p_{ij}^{42} & p_{ij}^{43} & p_{ij}^{44} & p_{ij}^{45} & 0 \\ p_{ij}^{50} & p_{ij}^{51} & p_{ij}^{52} & p_{ij}^{53} & p_{ij}^{54} & p_{ij}^{55} & 0 \\ n_j & 0 & 0 & 0 & 0 & 0 & 0 \end{vmatrix}, \tag{5.111}$$

where

$$|p| = \begin{vmatrix} p_{ij}^{00} & p_{ij}^{01} & p_{ij}^{02} & p_{ij}^{03} & p_{ij}^{04} & p_{ij}^{05} \\ p_{ij}^{10} & p_{ij}^{11} & p_{ij}^{12} & p_{ij}^{13} & p_{ij}^{14} & p_{ij}^{15} \\ p_{ij}^{20} & p_{ij}^{21} & p_{ij}^{22} & p_{ij}^{23} & p_{ij}^{24} & p_{ij}^{25} \\ p_{ij}^{30} & p_{ij}^{31} & p_{ij}^{32} & p_{ij}^{33} & p_{ij}^{34} & p_{ij}^{35} \\ p_{ij}^{40} & p_{ij}^{41} & p_{ij}^{42} & p_{ij}^{43} & p_{ij}^{44} & p_{ij}^{45} \\ p_{ij}^{50} & p_{ij}^{51} & p_{ij}^{52} & p_{ij}^{53} & p_{ij}^{54} & p_{ij}^{55} \end{vmatrix}.$$

The results of the viscosity calculation by Eq. (5.111) are represented in Fig. 5.2. For comparison, data of the other authors and the results of calculation by the classical model are also shown in [10–14].

The results of calculations using the Boltzmann kinetic equation by Chapman–Enskog approach the best way match to the experimental data than the results of the calculation by the classical model.

5.5 Determination the Coefficients of Diffusion, Thermal Diffusion, and Thermal Conductivity

5.5.1 Calculation of Diffusion and Thermal Diffusion Coefficients

The diffusion occurs in a gas mixture if components of the mixture have the different average speed, $\bar{\mathbf{v}}_1 - \bar{\mathbf{v}}_2 = \bar{\mathbf{V}}_1 - \bar{\mathbf{V}}_2 \neq 0$. This implies that

$$\bar{\mathbf{V}}_1 - \bar{\mathbf{V}}_2 = n_1^{-1} \int f_1 \mathbf{V}_1 d\mathbf{v}_1 - n_2^{-1} \int f_2 \mathbf{V}_2 d\mathbf{v}_2 \neq 0. \quad (5.112)$$

Using the function of the first-order approximation (5.69), (5.70), and $n_1^{-1} \int f_1^{(0)} \mathbf{V}_1 d\mathbf{v}_1 + n_2^{-1} \int f_2^{(0)} \mathbf{V}_2 d\mathbf{v}_2 = 0$, the velocity of diffusion in one direction can be determined by equations

$$\begin{aligned}
\bar{\mathbf{v}}_d = \frac{1}{3}(\bar{\mathbf{V}}_1 - \bar{\mathbf{V}}_2) &= -\frac{1}{3}\left[\left\{n_1^{-1} \int f_1^{(0)} \mathbf{V}_1 \cdot \mathbf{D}_1 d\mathbf{v}_1 - n_2^{-1} \int f_2^{(0)} \mathbf{V}_2 \cdot \mathbf{D}_2 d\mathbf{v}_2 \right\} \mathbf{d}_{12}\right. \\
&\quad \left. + \left\{n_1^{-1} \int f_1^{(0)} \mathbf{V}_1 \cdot \mathbf{A}_1 d\mathbf{v}_1 - n_2^{-1} \int f_2^{(0)} \mathbf{V}_2 \cdot \mathbf{A}_2 d\mathbf{v}_2 \right\} \nabla \ln T \right] \\
&= -\frac{1}{3} n [\{\mathbf{D}, \mathbf{D}\} \mathbf{d}_{12} + \{\mathbf{D}, \mathbf{A}\} \nabla \ln T].
\end{aligned} \quad (5.113)$$

For the case where the gas mixture does not have gradients of temperature and pressure, and there are no external forces acting on a molecule, the expression for the diffusion coefficient can be obtained from Eq. (5.113). At the same time, $\mathbf{d}_{12} = \nabla x_1 = n^{-1} \nabla n_1$ and $\nabla \ln(T) = 0$, so

$$\bar{\mathbf{v}}_d = \frac{1}{3}(\bar{\mathbf{V}}_1 - \bar{\mathbf{V}}_2) = -\frac{1}{3}\{\mathbf{D}, \mathbf{D}\} \nabla n_1. \quad (5.114)$$

Equation (5.114) can be written as

$$\bar{\mathbf{v}}_d = -D_{12}\left(n_1^{-1} \nabla n_1 - n_2^{-1} \nabla n_2\right) = -D_{12} \frac{n}{n_1 n_2} \nabla n_1, \quad (5.115)$$

where the proportional constant D_{12} is the diffusion coefficient. And the diffusion coefficient should be

$$D_{12} = \frac{n_1 n_2}{3n} \{\mathbf{D}, \mathbf{D}\}. \quad (5.116)$$

The presence of a temperature gradient means that the second term of Eq. (5.113) contains a thermal diffusion coefficient, that can be expressed as [1, 4, 15]

$$D_T = \frac{n_1 n_2}{3n}\{\mathbf{D}, \mathbf{A}\}. \tag{5.117}$$

The above equation allows us to rewrite (5.113) as

$$\bar{\mathbf{v}}_d = -\frac{n^2}{n_1 n_2}\{D_{12}\mathbf{d}_{12} + D_T \nabla \ln T\} = -\frac{n^2}{n_1 n_2} D_{12}\left\{\mathbf{d}_{12} + k_T \frac{\nabla T}{T}\right\}, \tag{5.118}$$

where $k_T = D_T / D_{12} \equiv \{\mathbf{D}, \mathbf{A}\}/\{\mathbf{D}, \mathbf{D}\}$.

5.5.2 *Determination of the Heat Flow*

Assume that the particles in the mixture can exchange with the kinetic energy (i.e., cannot exchange with the energy stored by the internal degrees of freedom), whereas the expression for heat flow may be written as [1, 4, 15]

$$\mathbf{q} = \int f_1 \frac{1}{2} m_1 V_1^2 \mathbf{V}_1 d\mathbf{v}_1 + \int f_2 \frac{1}{2} m_2 V_2^2 \mathbf{V}_2 d\mathbf{v}_2, \tag{5.119}$$

wherein

$$\begin{aligned}\frac{\mathbf{q}}{kT} - \frac{5}{2}(n_1\bar{\mathbf{V}}_1 + n_2\bar{\mathbf{V}}_2) &= -\frac{1}{3}\int f_1^{(0)}\left(W_1^2 - \frac{5}{2}\right)\{(\mathbf{V}_1 \cdot \mathbf{D}_1)\mathbf{d}_{12} + (\mathbf{V}_1 \cdot \mathbf{A}_1)\nabla \ln T\} d\mathbf{v}_1 \\ &\quad -\frac{1}{3}\int f_2^{(0)}\left(W_2^2 - \frac{5}{2}\right)\{(\mathbf{V}_2 \cdot \mathbf{D}_2)\mathbf{d}_{12} + (\mathbf{V}_2 \cdot \mathbf{A}_2)\nabla \ln T\} d\mathbf{v}_2 \\ &= -\frac{1}{3} n^2 [\{\mathbf{A}, \mathbf{D}\}\mathbf{d}_{12} + \{\mathbf{A}, \mathbf{A}\}\nabla \ln T].\end{aligned} \tag{5.120}$$

From (5.120) $\mathbf{q}$ can be obtained in the form of

$$\mathbf{q} = \frac{5}{2} kT(n_1\bar{\mathbf{V}}_1 + n_2\bar{\mathbf{V}}_2) - \frac{1}{3} kn^2 T[\{\mathbf{A}, \mathbf{D}\}\mathbf{d}_{12} + \{\mathbf{A}, \mathbf{D}\}\nabla \ln T]. \tag{5.121}$$

After the transformation of (5.121), we have

$$\begin{aligned}\mathbf{q} &= \frac{5}{2}kT(n_1\bar{\mathbf{V}}_1 + n_2\bar{\mathbf{V}}_2) + nkT(\bar{\mathbf{V}}_1 - \bar{\mathbf{V}}_2)\frac{\{\mathbf{A},\mathbf{D}\}}{\{\mathbf{D},\mathbf{D}\}} - \lambda\nabla T \\ &= -\lambda\nabla T + \frac{5}{2}kT(n_1\bar{\mathbf{V}}_1 + n_2\bar{\mathbf{V}}_2) + nkTk_T(\bar{\mathbf{V}}_1 - \bar{\mathbf{V}}_2),\end{aligned} \tag{5.122}$$

where

$$\lambda = \frac{1}{3}kn^2\left[\{\mathbf{A},\mathbf{A}\} - \frac{\{\mathbf{A},\mathbf{D}\}\{\mathbf{D},\mathbf{A}\}}{\{\mathbf{D},\mathbf{D}\}}\right] = \frac{1}{3}kn^2\{\tilde{\mathbf{A}},\tilde{\mathbf{A}}\}; \tag{5.123}$$

$$\{\tilde{\mathbf{A}},\tilde{\mathbf{A}}\} = \{\mathbf{A},\mathbf{A}\} - \frac{\{\mathbf{A},\mathbf{D}\}\{\mathbf{D},\mathbf{A}\}}{\{\mathbf{D},\mathbf{D}\}} = \{\mathbf{A},\mathbf{A}\} - 2k_T\{\mathbf{A},\mathbf{D}\} + k_T^2\{\mathbf{D},\mathbf{D}\}, \tag{5.124}$$

wherein

$$\tilde{\mathbf{A}}_1 = \mathbf{A}_1 - \frac{\{\mathbf{A},\mathbf{D}\}}{\{\mathbf{D},\mathbf{D}\}}\mathbf{D}_1 = \mathbf{A}_1 - k_T\mathbf{D}_1; \tag{5.125}$$

$$\tilde{\mathbf{A}}_2 = \mathbf{A}_2 - \frac{\{\mathbf{A},\mathbf{D}\}}{\{\mathbf{D},\mathbf{D}\}}\mathbf{D}_2 = \mathbf{A}_2 - k_T\mathbf{D}_2. \tag{5.126}$$

If we assume that diffusion does not occur (i.e., $\bar{\mathbf{V}}_1 = \bar{\mathbf{V}}_2 = 0$), we obtain $\mathbf{q} = -\lambda\nabla T$.

The first term on the right-hand side of Eq. (5.122) corresponds to the transfer of heat due to inequality of the mixture temperature, i.e., in the presence of a temperature gradient. The second term is the case because the heat flow is measured with respect to $\mathbf{v_0}$, this term corresponds to the heat flux carried by the molecules in the diffusion. The third term determines the effect of thermal diffusion.

The thermal conductivity can also be represented as

$$\lambda = \frac{2k^2T}{3mn}\int f^{(0)}W^2A(W)\left(W^2 - \frac{5}{2}\right)d\mathbf{v}. \tag{5.127}$$

5.5.3 *Defining Elements of the Bracket Integrals $\{\tilde{\mathbf{A}},\tilde{\mathbf{A}}\}$ and $\{\mathbf{A},\mathbf{D}\}$ Using the Sonine Polynomials*

To find the diffusion coefficient, thermal diffusion, and thermal conductivity, it is necessary to calculate the bracket integrals $\{\tilde{\mathbf{A}},\tilde{\mathbf{A}}\}$, $\{\mathbf{A},\mathbf{D}\}$, and $\{\mathbf{D},\mathbf{D}\}$. Using

Chapman–Enskog approach, the coefficients $\tilde{\mathbf{A}}_1$, $\tilde{\mathbf{A}}_2$, $\mathbf{D}_1$, and $\mathbf{D}_2$ can be expanded as a sum of the Sonine polynomials:

$$\tilde{\mathbf{A}}_1 = \sum_{p=-\infty}^{+\infty} a_p \mathbf{a}_1^{(p)}; \tilde{\mathbf{A}}_2 = \sum_{p=-\infty}^{+\infty} a_p \mathbf{a}_2^{(p)}; \mathbf{D}_1 = \sum_{p=-\infty}^{+\infty} d_p \mathbf{a}_1^{(p)}; \mathbf{D}_2 = \sum_{p=-\infty}^{+\infty} d_p \mathbf{a}_2^{(p)}, \tag{5.128}$$

where in accordance with the notations of Chapman and Cowling [1]

$$\mathbf{a}_1^{(p)} = \mathbf{a}_1^{(0)} = M_1^{1/2} \rho_2 \mathbf{W}_1 / \rho \qquad (p = 0); \tag{5.129}$$

$$\mathbf{a}_2^{(-p)} = a_2^{(0)} = -M_2^{1/2} \rho_1 \mathbf{W}_2 / \rho \qquad (p = 0); \tag{5.130}$$

$$\mathbf{a}_1^{(p)} = S_{3/2}^{(p)}(W_1^2)\mathbf{W}_1 \qquad (p > 0); \tag{5.131}$$

$$\mathbf{a}_2^{(p)} = 0 \qquad (p > 0); \tag{5.132}$$

$$\mathbf{a}_1^{(-p)} = 0 \qquad (p > 0); \tag{5.133}$$

$$\mathbf{a}_2^{(-p)} = S_{3/2}^{(p)}(W_2^2)\mathbf{W}_2 \qquad (p > 0). \tag{5.134}$$

For the determination of coefficients of the expansion a_p in (5.128), using (5.71), we obtain

$$\left\{\mathbf{A}, \mathbf{a}^{(q)}\right\} = \alpha_q, \tag{5.135}$$

wherein

$$n^2 \alpha_p = \int f_1^{(0)} \left(W_1^2 - \frac{5}{2} \right) \mathbf{V}_1 \cdot \mathbf{a}_1^{(q)} d\mathbf{v}_1 + \int f_2^{(0)} \left(W_2^2 - \frac{5}{2} \right) \mathbf{V}_2 \cdot \mathbf{a}_2^{(q)} d\mathbf{v}_2. \tag{5.136}$$

Integrating (5.136), we find that $\alpha_p = 0$ when $q \neq \pm 1$ and

$$\alpha_1 = -\frac{15}{4} \frac{n_1}{n^2} \sqrt{\frac{2kT}{m_1}}, \quad \alpha_{-1} = -\frac{15}{4} \frac{n_2}{n^2} \sqrt{\frac{2kT}{m_2}}. \tag{5.137}$$

Similarly, to determine the coefficients of the expansion d_p in (5.128) using (5.72), we obtain

$$\left\{\mathbf{D}, \mathbf{a}^{(q)}\right\} = \delta_q, \tag{5.138}$$

wherein

$$n^2\delta_q = x_1^{-1}\int f_1^{(0)}\mathbf{V}_1\cdot\mathbf{a}_1^{(q)}d\mathbf{v}_1 - x_2^{-1}\int f_2^{(0)}V_2\cdot\mathbf{a}_2^{(q)}d\mathbf{v}_2. \tag{5.139}$$

Integrating (5.139), we have

$$\delta_0 = \frac{3}{2n}\sqrt{\frac{2kT}{m_0}};\quad \delta_q = 0(q\neq 0). \tag{5.140}$$

By combining (5.128)–(5.134), (5.138)–(5.140), write the system of equations as

$$\sum_{p=-\infty}^{+\infty} d_p a_{pq} = \delta_q \quad (q = 0, \pm 1, \pm 2, \ldots, \pm\infty), \tag{5.141}$$

where

$$a_{pq} = \left\{\mathbf{a}^{(p)}, \mathbf{a}^{(q)}\right\} = a_{qp}. \tag{5.142}$$

If the values a_{pq} are known, all the values d_p can be found by solving a system of algebraic equations (5.141). Next, to determine the coefficients of the expansion a_p it is necessary to substitute (5.129)–(5.134) in (5.128). Since $\tilde{\mathbf{A}} = \mathbf{A} - k_T\mathbf{D}$ then

$$\left\{\tilde{\mathbf{A}}, \mathbf{a}^{(q)}\right\} = \left\{\mathbf{A}, \mathbf{a}^{(q)}\right\} - k_T\left\{\mathbf{D}, \mathbf{a}^{(q)}\right\} = \alpha_q - k_T\delta_q. \tag{5.143}$$

Since $\delta_q = 0$ $(q \neq 0)$, it can be written as

$$\sum_{\substack{p=-\infty\\ p\neq 0}}^{+\infty} a_p a_{pq} = \alpha_q \quad (q = \pm 1, \pm 2, \ldots, \pm\infty). \tag{5.144}$$

From the system of Eqs. (5.144) and (5.142) the expansion coefficients a_p can be obtained. Based on the above equations, it can be shown that

$$\{\mathbf{D},\mathbf{D}\} = d_0\delta_0;\ \{\mathbf{D},\mathbf{A}\} = d_1\alpha_1 + d_{-1}\alpha_{-1};\ \left\{\tilde{\mathbf{A}},\tilde{\mathbf{A}}\right\} = a_1\alpha_1 + a_{-1}\alpha_{-1}. \tag{5.145}$$

In accordance with the work of Chapman and Cowling [1] as well as designations adopted by them, we note that for a two-component gas mixture the diffusion, thermal diffusion coefficients, and thermal conductivity can be expressed in the form of approximation of any given order m through Sonine polynomials as follows:

$$[D_{12}]_m = \frac{1}{2} x_1 x_2 \sqrt{\frac{2kT}{m_0}} d_0^{(m)}; \tag{5.146}$$

$$[D_T]_m = -\frac{5}{4} x_1 x_2 \sqrt{\frac{2kT}{m_0}} \left(x_1 M_1^{-\frac{1}{2}} d_1^{(m)} + x_2 M_2^{-\frac{1}{2}} d_{-1}^{(m)} \right); \tag{5.147}$$

$$[\lambda]_m = -\frac{5}{4} kn \sqrt{\frac{2kT}{m_0}} \left(x_1 M_1^{-1/2} a_1^{(m)} + x_2 M_2^{-1/2} a_{-1}^{(m)} \right), \tag{5.148}$$

where $d_{-1}^{(m)}$, $d_0^{(m)}$, $d_1^{(m)}$ can be found from the system of Eqs. (5.141), $a_{-1}^{(m)}$, $a_1^{(m)}$—from (5.144).

5.5.4 The Bracket Integrals

To find the diffusion, thermal diffusion coefficients, and thermal conductivity it is necessary to determine the coefficients of expansion $d_{-1}^{(m)}$, $d_0^{(m)}$, $d_1^{(m)}$ and $a_{-1}^{(m)}$, $a_1^{(m)}$. For this purpose, it is necessary to calculate the bracket integrals, that are represented in (5.74) for $\{\mathbf{a}^{(p)}, \mathbf{a}^{(q)}\}$. In turn, it is necessary to calculate integrals in square brackets (5.75), (5.76), i.e., $\left[\mathbf{a}_1^{(p)}, \mathbf{a}_1^{(q)}\right]_1$, $\left[\mathbf{a}_1^{(p)}, \mathbf{a}_1^{(q)}\right]_{12}$, and $\left[\mathbf{a}_1^{(p)}, \mathbf{a}_2^{(q)}\right]_{12}$. Integrating by all the variables defining the parameters of the collision, and information about the intermolecular interaction potential are also required.

Using the property of orthogonal Sonine polynomials, a number of equations containing the matrix coefficients can be obtained, they are necessary for the determination of solutions. Right side of the equations contain constants d_q, α_q, and the left—a combination of bracket integrals corresponding to the required elements of the matrix a_{pq} that is used to determine the coefficients of the expansion d_p, a_p.

Using matrix notation this system of equations can be written as

$$\mathbf{Dd} = \boldsymbol{\delta} \quad \text{and} \quad \mathbf{Aa} = \boldsymbol{\alpha}, \tag{5.149}$$

where

$$\mathbf{D} = \begin{bmatrix} a_{-m-m} & \cdots & a_{-m-1} & a_{-m0} & a_{-m1} & \cdots & a_{-mm} \\ \vdots & \ddots & \vdots & \vdots & \vdots & \ddots & \vdots \\ a_{-1-m} & \cdots & a_{-1-1} & a_{-10} & a_{-11} & \cdots & a_{-1m} \\ a_{0-m} & \cdots & a_{0-1} & a_{00} & a_{01} & \cdots & a_{0m} \\ a_{1-m} & \cdots & a_{1-1} & a_{10} & a_{11} & \cdots & a_{1m} \\ \vdots & \ddots & \vdots & \vdots & \vdots & \ddots & \vdots \\ a_{m-m} & \cdots & a_{m-1} & a_{m0} & a_{m1} & \cdots & a_{mm} \end{bmatrix}; \quad \mathbf{d} = \begin{bmatrix} d_{-m} \\ \vdots \\ d_{-1} \\ d_0 \\ d_1 \\ \vdots \\ d_m \end{bmatrix};$$

$$\boldsymbol{\delta} = \begin{bmatrix} 0 \\ \vdots \\ 0 \\ \delta_0 \\ 0 \\ \vdots \\ 0 \end{bmatrix}; \tag{5.150}$$

$$\mathbf{A} = \begin{bmatrix} a_{-m-m} & \cdots & a_{-m-1} & a_{-m1} & \cdots & a_{-mm} \\ \vdots & \ddots & \vdots & \vdots & \ddots & \ddots \\ a_{-1-m} & \cdots & a_{-1-1} & a_{-11} & \cdots & a_{-1m} \\ a_{1-m} & \cdots & a_{1-1} & a_{11} & \cdots & a_{1m} \\ \vdots & \ddots & \vdots & \vdots & \ddots & \vdots \\ a_{m-m} & \cdots & a_{m-1} & a_{m1} & \cdots & a_{mm} \end{bmatrix}; \quad \mathbf{a} = \begin{bmatrix} a_{-m} \\ \vdots \\ a_{-1} \\ a_1 \\ \vdots \\ a_m \end{bmatrix}; \tag{5.151}$$

$$\boldsymbol{\alpha} = \begin{bmatrix} 0 \\ \vdots \\ \alpha_{-1} \\ \alpha_1 \\ \vdots \\ 0 \end{bmatrix}.$$

In (5.150), (5.151) superscript (m) for the expansion coefficients $d_p^{(m)}$ and $a_p^{(m)}$ was omitted, as it was done by Chapman and Cowling.

Thus, for obtaining the expansion coefficients $d_{-1}^{(m)}$, $d_0^{(m)}$, $d_1^{(m)}$ and $a_{-1}^{(m)}$, $a_1^{(m)}$ and the coefficients of diffusion, thermal diffusion, and thermal conductivity, respectively, in this order of decomposition, it is necessary only to form matrices $(2m+1) \times (2m+1)$ and $2m \times 2m$ and invert them. Elements of the matrix a_{pq} in (5.151) are obtained by combining the bracket integrals containing relevant Sonine polynomials. As for any p, q it is easy to show that $a_{pq} = a_{qp}$, then for a_{pq} it can be written [1, 4, 15] as

$$a_{pq} = a_{qp} = x_1^2\left[S_{3/2}^{(p)}(W_1^2)\mathbf{W}_1,\, S_{3/2}^{(q)}(W_1^2)\mathbf{W}_1\right]_1 + x_1x_2\left[S_{3/2}^{(p)}(W_1^2)\mathbf{W}_1,\, S_{3/2}^{(q)}(W_1^2)\mathbf{W}_1\right]_{12}; \quad (5.152)$$

$$a_{p-q} = a_{-qp} = x_1x_2\left[S_{3/2}^{(p)}(W_1^2)\mathbf{W}_1,\, S_{3/2}^{(q)}(W_2^2)\mathbf{W}_2\right]_{12}; \quad (5.153)$$

$$a_{-pq} = a_{q-p} = x_1x_2\left[S_{3/2}^{(p)}(W_2^2)\mathbf{W}_2,\, S_{3/2}^{(q)}(W_1^2)\mathbf{W}_1\right]_{21}; \quad (5.154)$$

$$a_{-p-q} = a_{-q-p} = x_2^2\left[S_{3/2}^{(p)}(W_2^2)\mathbf{W}_2,\, S_{3/2}^{(q)}(W_2^2)\mathbf{W}_2\right]_2 + x_1x_2\left[S_{3/2}^{(p)}(W_2^2)\mathbf{W}_2,\, S_{3/2}^{(q)}(W_2^2)\mathbf{W}_2\right]_{21}. \quad (5.155)$$

Analyzing the expression (5.152)–(5.155), it can be noted that (5.154) and (5.155) are identical to (5.153) and 5.152) and have the difference only in the indexes that indicate the components of mixture. Thus, in general, for a complete solution of the BKE using Chapman–Enskog approach for determination of the diffusion, thermal diffusion coefficients, and thermal conductivity of a two-component gas mixture, the determination of only three types of bracket integrals is required:

$$\left[S_{3/2}^{(p)}(W_1^2)\mathbf{W}_1,\, S_{3/2}^{(q)}(W_1^2)\mathbf{W}_1\right]_1; \quad (5.156)$$

$$\left[S_{3/2}^{(p)}(W_1^2)\mathbf{W}_1,\, S_{3/2}^{(q)}(W_2^2)\mathbf{W}_2\right]_{12}; \quad (5.157)$$

$$\left[S_{3/2}^{(p)}(W_1^2)\mathbf{W}_1,\, S_{3/2}^{(q)}(W_1^2)\mathbf{W}_1\right]_{12}. \quad (5.158)$$

Let us write the final equations for determination of the diffusion, thermal diffusion coefficients, and thermal conductivity. We obtain

$$\tilde{Q}_{ij}^{pq} = Q_{ij}^{pq} - \frac{n_j}{n_i}\sqrt{\frac{m_j}{m_i}}Q_{ii}^{pq}\delta_{p0}\delta_{q0};\, \overline{\overline{W_i}} = W_i, \quad (5.159)$$

where Q_{ij}^{pq} is defined as [10, 16]

$$Q_{ij}^{pq} = \sum_{l=1}^{v} n_i n_l\Big\{\delta_{ij}\left[S_n^{(p)}(W_i^2)\overline{\overline{W_i}};\, S_n^{(q)}(W_i^2)\overline{\overline{W_i}}\right]_{il} + \delta_{jl}\left[S_n^{(p)}(W_i^2)\overline{\overline{W_i}};\, S_n^{(q)}(W_l^2)\overline{\overline{W_l}}\right]_{il}. \quad (5.160)$$

Formulas of the bracket integrals in determining the diffusion coefficient, thermal diffusion, electrical conductivity, and thermal conductivity of the plasma are given in Appendices D and E [4, 15].

The elements of the matrix will look like

$$q_{ij}^{pq} = \sqrt{\frac{2\pi m_i}{kT}} \tilde{Q}_{ij}^{pq}. \tag{5.161}$$

We write the expression for the definition of thermal conductivity for a sixth-order approximation.

$$[\lambda]_6 = -\frac{75k}{8}\frac{(2\pi kT)^{1/2}}{|q|}\begin{vmatrix} q_{ij}^{00} & q_{ij}^{01} & q_{ij}^{02} & q_{ij}^{03} & q_{ij}^{04} & q_{ij}^{05} & 0 \\ q_{ij}^{10} & q_{ij}^{11} & q_{ij}^{12} & q_{ij}^{13} & q_{ij}^{14} & q_{ij}^{15} & n_i \\ q_{ij}^{20} & q_{ij}^{21} & q_{ij}^{22} & q_{ij}^{23} & q_{ij}^{24} & q_{ij}^{25} & 0 \\ q_{ij}^{30} & q_{ij}^{31} & q_{ij}^{32} & q_{ij}^{33} & q_{ij}^{34} & q_{ij}^{35} & 0 \\ q_{ij}^{40} & q_{ij}^{41} & q_{ij}^{42} & q_{ij}^{43} & q_{ij}^{44} & q_{ij}^{45} & 0 \\ q_{ij}^{50} & q_{ij}^{51} & q_{ij}^{52} & q_{ij}^{53} & q_{ij}^{54} & q_{ij}^{55} & 0 \\ 0 & n_j/\sqrt{m_j} & 0 & 0 & 0 & 0 & 0 \end{vmatrix}, \tag{5.162}$$

where

$$|q| = \begin{vmatrix} q_{ij}^{00} & q_{ij}^{01} & q_{ij}^{02} & q_{ij}^{03} & q_{ij}^{04} & q_{ij}^{05} \\ q_{ij}^{10} & q_{ij}^{11} & q_{ij}^{12} & q_{ij}^{13} & q_{ij}^{14} & q_{ij}^{15} \\ q_{ij}^{20} & q_{ij}^{21} & q_{ij}^{22} & q_{ij}^{23} & q_{ij}^{24} & q_{ij}^{25} \\ q_{ij}^{30} & q_{ij}^{31} & q_{ij}^{32} & q_{ij}^{33} & q_{ij}^{34} & q_{ij}^{35} \\ q_{ij}^{40} & q_{ij}^{41} & q_{ij}^{42} & q_{ij}^{43} & q_{ij}^{44} & q_{ij}^{45} \\ q_{ij}^{50} & q_{ij}^{51} & q_{ij}^{52} & q_{ij}^{53} & q_{ij}^{54} & q_{ij}^{55} \end{vmatrix}.$$

Figure 5.3 shows the results of the thermal conductivity calculation by solving the Boltzmann kinetic equation using Chapman–Enskog approach, the data of other authors as well (for comparison) were calculated using the classical model using the expression obtained in [13]. Figure 5.4 shows the results for the components of the thermal conductivity calculations using the classical model and by solving BKE by Chapman–Enskog approach. Calculation by the classical model was obtained using the equations from [13].

The total thermal conductivity is the sum of the components:

$$\lambda_{\text{total}} = \lambda_{\text{a}} + \lambda_{\text{i}} + \lambda_{\text{e}} + \lambda_{\text{R}} = \lambda_{\text{c}} + \lambda_{\text{R}}, \tag{5.163}$$

where λ_{a}, λ_{i}, λ_{e}, λ_{R}—are, respectively, atomic, ionic, and reactive components of the thermal conductivity; $\lambda_{\text{c}} = \lambda_{\text{a}} + \lambda_{\text{i}} + \lambda_{\text{e}}$—the contact thermal conductivity.

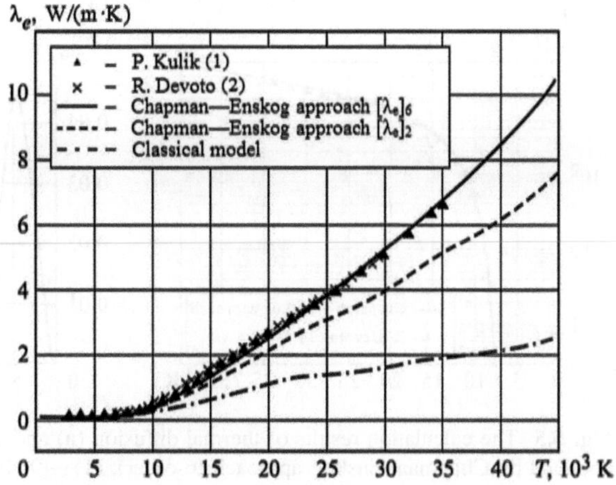

Fig. 5.3 The calculation results of the thermal conductivity for electronic in the Argon by Chapman–Enskog approach (2- and 6-order) and comparison with calculation results using the classical model and experimental and theoretical data of other authors: (1)—[11]; (2)—[9, 10]

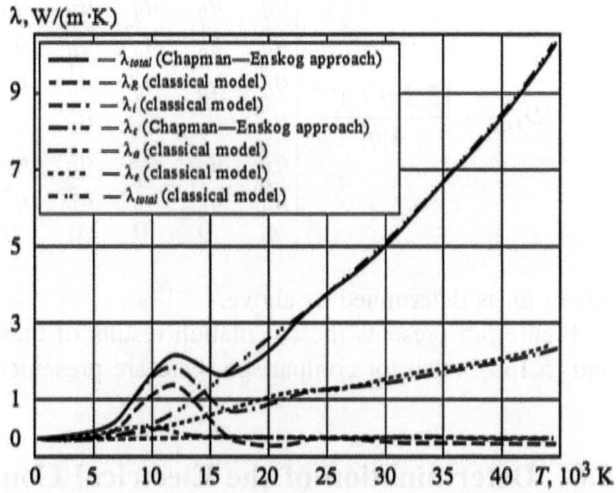

Fig. 5.4 Comparison of the thermal conductivity calculation results for argon components obtained by the classical model and by Chapman–Enskog approach (6-order)

Here is the equation for determination of diffusion and thermal diffusion coefficients of the sixth-order approximation [1, 2, 4, 9, 10, 15].

$$[D_{ij}]_6 = \frac{3\rho(2\pi kT_i)^{1/2}}{2n|q|}\begin{vmatrix} q_{ij}^{00} & q_{ij}^{01} & q_{ij}^{02} & q_{ij}^{03} & q_{ij}^{04} & q_{ij}^{05} & n_i/\sqrt{m_i} \\ q_{ij}^{10} & q_{ij}^{11} & q_{ij}^{12} & q_{ij}^{13} & q_{ij}^{14} & q_{ij}^{15} & 0 \\ q_{ij}^{20} & q_{ij}^{21} & q_{ij}^{22} & q_{ij}^{23} & q_{ij}^{24} & q_{ij}^{25} & 0 \\ q_{ij}^{30} & q_{ij}^{31} & q_{ij}^{32} & q_{ij}^{33} & q_{ij}^{34} & q_{ij}^{35} & 0 \\ q_{ij}^{40} & q_{ij}^{41} & q_{ij}^{42} & q_{ij}^{43} & q_{ij}^{44} & q_{ij}^{45} & 0 \\ q_{ij}^{50} & q_{ij}^{51} & q_{ij}^{52} & q_{ij}^{53} & q_{ij}^{54} & q_{ij}^{55} & 0 \\ 1/m_j & 0 & 0 & 0 & 0 & 0 & 0 \end{vmatrix}; \quad (5.164)$$

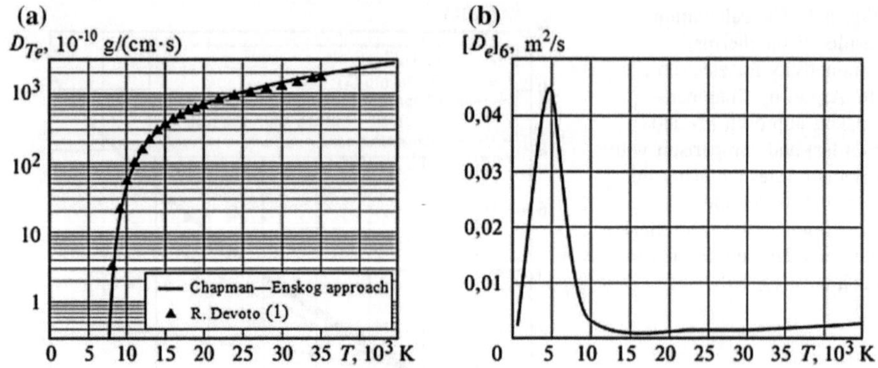

Fig. 5.5 The calculation results of thermal diffusion (**a**) and diffusion (**b**) coefficients for Argon, obtained by Chapman–Enskog approach (6-order): (1)—[9, 10]

$$[D_{Ti}]_6 = \frac{15(2\pi kT_i)^{1/2}}{4|q|}\begin{vmatrix} q_{ij}^{00} & q_{ij}^{01} & q_{ij}^{02} & q_{ij}^{03} & q_{ij}^{04} & q_{ij}^{05} & 0 \\ q_{ij}^{10} & q_{ij}^{11} & q_{ij}^{12} & q_{ij}^{13} & q_{ij}^{14} & q_{ij}^{15} & n_i\sqrt{m_i} \\ q_{ij}^{20} & q_{ij}^{21} & q_{ij}^{22} & q_{ij}^{23} & q_{ij}^{24} & q_{ij}^{25} & 0 \\ q_{ij}^{30} & q_{ij}^{31} & q_{ij}^{32} & q_{ij}^{33} & q_{ij}^{34} & q_{ij}^{35} & 0 \\ q_{ij}^{40} & q_{ij}^{41} & q_{ij}^{42} & q_{ij}^{43} & q_{ij}^{44} & q_{ij}^{45} & 0 \\ q_{ij}^{50} & q_{ij}^{51} & q_{ij}^{52} & q_{ij}^{53} & q_{ij}^{54} & q_{ij}^{55} & 0 \\ n_j & 0 & 0 & 0 & 0 & 0 & 0 \end{vmatrix}, \tag{5.165}$$

where $|q|$ is determined as above.

Figure 5.5 presents the calculation results of these coefficients by Eqs. (5.164) and (5.165). Also for comparison, here are presented the data of [9, 10].

5.6 Determination of the Electrical Conductivity

Electrical conductivity is one of the main parameters that affect the properties of the plasma. The electrical conductivity is determined by the scattering of electrons on atoms, ions, and electrons, and can be found using the diffusion coefficients [1, 2]. We write the expression for calculating the electric conductivity [9, 10] as

$$[\sigma_{ij}]_6 = \frac{3e^2\sqrt{\frac{2\pi}{kT_i}}}{2|q|}\begin{vmatrix} q_{ij}^{00} & q_{ij}^{01} & q_{ij}^{02} & q_{ij}^{03} & q_{ij}^{04} & q_{ij}^{05} & n_i/\sqrt{m_i} \\ q_{ij}^{10} & q_{ij}^{11} & q_{ij}^{12} & q_{ij}^{13} & q_{ij}^{14} & q_{ij}^{15} & 0 \\ q_{ij}^{20} & q_{ij}^{21} & q_{ij}^{22} & q_{ij}^{23} & q_{ij}^{24} & q_{ij}^{25} & 0 \\ q_{ij}^{30} & q_{ij}^{31} & q_{ij}^{32} & q_{ij}^{33} & q_{ij}^{34} & q_{ij}^{35} & 0 \\ q_{ij}^{40} & q_{ij}^{41} & q_{ij}^{42} & q_{ij}^{43} & q_{ij}^{44} & q_{ij}^{45} & 0 \\ q_{ij}^{50} & q_{ij}^{51} & q_{ij}^{52} & q_{ij}^{53} & q_{ij}^{54} & q_{ij}^{55} & 0 \\ n_j & 0 & 0 & 0 & 0 & 0 & 0 \end{vmatrix}, \tag{5.166}$$

where $|q|$ is determined as above.

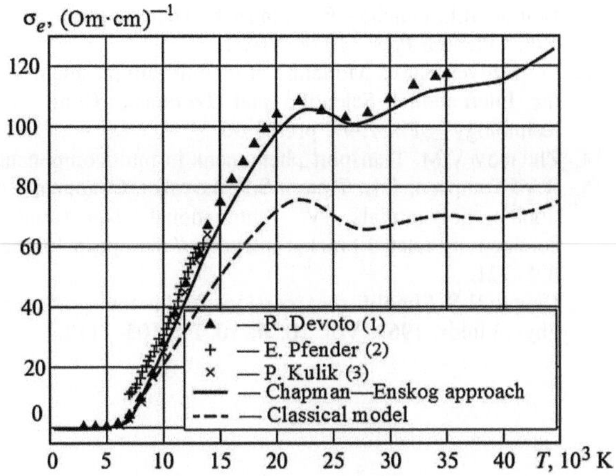

Fig. 5.6 The calculation results of the electrical conductivity of argon, obtained by Chapman–Enskog approach (6-order) and the classical model, and compare with those of other authors: (1)—[9, 10]; (2)—[12]; (3)—[11]

Matrix elements are calculated in the same way as for the coefficient of diffusion, thermal diffusion, and thermal conductivity (5.159–5.162). For comparison, calculation was performed using equations of classical model [13, 14]. Figure 5.6 shows the results of calculation of the electrical conductivity of the Argon plasma by Chapman–Enskog approach and their comparison with data of other authors.

References

1. Chapman S., Cowling T.G. The Mathematical Theory of Non-Uniform Gases. Cambridge: Cambridge University Press, 1970.
2. J. O. Hirschfelder, Ch. F. Curtiss, R. B. Bird. Molecular theory of gases and liquids / John Wiley and Sons, New York. 2nd Edition: 1964.
3. Ferzinger J.H., Kaper H.G. Mathematical Theory of Transport Processes in Gases, North-Holland. Amsterdam, 1972.
4. Loyalka S.K., Tipton E.L., Tompson R.V. Chapman-Enskog solutions to arbitrary order in Sonine polynomials I: Simple, rigid-sphere gas // Phys. Rev. A. 2007. Vol. 379. PP. 417–435.
5. Transport properties in a two temperature plasma: theory and application / Rat V. et al. // Phys. Rev. E. 2001. Vol. 64.
6. Rat V., Andre P., Aubreton J. et al. // Phys. D: Applied Phys. 2001. Vol. 34. P. 2191–2204.
7. Transport coefficients including diffusion in a two-temperature argon plasma / V. Rat, et al. // J. Phys. D: Appl. Phys. 2002. Vol. 35. P. 981–991.
8. R.V. Tompson, E.L. Tipton, S.K. Loyalka. Chapman–Enskog solutions to arbitrary order in Sonine polynomials V: Summational expressions for the viscosity-related bracket integrals // European Journal of Mechanics B/Fluids 29 (2010) 153–179.
9. Devoto R.S. Transport coefficients of ionized argon // Phys. Fluids. 1973. Vol. 16. № 5. P. 616–623.
10. Devoto R.S. Transport coefficients of partially ionized argon // Phys. Fluids. 1967. Vol. 10. № 2. PP. 354–364.
11. Kulik P.P. Essays on physics and chemistry of Low-temperature plasma / Ed. by L.S. Polak. M.: Nauka, 1971.

12. Boulos M.I., Fauchais P., Pfender E. Thermal plasmas, Fundamentals and Applications. New York: Plenum Press, 1994. Vol. 1.
13. Shi Nguyen-Kuok. Modeling of equilibrium plasma in RF and Arc plasma torches // Proc. of the International Scientific and Technical. Conf. "Electrophysical and electrochemical technology" SPb. 1997. pp. 63-66.
14. Zhdanov V.M. Transport phenomena in multicomponent plasma. M.: Energoizdat 1982.
15. R.V. Tompson, E.L. Tipton, S.K. Loyalka. Chapman–Enskog solutions to arbitrary order in Sonine polynomials IV: Summational expressions for the diffusion- and thermal conductivity-related bracket integrals // European Journal of Mechanics B/Fluids 28 (2009), 695–721.
16. Devoto R.S. Simplified expressions for the transport properties of ionized monatomic gases // Phys. Fluids. 1967. Vol. 10. № 10. P. 2105–2112.

Chapter 6
Numerical Methods of the Plasma Physics

6.1 Basis of the Numerical Methods of the Plasma Physics

6.1.1 Solution of the Generalized Differential Equation

Problems of mathematical simulation of plasma processes are the most difficult tasks among of the plasma research problems. Numerous physical processes such as the exchange of energy and momentum, transformation of particles, ionization and recombination, dissociation of molecules, radiative processes, chemical reactions of particles activation and deactivation, and others occur simultaneously in plasma. All of them usually happen with the active interaction of plasma with an external electromagnetic field.

Numerical methods are the main tools for solving the problems of mathematical simulation. Effective using of them is possible only in the case of the best-fit of mathematical models and physical processes occurring in the plasma.

The model description of plasma and plasma processes (Chap. 1) is often expressed using nonlinear differential equations in partial derivatives. These equations have a similar structure, and they can be written in the same form of a Generalized Differential Equations (GDE). For low-temperature plasma the stationary case is of the greatest interest, when the flow of plasma maintains its state for a period of time greater than its relaxation time. In this case, the GDE has the form

$$\mathrm{div}(\rho\vec{v}\Phi) = \mathrm{div}(\Gamma \,\mathrm{grad}\,\Phi) + S, \tag{6.1}$$

where Φ—a generalized variable; Γ—coefficients that is included in the flow $\Gamma\,\mathrm{grad}\,\Phi$—that can be called as the diffusion coefficient; S—some source function.

The GDE consists of three groups of members: $\mathrm{div}(\rho\vec{v}\Phi)$—convection; $\mathrm{div}(\Gamma\,\mathrm{grad}\,\Phi)$—diffusion; S—the source function. If there are other members in the

S. Nguyen-Kuok, *Theory of Low-Temperature Plasma Physics*,
Springer Series on Atomic, Optical, and Plasma Physics 95,
DOI 10.1007/978-3-319-43721-7_6

differential equation that cannot be expressed in this form, they can be combined into a common of the source function.

Most designs of the plasma torches channels have a symmetrical cylindrical shape and it is convenient to solve GDE (6.1) using the two-dimensional cylindrical coordinate system and this equation can be written in a form

$$\frac{\partial(\rho v_z \Phi)}{\partial z} + \frac{1}{r}\frac{\partial}{\partial r}(r\rho v_r \Phi) = \frac{\partial}{\partial z}\left(\Gamma\frac{\partial\Phi}{\partial z}\right) + \frac{1}{r}\frac{\partial}{\partial r}\left(r\gamma\frac{\partial\Phi}{\partial r}\right) + S. \qquad (6.2)$$

Presentation the equations of the plasma model in the form of GDE (6.1) or (6.2) allows the development of general mathematical and software algorithms for solving this equation, as well as the design of the computational procedures for modeling of plasma.

The Analytical Solution of One-Dimensional GDE Let us analyze the simplest solution to the GDE for example on one-dimensional problem of steady processes of the cold gas purging and heating in a half-space of heated plasma (Fig. 6.1).

Using the model of equilibrium plasma (Sect. 1.2) one-dimensional energy balance equation describing the heating of the gas has the form

$$\frac{\mathrm{d}(\rho v_z T)}{\mathrm{d}z} = \frac{\mathrm{d}}{\mathrm{d}z}\left(\frac{\lambda}{c_p}\frac{dT}{\mathrm{d}z}\right) + \frac{1}{c_p}\left(\sigma E^2 - U_{\mathrm{rad}}\right). \qquad (6.3)$$

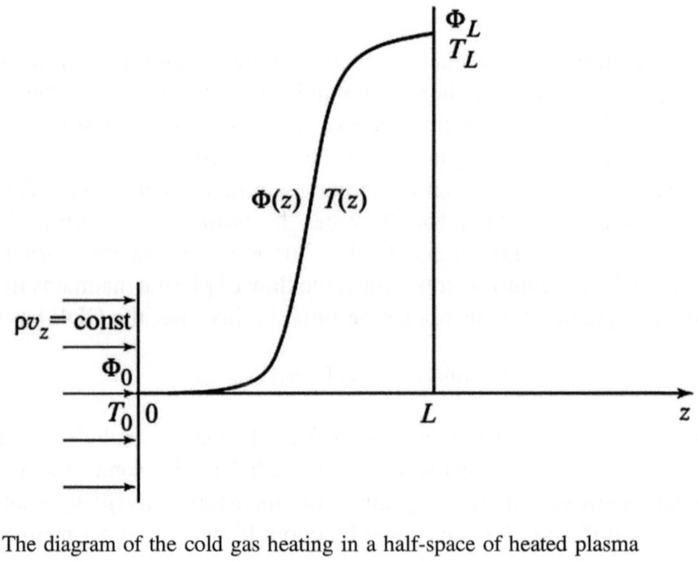

Fig. 6.1 The diagram of the cold gas heating in a half-space of heated plasma

The one-dimensional GDE with one-dimensional gas flow takes the form

$$\frac{\mathrm{d}(\rho v_z \Phi)}{\mathrm{d}z} = \frac{\mathrm{d}}{\mathrm{d}z}\left(\Gamma \frac{\mathrm{d}\Phi}{\mathrm{d}z}\right) + S. \tag{6.4}$$

The one-dimensional continuity equation of the gas is written as

$$\frac{\mathrm{d}}{\mathrm{d}z}(\rho v_z) = 0 \,\text{or}\, \rho v_z = \text{const.} \tag{6.5}$$

The exact analytical solutions of one-dimensional energy balance equation and GDE may be obtained only in the case where λ/c_p = const and Γ = const, and with a relatively simple form of the source function. For a simple analysis of the solution (6.3) and (6.4) we assume that the source function is equal to zero. Divide the region of gas heating into plurality sub-domains with the length L (Fig. 6.1). Because of the smallness of each subregion it can be assumed that the factors λ/c_p = const and Γ = const. We assume that the unknown variables at the beginning and in the end of each subregion with length L will be T_0, T_L and Φ_0, Φ_L respectively (Fig. 6.1), and then changes $T(z)$ and $\Phi(z)$ in each subregion with such boundary condition can be obtained from the analytical solutions

$$\left.\begin{aligned} T(z) &= T_0 + (T_L - T_0)\frac{\exp(Pe\Delta z/L) - 1}{\exp Pe - 1}, (0 \le \Delta z \le L); \\ \frac{\mathrm{d}^n T}{\mathrm{d}z^n} &= \frac{T_L - T_0}{\exp Pe - 1}\left(\frac{Pe}{L}\right)^n \exp\left(\frac{Pe\Delta z}{L}\right); \end{aligned}\right\} \tag{6.6}$$

$$\left.\begin{aligned} \Phi(z) &= \Phi_0 + (\Phi_L - \Phi_0)\frac{\exp(Pe\Delta z/L) - 1}{\exp Pe - 1}; \\ \frac{\mathrm{d}^n \Phi}{\mathrm{d}z^n} &= \frac{\Phi_L - \Phi_0}{\exp Pe - 1}\left(\frac{Pe}{L}\right)^n \exp\left(\frac{Pe\Delta z}{L}\right), \end{aligned}\right\} \tag{6.7}$$

where Pe—the grid Peclet number, defined as the ratio of the intensities of the convective F and diffusive D fluxes: $Pe = \frac{F}{D} = \frac{\rho v_z}{\Gamma/L}$. Dependencies $\Phi(z)$ for different Pe at constant ratios are shown in Fig. 6.2.

The resulting one-dimensional analytical solution of the GDE with constant coefficients Γ in the form (6.7) can then be used to describe the change in the target variable between the nodes of the computational grid.

The method of control volume is known as a modern method that combines the advantages of two well-known techniques that are commonly used in solving differential equations in partial derivatives: method of the finite elements and method of the finite difference. As in method of the finite elements, representation of the target variable between the nodes of an elementary volume in the form of a polynomial is used at method of the control volume. However, unlike the method of finite element, this functional relationship is not arbitrary, and it is obtained by

Fig. 6.2 Dependencies $\Phi(z)$ for different values of the grid Peclet number *Pe*

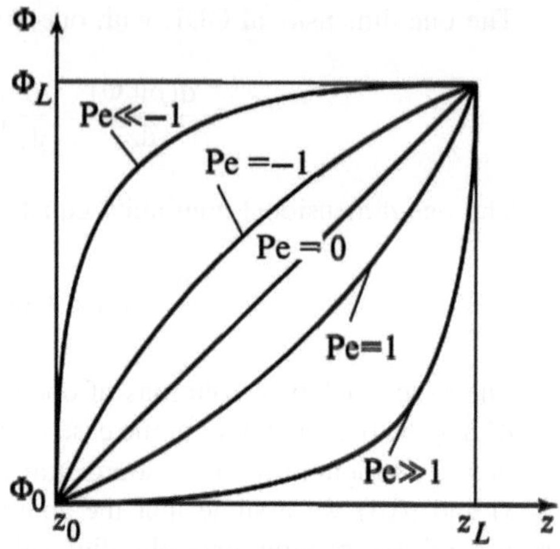

solving a physical problem, such as (6.7). At the same time, the functional relationship is used to determine the appropriate flow through the faces of the elementary volume. As a result, we obtain a finite-difference equation in the form of discrete analogue for the unknown variables in the nodes, as it was done in all methods of the finite difference.

To use the result of a one-dimensional analytical solutions of GDE in the form (6.7), we apply the certain grid to computational domain (as in the method of finite differences), dividing the study area into a plurality of disjoint sub-domains so that each sub-domain contains only one estimated point. These sub-domains can be called as the control volume. Thereafter, for each control volume it is necessary to make a balance of physical flows by integrating the corresponding equation for the control volume. The result of this operation will be the discrete analogue that expresses the relevant law of conservation for this control volume at the same way as a differential equation expresses the law of conservation for the infinitesimal volume.

Thus, an important feature of the method of control volume is that it contains the foundations of exact integral conservation of such quantities as mass, momentum, and energy in any group of the control volume and therefore everywhere in the studied area, that is very important in numerical simulations of plasma processes. This property can be manifested using any quantity of the grid points, but not only at a very large number of them. The solution of equation using an approximate grid may correspond to the exact integral balances.

The Numerical Solution of the One-Dimensional GDE Consider the one-dimensional GDE and continuity equation in the form (6.4) and (6.5). One-dimensional control volume and the location of the grid points are shown in Fig. 6.3.

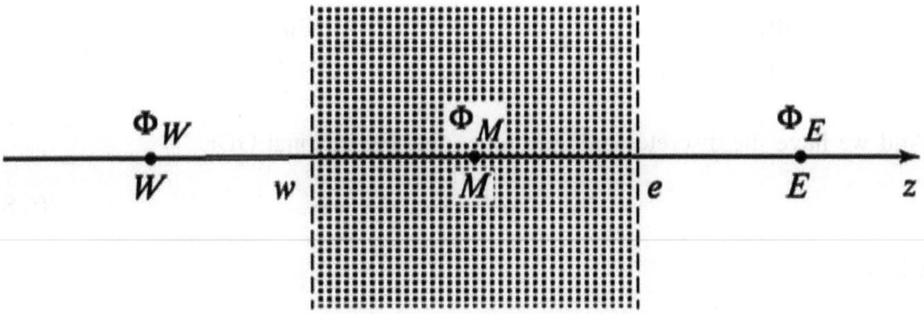

Fig. 6.3 The control volume and grid points for the numerical solution of one-dimensional GDE

The differential equation in the form (6.4) expresses the corresponding law of conservation in the infinitesimal volume, and to obtain an appropriate balance in the finite control volume (Fig. 6.3) it is necessary to integrate it across all the control volume:

$$\int_{z_w}^{z_e} \frac{\mathrm{d}(\rho v_z \Phi)}{\mathrm{d}z}\mathrm{d}z = \int_{z_w}^{z_e} \frac{\mathrm{d}}{\mathrm{d}z}\left(\Gamma \frac{\mathrm{d}\Phi}{\mathrm{d}z}\right)\mathrm{d}z + \int_{z_w}^{z_e} S\mathrm{d}z;$$

$$(\rho v_z)_e \Phi_e - (\rho v_z)_w \Phi_w = \Gamma_e\left(\frac{\mathrm{d}\Phi}{\mathrm{d}z}\right)_e - \Gamma_w\left(\frac{\mathrm{d}\Phi}{\mathrm{d}z}\right)_w + S_M(z_e - z_w),$$

Here the indices e, w correspond to the values on the respective faces of the control volume; it is assumed that the source function S in the control volume takes an average value at the point M.

Substituting the analytical solution (6.7) in the last equation we get

$$(\rho v_z)_e\left[\Phi_M + (\Phi_E - \Phi_M)\frac{\exp\left(\frac{Pe_e(z_e - z_M)}{z_E - z_M}\right) - 1}{\exp Pe_e - 1}\right]$$
$$-\frac{\Gamma_e(\Phi_E - \Phi_M)}{z_E - z_M}\exp\left(\frac{Pe_e(z_e - z_M)}{z_E - z_M}\right)\frac{Pe_e}{\exp(Pe_e) - 1}$$
$$-(\rho v_z)_w\left[\Phi_W + (\Phi_M - \Phi_W)\frac{\exp\left(\frac{Pe_w(z_w - z_W)}{z_M - z_W}\right) - 1}{\exp Pe_w - 1}\right]$$
$$+\frac{\Gamma_w(\Phi_M - \Phi_W)}{z_M - z_W}\exp\left(\frac{Pe_w(z_w - z_W)}{z_M - z_W}\right)\frac{Pe_w}{\exp Pe_w - 1} = S_M(z_e - z_w).$$

Designating $F_e \equiv (\rho v_z)_e, F_w = (\rho v_z)_w, D_e = \frac{\Gamma_e}{z_E - z_M}, D_w = \frac{\Gamma_w}{z_M - z_W}, Pe_e = \frac{F_e}{D_e}, Pe_w = \frac{F_w}{D_w}$, Last equation can be written as

$$F_e\left[\Phi_M + \frac{\Phi_M - \Phi_E}{\exp Pe_e - 1}\right] - F_w\left[\Phi_M - (\Phi_M - \Phi_W) - \frac{\Phi_M - \Phi_W}{\exp Pe_w - 1}\right] = S_M(z_e - z_w).$$

And we have the discrete analogue for a one-dimensional GDE

$$a_M\Phi_M = a_E\Phi_E + a_W\Phi_W + b_M, \tag{6.8}$$

where

$$\begin{cases} a_E = D_e A(Pe_e); & a_W = D_w A(Pe_w) + F_w; \\ a_M = a_E + a_W; & b_M = S_M(z_e - z_w). \end{cases}$$

The function $A(Pe) = \frac{Pe}{\exp Pe - 1}$ features the approximating method for the necessary variable in the control volume using this exponential computational scheme.

In the derivation of discrete analogue the continuity equation is used as $F_e = F_w = \rho v_z$ = const. As a result of the approximate solution using an iterative procedure, this condition can be fulfilled only in the final decision. However, using this condition makes it possible to obtain the discrete analogue (6.8) with positive coefficients satisfying the criterion of stability (6.49). In this case, the coefficients $D_e > 0, D_w > 0, F_e = F_w > 0$ and the requirement of the positivity of factors in discrete analogue [the stability criterion (6.49) will be considered the bottom] makes it necessary to maintain the function $A(Pe)$ as positive.

The Numerical Solution of the Two-Dimensional GDE Consider the algorithm for solving the two-dimensional GDE (6.2) using the method of control volume. A typical two-dimensional control volume for the solution of the GDE (6.2) and the location of the grid points of the grid are shown in Fig. 6.4.

As it was noted, the GDE (6.2) expresses the corresponding law of conservation at the infinitesimal volume, and to obtain an appropriate balance in the finite control volume it should be integrated through all the control volume (Fig. 6.4)

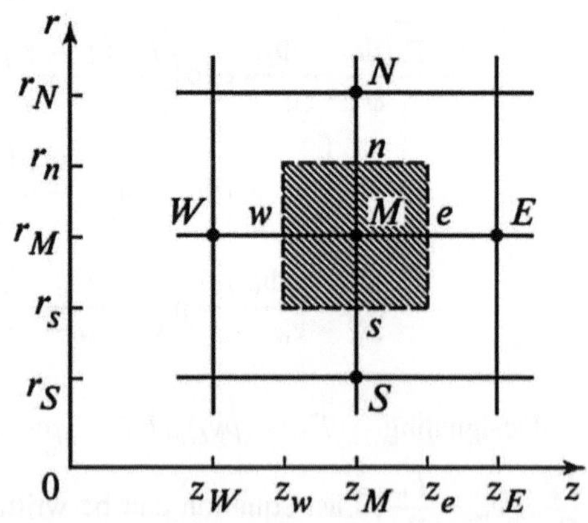

Fig. 6.4 Control volume and the grid points for solving the two-dimensional GDE

$$\int_{r_s}^{r_n}\int_{z_w}^{z_e}\frac{\partial(\rho v_z\Phi)}{\partial z}\mathrm{d}z r\mathrm{d}r+\int_{z_w}^{z_e}\int_{r_s}^{r_n}\frac{1}{r}\frac{\partial}{\partial r}(r\rho v_r\Phi)r\mathrm{d}r\mathrm{d}z-\int_{r_s}^{r_n}\int_{z_w}^{z_e}\frac{\partial}{\partial z}\left(\Gamma\frac{\partial\Phi}{\partial z}\right)\mathrm{d}z r\mathrm{d}r$$
$$-\int_{z_w}^{z_e}\int_{r_s}^{r_n}\frac{1}{r}\frac{\partial}{\partial r}\left(r\Gamma\frac{\partial\Phi}{\partial r}\right)r\mathrm{d}r\mathrm{d}z=\int_{z_w}^{z_e}\int_{r_s}^{r_n}Sr\mathrm{d}r\mathrm{d}z$$

The elementary control volume $\mathrm{d}V=\mathrm{d}r\mathrm{d}z r\mathrm{d}\varphi$; $\mathrm{d}\varphi$—the single rotation angle. In the cylindrical axial symmetry the variable Φ is independent of the angle φ, so the integration in this elementary volume $\mathrm{d}V$ both sides of equation reduces to $\int \mathrm{d}\varphi$. The resulting integral equation can be divided into three groups

– for the convective fluxes

$$I_{c1}=\int_{r_s}^{r_n}\int_{z_w}^{z_e}\frac{\partial}{\partial z}(\rho v_z\Phi)\mathrm{d}z r\mathrm{d}r;\quad I_{c2}=\int_{z_w}^{z_e}\int_{r_s}^{r_n}\frac{1}{r}\frac{\partial}{\partial r}(r\rho v_r\Phi)r\mathrm{d}r\mathrm{d}z;$$

– for the diffusion fluxes

$$I_{d1}=-\int_{r_s}^{r_n}\int_{z_w}^{z_e}\frac{\partial}{\partial z}\left(\Gamma\frac{\partial\Phi}{\partial z}\right)\mathrm{d}z r\mathrm{d}r;\quad I_{d2}=-\int_{z_w}^{z_e}\int_{r_s}^{r_n}\frac{1}{r}\frac{\partial}{\partial r}\left(r\Gamma\frac{\partial\Phi}{\partial r}\right)r\mathrm{d}r\mathrm{d}z;$$

– for the source function

$$I_S=\int_{z_w}^{z_e}\int_{r_s}^{r_n}Sr\mathrm{d}r\mathrm{d}z.$$

Further integration is carried out separately for convective and diffusive fluxes through the corresponding verges of the control volume *e, w, n, s*. Note that the differential of the same order is situated under the first integral of the equations for convective and diffusive fluxes, so these integrals can only be exactly once integrated. We obtain four subgroups of convective and diffusive fluxes through the verges of the control volume *e, w, n, s* in the form:

$$I_{ce} = \int_{r_s}^{r_n} (\rho v_z)_e \Phi_e r dr; \quad I_{cw} = -\int_{r_s}^{r_n} (\rho v_z)_w \Phi_w r dr;$$

$$I_{cn} = \int_{z_w}^{z_e} (r\rho v_r)_n \Phi_n dz; \quad I_{cs} = -\int_{z_w}^{z_e} (r\rho v_r)_s \Phi_s dz;$$

$$I_{de} = -\int_{r_s}^{r_n} \left(\Gamma \frac{\partial \Phi}{\partial z}\right)_e r dr; \quad I_{dw} = \int_{r_s}^{r_n} \left(\Gamma \frac{\partial \Phi}{\partial z}\right)_w r dr;$$

$$I_{dn} = -\int_{z_w}^{z_e} \left(r\Gamma \frac{\partial \Phi}{\partial r}\right)_n dz; \quad I_{ds} = \int_{z_w}^{z_e} \left(r\Gamma \frac{\partial \Phi}{\partial r}\right)_s dz.$$

Integrands with subscripts *e, w, n, s* should be computed along each corresponding verge of the control volume. In most of plasma processes associated with convective and diffusive fluxes the integrands are smooth. And for each of them at its range of integration the average value can be found when the integral is equal to the product of the average value and the range of integration. The average value of the integrand along each side *e, w, n, s* is accepted as an approximation. For convenience, we use the same notations *e, w, n, s* for the sides of the control volume and intersection points of the grid lines (Fig. 6.4). The recent subgroup integrals thus can be written as

$$\begin{aligned} I_{ce} &= (\rho v_z)_e \Phi_e \tfrac{(r_n - r_s)(r_n + r_s)}{2}; & I_{cw} &= -(\rho v_z)_w \Phi_w \tfrac{(r_n - r_s)(r_n + r_s)}{2}; \\ I_{cn} &= (\rho v_r)_n r_n \Phi_n (z_e - z_w); & I_{cs} &= -(\rho v_r)_s r_s \Phi_s (z_e - z_w); \\ I_{de} &= -\Gamma_e \left(\tfrac{\partial \Phi}{\partial z}\right)_e \tfrac{(r_n - r_s)(r_n + r_s)}{2}; & I_{dw} &= \Gamma_w \left(\tfrac{\partial \Phi}{\partial z}\right)_w \tfrac{(r_n - r_s)(r_n + r_s)}{2}; \\ I_{dn} &= -\Gamma_n \left(\tfrac{\partial \Phi}{\partial r}\right)_n r_n (z_e - z_w); & I_{ds} &= \Gamma_s \left(\tfrac{\partial \Phi}{\partial r}\right)_s r_s (z_e - z_w). \end{aligned}$$

Note:

$$\left. \begin{aligned} F_e &= (\rho v_z)_e \tfrac{(r_n - r_s)(r_n + r_s)}{2}; & F_w &= (\rho v_z)_w \tfrac{(r_n - r_s)(r_n + r_s)}{2}; \\ F_n &= (\rho v_r)_n r_n (z_e - z_w); & F_s &= (\rho v_r)_s r_s (z_e - z_w). \end{aligned} \right\} \tag{6.9}$$

$$\begin{aligned} D_e &= \tfrac{\Gamma_e}{z_E - z_M} \tfrac{(r_n - r_s)(r_n + r_s)}{2}; & D_w &= \tfrac{\Gamma_w}{z_M - z_W} \tfrac{(r_n - r_s)(r_n + r_s)}{2}; \\ D_n &= \tfrac{\Gamma_n}{r_N - r_M} r_n (z_e - z_w); & D_s &= \tfrac{\Gamma_s}{r_M - r_S} r_s (z_e - z_w). \end{aligned} \tag{6.10}$$

$$Pe_e = \frac{F_e}{D_e}; \quad Pe_w = \frac{F_w}{D_w}; \quad Pe_n = \frac{F_n}{D_n}; \quad Pe_s = \frac{F_s}{D_s}, \tag{6.11}$$

Here F_e, F_w, F_n, F_s—the mass flow through the corresponding verges e, w, n, s of the control volume (intensities of the convective flux); D_e, D_w, D_n, D_s—intensities of the diffusion flux through the same verge (always positive); Pe_e, Pe_w, Pe_n, Pe_s—the grid Peclet numbers at the verges of control volume e, w, n, s, respectively.

Then it can be written

$$I_{ce} = F_e\Phi_e; \quad I_{cw} = -F_w\Phi_w; \quad I_{cn} = F_n\Phi_n; \quad I_{cs} = -F_s\Phi_s;$$
$$I_{de} = -D_e(z_E - z_M)\left(\frac{\partial\Phi}{\partial z}\right)_e; \quad I_{dw} = D_w(z_M - z_W)\left(\frac{\partial\Phi}{\partial z}\right)_w;$$
$$I_{dn} = -D_n(r_N - r_M)\left(\frac{\partial\Phi}{\partial r}\right)_n; \quad I_{ds} = D_s(r_M - r_S)\left(\frac{\partial\Phi}{\partial r}\right)_s.$$

The further compilation of balance fluxes through the control volume is produced by adding the appropriate convective and diffusive fluxes through every verge of the control volume. The assumption of the constancy of some variables on the verges of the control volume provides an opportunity to apply the result of a one-dimensional analytical solution of GDE (6.7) for the definitions of these fluxes. We introduce a function of the grid Peclet number Pe that characterizes the calculation circuit [1]

$$A(Pe) \equiv \frac{Pe}{\exp Pe - 1} = A(|Pe|) + \max(-Pe, 0). \tag{6.12}$$

Then the amount of fluxes through the verges of control volume will be

$$\begin{cases} I_{ce} + I_{de} \equiv F_e\Phi_M + (\Phi_M - \Phi_E)D_eA(Pe_e); \\ I_{cw} + I_{dw} \equiv -F_w\Phi_M + (\Phi_M - \Phi_W)[D_wA(Pe_w) + F_w]; \\ I_{cn} + I_{dn} \equiv F_n\Phi_M + (\Phi_M - \Phi_N)D_nA(Pe_n); \\ I_{cs} + I_{ds} \equiv -F_s\Phi_M + (\Phi_M - \Phi_S)[D_sA(Pe_s) + F_s]. \end{cases} \tag{6.13}$$

The source function is constant inside the control volume. It takes the value $S = \text{const} = S_M$ in the central grid M, i.e.,

$$I_S = \int_{z_w}^{z_e}\int_{r_s}^{r_n} Srdrdz = S_M\Delta V, \tag{6.14}$$

where $\Delta V = \frac{(r_n - r_s)(r_n + r_s)}{2}(z_e - z_w)$—the volume of the two-dimensional control volume.

The corresponding balance for the control volume is the sum of all incoming and outgoing fluxes through the verges of control volume (6.13) and the source function (6.14)

$$I_{ce} + I_{de} + I_{cw} + I_{dw} + I_{cn} + I_{dn} + I_{cs} + I_{ds} = I_S$$

Substituting (6.13) and (6.14) into the last balance equation, we have

$$\begin{aligned}&(\Phi_M-\Phi_E)D_eA(P_e)+(\Phi_M-\Phi_W)(D_wA(P_w)+F_w)\\&+(\Phi_M-\Phi_N)D_nA(P_n)+(\Phi_M-\Phi_S)(D_sA(P_s)+F_s)\\&+(F_e-F_w+F_n-F_s)\Phi_M=S_M\Delta V.\end{aligned}\tag{6.15}$$

On the basis of (6.15), some discrete analogue of GDE (6.2) can already be obtained. However it does not guarantee the performance of the stability criterion (6.49) and this fact considerably restricts its application. To display the discrete analogue of GDE the continuity equation and the condition of the fulfillment of the law of mass conservation can be used inside the control volume.

The two-dimensional continuity equation in cylindrical coordinates has the form

$$\frac{\partial}{\partial z}(\rho v_z)+\frac{1}{r}\frac{\partial}{\partial r}(r\rho v_r)=0.$$

The mass balance in the continuity equation can be fulfilled using the integration inside the control volume (Fig. 6.4).

Analogically to the procedure of integration of GDE (6.2) we have

$$\begin{aligned}&\int_{r_s}^{r_n}\int_{z_w}^{z_e}\frac{\partial}{\partial z}(\rho v_z)\mathrm{d}z r\mathrm{d}r+\int_{z_w}^{z_e}\int_{r_s}^{r_n}\frac{1}{r}\frac{\partial}{\partial r}(r\rho v_r)r\mathrm{d}r\mathrm{d}z\\&=\int_{r_s}^{r_n}\left[(\rho v_z)_e-(\rho v_z)_w\right]r\mathrm{d}r+\int_{z_w}^{z_e}\left[(r\rho v_r)_n-(r\rho v_r)_s\right]\mathrm{d}z=0.\end{aligned}$$

Using the assumption of the average value of the integrands and of the constancy of the fluxes through the verges of control volume *e*, *w*, *n*, *s* we obtain

$$\begin{aligned}&(\rho v_z)_e\frac{(r_n-r_s)(r_n+r_s)}{2}-(\rho v_z)_w\frac{(r_n-r_s)(r_n+r_s)}{2}\\&+(\rho v_r)_n(z_e-z_w)r_n-(\rho v_r)_s(z_e-z_w)r_s=0,\end{aligned}$$

or taking into account (6.9) we have

$$F_e-F_w+F_n-F_s=0.\tag{6.16}$$

Equation (6.16) is the continuity equation in an integral form for this control volume. Taking into account (6.16) the discrete analogue of GDE (6.2) can be obtained from the balance Eq. (6.15), it has the following form:

$$a_M\Phi_M=a_E\Phi_E+a_W\Phi_W+a_N\Phi_N+a_S\Phi_S+b_M,\tag{6.17}$$

where

$$\begin{cases} a_E = D_e A(|Pe_e|) + \max(-F_e, 0) \geq 0; & a_W = D_w A(|Pe_w|) + \max(F_w, 0) \geq 0; \\ a_N = D_n A(|Pe_n|) + \max(-F_n, 0) \geq 0; & a_S = D_s A(|Pe_s|) + \max(F_s, 0) \geq 0; \\ a_M = a_E + a_W + a_N + a_S > 0; & b_M = S_M \Delta V. \end{cases}$$

Here the function $A(Pe)$ (6.12) was used, and the corresponding computational scheme is called exponential and it is based on the analytical solution (6.7). Calculation of the exponential function in (6.12) is time-consuming, so the other scheme based on the approximation of the exponential function in the form [1] can be used

$$A(Pe) = \frac{Pe}{\exp Pe - 1} \approx \max[0; (1 - 0.1\, Pe)^5]. \tag{6.18}$$

It should be noted that the implementation of the law of mass conservation in an integrated form (6.16) makes it possible to guarantee the positivity of the coefficients of the discrete analogue in (6.17), i.e., $a_E, a_W, a_N, a_S, a_M \geq 0$ and $\frac{a_M}{a_E + a_W + a_N + a_S} = 1$ that is necessary for compliance of the stability criterion (6.49).

The obtained discrete analogue of GDE (6.17) expresses the appropriate balance for the finite control volume exactly the same as the original differential equation for the infinitesimal volume.

The Numerical Solution of the Three-Dimensional GDE Generalized differential Eq. (6.1) in a three-dimensional cylindrical coordinate system is

$$\frac{\partial(\rho v_z \Phi)}{\partial z} + \frac{1}{r}\frac{\partial}{\partial r}(r\rho v_r \Phi) + \frac{1}{r}\frac{\partial(\rho v_\varphi \Phi)}{\partial \varphi} \\ = \frac{\partial}{\partial z}\left(\Gamma \frac{\partial \Phi}{\partial z}\right) + \frac{1}{r}\frac{\partial}{\partial r}\left(r\Gamma \frac{\partial \Phi}{\partial r}\right) + \frac{1}{r}\frac{\partial}{\partial \varphi}\left(\frac{\Gamma}{r}\frac{\partial \Phi}{\partial \varphi}\right) + S. \tag{6.19}$$

The corresponding balance for the control volume is obtained by integrating (6.19) in the three-dimensional control volumes (Fig. 6.4 with additional grid points T, B and verges t and b in the direction φ).

$$\int_{\varphi_b}^{\varphi_t}\int_{r_s}^{r_n}\int_{z_w}^{z_e} \frac{\partial(\rho v_z \Phi)}{\partial z} dzrdrd\varphi - \int_{\varphi_b}^{\varphi_t}\int_{r_s}^{r_n}\int_{z_w}^{z_e} \frac{\partial}{\partial z}\left(\Gamma \frac{\partial \Phi}{\partial z}\right) dzrdrd\varphi \\ + \int_{z_w}^{z_e}\int_{\varphi_b}^{\varphi_t}\int_{r_s}^{r_n} \frac{1}{r}\frac{\partial}{\partial r}(r\rho v_r \Phi) rdrd\varphi dz - \int_{z_w}^{z_e}\int_{\varphi_b}^{\varphi_t}\int_{r_s}^{r_n} \frac{1}{r}\frac{\partial}{\partial r}\left(r\Gamma \frac{\partial \Phi}{\partial r}\right) rdrd\varphi dz$$

$$+\int_{r_s}^{r_n}\int_{z_w}^{z_e}\int_{\varphi_b}^{\varphi_t}\frac{1}{r}\frac{\partial(\rho v_\varphi\Phi)}{\partial\varphi}rd\varphi dzdr-\int_{r_s}^{r_n}\int_{z_w}^{z_e}\int_{\varphi_b}^{\varphi_t}\frac{1}{r}\frac{\partial}{\partial\varphi}\left(\frac{\Gamma}{r}\frac{\partial\Phi}{\partial\varphi}\right)rd\varphi dzdr$$
$$=\int_{\varphi_b}^{\varphi_t}\int_{z_w}^{z_e}\int_{r_s}^{r_n}Srdrdzd\varphi.$$

The integration procedure (6.19) is similar to the process of integration of two-dimensional GDE, and the discrete analogue is

$$a_M\Phi_M=a_E\Phi_E+a_W\Phi_W+a_N\Phi_N+a_S\Phi_S+a_T\Phi_T+a_B\Phi_B+b_M, \tag{6.20}$$

where

$$\begin{cases} a_E=D_eA(|Pe_e|)+\max(-F_e,0)\geq 0; \quad a_W=D_wA(|Pe_w|)+\max(F_w,0)\geq 0;\\ a_N=D_nA(|Pe_n|)+\max(-F_n,0)\geq 0; \quad a_S=D_sA(|Pe_s|)+\max(F_s,0)\geq 0;\\ a_T=D_tA(|Pe_t|)+\max(-F_t,0)\geq 0; \quad a_B=D_bA(|Pe_b|)+\max(F_b,0)\geq 0;\\ a_M=a_E+a_W+a_N+a_S+a_T+a_B>0; \quad b_M=S_M\Delta V;\\ \Delta V=\frac{(r_n-r_s)(r_n+r_s)}{2}(z_e-z_w)(\varphi_t-\varphi_b). \end{cases}$$

Fluxes through the verges of the control volume are determined using the equations

$$\left.\begin{aligned} F_e&=(\rho v_z)_e\tfrac{(r_n-r_s)(r_n+r_s)}{2}(\varphi_t-\varphi_b);\\ F_w&=(\rho v_z)_w\tfrac{(r_n-r_s)(r_n+r_s)}{2}(\varphi_t-\varphi_b);\\ F_n&=(\rho v_r)_n r_n(\varphi_t-\varphi_b)(z_e-z_w);\\ F_s&=(\rho v_r)_s r_s(\varphi_t-\varphi_b)(z_e-z_w);\\ F_t&=(\rho v_\varphi)_t(z_e-z_w)(r_n-r_s);\\ F_b&=(\rho v_\varphi)_b(z_e-z_w)(r_n-r_s). \end{aligned}\right\} \tag{6.21}$$

$$\left.\begin{aligned} D_e&=\tfrac{\Gamma_e}{z_E-z_M}\tfrac{(r_n-r_s)(r_n+r_s)}{2}(\varphi_t-\varphi_b);\\ D_w&=\tfrac{\Gamma_w}{z_M-z_W}\tfrac{(r_n-r_s)(r_n+r_s)}{2}(\varphi_t-\varphi_b);\\ D_n&=\tfrac{\Gamma_n}{r_N-r_M}r_n(\varphi_t-\varphi_b)(z_e-z_w);\\ D_s&=\tfrac{\Gamma_s}{r_M-r_S}r_s(\varphi_t-\varphi_b)(z_e-z_w)\\ D_t&=\tfrac{\Gamma_t}{r_M(\varphi_T-\varphi_M)}(z_e-z_w)(r_n-r_s);\\ D_b&=\tfrac{\Gamma_b}{r_M(\varphi_M-\varphi_B)}(z_e-z_w)(r_n-r_s). \end{aligned}\right\} \tag{6.22}$$

$$Pe_e=\frac{F_e}{D_e};\ Pe_w=\frac{F_w}{D_w};\ Pe_n=\frac{F_n}{D_n};\ Pe_s=\frac{F_s}{D_s};\ Pe_t=\frac{F_t}{D_t};\ Pe_b=\frac{F_b}{D_b}. \tag{6.23}$$

The function $A(|Pe|)$ can be determined as (6.18).

6.1.2 Solution of the Momentum Equations and the Continuity Equation of the Gas Flux

The Main Methods for Solving the Momentum Equations of the Gas Flux The momentum equations of the plasma flux for the velocity components v_z, v_r, v_φ in a cylindrical coordinate system can be written in a convenient form for decision

$$\begin{cases} \operatorname{div}(\rho\vec{v}v_z) = \operatorname{div}(2\mu \operatorname{grad} v_z) - \frac{\partial p}{\partial z} + S_z; \\ \operatorname{div}(\rho\vec{v}v_r) = \operatorname{div}(2\mu \operatorname{grad} v_r) - \frac{\partial p}{\partial r} + S_r; \\ \operatorname{div}(\rho\vec{v}v_\varphi) = \operatorname{div}(\mu \operatorname{grad} v_\varphi) + S_\varphi, \end{cases} \tag{6.24}$$

where $-\frac{\partial p}{\partial z}$, $-\frac{\partial p}{\partial r}$—the components of the static pressure gradient; $S_z = F_{Bz} + \rho g_z + F_{\mu z}$; $S_r = F_{Br} + F_{\mu r}$; $S_\varphi = F_{\mu\varphi}$—the source functions in the continuity equation; F_{Bz}, F_{Br}—the components of electromagnetic force $\vec{F}_B = \vec{j} \times \vec{B}$; ρg_z—the gravity force; $F_{\mu z}, F_{\mu r}, F_{\mu\varphi}$—the additional viscosity members.

$$\left.\begin{aligned} F_{\mu z} &= \frac{1}{r}\frac{\partial}{\partial r}\left[r\mu\left(\frac{\partial v_r}{\partial z} - \frac{\partial v_z}{\partial r}\right)\right] - \frac{2}{3}\frac{\partial}{\partial z}\left[\mu\left(\frac{\partial v_z}{\partial z} + \frac{1}{r}\frac{\partial (rv_r)}{\partial r}\right)\right]; \\ F_{\mu r} &= \frac{\partial}{\partial z}\left[\mu\left(\frac{\partial v_z}{\partial r} - \frac{\partial v_r}{\partial z}\right)\right] - \frac{2}{3}\frac{\partial}{\partial r}\left[\mu\left(\frac{\partial v_z}{\partial z} + \frac{1}{r}\frac{\partial (rv_r)}{\partial r}\right)\right] - 2\mu\frac{v_r}{r^2}; \\ F_{\mu\varphi} &= -\frac{v_\varphi}{r}\left(\frac{\mu}{r} + \frac{\partial \mu}{\partial r} + \rho v_r\right). \end{aligned}\right\}$$

The momentum Eq. (6.24) can be written in the similar form of GDE (6.1), i.e., the discrete analogue of the GDE (6.17) or (6.20) can be used to solve them. Thus, the moment equation for the gas flux in the cylindrical coordinate system can be represented as

$$\frac{\partial}{\partial z}(\rho v_z v_z) + \frac{1}{r}\frac{\partial}{\partial r}(r\rho v_r v_z) = \frac{\partial}{\partial z}\left(2\mu\frac{\partial v_z}{\partial z}\right) + \frac{1}{r}\frac{\partial}{\partial r}\left(r2\mu\frac{\partial v_z}{\partial r}\right) - \frac{\partial P}{\partial z} + S_z; \tag{6.25}$$

$$\frac{\partial}{\partial z}(\rho v_z v_r) + \frac{1}{r}\frac{\partial}{\partial r}(r\rho v_r v_r) = \frac{\partial}{\partial z}\left(2\mu\frac{\partial v_r}{\partial z}\right) + \frac{1}{r}\frac{\partial}{\partial r}\left(r2\mu\frac{\partial v_r}{\partial r}\right) - \frac{\partial P}{\partial r} + S_r; \tag{6.26}$$

$$\frac{\partial}{\partial z}(\rho v_z v_\varphi) + \frac{1}{r}\frac{\partial}{\partial r}(r\rho v_r v_\varphi) = \frac{\partial}{\partial z}\left(\mu\frac{\partial v_\varphi}{\partial z}\right) + \frac{1}{r}\frac{\partial}{\partial r}\left(r\mu\frac{\partial v_\varphi}{\partial r}\right) + S_\varphi. \tag{6.27}$$

If the gradient of static pressure is taken into account through a common source function, the momentum Eqs. (6.25–6.27) are written in the form of GDE (6.19) and to solve them we can use the discrete analogue (6.20).

However, the difficulty in the solutions of the momentum Eqs. (6.25) and (6.26) is the unknown pressure field p (for its definition there is no clear differential equation). In computational mathematics, there are two ways to solve this problem and, accordingly, two main methods of the momentum equations solutions. The first method is based on the elimination of pressure from the momentum equation. Thus, in the case of two-dimensional problems pressure is eliminated from the two momentum equations through the cross differentiation of each equation using the vortex intensity ω/r and the current function ψ. For two-dimensional fluxes of gas flow, this method is based on the known method of solving of the momentum equation with variable current function and the vortex intensity, described, for example, in [2–4].

This technique has several features. The pressure is not one of the dependent variables, instead of two momentum equations and continuity equation for the gas it is necessary to solve two equations for the current function ψ and the vortex intensity ω/r. Conditions for the vortex intensity on the wall of the plasma torch are very hard, and because of its approximate representation it is often difficult to obtain convergent solutions. The pressure has been so successfully excluded from the formulation of the problem; it is often final desired or intermediate and it is necessary to calculate the thermal properties of the plasma. In these cases, the difficulties encountered in determining the pressure field shifted the benefits of using the variables of the current function and the vortex intensity. In addition, the main disadvantage is the inability to apply it in the case of three-dimensional problems where there is no the current function ψ.

The second method is based on direct solution of the momentum equations and the continuity equation with physical variables—components of the velocity and pressure. The pressure field is determined from the continuity equation (integrated form) after substituting to it the discrete analogues of the momentum equations. Next, the resulting pressure field is used for solution of the momentum equations and calculation of the plasma composition, its transfer coefficients and thermodynamic properties. Thus, the momentum equations and continuity equation are solved together, the pressure plays a linking role for the parameters of the momentum and continuity equations, and the resulting velocity field automatically satisfies the two laws of conservation: of momentum and mass. This method is described in some of papers [1, 3, 5].

Using the continuity equation to determine the pressure has some difficulties. In many problems the pressure is known only at one boundary—at the exit from the plasma torch, or the reactor chamber. On the other boundaries (on the wall, at the entrance and at the axis of symmetry of the plasma torch) the pressure is generally unknown.

However, components of the velocity, temperature, gas or gradients normal to these boundaries are usually known at these boundaries. Thus, in the entrance of plasma torch or gas inlet chamber the gas flow velocity can be measured or determined from the total gas flow Q. On the wall of the plasma torch gas as it "sticks", and its speed can be equated to zero. In the plasma torch axis of the radial component of the plasma velocity is equal to zero by the condition of axial symmetry. The temperature and density of the gas at these boundaries are easily measured or can be determined by simple models. Later, we will show that the pressure on the borders of these can be found with the defined components of the velocity and density of the gas, if used in solving shifted chess grid.

Consequently, both of the above methods require joint solutions of momentum equations and continuity equation (to exclude or determine the pressure). The second method allows directly obtaining the pressure field required for finding the composition and properties of the plasma. It describes the physical processes more accurately and has greater clarity than the first method. In the future, we focus on the second method, used for the development of algorithms for solving of momentum equations and continuity equation for the plasma flux. Note that this method is more time-consuming. It requires performing the additional calculations, compared with the first method. However, its advantage in the direct determination of the pressure is it compensates the additional complexity.

The Structure of the Shifted Chess Grids To implement the numerical methods it is not necessarily to solve all the equations, using the same grid, but each equation can be solved using its own grid. The boundary conditions for the equations are set differently on the borders and to use them effectively we need to apply the different grids.

Therefore it has been developed, the so-called shifted chess grid in which the points are displaced relative to each other and are staggered. For example, the grid points for determination of components of the velocity and pressure may be shifted from each other for a half pitch and the velocity components can be obtained at the verges of the control volume during the determination of the pressure (Fig. 6.5).

The main advantage of the shifted chess grid structure compared with a grid for solving of all the equations is to eliminate the need to determine the pressure directly on the borders, where it is not known (at the entrance, the wall, and the symmetry axis of the plasma torch). It offers the possibility of finding its value in terms of known components of the gas velocity.

Shifted chess grid allows obtaining the pressure difference that is necessary for the calculation of the pressure gradient, using the points on the verges for each component of the velocity (Fig. 6.5). It does not require a special algorithm for determining or approximating, as needed in the case of solutions of equations using the same grid. Similarly, for obtaining the solution of the continuity equation is not necessary to perform additional approximation on the faces of the velocity components of the control volume; they were obtained for these verges (Fig. 6.5) using its grids. In particular, the use of shifted chess grids lies at the basis of the developed solution procedures of the momentum equations SIMPLE and SIMPLER [1].

To solve the momentum equation and the continuity equation for the components of the velocity v_z, v_r and the pressure p the developed shifted grid that is shown in Fig. 6.5 is used, and the typical of the control volume are shown at Figs. 6.6, 6.7 and 6.8.

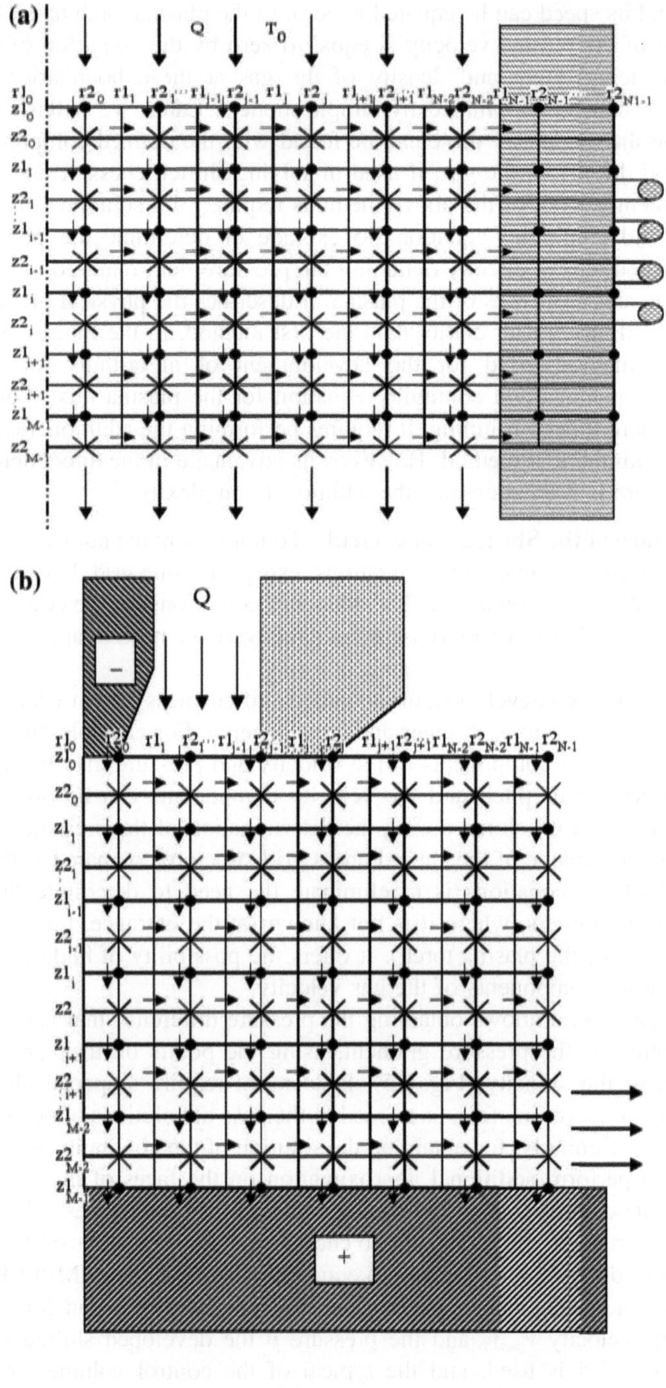
(a)
Q
T0
(b)
Q
−
+

◄ **Fig. 6.5** The structure of the shifted chess grids for simulation RF-plasma torches and Arc in the channel (**a**), and as well as the free Arc (**b**): •, grid points [$z1$, $r2$] to solve the equation of energy balance; →, grid points [$z2$, $r1$] for solution of momentum equation for vr; ↓, grid point [$z1$, $r2$] for solution of momentum equation for v_z; ×, grid point [$z2$, $r2$] for solving the continuity equation and determine the pressure p; grid point [$z1$, $r1$] is used to solve the momentum equations for v_φ and Maxwell's equations; three-dimensional momentum equations for v_z, v_r, v_φ are solving in the grid point [$z1$, $r2$, $\varphi2$], [$z2$, $r1$, $\varphi2$], [$z2$, $r2$, $\varphi1$] respectively

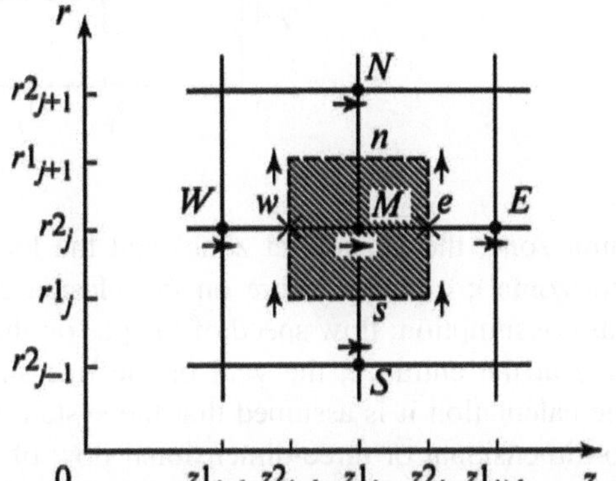

Fig. 6.6 The control volume and the grid [$z1$, $r2$] for solving the momentum equation (for v_z) and the equation of the energy balance: →, points for determination of the component v_z; ↑, the same for the component v_r; ×, the same for the pressure p

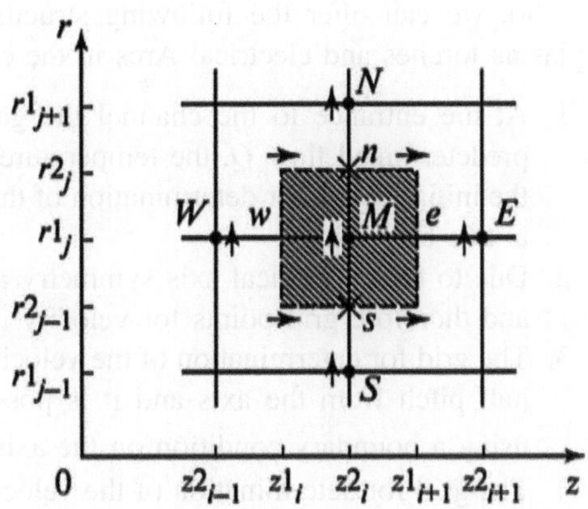

Fig. 6.7 The control volume and the grid [$z2$, $r1$] for solving the momentum equation (for v_r): the notations are the same as in Fig. 6.6

In solving the momentum equations and continuity equation for RF- and Arc plasma torches are generally known the following values: the geometric dimensions of the channel of the plasma torch; field of study of plasma processes (the inductor

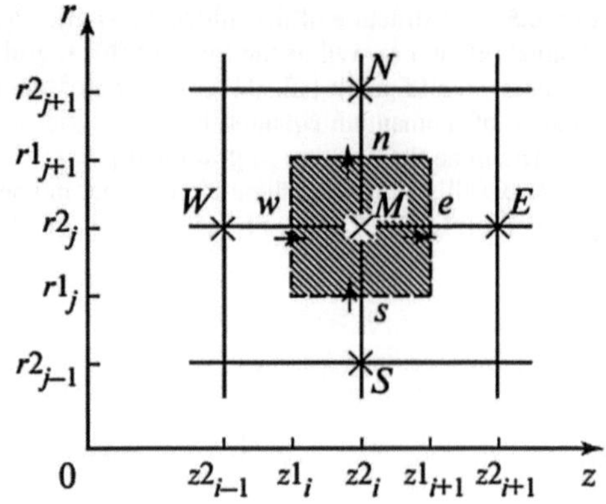

Fig. 6.8 The control volume and grid [z2, r2] for solving the continuity equation (for p): the notations are the same as in Fig. 6.6

zone, the Arc column zone, the plasma jet zone) and the location of the plasma torches (vertical, horizontal); outlet pressure on the plasma channel, chamber or reactor; kind and gas consumption; flow speed of the gas on the inlet of the plasma torch; its temperature at the entrance, the wall or the free surface of the plasma torch. Usually in the calculation it is assumed that the system has cylindrical axial symmetry with two-dimensional or three-dimensional flow of the gas stream.

In view of the above, a system of shifted grids can be built for the effectively use of the information about the boundary conditions of the problem and the regime of the plasma torch operation. In constructing the shifted grids we will be guided by the following: boundary points to determine the unknown variables will be located directly on the border, where their values are known, or set. For unknown quantities, or values for a known gradient at the boundaries of these points, because of the shifted grid would move inward by a half pitch calculation area.

So, we can offer the following structure shifted grids for simulation of RF-plasma torches and electrical Arcs in the channels (Fig. 6.5a).

1. At the entrance to the channel the gas velocity can be determined from its predetermined flow Q, the temperature of the input stream is often known, so the initial points for determination of the speed v_z and temperature T are located at this boundary.
2. Due to the cylindrical axis symmetry at the center of the plasma torch $v_r = 0$, and therefore grid points for velocity v_r are on the symmetry axis.
3. The grid for determination of the velocity v_z and temperature T is displaced by a half pitch from the axis and it is possible to find the corresponding streams using a boundary condition on the axis $v_r = v_\varphi = \frac{\partial v_z}{\partial r} = \frac{\partial T}{\partial r} = 0$.
4. The grid for determination of the velocity v_r is shifted by a half pitch down the z axis.
5. The grid for determination of the pressure p is shifted by a half step down the z axis compared to the grid for v_z and a half step on the r axis compared with grid

for v_r, thereby there is an opportunity to find the components of the velocity on the verges of the discrete analogue for solving the continuity equation.

6. The pressure at the exit from channel of the plasma torch is commonly known, and the output grid points for determination of the pressure p in the channel are located directly on this boundary.
7. Consider that gas «sticks» to the wall, i.e., its velocity at the wall of the plasma torch is equal to zero. The grid for determination of v_r is constructed so that the points are located on the wall.
8. The gas temperature T on the wall of the plasma torch is usually defined or specified on the inside or the outside. The latter condition is more accurate, and the grid for the decision of energy balance equation is structured to cover the wall of the plasma torch and had the grid points on its outside. The decision of energy balance equation may give the additional information about the temperature distribution inside the wall of the plasma torch and its degree of heating.
9. Since virtually every equation is solved with the use of a separate grid for calculating the coefficients of the discrete analogue it should take the appropriate approximation of the temperature T, the velocity components v_z and v_r using the grids, other than the main one.
10. Some of the convenience for the approximation and determination of the discrete analogue coefficients can be achieved in the solution of energy balance equation and the momentum equation for v_z using the same grid.

The structure of a shifted grid for the simulation of a free electric Arc (Chap. 8) using its boundary conditions is represented on the Fig. 6.5b. Below, we consider the use of the structure shifted chess grid for discrete analogue of the momentum equation and the continuity equation of the plasma flow.

The Discrete Analogue of the Momentum Equations The procedure for obtaining the discrete analogue of the momentum Eqs. (6.25–6.27) is similar to the procedure for the obtaining a discrete analogue of the GDE (6.2) with a difference of integration and static pressure gradients $-\frac{\partial p}{\partial z}$ and $-\frac{\partial p}{\partial r}$. The discrete analogue for them can be written as (6.17)

$$\left.\begin{aligned} a_M v_{zM} &= a_E v_{zE} + a_W v_{zW} + a_N v_{zN} + a_S v_{zS} + b_{zM} + (p_w - p_e)\Delta S_r; \\ \Delta S_r &= (r_n - r_s)(r_n + r_s)/2; \\ a_M v_{rM} &= a_E v_{rE} + a_W v_{rW} + a_N v_{rN} + a_S v_{rS} + b_{rM} + (p_s - p_n)\Delta S_z; \\ \Delta S_z &= (z_e - z_w)(r_n + r_s)/2; \\ a_M v_{\varphi M} &= a_E v_{\varphi E} + a_W v_{\varphi W} + a_N v_{\varphi N} + a_S v_{\varphi S} + b_{\varphi M}. \end{aligned}\right\} \quad (6.28)$$

The coefficients $a_E, a_W, a_N, a_S, a_M, b_{zM}, b_{rM}, b_{\varphi M}$ are determined using their grids by analogy with coefficients of the discrete analogue (6.17); $(p_w - p_e)$ and $(p_s - p_n)$—the difference of the pressure in grid points to obtain v_z and v_r, respectively; $(p_w - p_e)\Delta S_r$ and $(p_s - p_n)\Delta S_z$—the pressure forces on the surfaces with the areas ΔS_r and ΔS_z respectively.

For further change it is convenient to introduce the following notations:

$$\begin{cases} a_E v_{zE} + a_W v_{zW} + a_N v_{zN} + a_S v_{zS} = \sum a_m v_{zm}; \\ a_E v_{rE} + a_W v_{rW} + a_N v_{rN} + a_S v_{rS} = \sum a_m v_{rm}; \\ a_E v_{\varphi E} + a_W v_{\varphi W} + a_N v_{\varphi N} + a_S v_{\varphi S} = \sum a_m v_{\varphi m}; \\ \frac{\Delta S_r}{a_M} = d_{zM}; \ \frac{\Delta S_z}{a_M} = d_{rM}, \end{cases} \tag{6.29}$$

Here the summation is over all the nearest mashes of the central point M of this control volume, $m = E, W, N, S$. The discrete analogues of the momentum Eq. (6.28) can be written as

$$\begin{cases} v_{zM} = \frac{\sum a_m v_{zm} + b_{zM}}{a_M} + (p_w - p_e) d_{zM}; \\ v_{rM} = \frac{\sum a_m v_{rm} + b_{rM}}{a_M} + (p_s - p_n) d_{rM}; \\ v_{\varphi M} = \frac{\sum a_m v_{\varphi m} + b_{\varphi M}}{a_M}. \end{cases} \tag{6.30}$$

Note that the notations are strictly defined for each respective grid. Note

$$\hat{v}_{zM} = \frac{\sum a_m v_{zm} + b_{zM}}{a_M}; \ \hat{v}_{rM} = \frac{\sum a_m v_{rm} + b_{rM}}{a_M}. \tag{6.31}$$

Here $\hat{v}_{zM}$ and $\hat{v}_{rM}$, the partial rate velocities for v_z and v_r, accordingly, and there is no the pressure difference in the equations for determination of them. Discrete analogues of the momentum equations for v_z and v_r can also be written in the form convenient to insert into the continuity equation

$$\begin{cases} v_{zM} = \hat{v}_{zM} + (p_w - p_e) \mathrm{d}_{zm}; \\ v_{rM} = \hat{v}_{rM} + (p_s - p_n) \mathrm{d}_{rm}. \end{cases} \tag{6.32}$$

The Discrete Analogue of the Continuity Equation for the Pressure According to a second method of solving the momentum equations, the pressure is determined from the decision of the continuity equation for all the gas flow. At the same time, the following information can be used: the correct pressure field after the substitution to the momentum equation (equation of the momentum conservation) gives the velocity field that would satisfy the continuity equation (the law of mass conservation). For this reason, the momentum equation and the continuity equation should be solved jointly. The two-dimensional continuity equation in cylindrical coordinates is

$$\frac{\partial(\rho v_z)}{\partial z} + \frac{1}{r}\frac{\partial}{\partial r}(r \rho v_r) = 0. \tag{6.33}$$

The continuity equation for determination of the pressure is solved using grid [$z2$, $r2$] (Fig. 6.5). The mass balance for the control volume that is shown in Fig. 6.8, can be carried out by integrating (6.33) for this control volume

$$\int_{r_s}^{r_n}\int_{z_w}^{z_e}\frac{\partial(\rho v_z)}{\partial z}dzdr+\int_{z_w}^{z_e}\int_{r_s}^{r_n}\frac{1}{r}\frac{\partial}{\partial r}(r\rho v_r)rdrdz=0.$$

$$\left[(\rho v_z)_e-(\rho v_z)_w\right]\frac{(r_n-r_s)(r_n+r_s)}{2}+\left[r_n(\rho v_r)_n-r_s(\rho v_r)_s\right](z_e-z_w)=0. \quad (6.34)$$

Using (6.32), components of the velocity v_z and v_r can be represented on the verges of the control volume (Fig. 6.8) in the form

$$\begin{cases} v_{ze}=\hat{v}_{ze}+(p_M-p_E)d_{ze};\ v_{zw}=\hat{v}_{zw}+(p_W-p_M)d_{zw};\\ v_{rn}=\hat{v}_{rn}+(p_M-p_N)d_{rn};\ v_{rs}=\hat{v}_{rs}+(p_S-p_M)d_{rs}, \end{cases} \quad (6.35)$$

Here the partial rate velocities and ratios are defined as

$$\begin{cases} \hat{v}_{ze}=\dfrac{\sum a_m v_{zm}+b_{ze}}{a_e};\ \hat{v}_{zw}=\dfrac{\sum a_m v_{zm}+b_{zw}}{a_w};\\ \hat{v}_{rn}=\dfrac{\sum a_m v_{rm}+b_{rn}}{a_n};\ \hat{v}_{rs}=\dfrac{\sum a_m v_{rm}+b_{rs}}{a_s}. \end{cases} \quad (6.36)$$

$$\begin{cases} d_{ze}=\dfrac{(r_n-r_s)(r_n+r_s)}{2a_e};\ d_{zw}=\dfrac{(r_n-r_s)(r_n+r_s)}{2a_w};\\ d_{rn}=\dfrac{(z_e-z_w)(r_N+r_M)}{2a_n};\ d_{rs}=\dfrac{(z_e-z_w)(r_S+r_M)}{2a_s}. \end{cases} \quad (6.37)$$

All the neighboring points in the corresponding grid for determination of v_z (grid [$z1$, $r2$]) and v_r (grid [$z2$, $r1$]) are marked using the index «m». It depends on the location of these grids points e, w, n, s of the control volume for determination of the pressure p (grid [$z2$, $r2$]). Substituting (6.35) into (6.34), we obtain a discrete analogue of the continuity equation for the pressure p

$$a_M p_M = a_E p_E + a_W p_W + a_N p_N + a_S p_S + b_M,$$

where

$$\begin{cases} a_E=\rho_e d_{ze}\frac{(r_n-r_s)(r_n+r_s)}{2};\quad a_W=\rho_w d_{zw}\frac{(r_n-r_s)(r_n+r_s)}{2};\\ a_N=\rho_n d_{rn}(z_e-z_w)r_n;\quad a_S=\rho_s d_{rs}(z_e-z_w)r_s;\\ a_M=a_E+a_W+a_N+a_S;\\ b_M=(\rho_w\hat{v}_{zw}-\rho_e\hat{v}_{ze})\frac{(r_n-r_s)(r_n+r_s)}{2}+(r_s\rho_s\hat{v}_{rs}-r_n\rho_n\hat{v}_{rn})(z_e-z_w). \end{cases} \quad (6.38)$$

Thus, with the help of the shifted chess grid by substituting the solutions of the continuity equation for the velocity components v_z and v_r into the integrated continuity equation we obtain a new kind of discrete analogue of the continuity equation for the pressure (6.38). Together with (6.28) we have four systems of the discrete analogues of the momentum equations and the continuity equation for all the gas flow for the four unknown variables v_z, v_r, v_φ and p.

The algorithm of the joint solution of the momentum equations and the continuity equation is

1. The shifted chess grids for the solution of the momentum equation and the continuity equation are built using the boundary conditions and the features of this problem.
2. Some suggested initial velocity is introduced in dependence on the coordinates of the computational grid. In many cases, for a rapid convergence the zero velocity is selected as the initial velocity (except the boundary points).
3. The coefficients of the momentum equation are calculated using their grids and then the partial velocities are calculated (6.31) by substituting the values of the velocities in neighboring points.
4. The coefficients of discrete analogue of continuity Eq. (6.38) are calculated and (6.38) is solved for the pressure.
5. The system of discrete analogs of the momentum equation is solved using the pressure (6.28) and a new velocity is found.
6. The calculations are repeated from position 3 as long as there is no convergence of the solution.

6.1.3 Conversion of Different Equations in Generalized Form

Equation of the Energy Balance for the Equilibrium Plasma If we assume that the total specific heat capacity of the plasma c_p in the control volume is constant, the equation of the energy balance for the equilibrium plasma can be written in the form of GDE (6.1) as

$$\mathrm{div}(\rho\vec{v}T) = \mathrm{div}\left(\frac{\lambda}{c_p}\mathrm{grad}T\right) + \frac{1}{c_p}\left[\sigma E^2 - U_{\mathrm{rad}}\right]. \tag{6.39}$$

There is another form of the equation of the energy balance using enthalpy H

$$\mathrm{div}(\rho\vec{v}H) = \mathrm{div}(\lambda\mathrm{grad}T) + \sigma E^2 - U_{\mathrm{rad}}. \tag{6.40}$$

For quasi-neutral atomic plasma with single, double, and triple ionization the electron density, the plasma density, and enthalpy are determined by the equations

$$n_{\mathrm{e}} = n_{\mathrm{i}}^{+} + 2n_{\mathrm{i}}^{++} + 3n_{\mathrm{i}}^{+++};\ \rho = m_{\mathrm{a}}\left(n_{\mathrm{a}} + n_{\mathrm{i}}^{+} + n_{\mathrm{i}}^{++} + n_{\mathrm{i}}^{+++}\right);$$

$$\begin{aligned} H &= \frac{1}{\rho}\left[\frac{5}{2}kTn_{\mathrm{e}} + E_{\mathrm{i}}^{+}n_{\mathrm{i}}^{+} + E_{\mathrm{i}}^{++}n_{\mathrm{i}}^{++} + E_{\mathrm{i}}^{+++}n_{\mathrm{i}}^{+++}\right. \\ &\quad \left. + \frac{5}{2}kT\left(n_{\mathrm{a}} + n_{\mathrm{i}}^{+} + n_{\mathrm{i}}^{++} + n_{\mathrm{i}}^{+++}\right)\right] \\ &= \frac{1}{\rho}\left(\frac{5}{2}kTn_{\mathrm{e}} + E_{\mathrm{i}}^{+}n_{\mathrm{i}}^{+} + E_{\mathrm{i}}^{++}n_{\mathrm{i}}^{++} + E_{\mathrm{i}}^{+++}n_{\mathrm{i}}^{+++}\right) + \frac{5k}{2m_{\mathrm{a}}}T. \end{aligned} \tag{6.41}$$

Therefore, the equation of the energy balance (6.40) for the atomic plasma has form:

$$\begin{aligned} \operatorname{div}(\rho\vec{v}T) &= \operatorname{div}\left(\frac{2m_{\mathrm{a}}}{5k}\lambda \operatorname{grad}T\right) + \frac{2m_{\mathrm{a}}}{5k} \\ &\times \left\{\sigma E^{2} - U_{\mathrm{rad}} - \operatorname{div}\left[\left(\frac{5}{2}kTn_{\mathrm{e}} + E_{\mathrm{i}}^{+}n_{\mathrm{i}}^{+} + E_{\mathrm{i}}^{++}n_{\mathrm{i}}^{++} + E_{\mathrm{i}}^{+++}n_{\mathrm{i}}^{+++}\right)\vec{v}\right]\right\}. \end{aligned} \tag{6.42}$$

Note that the solution of (6.42), there is no need to determine the specific heat capacity c_p.

Equation of the Energy Balance for the Electron Gas in Two-Temperature Plasma Consider the equation of the energy balance for the electron gas of atomic neutral plasma with single, double, and triple ionization. Using the continuity equation in the form $\operatorname{div}\left[m_{\mathrm{a}}\left(n_{\mathrm{a}} + n_{\mathrm{i}}^{+} + n_{\mathrm{i}}^{++} + n_{\mathrm{i}}^{+++}\right)\vec{v}\right] = 0$, we get the following relations:

$$\begin{cases} \operatorname{div}\left[\left(n_{\mathrm{a}} + n_{\mathrm{i}}^{++} + n_{\mathrm{i}}^{+++}\right)\vec{v}\right] = -\operatorname{div}\left(n_{\mathrm{i}}^{+}\vec{v}\right); \\ \operatorname{div}\left[\left(n_{\mathrm{a}} + n_{\mathrm{i}}^{+} + n_{\mathrm{i}}^{+++}\right)\vec{v}\right] = -\operatorname{div}\left(n_{\mathrm{i}}^{++}\vec{v}\right); \\ \operatorname{div}\left[\left(n_{\mathrm{a}} + n_{\mathrm{i}}^{+} + n_{\mathrm{i}}^{++}\right)\vec{v}\right] = -\operatorname{div}\left(n_{\mathrm{i}}^{+++}\vec{v}\right). \end{cases}$$

Consequently,

$$\begin{aligned} &\operatorname{div}(\rho\vec{v}H_{e}) + \operatorname{div}\left\{\left[\frac{5}{2}kT_{\mathrm{e}}\left(n_{\mathrm{a}} - n_{\mathrm{i}}^{++} - 2n_{\mathrm{i}}^{+++}\right) - E_{\mathrm{i}}^{+}n_{\mathrm{i}}^{+} - E_{\mathrm{i}}^{++}n_{\mathrm{i}}^{++} - E_{\mathrm{i}}^{++}n_{\mathrm{i}}^{+++}\right]\vec{v}\right\} \\ &= \operatorname{div}\left\{\left[\frac{5}{2}kT_{e}n_{\mathrm{e}} + E_{\mathrm{i}}^{+}n_{\mathrm{i}}^{+} + E_{\mathrm{i}}^{++}n_{\mathrm{i}}^{++} + E_{\mathrm{i}}^{+++}n_{\mathrm{i}}^{+++} + \frac{5}{2}kT_{\mathrm{e}}\left(n_{\mathrm{a}} - n_{\mathrm{i}}^{++} - 2n_{\mathrm{i}}^{+++}\right)\right.\right. \\ &\quad \left.\left. + E_{\mathrm{i}}^{+}\left(n_{\mathrm{a}} + n_{\mathrm{i}}^{++} + n_{\mathrm{i}}^{+++}\right) + E_{\mathrm{i}}^{++}\left(n_{\mathrm{a}} + n_{\mathrm{i}}^{+} + n_{\mathrm{i}}^{+++}\right) + E_{\mathrm{i}}^{+++}\left(n_{\mathrm{a}} + n_{\mathrm{i}}^{+} + n_{\mathrm{i}}^{++}\right)\right]\vec{v}\right\} \\ &= \operatorname{div}\left[\left(\frac{5}{2}kTn_{\mathrm{e}} + E_{\mathrm{i}}^{+} + E_{\mathrm{i}}^{++} + E_{\mathrm{i}}^{+++}\right)\left(n_{\mathrm{a}} + n_{\mathrm{i}}^{+} + n_{\mathrm{i}}^{++} + n_{\mathrm{i}}^{+++}\right)\vec{v}\right] = \frac{5k}{2m_{\mathrm{a}}}\operatorname{div}(\rho\vec{v}T_{\mathrm{e}}). \end{aligned}$$

The equation of energy balance for the electron gas (1.34) for the stationary case and $\vec{v}_e = \vec{v}$ can be written in the form of GDE (6.1)

$$\operatorname{div}(\rho\vec{v}T_e) = \operatorname{div}\left(\frac{2m_a}{5k}\lambda_e \operatorname{grad} T_e\right) + \frac{2m_a}{5k}\{[\sigma_e E^2 - U_{rad} - b(T_e - T_a)] \\ -\operatorname{div}\left[\left(\frac{5}{2}kT_e(n_a - n_i^{++} - 2n_i^{+++}) - E_i^+ n_i^+ - E_i^{++} n_i^{++} - E_i^{+++} n_i^{+++}\right)\vec{v}\right]\right\} \tag{6.43}$$

Here $b = \frac{3}{2}k\frac{2m_e}{m_a}n_e\sqrt{\frac{8kT_e}{\pi m_e}}(n_a Q_{ea} + n_i^+ Q_{ei}^+ + n_i^{++} Q_{ei}^{++} + n_i^{+++} Q_{ei}^{+++})$—the coefficient of the energy exchange of electrons with atoms and ions; $Q_{ea}, Q_{ei}^+, Q_{ei}^{++}, Q_{ei}^{+++}$—averaged effective cross sections of the electron with atoms, single, double, and triple ions, respectively (Sects. 3.4 and 3.5).

Equation of the Energy Balance for Atom–Ion Gas in Two-Temperature Plasma For the atom–ion gas we have enthalpy

$$H_{ai} = \frac{1}{\rho}\left[\frac{5}{2}kT_a(n_a + n_i^+ + n_i^{++} + n_i^{+++})\right] = \frac{5k}{2m_a}T_a.$$

The equation of energy balance for atom–ion gas (1.37) for the stationary case can be written in the form of GDE (6.1) as

$$\operatorname{div}(\rho\vec{v}T_a) = \operatorname{div}\left(\frac{2m_a}{5k}\lambda_{ai} gradT_a\right) + \frac{2m_a}{5k}b(T_e - T_a). \tag{6.44}$$

The Continuity Equation of the Chemical Elements and the Equation of the Ionization Equilibrium In general, the steady continuity equation, or the equation of mass conservation of the *i*th chemical elements has the form GDE (6.1) as

$$\operatorname{div}(\rho\vec{v}x_i) = \operatorname{div}(\Gamma_i \operatorname{grad} x_i) + S_i, \tag{6.45}$$

Here $x_i = \rho_i/\rho$—the mass concentration of *i*th chemical element; Γ_i, S_i—the diffusion coefficient and the source function of the *i*th chemical element.

Stationary equation of ionization equilibrium for atomic plasma is written as:

$$\operatorname{div}(n_e\vec{v}) = \operatorname{div}(D_{amb} \operatorname{grad} n_e) + k_r n_e n_a\left(\frac{n_e^{*2}}{n_a^*} - \frac{n_e^2}{n_a}\right), \tag{6.46}$$

Here D_{amb}—the bipolar diffusion coefficient; k_r—the coefficient of the triple (impact-radiative) recombination; n_e^*, n_a^*—the concentration of electrons and atoms of equilibrium plasma, determined by the Saha equation.

Equation (6.46) can be written in the form of GDE (6.1) as

$$\operatorname{div}(\rho\vec{v}x_e) = \operatorname{div}(\rho D_{amb} \operatorname{grad} x_e) + k_r m_e n_e n_a\left(\frac{n_e^{*2}}{n_a^*} - \frac{n_e^2}{n_a}\right). \tag{6.47}$$

Taking into account, boundary conditions for the solution of the equation of ionization equilibrium, and a link between the particle concentration and temperature in the determination of the transfer coefficients and the thermodynamic properties of the plasma, it is desirable to solve the equation of ionization equilibrium (6.47) using the same grid [$z1$, $r2$] (Fig. 6.5), like the equation of energy balance.

For the convenience, the parameters of different equations in the form of GDE (6.1) are shown in Table 6.1.

Note that the static pressure gradient is included to the source function of the momentum equation for v_z and v_r, this gradient is then substituted into the continuity equation of the gas flux. The momentum equations have the discrete analogue (6.28), and the continuity equation has the discrete analogue (6.38). For all other equations in the form of GDE (6.1) the discrete analogue can be applied the form

Table 6.1 The equations in the form of GDE (6.1)

Equations	Parameters		
	Φ	Γ	S
Equation of the energy balance for the equilibrium plasma model:			
Taking into the specific heat	T	λ/c_p	$S=(\sigma E^2-U_{\text{rad}})/c_p$
Without the specific heat	T	$\frac{2m_a}{5k}\lambda$	$\frac{2m_a}{5k}\left\{\sigma E^2-U_{\text{rad}}-\text{div}\left[\left(\frac{5}{2}kTn_e+E_i^{+}n_i^{+}+E_i^{++}n_i^{++}+E_i^{+++}n_i^{+++}\right)\vec{v}\right]\right\}$
Equation of the energy balance for the electron gas	T_e	$\frac{2m_a}{5k}\lambda_e$	$\frac{2m_a}{5k}\{\sigma_e E^2-U_{\text{rad}}-b(T_e-T_a)-\text{div}\left[\left(\frac{5}{2}kT_e(n_a-n_i^{++}-2n_i^{+++})-E_i^{+}n_i^{+}-E_i^{++}n_i^{++}-E_i^{+++}n_i^{+++}\right)\vec{v}\right]\}$
Equation of the energy balance for the atom–ion gas	T_a	$\frac{2m_a}{5k}\lambda_{ai}$	$\frac{2m_a}{5k}b(T_e-T_a)$
Momentum equation for v_z	v_z	2μ	$-\frac{\partial p}{\partial z}+[\vec{j}\times\vec{B}]_z+\frac{1}{r}\frac{\partial}{\partial r}\left[r\mu\left(\frac{\partial v_r}{\partial z}-\frac{\partial v_z}{\partial r}\right)\right]-\frac{2}{3}\frac{\partial}{\partial z}\left[\mu\left(\frac{\partial v_z}{\partial z}+\frac{1}{r}\frac{\partial(rv_r)}{\partial r}\right)\right]+\rho g_z$
Momentum equation for v_r	v_r	2μ	$-\frac{\partial p}{\partial r}+[\vec{j}\times\vec{B}]_r+\frac{\partial}{\partial z}\left[\mu\left(\frac{\partial v_z}{\partial r}-\frac{\partial v_r}{\partial z}\right)\right]-\frac{2}{3}\frac{\partial}{\partial r}\left[\mu\left(\frac{\partial v_z}{\partial z}+\frac{1}{r}\frac{\partial(rv_r)}{\partial r}\right)\right]-2\mu\frac{v_r}{r^2}$
Momentum equation for v_φ	v_φ	μ	$-\frac{v_\varphi}{r}\left(\frac{\mu}{r}+\frac{\partial\mu}{\partial r}+\rho v_r\right)$
Continuity equation for all the gas flux	1	0	0
Continuity equation for the ith chemical elements	x_i	Γ_i	S_i
Equation of ionization equilibrium	x_e	ρD_{amb}	$k_r m_e n_e n_a\left(\frac{n_e^{*2}}{n_a^*}-\frac{n_e^2}{n_a}\right)$

(6.17) or (6.20). The above transfer equations of the turbulent flow of the models $k-\varepsilon$ and $k-\omega$ (Sect. 1.5) have also the form of GDE (6.1), their basic coefficients are presented in Tables 1.1, 1.2.

6.2 The Stability of the Difference Methods and Computational Procedures

6.2.1 Stability of Difference Methods

Property of the difference methods that is able to cause a computational system to a single final decision under certain perturbations of the input data is called stability.

If Φ_h—the solution of *k*th differential equations and ξ_h—the input data for some difference method. These parameters depend on the integration step or the grid step Δh. And if we change Δh we get a sequences of decisions $\{\Phi_h\}$ and input data $\{\xi_h\}$. In mathematical physics it is accepted that the problem and its difference method of solution are correct if the following conditions are carried out:

1. The problem is uniquely solvable and the solution exists and is unique for all the input data ξ_h from some admissible family (the region of stability).
2. The solution of the difference problem continuously depends on the initial data ξ_h, and this dependence can be uniform for the grid steps Δh.

The condition (2) implies that a constant value $M > 0$ exists and it is independent of Δh, and for sufficiently small step $|\Delta h| \leq h_0$, the inequality [6] occurs

$$\left\|\Phi'_h - \Phi_h\right\|_{(1h)} \leq M\left\|\xi'_h - \xi_h\right\|_{(2h)}, \tag{6.48}$$

where Φ'_h—the solution of the difference problem with the input data ξ'_h, and $(1h)$ and $(2h)$—the norms in a variety of gridlines functions defined on this grid.

The property of continuous dependence of the solution of the difference problem on the input data expressed by inequality (6.48) is called the stability of difference methods for input data, or simply stability. Thus, if the difference method is stable, it converges; and the establishing of the convergence and accuracy of the solution of the problem are connected with the choice of the approximation method and stability of the system.

Stability Criterion After the sampling of the differential equation using a finite-difference method a system of discrete analogues can be obtained in the form: $a_M\Phi_M = \sum a_m\Phi_m + b_M$. Then a sufficient condition for the convergence of solutions of such a system in the iterative methods of Gauss–Seidel is a diagonal dominance, i.e.,:

$$|a_M| \geq \sum |a_m| \text{ for all equations;}$$
$$|a_M| > \sum |a_m| \text{ at least, for one of the equations.} \quad (6.49)$$

Conventionally, we call the difference method for obtaining the discrete analogues with the property (6.49) as stable. Note that the stability criterion (6.49) exists only in the solution of linear algebraic equations. In solving the nonlinear systems, the stability criterion does not exist, and the condition (6.49) is not sufficient for convergence of the solutions.

Stability criterion (6.49) has the following features:

1. The criterion (6.49) is a sufficient but not a necessary condition. Therefore the convergence may be even at the partial violation of (6.49).
2. At the recording of a discrete analogue in a form $\Phi_M = \frac{\sum a_m \Phi_m + b_M}{a_M}$, it follows that if the value a_M higher, the value Φ_M is the more slowly changing from iteration to iteration, and the less it depends on the values of Φ_m in neighboring nodes, i.e., on the disturbances of the iterative procedure, it contributes to the sustainability of solutions.
3. If $a_M = \sum a_m$, as is typical for most of the difference methods, for the implementation of the stability conditions (6.49) the requirement that all the coefficients of the discrete analogue have the same sign should be complied. In this case, the second condition for the stability criterion is automatically met if at least one value of the required variable is known at the border, i.e., $a_M = 1$, $\sum a_m = 0$.

The Convergence Condition Suppose that Φ_M^* is the value of the unknown variable in the node M, obtained in the previous iteration, and Φ_M—the same as found in the current iteration, then the condition for solutions convergence in an iterative procedure of solving the inequality:

$$\max\left|\frac{\Phi_M - \Phi_M^*}{\Phi_M}\right| \leq \varepsilon \text{ for all } M. \quad (6.50)$$

Here $\varepsilon = 10^{-5} - 10^{-3}$—some small value, depending on the conditions of the problem. If the value of Φ have a large gap, the following condition of convergence can be used:

$$\max\frac{|\Phi_M - \Phi_M^*|}{\max|\Phi_M|} \leq \varepsilon \text{ for all } M. \quad (6.51)$$

A very slow convergence of (6.50) and (6.51) may prove to be incorrect and insufficient to stop calculating and making the intermediate value as a final solution. It so happens that the solution is far from steady-state value, but the conditions (6.50) or (6.51) already occurs. In this case, it is necessary to define some residual R for the solution of differential equation and sufficient condition for the convergence of solutions in this case is an extremely small value R.

To analyze the stability of difference methods, applicable to the solution of problems of gas dynamics and heat transfer in plasma using numerical solution of one-dimensional GDE in the form (6.8).

Using the function $A(Pe)$ in (6.8) allows the compact construction of all the numerical method in one record for their analysis without changing the type of discrete analog. For example, the exponential scheme $A(Pe) = \frac{Pe}{\exp Pe - 1}$ for central difference scheme $A(Pe) = 1 - 0.5Pe$ for the scheme upstream $A(Pe) = 1$, etc.

In sampling the condition $F_e = F_w = \rho v_z = \text{const}$ is used. Because of the approximate solution this condition cannot be satisfied immediately in an iterative process. However, its use makes it possible to obtain the discrete analogue (6.8) with coefficients satisfying stability criterion (6.49). In this case, the coefficients $D_e > 0, D_w > 0, F_e = F_w > 0$, so the requirement of positiveness of the coefficients in the discrete analogue[the 3-d consequence of the stability criterion (6.49)] leads to the requirement positiveness for the function $A(Pe)$.

It is easy to note that for $|Pe| > 2$ the using of a central difference scheme $A(Pe) < 0$ (Fig. 6.9), we get the negative coefficients of the discrete analogue (6.8) and violation of the stability criterion (6.49). Consequently, the central difference scheme in this case is unstable and its use is not desirable.

To prevent $A(Pe) < 0$ by approximation using the central difference scheme, a combined scheme $[A(Pe) = \max(0; 1 - 0.5Pe)]$ can be used.

The scheme «upstream» is stable if $A(Pe) > 0$ for all Pe, but it is not accurate for small values of $|Pe|$. In this case, according to the «up flow» the average value of the unknown variable on the verges of the control volume still has the value of this variable in the near node down the stream, that is true only for a large predominance of convective flow under the diffusion one, i.e., at the large value of Pe (Fig. 6.2).

For the exponential scheme $A(Pe) > 0$ for all Pe (Fig. 6.9), and all the coefficients of the obtained discrete analogue are positive that satisfies the stability criterion (6.49). In terms of accuracy and stability, an exponential scheme can be used to construct the discrete analogue. However, the exponential calculation usually requires more computation time, so this scheme is often approximated by different polynomials, one of them is available in [1]—the scheme with a power-law (Fig. 6.9): $A(Pe) = \dfrac{Pe}{\exp Pe - 1} \approx \max[0; (1 - 0.1Pe)^5]$. The exponential scheme is accurately approximated using this function by and it guarantees the positive values of the coefficients of the discrete analogue.

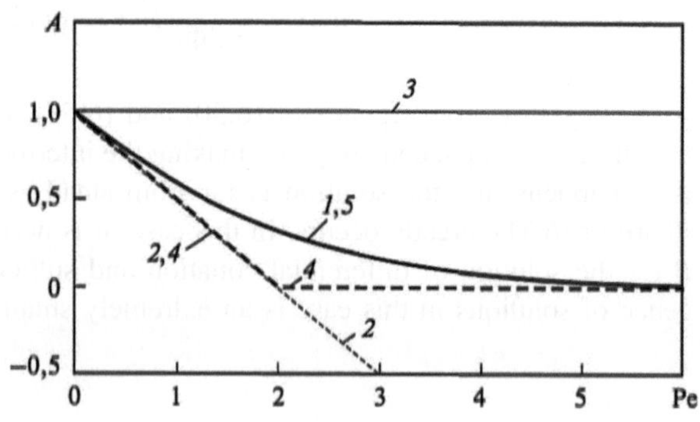

Fig. 6.9 The function $A(Pe)$ for different computational schemes: *1* the exponential scheme; *2* the central difference scheme; *3* the scheme «upstream» ; *4* the combined scheme; *5* the scheme with a power-law

6.2.2 Analysis of the Computational Procedures Stability

From the above calculation schemes (Sect. 6.2.1), the best sustainable ones can be selected. However, these schemes are stable only in the solution of linear problems (i.e., with constant coefficients). For nonlinear stability problem solving procedure is not guaranteed in the application of even the most stable of the scheme. A divergence or oscillation process of solutions with increasing amplitude is usually the sign of the instability of the computational procedure.

Based on the fact that the object modeling—the low-temperature plasma is in a stable state, it can be assumed that the stability of calculation procedures may be lost due to the inconsistency of the structural model of computational procedures with natural processes, i.e., due to loss of information. Stability of the computational procedure is particularly disturbed at the decision purely nonlinear differential equations, where the loss of information is essential and cannot be recovered or compensated.

In practice the iterative methods are often used as the main tool of exclusion of the instability of computational procedures for solving nonlinear problems. Using iterations greatly simplifies the calculations, instead of the simultaneous solution of all the equations their decision is consistently carried out. Starting with some initial decisions their new approximate solution can be found. The old decision should be replaced by the new one, and the process should be repeated until the convergence is reached. Links of the physical processes that cannot be established at this stage of solution are recovered in solution of equations with repeated clarifications and translation of the relevant factors.

This structure (iterative model) of solutions of differential equations as a result of iterative procedures is shown in Fig. 6.10, here there are two levels of iterative procedures—local for solving of each equation and global for consistent solutions of equations (to clarify and enumeration of relevant factors, i.e., to restore the connection between the equations). The coefficient α is the coefficient of external feedback and it is implemented using the known algorithms of top or bottom relaxation.

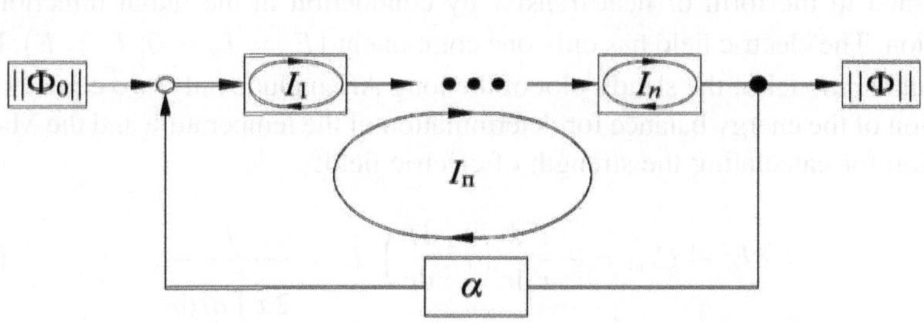

Fig. 6.10 The structure of the sequential solution of differential equations as a result of iterative procedures: $I_1, I_2, \ldots, I_n$, local iterative procedures; I_{II}, global iterative procedure

Despite the popularity of the use of iterative procedures, they have a significant drawback: the information about the connection of some physical processes (between the laws of conservation) may be only partially restored at the end of new iteration and is not reflected in the solution of the equation. The coefficients of each equation are determined and corrected only in the transition from decision to decision, and in the process of solving each equation remain unchanged. Therefore, in the computational procedure of the simulation model, there is always some delay due to the processes of simulation compared to actual physical processes. Moreover, if the physical connection is simulated for local iterative procedures, the time delay is equal to the working time of one local procedure, and if the connections between the physical processes exist at the global procedure, the time lag even further—it is equal to the time of one global calculation procedure.

The harder the physical processes are associated, the greater the delay in the management of computational procedures becomes and the greater the distortion of the simulation model becomes, compared with the mathematical model. In the case where the external feedback circuit cannot compensate the delay in the control, the calculation process becomes unstable or diverging.

The successful use of traditional iterative structures for modeling of some of nonlinear physical processes only notes that for these processes, the natural communications are not so strong, and the information that restores by the iterative procedures is sufficient to compensate the lost information inside the solution. At this structural representation, the nonlinear and rigid connections of the physical processes cannot yet lead to the instability of the system. Therefore, to ensure the stability of the calculation procedures, it is necessary to simulate and determine not only the external but also the internal communication of the physical processes in the computational structure.

For example, analyze the stability of the modeling procedures for established long plot Arc in the channel. For a long Arc, where the radius of the channel that is much less than its length, the established section occupies a large space of the channel and has the parameters that almost do not change along the channel $\left(\frac{\partial T}{\partial z} = 0\right)$. The motion of the gas is one-dimensional $\left(v_r = v_\varphi = 0\right)$. Isolation of energy σE^2 is consumed in the form of heat transfer by conduction in the radial direction, and radiation. The electric field has only one component $\left(E_\varphi = E_r = 0;\ E_z = E\right)$. In this case, in the model of the steady plot of the long Arc include only two equations: the equation of the energy balance for determination of the temperature and the Maxwell equation for calculating the strength of electric field:

$$\sigma E_z^2 - U_{\mathrm{rad}} = -\frac{1}{r}\frac{\mathrm{d}}{\mathrm{d}r}\left(r\lambda\frac{\mathrm{d}T}{\mathrm{d}r}\right); E_z = \frac{I}{2\pi\int\limits_0^R \sigma r \mathrm{d}r}, \tag{6.52}$$

Here *I*—current of the Arc; *R*—the radius of the Arc channel.

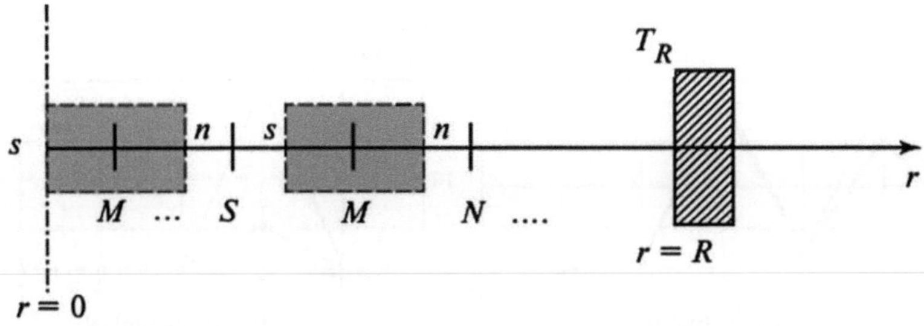

Fig. 6.11 The control volumes and grid for solution of the one-dimensional equation of the energy balance of the Arc in the channel

The solution is carried out by method of the control volume. The balance of power and the discrete analogue for the equation of energy balance are obtained by the integrating of the equation of energy balance in the control volume (Fig. 6.11).

Discrete analoguefor the equation of energy balance (6.52) takes the following form:

$$a_M T_M = a_N T_N + a_S T_S + b_M, \tag{6.53}$$

where

$$\begin{cases} a_N = \frac{r_n \lambda_n}{r_N - r_M} > 0; a_S = \frac{r_s \lambda_s}{r_M - r_S} > 0; a_M = a_N + a_S > 0; \\ b_M = 0.5 S_M (r_n - r_s)(r_n + r_s); S_M = \left(\sigma E_z^2 - U_{\text{rad}}\right)_M. \end{cases}$$

The standard solving procedure consists of the following:

1. Some grid is constructed and the initial temperature field is given on this grid for the first iteration, for example, T = 10,000 K.
2. The strength of electric field E, the released (specific) power σE^2 and coefficients of the discrete analogue are determined by (6.53).
3. The system of discrete analogs (6.53) (with constant coefficients) is solved in order to find a new field of temperature.
4. The convergence condition is checking (6.50).
5. If the convergence condition is satisfied, then the solution is over, if not, a new field of temperature is adopted as the current and the calculation is continued, starting from position 2.

Solution using this procedure and grid with the steps $\Delta r = 0.001 - 0.1$ mm is unstable. The dynamics of the iterative procedure (a change of temperature and released power from iteration to iteration) is shown in Fig. 6.12. The rare changes in temperature and released power from iteration to iteration indicate the instability of computational procedure.

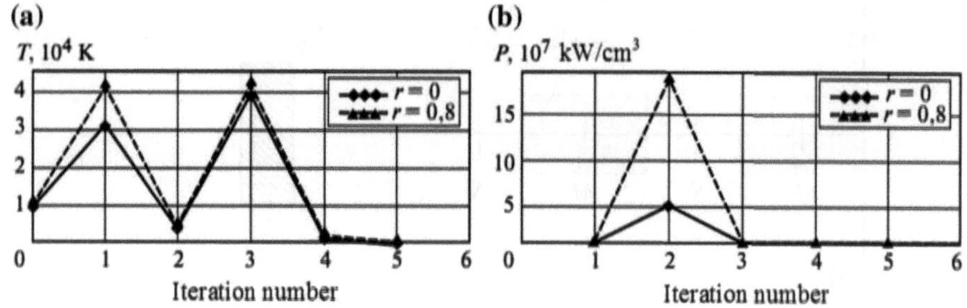

Fig. 6.12 The change in temperature T (**a**) and released power $P = \sigma E^2$ (**b**) in the Arc from iteration to iteration at the unstable solution for different points ($r = 0$ and $r = 0.8$ mm) of the Arc channel with radius $R = 1.5$ mm and current $I = 90$ A

This iterative procedure has the following features: the coefficients of the discrete analogue remain constant inside the iterations (position 3) regardless of its decision; conversion of the coefficients is performed after obtaining of the new field of temperature. Links between the various physical processes and the field temperature (i.e., by the decision of the equation of the energy balance) are corrected or restored only after solving of the discrete analog, so there is a delay in the administration of the simulation model for the computational procedure for some duration, at least, one iteration in comparison with natural processes .

This equation of energy balance consists of three members: the transfer of power by heat conduction $\frac{1}{r}\frac{\mathrm{d}}{\mathrm{d}r}\left(r\lambda\frac{\mathrm{d}T}{\mathrm{d}r}\right)$; the losses of power due to radiation U_{rad} and the emitted power σE^2. Beside the established link of power transmitted by heat conduction with the temperature in the discrete analogue (6.53) the knowledge of the relations between the losses of power due to radiation $U_{\mathrm{rad}} = \mathrm{const}$ and emitted power σE^2 with temperature is required. These links are not established inside the iteration. They are restored after each of iterations through the conversion of the strength of electric field, emitted power, and the losses of power due to radiation using the resulting field of temperature.

It is difficult to set the links between U_{rad} and σE^2 and temperature inside the iteration, because they depend on many factors. However, the rare change in the decisions of equation of the energy balance and released power from iteration to iteration (Fig. 6.12) indicates a rigid connection between the released power and the temperature of plasma. The exact dependence σE^2 (or $\sigma E^2 - U_{\mathrm{rad}}$) on the temperature cannot be determined, but based on physical reasoning it can be assumed that it has a decreasing character, as well as for any sustainable process.

Indeed, if it is not so, then any perturbation (raising or lowering of temperature) may cause an additional increase or respectively decrease of the emitted power, and it contributes to further increase or decrease of the temperature. This process does not occur in nature as the increase or decrease in the temperature cannot continue indefinitely, always in nature, particularly in plasma, there is a negative feedback mechanism for the stable existence.

Thus, to eliminate the delays in the administration of the simulation model, as compared to the physical processes beside the iterative procedure, it is necessary to establish a special internal communication directly inside the solutions each of equation. This connection can be realized with the help of some dependence between the source function S and the unknown variable Φ.

6.3 The Structural Organization of Computational Procedures

6.3.1 *The External Feedback in the Structure of Computational Procedures*

Most of the tasks at the simulation of plasma processes are solved using two iteration loops: the internal (local) for solving the individual equations and the external (global) for the consistent solutions of equations (Fig. 6.10).

The traditional structure of the local iterative procedure with constant coefficients is shown in Fig. 6.13. It has the following features:

1. The decision of any equation begins with some initial (approximate) field $\|\Phi_M^0\|$.
2. In solving each of equations the solution of other equations remains unchanged. Besides the physical processes described in this equation, the communications of the other processes and the laws of conservation are not established in its decision.
3. The stability condition of the iterative procedure (6.49) exists only for the constant coefficients $a_M, a_E, a_W, a_N, a_S, b_M$ and the block of definitions of these factors is beyond the scope of an iterative procedure. The coefficients of a discrete analogue are determined again after each of the global iteration cycle, taking into account the decisions of other equations. Thus, the dependence of some physical processes can be simulated only in the external (global) feedback loop (Fig. 6.10).

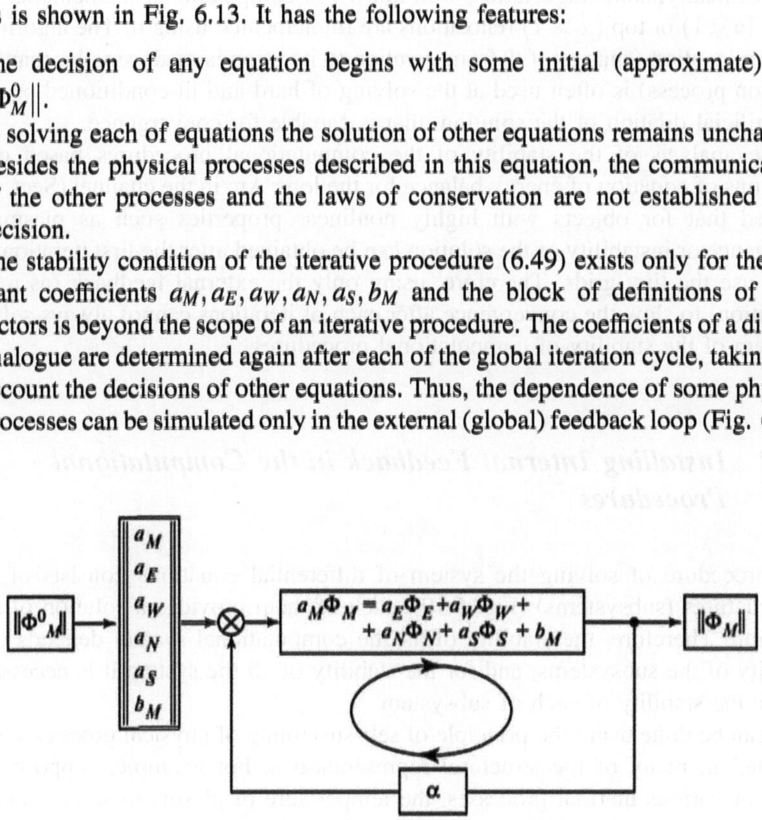

Fig. 6.13 The local iterative procedure with the external determination of the coefficients

4. The new solution after each of iteration cycles replaces the old one. The solving process is repeated until the convergence condition is satisfied (6.50).
5. The recursive solution is carried out for all grid points using the discrete analogue of the differential equation. For multidimensional problems the rapid convergence of solutions for the system of algebraic equations can be achieved through a combination of two methods: direct in the same direction of coordinates and recursive in other directions (the method of alternating directions).
6. Repeat of the solutions is realized by the circuit with the external feedback using the factor α, the regulation of this factor can be slowed down or speeded up the convergence of solutions, if it is necessary.

Control of the iterative procedure using external feedback is implemented using the following algorithm:

$$\Phi_M = \alpha\Phi_M^{(n)} + (1-\alpha)\Phi_M^{(n-1)}, \quad (6.54)$$

Here $\Phi_M^{(n)}$ and $\Phi_M^{(n-1)}$—the solutions of the point M in two nearest iterations n and $n-1$.

Specifically, some algorithms, well-known in computational mathematic, of the lower ($\alpha < 1$) or top ($\alpha > 1$) relaxations are implemented using α. The algorithm of lower relaxation (change of Φ from iteration to iteration is made smaller than in the iteration process) is often used at the solving of hard and ill-conditioned problems for artificial dilation of the solution, that is capable for convergence.

The analysis of the stability of the computational procedures based on the decisions of equation of energy balance for the long Arc in the channel (Sect. 6.2.2) showed that for objects with highly nonlinear properties such as plasma, the divergence or instability of the solution can be obtained after the first iteration, even if we use the tiny grids. Therefore, using only the external feedback (as a lower relaxation) to slow the convergence after each of iterations cannot always solve the problem of the stability of computational procedures.

6.3.2 *Installing Internal Feedback in the Computational Procedures*

The procedure of solving the system of differential equations consists of many substructures (subsystems) (Fig. 6.10), each of them provides a solution of single equation. Therefore, the stability of all the computational system depends on the stability of the subsystems, and for the stability of all the system it is necessary to ensure the stability of each of subsystem.

It can be done using the principle of self-sustaining of physical processes, that is reflected in many of the structural representations. For example, suppose that a result of various thermal processes, the temperature of plasma in some volume of the plasma torch channel increases. Then for the steady state of the plasma the release of energy in the form of Joule heat σE^2 in this volume should be reduced, and at the same time the flow of energy transfer from this volume the other one have to increase due to different dissipative processes. As a result, the temperature in the volume cannot be increased indefinitely. The stable plasma, due to internal reverse (negative) links, is regulated itself by initiating the various feedback processes that may lead to the counter to any disturbance.

A similar picture can be obtained by considering the changes in gas-dynamic, thermodynamic, electromagnetic, and other physical processes. We see that everywhere and for every object the internal feedback exists for self-regulation and maintaining of the steady state of the object. Thus, in simulations these connections must be established in the model as a negative feedback.

Based on the principle of sustainability and self-regulation of natural object the internal feedback of different physical processes can be restored and set in a decreasing dependence of source function S of the target variable Φ. In various studies [1, 7] it is proposed to establish this relationship inside the procedure of solving of each equation as a linear source function S:

$$S = S_c + S_p\Phi, \tag{6.55}$$

where the coefficients S_c, S_p depend on the coordinates and on Φ. After each iteration cycle, S_c and S_p and be converted with other factors of a discrete analogue using the new values of Φ.

The dependence of the source function S on Φ (6.55) is set inside the iterative solving procedure and it is an internal feedback for output value. Linearization of the source function is performed on the grounds that, first, the system of the discrete analogues allows only the formal linear relationship, and second, for the iterative procedure the depending of the source function S on Φ can be formed approximately. Mistakes in the approximation of this function can be eliminated with the help of external feedback of iterative procedure. Third, the structure of the GDE (6.1) shows that the inner connection of various physical processes can be set in the discrete analogue only using the source function, and other members have already been successfully used in the sampling of the differential equation.

Although the adoption of a linear dependence $S(\Phi)$ allows us to describe some connection inside the iterative procedure, a method of determination of its coefficients is unknown. In addition to the trivial cases when the source function can be explicitly expressed as a function of the independent variable, for the linearization of the source function for most of the equations for description of plasma when only the values S and Φ from the previous iteration are usually known. For the construction of the linear dependence it is necessary to have additional information. This creates some difficulties in the formation of such a function.

The proposal about the maximum possible range of the target variable Φ [1] very often leads to a small value of the coefficient S_p, i.e., it fails to effectively control of the course of solutions that in some cases can negate the effect of feedback in solving of stiff systems of equations. In this connection, we will offer the following algorithm to establish the connection and control the speed of solving the equations:

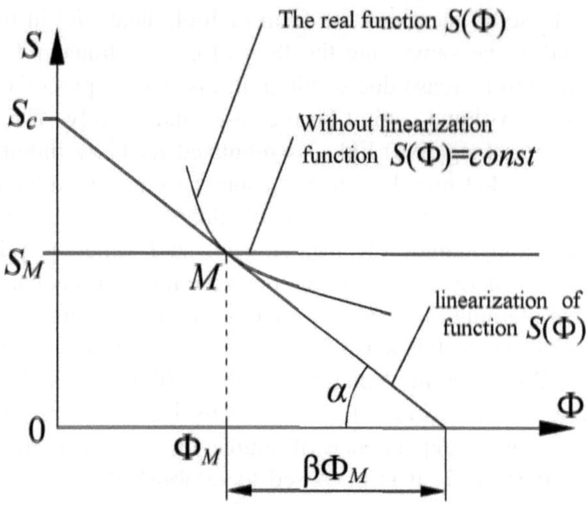

Fig. 6.14 Illustration of linearization for the source function $S = S_c + S_p\Phi$ ($S = \text{const}$ when $\beta = \infty, S_p = 0$)

because of the possible approximate solution in an iterative procedure the relationship can be constructed where the solution obtained in the following procedure does not exceed the some certain value. If the source function in the point M takes the value S_M for Φ_M in this iteration, it can be requested that in the next iteration the resulting solution does not exceed the value $\beta\Phi_M$, where β—some coefficient (Fig. 6.14). This can be achieved using the following algorithm [7, 8]:

$$S_p = -\left|\frac{S}{\beta\Phi}\right|; \quad S_c = S - S_p\Phi. \tag{6.56}$$

Coefficient β determines the speed of change of the function S, depending on Φ. It can be called the coefficient of internal feedback in solving of this equation. Basically the coefficient β varies over a wide range ($\beta = 0 - \infty$) and it is determined depending on specific conditions of the problem. Note that always $S_p \leq 0$. It is in accordance with the above principle of the stability of the physical process.

The meaning of the input coefficient β can be identified by (6.55) and (6.56). The smaller β is, the greater the angle of the dependence $S(\Phi)$ is and greater the source function S reacts to a change in the solution of the equation and internal feedback effect is more; wherein, the relationship $S(\Phi)$ is not the same for all control volumes and it is determined by the values of the source function S and the variable Φ at each point M, obtained in the previous iteration. Using the proposed the linearization algorithm for the function $S(\Phi)$ and the installation of internal feedback (6.56) in contrast to the algorithms [1] there is no need to predetermine the maximum range of variation of the target variable Φ. This algorithm is universal, effective, and useful in solving the equations with any possible range of variation for the variables Φ, so the changes of the target variable from iteration to iteration can be effectively controlled.

In this case, the coefficient β is introduced to establish the internal negative feedback between the source function and the variable Φ, but given its regulatory role, we will be able to artificially mimic the relationship (on the principle of sustainability of the physical processes) not only for the source function, but also for other physical processes with variable Φ. Thus, with the introduction of the internal feedback coefficient β of we were able to fully restore the information lost as a result of the assumption of the constancy of the coefficients in iterative procedures, simulating the natural physical processes in computational procedures, making them identical, and thus achieve the stability of these procedures.

Dynamics of the Iterative Procedures with Internal Feedback Results of the iterative procedures are approximate and vary from iteration to iteration, if the iterative procedures are stable, they, as the results of the computational processes, can satisfy the mathematical model and the gradient of the steady solution in time (actually in iteration) is equal to zero.

Thus, the models of iterative procedures are always unsteady, even for stationary physical processes. If the real processes are sustainable, the iterative procedures still may be unstable. If we take one of iteration per unit time, the change of the solution in the computational procedure by iterations and the real time of the processes are identical. In this case, the iterative procedure simulates the time-dependent model of physical processes and the equation for them has the form of GDE

$$\frac{\partial(\rho\Phi)}{\partial t} + \mathrm{div}(\rho\vec{v}\Phi) = \mathrm{div}(\Gamma \mathrm{grad}\Phi) + S. \tag{6.57}$$

Here the generalized variable Φ is regarded as the vector of state for all unknown variables and the generalized value S—as vector of state for all the source functions. Therefore, the linearization of the generalized value S is equivalent to the linearization of the source function, and the installation of internal feedback loop of the global iterative procedure is equivalent to its installation at all local iterative procedures.

Equation (6.57) is the general equation for describing the dynamics of plasma processes and the global iterative procedure. In particular, at the convergence of the solutions $\frac{\partial\Phi}{\partial t} = 0$, and the set values of the variables are the parameters in stationary processes of the Eq. (6.1).

Equation (6.57) describing the dynamics of iterative procedures using internal feedback (6.55) and (6.56) can be represented as

$$\frac{\partial(\rho\Phi)}{\partial t} - S_p\Phi = \mathrm{div}(\Gamma \mathrm{grad}\Phi) - \mathrm{div}(\rho\vec{v}\Phi) + S_c. \tag{6.58}$$

Character of the transients for iterative procedure from the initial decision to installed one in the system is determined by the view of the left side of Eq. (6.58).

At the negative factor S_p all the coefficients in the left side of (6.58) are positive, i.e., the equation satisfies the Hurwitz stability criterion [9]. It follows that the linearization of the source function (6.55) using the algorithm (6.56) can lead the system to stability. Note also that at the large negative coefficient S_p, the transition (iterative) process is faster, i.e., the convergence condition (6.50) is performed faster due to the exponential function $\Phi(t) = A\,\exp(S_p/\rho) + C$, where A and C—are some constants.

To analyze the dynamics of iterative procedures with internal feedback a simple model of new long Arc in the channel is used (6.52). Recall that the traditional structure of iterative procedures with constant coefficients without installing of the internal feedback (6.53) led to a divergence of solutions already almost after the first iteration (Fig. 6.12). For this case, individual feedbacks of the released power σE^2 and power losses due to radiation U_{rad} with temperature can be set. However, for simplicity, combine them in a common source function: $S = \sigma E^2 - U_{\text{rad}}$, the internal feedback $S(\Phi)$ is implemented using the algorithm (6.55 and 6.56).

The discrete analoguefor the equation of energy balance (6.52) with internal feedback is the following:

$$a_M T_M = a_N T_N + a_S T_S + b_M,$$

where

$$\begin{cases} a_N = \frac{r_n \lambda_n}{r_N - r_M} > 0; a_S = \frac{r_s \lambda_s}{r_M - r_S} > 0; \\ a_M = a_N + a_S - 0.5 S_{pM}(r_n - r_s)(r_n + r_s) > 0; \\ b_M = 0.5 S_{cM}(r_n - r_s)(r_n + r_s); S_M = \left(\sigma E_z^2 - U_{\text{rad}}\right)_M. \end{cases} \quad (6.59)$$

The dynamics of an iterative procedure for the solution of (6.59) with different coefficients of internal feedback β and the initial approximations for the same grid are shown in Fig. 6.15. The various initial conditions were selected for analysis (T_0 = 8000, 10,000, 13,000 K) with appropriate coefficients of internal feedback β = 0.1; 0.5. The computational grid with steps Δr = 0.01—0.1 mm was used.

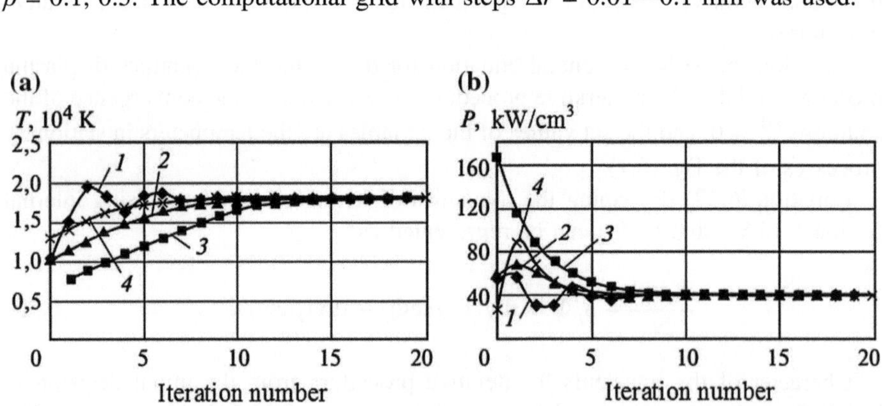

Fig. 6.15 The change of temperature T (**a**) and released power σE^2 (**b**) from iteration to iteration for the stable solutions with internal feedback for the point on the axis (r = 0) of the Arc channel (I = 90 A, R = *1.5* mm): *1* β = 0.5, T_0 = 10,000 K; *2* β = 0.1, T_0 = 10,000 K; *3* β = 0.1, T_0 = 8000 K; *4* β = 0.1, T_0 = 13,000 K

The results (Fig. 6.15a) show that for the procedure with internal feedback the convergent solutions can be obtained on the computational grid with the steps Δr = 0.01–0.1 mm with a coefficient of internal feedback $\beta \leq 0.5$. If $\beta = 0.5$, convergence is provided for all the points but with some hesitation (the two-way convergence), hence $\beta = 0.5$—a boundary point of the stability for the procedure. If $\beta = 0.1$ the solution converges without hesitation (unilateral convergence) for all central points with different initial conditions. Small fluctuations, as can be seen, can be obtained for the points in the area of sharp temperature gradient with the initial condition T_0 = 13,000 K. Thus, the dynamics of the iterative procedure depends on its structure—the coefficient of the internal feedback and the input data of the problem: current and power of the Arc, its geometry, type of gas in plasma, etc., as well as on the initial conditions and the step of the computational grid. All of the solutions are stable, however, the rapid convergence is obtained by the unilateral convergence $\beta = 0.1$ and the initial condition T_0 = 10,000 K.

Figure 6.15b shows the variation of the released power by iteration. It is seen that due to the inputted coefficient β and obtained negative feedback $(S_p < 0)$, the value of the released power vary from iteration to iteration at reasonable limits, it promotes the convergence of solution.

This example clearly shows that the new structure of iterative procedures with internal feedback has the property of stability compared to the conventional structure (without internal connections). The introduction of negative feedback on the output values provides the additional information about the nonlinearity and interdependence of some physical processes inside the procedure of the iterative solving of equation that is not in the traditional structure. From the point of view of the theory of automatic control the feedback for the time derivative with the output value is a powerful tool to ensure the stability [10].

6.3.3 The Mathematical Algorithms with Internal Feedback

As shown above (Sect. 6.3.2), the installation of additional internal feedback loop allows recovering of the lost information inside the iterative procedure and ensuring the manageability and stability of the system as a whole. Consider the discrete analogue of GDE (6.1) considering internal feedback loop and linearization algorithms of the source function for concrete equations of plasma.

The Discrete Analogue of the GDE Based on Internal Feedback Generalized differential Eq. (6.1) using the linearization of the source function (6.55) has the following form:

$$\operatorname{div}(\rho\vec{v}\Phi) = \operatorname{div}(\Gamma\operatorname{grad}\Phi) + S_c + S_p\Phi. \tag{6.60}$$

Using the method of the control volume and the corresponding computational scheme (Sects. 6.1.1 and 6.2.1), the discrete analogue of the two-dimensional GDE

(6.2) for the control volume (Fig. 6.2) with the internal feedback (6.55) and (6.56) is

$$a_M\Phi_M = a_E\Phi_E + a_W\Phi_W + a_N\Phi_N + a_S\Phi_S + b_M,$$

where

$$\begin{cases} a_E = D_eA(|Pe_e|) + \max(-F_e, 0); a_W = D_wA(|Pe_w|) + \max(F_w, 0); \\ a_N = D_nA(|Pe_n|) + \max(-F_n, 0); a_S = D_sA(|Pe_s|) + \max(F_s, 0); \\ a_M = a_E + a_W + a_N + a_S - S_{pM}\Delta V; b_M = S_{cM}\Delta V; \\ \Delta V = (z_e - z_w)\frac{(r_n - r_s)(r_n + r_s)}{2}. \end{cases} \tag{6.61}$$

Similarly the discrete analogue for three-dimensional GDE (6.19) has the form

$$a_M\Phi_M = a_E\Phi_E + a_W\Phi_W + a_N\Phi_N + a_S\Phi_S + a_T\Phi_T + a_B\Phi_B + b_M,$$

where

$$\begin{cases} a_E = D_eA(|Pe_e|) + \max(-F_e, 0) \geq 0; a_W = D_wA(|Pe_w|) + \max(F_w, 0) \geq 0; \\ a_N = D_nA(|Pe_n|) + \max(-F_n, 0) \geq 0; a_S = D_sA(|Pe_s|) + \max(F_s, 0) \geq 0; \\ a_T = D_tA(|Pe_t|) + \max(-F_t, 0) \geq 0;\ a_B = D_bA(|Pe_b|) + \max(F_b, 0) \geq 0; \\ a_M = a_E + a_W + a_N + a_S + a_T + a_B - S_{pM}\Delta V > 0; b_M = S_{cM}\Delta V; \\ \Delta V = (z_e - z_w)\frac{(r_n - r_s)(r_n + r_s)}{2}(\varphi_t - \varphi_b). \end{cases} \tag{6.62}$$

In the calculation of the coefficients for the discrete analogues of (6.61), (6.62) calculating the Eqs. (6.9)–(6.11), (6.18), (6.21)–(6.23) should be used.

Algorithms of the Installation of the Internal Feedback Loop The first linearization algorithm (6.56) for the absolute values of the variables is illustrated in Fig. 6.14. The increment of the target variable is formed in absolute value of Φ_M, multiplied by β, that is set based on the analysis of the possible increments Φ in iteration by simultaneously solving the equations. This algorithm provides a greater increment for large absolute values of variables, and vice versa. This pattern is logically justified and is particularly effective in solving the equations and the value of variables may vary over a wide range, for example, the value of the temperature or the mass concentration of particles in the plasma.

However, this algorithm has some flaw: in the case when the absolute values of the variables inside the iterative procedure are small enough, and the source function takes a very high value, the convergence of the solution can be achieved very slowly because of the large coefficient S_p. Moreover, when the current values of the variables are equal to zero, the algorithm cannot work because of the division by zero. For example, in solving the momentum equations when the components of the plasma velocity at some points may be equal to zero, or very close to zero. Thus it is possible to adopt $S_p = 0$, $S_c = S$, as well as in the traditional iterative procedure without internal feedback loop. However, the experience shows that the simulation of highly nonlinear objects such as plasma and plasma processes, even the use of a structure without internal feedback loop in one equation at a single iteration can lead to loss of stability of the computational procedures.

At the same time in many cases, it is difficult to find the initial values in the nodes that are sufficiently close to the final values. It is convenient to put some of the main variables (e.g., temperature) as constant, and the other equal to zero, especially when they can take both positive and negative values (for example, the components of the gas velocity). In this case, a second algorithm can be recorded by for the linearization of the source function not for the absolute value of the variable, but for a given their increment $\Delta\Phi$ (Fig. 6.16) [8].

$$S_p = -\left|\frac{S}{\Delta\Phi}\right|; \; S_c = S_M - S_p\Phi. \tag{6.63}$$

This linearization algorithm of the source function allows a fully getting rid of the first algorithm lack (6.56) and enabling the zero crossing for the variable inside the iteration. The source function is linearized for all possible values of the variables, without exception, in the formation of its coefficients not the absolute values of the variables, but their increments ($\Delta\Phi \neq 0$) are used. Note that because of constancy $\Delta\Phi$ the algorithm (6.63) may not be effective for the equations where the variables can take values over a wide range.

Thus, the third linearization algorithm of the source function can be obtained by combining the first two algorithms (6.56) and (6.63):

$$S_p = -\left|\frac{S}{\beta\Phi}\right| \text{ for } |\Phi| \geq \Delta\Phi; \; S_p = -\left|\frac{S}{\Delta\Phi}\right| \text{ for } |\Phi| < \Delta\Phi; \; S_c = S - S_p\Phi. \tag{6.64}$$

If the source function is explicitly analytic function $S(\Phi)$, then there are many variants of linearization for the source function $S = S_c + S_p\Phi$. The main thing is

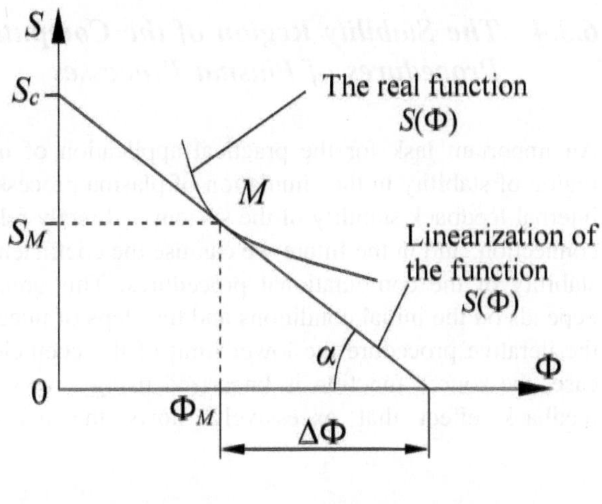

Fig. 6.16 Linearization of the source function using the given increment of variables

that it is necessary to choose the appropriate terms S_c, S_p depending on the physical formulation of the problem. For example, if the source function can be written as $S = S_1 - S_2\Phi$, then the linearization can be carried out according to the algorithm

$$S_p = -\beta S_2; S_c = S - S_p\Phi. \tag{6.65}$$

In the fourth linearization algorithm of the source function (6.65) the coefficient β plays the role of internal feedback and is regulated. It shows how many times the real coefficient S_p effectively increases. For example, the source function

For the equation of the energy balance of atom–ion gas in the two-temperature plasma model has the form

$$S = \frac{6m_e}{5} n_e \sqrt{\frac{8kT_e}{\pi m_e}} (n_a Q_{ea} + n_i Q_{ei})(T_e - T_a).$$

The linearization of the source function can be performed as

$$S_p = -\beta \frac{6m_e}{5} n_e \sqrt{\frac{8kT_e}{\pi m_e}} (n_a Q_{ea} + n_i Q_{ei});\ S_c = S - S_p T_a.$$

The linearization of the source function is possible even when it is equal to zero. When using the above four algorithms the linearization of the source function fails only on condition that if the source function $S \neq 0$. If $S = 0$ or it is close to zero, it is possible to artificially create an internal feedback, expanding the source function into two equal parts $S = S^* - S^*$, where S^* should be selected on the basis of physical considerations for the particular case. The first function S^* is further a subject for linearization $\left(S^* = S_c + S_p\Phi\right)$ using one of the above algorithms.

All of the above algorithms have been tested and successfully implemented in the simulation of different plasma processes.

6.3.4 The Stability Region of the Computational Procedures of Plasma Processes

An important task for the practical application of internal feedback is to find a region of stability in the simulation of plasma processes. In a closed structure with internal feedback stability of the system is directly related to the installation of this connection, and in the future we can use the coefficient β to determine the region of stability of the computational procedures. This area has the closed borders and depends on the initial conditions and the steps of integration (steps of the grid). For the iterative procedure the lower limit of the coefficient β is close to zero. In this case, the source function is linearized using a very large coefficient S_p with the feedback effect that excessively slows the decision and achieving of the convergence. The upper limit of β depends on the conditions of a specific task. The region of stability for computational procedures can be determined as the multi-dimensional enclosed space, whose dimension is determined by the number of equations, included in the model, and each coordinate of the space is the range of variation for the corresponding coefficient of internal feedback β.

As a result of the simulation of plasma processes the information about the region of calculation procedures stability at the solving of the basic equations for plasma in a wide range of input date for RF- and Arc plasma torches is obtained. Thus, in the solution of the equation of the energy balance it was obtained: $P = 0.5$–$1\ \text{kW/cm}^3$—for RF-plasma torches, $P = 1$–$10\ \text{kW/cm}^3$—for free Arc column and most Arcs in the channel and the anode Arc spots, $P = 50$–$100\ \text{kW/cm}^3$—for the compressed Arc in the channel and for the cathode Arc spots. Similarly, at the decision of the equation of momentum $F = 0.005$–$1\ \text{N/cm}^3$, at the decision of the continuity equation $\Delta p = 20$–5000 Pa and at the decision of the equation of the energy balance for the electron gas $P = 0.5$–$100\ \text{kW/cm}^3$. Steps of the grids are regulated in the range $\Delta h = 0.001$–10 mm for the power of modern computers. Figures 6.17, 6.18, 6.19 and 6.20 show the stability regions for the simultaneously solving the different equations in various models of plasma.

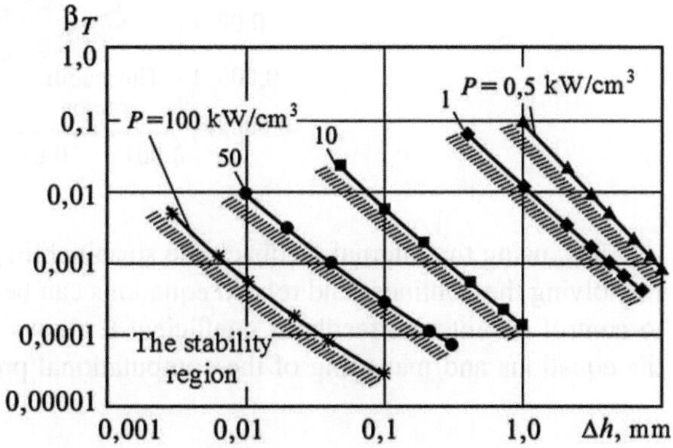

Fig. 6.17 The stability region of iterative procedure at solving the equation of the balance energy together with other equations for the model of equilibrium plasma (analogue *D*-decomposition of Y. Neumark [10])

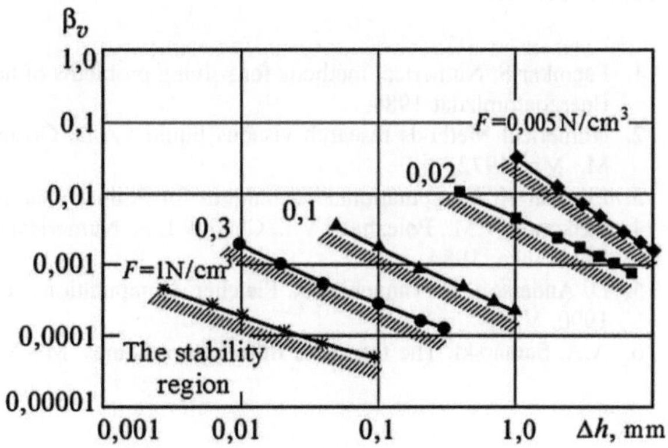

Fig. 6.18 The stability region of the iterative procedure at solving the momentum equation together with other equations of equilibrium plasma

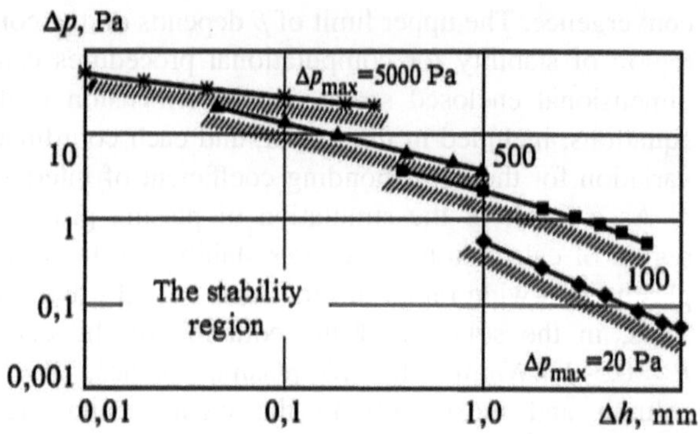

Fig. 6.19 The stability region of the iterative procedure at solving the continuity equation together with other equations of equilibrium plasma

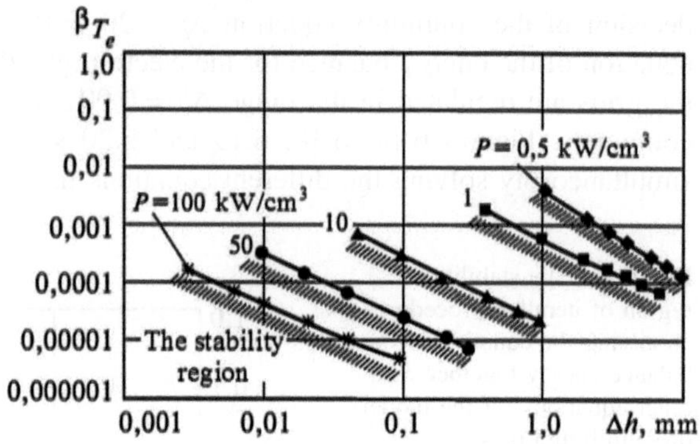

Fig. 6.20 The stability region of the iterative procedure at solving the equation of the balance energy for the electron gas together with other equations of nonequilibrium plasma

Thus, using the internal feedback the sustainability of computational procedures for solving the nonlinear and related equations can be fully achieved, and the ability to control the internal feedback coefficient β allows the monitoring the solving of the equations and managing of the computational processes.

References

1. Patankar S. Numerical methods for solving problems of heat transfer and fluid dynamics: M.: Energoatomizdat 1984.
2. Numerical Methods research viscous liquid / A.D. Gosmen, V.M. Pan, A.K. Ranchen, etc. M.: Mir, 1972.
3. Fletcher K. Computational Techniques for Fluid Dynamics. M.: Mir, 1991. V. 1, 2.
4. Paskonov V.M., Polezhaev V.I., Chudov L.A. Numerical modeling of heat and mass transfer. M.: Nauka, 1984.
5. D. Anderson, D. Tannehill, R. Pletcher. Computational Fluid Mechanics and Heat. M.: Mir, 1990. V. 1.
6. A.A. Samarski. The theory of difference schemes. M.: Nauka, 1989.

7. Shi Nguyen-Kuok. Ensuring the sustainability of the computational process in the simulation of heat and gas-dynamic systems with distributed parameters // Proc. of the International Scientific and Technical. Conf. "Information technologies in modeling and management." SPb. 1996. PP. 68–70.
8. Shi Nguyen-Kuok. A stable solution of system of differential equations for modeling the Low-temperature plasma processes // Proc. of the I International scientific-practical conference. Conf. "Differential Equations and Applications". SPb. 1996. PP. 234–236.
9. Besekersky V.A., Popov E.P. The theory of automatic control systems. M.: Nauka. 1972.
10. The theory of automatic control / Ed. by A.V. Netushin. M.: Higher School. 1976.

Chapter 7
The RF Plasma Torches

7.1 Characteristics of the RF Plasma Torches

RF plasma torches—plasma devices that are widely used in science and technology. RF discharge does not require the electrodes, there is a ring current is one of the few sources that generate clean plasma electrode materials are not contaminated. These plasma torches have a relatively large space plasma flow and have a virtually unlimited service life in any gaseous medium. RF plasma torches (Fig. 7.1) allow obtaining the pure plasma with a temperature of 7000–11,000 K at atmospheric pressure. The heating of the plasma without electrodes in the RF plasma torches of the induction type is described using the law of induction and conditions of the heating in a conducting medium and in a high-frequency alternating electromagnetic field. The conducting medium, in this case, is ionized gas—plasma. High-frequency magnetic field of the inductor generates an alternating magnetic flux $d\Phi/dt$ variable and induces an electromotive force (EMF), whereby currents having annular i_φ differing in amplitude and phase and having a spatial distribution of the plasma in the induction.

There are the following types of discharge chambers of the RF plasma torches (Fig. 7.2): the quartz chamber that is not cooled by water; the quartz water-cooled chamber; the metal chamber with cuts; the porous chamber.

The stabilization of the plasma in the discharge chamber of the RF plasma torches is generally carried out by specially formed gas flows. There are three ways of such stabilization: the axial gas stabilization; the gas vortex stabilization; and stabilization by unformed gas flow.

The temperature of the RF plasma torches The investigations of the temperature distribution in RF plasma were carried out by many authors [1–11]. The spatial distribution of temperature and electrical conductivity for the RF Argon plasma are shown in Fig. 7.3a. The radial temperature distributions in argon plasma and in plasma of molecular gases in the area of maximum temperature, according to different authors, are represented at Fig. 7.3b, c. The results shown in Fig. 7.3 are

S. Nguyen-Kuok, *Theory of Low-Temperature Plasma Physics*,
Springer Series on Atomic, Optical, and Plasma Physics 95,
DOI 10.1007/978-3-319-43721-7_7

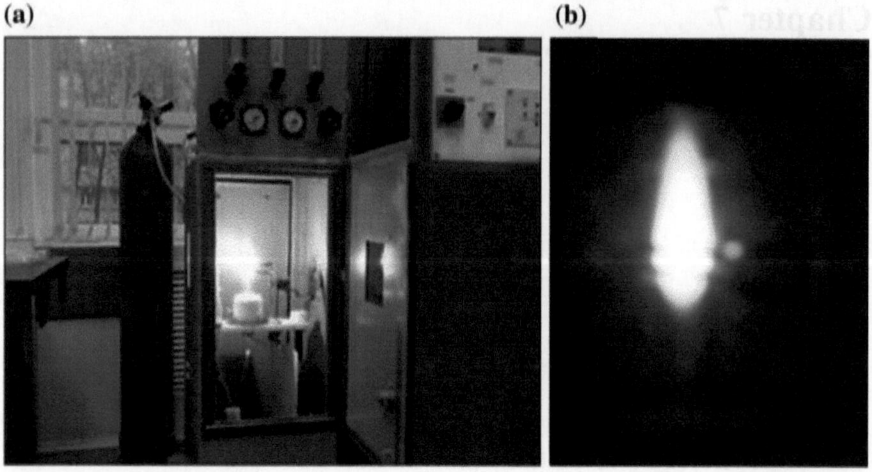

Fig. 7.1 The installation of the RF plasma torch (f = 27.12 MHz, the discharge power P_2 = 3 kW) (**a**) and an Argon plasma in the inductor with gas consumption Q_{Ar} = 30 l/min (**b**)

obtained for the plasma torches with the quartz chamber when the gas vortex stabilization of the RF plasma. The average temperatures in the central part of plasma depending on the power of plasma discharge and the gas consumption are shown in Fig. 7.4.

The results of the general analysis of temperature represented in Figs. 7.3 and 7.4 show that the maximum of temperature in the RF plasma torches is located not on the axis of the plasma torch, but in the zone of maximum current density and the released power. The values of temperature obtained by the different authors in various conditions at certain frequencies of generator, diameters of plasma torches, the powers of discharge are not very different to each other. Thus, for RF Argon plasma the maximum of temperature is 9000–11,000 K, for Air and Nitrogen 6500–8500 K, for oxygen 8500–9500 K. Such a coincidence of the results is due to the fact that in all experiments, the power per unit of plasma volume is changed slightly.

Thermal nonequilibrium plasma in the RF plasma torches is found by researchers [12–14]. The first attempt to quantitatively measure the effect of nonequilibrium was made by author in [12]. The electron temperature in Argon was measured by the absolute intensity of the recombination spectrum of ArI in the wavelength λ = 430–480 μm. The measurements were carried out (on the cut) of the RF plasma torch with a tube diameter of 28 mm with a power of discharge in plasma 7–8 kW and the frequency of generator 17 MHz.

The temperature of the atom–ion gas was determined by hot-wire anemometer, i.e., the heat flow going to the water-cooled copper cross-flow tube with diameter of 1.35 mm. These measurements of the total heat flow via a tube from the center to the edge of the plasma jet by means of conversion were leaded to the values of the heat flux in the local point. Heat flow in accordance with the theory of heat transfer

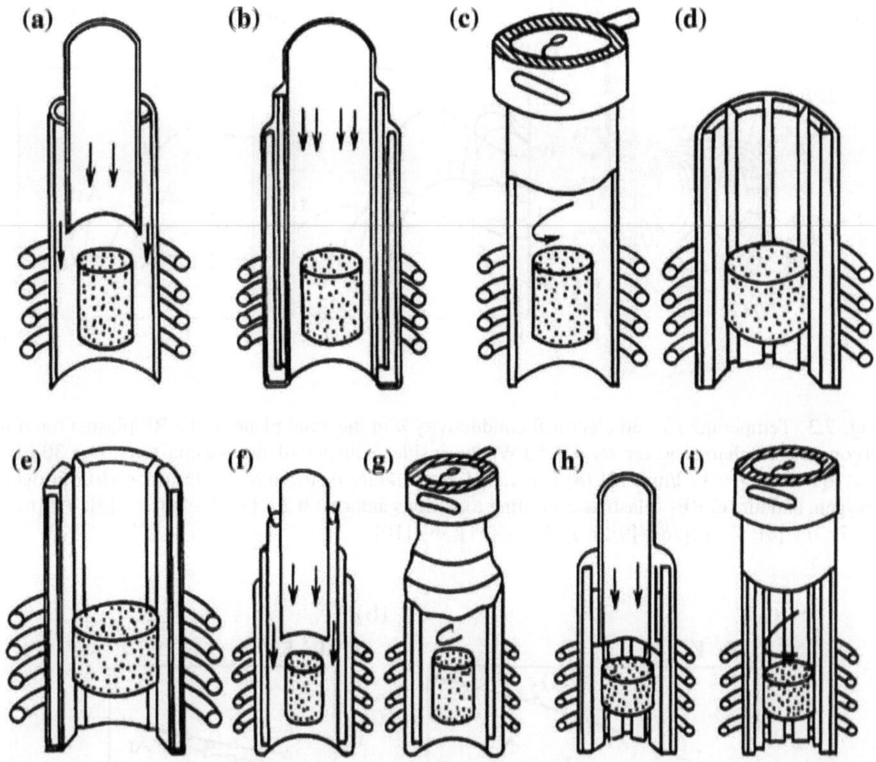

Fig. 7.2 The main types of discharge chambers in the RF plasma torches [1]: **a** the quartz chamber that is not cooled by water (axial gas stabilization); **b** the quartz water-cooled chamber (unformed axial flow); **c** the quartz chamber without water cooling (vortex stabilization); **d** the metal cooled split chamber with a quartz sheath; **e** the metal cooled chamber with refractory inserts; **f** the quartz water-cooled chamber (axial stabilization); **g** the same (vortex stabilization); **h** the metal split cooling chamber (axial stabilization); **i** the same (vortex stabilization)

is related to the enthalpy (i.e., the temperature of the heavy component) and it allows obtaining of the value T_e and *Ta* in the same local point. It was found that on the axis of the plasma torch the temperature of electrons is not equal to the temperature of the ions and atoms: (T_e = 10,200 K; T_a = 7000–8000 K) [12].

In [13] the relaxation method for measuring of T_e and T_a was used. It is based on the temperature difference between the relaxation times of the atom–ion and electron gases. It was estimated [13], the relaxation time for T_e is $\tau_e = 5 \times 10^{-6}$ s and for $T_a - \tau_a = 10^{-2}$ s. Measurements of T_e and T_a in such method [13] were made at the following conditions: frequency f = 26 MHz, the plasma discharge power of 3–6 kW, the diameter of quartz chamber $d = 2R$ = 20 mm. The grid modulation scheme (quenching) of the high voltage and automatic recording of the intensity of the spectral lines and the continuum on a fast spectrograph were applied. The results of temperature measurements by this method are presented in Table 7.1.

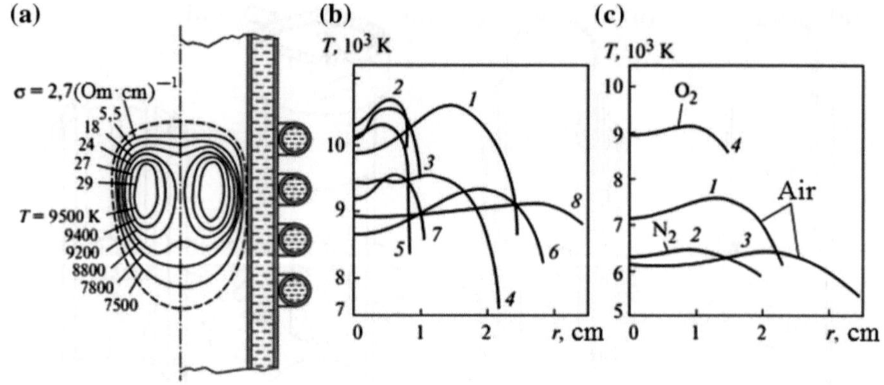

Fig. 7.3 Temperature T and electrical conductivity σ in the axial plane of the RF plasma torch in argon at a discharge power P_2 = 2.5 kW, the inside diameter of the plasma torch d = 30 mm, f = 26 MHz, Q = 15 l/min [2] (**a**) and radial temperature distribution in an argon (**b**), nitrogen, oxygen, and air (**c**) RF—plasmas according to various authors: **b** *1*—[3]; *2*—[4]; *3*—[5]; *4*—[6]; *5*—[7]; *6*—[8]; *7*—[2]; *8*—[9]; **c** *1, 2, 3*—[11]; *4*—[10]

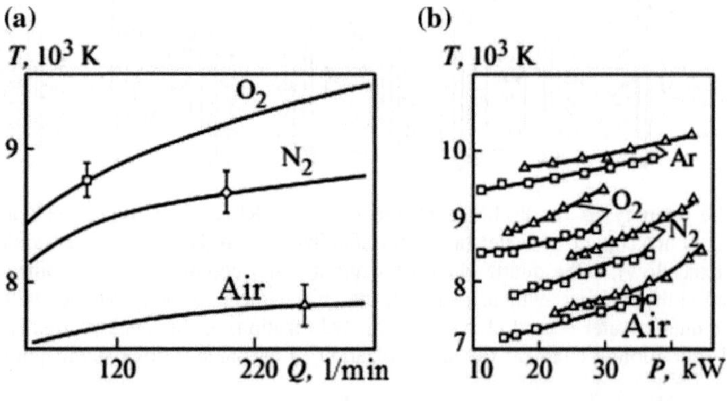

Fig. 7.4 The average temperatures in the RF plasma torches from the gas consumption Q (*a*) (P = 30 kW) and from the power of plasma P (*b*): *triangle* Q = 250 l/min; *square* Q = 75 l/min; the inner diameter of the quartz plasma torch 60 mm; f = 5.28 MHz

Table 7.1 The results of measurements of the electron and atom–ion temperatures using the gas relaxation method [13]

The coordinate r/R	T_e (10^3 K)	T_a (10^3 K)	$T_e - T_a$ (10^3 K)
0.9	9.3	6.65	2.63
0.5	10	8.3	1.7
0	10	8.7	1.3

One of the carefully the temperature of electrons and ion–atom gas in the RF Argon plasma was measured by authors in [14]. In this work, in addition to hot-wire anemometer, the method of enthalpy probe was used. RF discharge was excited in

Table 7.2 The values of T_e and T_a in induction argon plasma at the exit from the plasma torch according to [14]

U_1 (kV)	Q (l/min)	T_e (10^3 K)	T_a (10^3 K)	P_2 (kW)
3	5	9.5	9.0	13.5
3	10	9.5	6.5	15.8
3	15	9.5	5.0	19.0
3	20	10.0	4.0	19.5
3	30	10.5	4.0	20.5
2	5	8.5	5.5	6.0
2	10	9.0	4.0	8.2
2	15	9.5	3.0	9.4
2	20	10.0	3.0	9.5
2	30	10.5	3.0	9.5

the metal water-cooled chamber with an internal diameter of 32 mm. Generator with the frequency of 6 MHz was used, power dissipated in plasma was about 4–6 kW. The plasma-forming gas was axially supplied into the discharge and its consumption was varied from 1 to 30 l/min.

The measurement results for T_e and T_a for axial point $r = 0$ in various regimes are shown in Table 7.2.

It also shows the total power of the induction plasma torch P_2. Values of T_e at the axis of plasma torch changing with the changing P_2 and Q vary slightly in the wide diapason. This is confirmed by many measurements of different authors, and the temperature of the ion–atom gas is strongly dependent on consumption of argon and voltage in the inductor (Fig. 7.5a). With increasing consumption Q, the temperature of atoms and ions falls sharply. For example, for Q = 20–30 l/min, T_a is about 3 times less than T_e. At low consumptions (1–5 l/min), when U_1 = 2.5–3 kV temperature T_e at the axis coincides with T_a. However, at the voltage across the inductor U_1 = 1.6–2 kV, T_a, even at low flow consumptions T_e is well below.

At Fig. 7.5b radial temperature distribution in the measured cross section for two cases is shown: (1) on an axis $T_e = T_a$ (Q = 5 l/min); (2) $T_e \neq T_a$, (Q = 20 l/min). The figure shows that even in the case where the plasma at the axis of the discharge is equilibrium, in the periphery region there is a significant discrepancy of T_e, and T_a.

From the data [15–17] it follows that the primary mechanism of disorders of the thermal equilibrium is the penetration of a "cold" plasma gas in plasma clot. The "cold" gas incoming into plasma is almost no change the electron temperature, but significantly reduces the atom–ion temperature, but the consumption of gas, "captured" by plasmoid is exclusively determined by the value of the magnetic overpressure (magnetic pinch effect), and very weakly depends on the consumption of gas supplied to the plasma torch. Gas that is not captured by plasmoid wraps around its front part. In the measurement results [12–14] it is shown that a large temperature difference between the electrons and the atom–ion gas is observed only in argon plasma. In the plasma of molecular gases T_e and T_a have only the slightly difference.

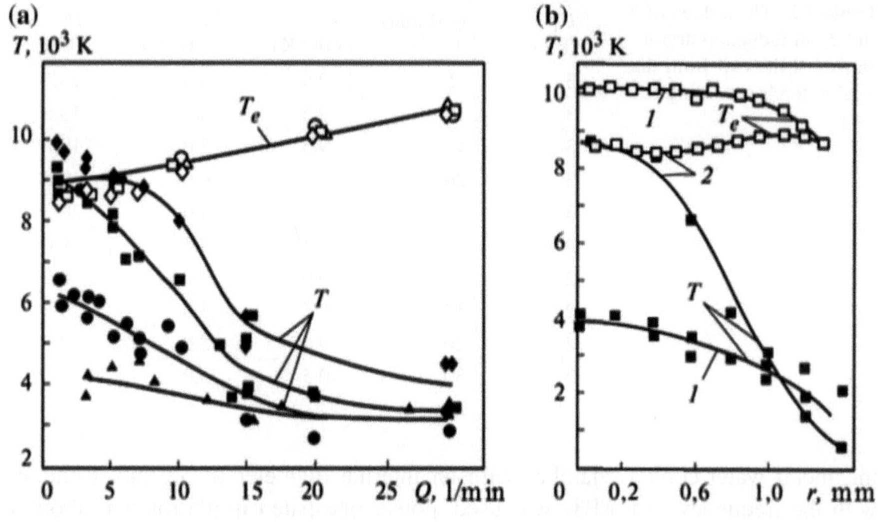

Fig. 7.5 The electron and atom–ion temperatures (T_e and T) at the cut of the RF plasma torch on the various voltages of inductor U_1 (**a**) and gas consumption Q (**b**): **a** *open diamond, filled diamond*—U_1 = 3 kV; *open square, filled square*—U_1 = 2,5 kV; *open circle, filled circle*—U_1 = 2 kV; *open triangle, filled triangle*—U_1 = 1,6 kV; **b** *1*—Q = 5 l/min; 2—Q = 20 l/min

Gas dynamics in the RF plasma torches In numerous studies of gas dynamics of the RF plasma torches [12, 18–22], the speed of the plasma jet, flowing into the open space, was measured. These experiments are interesting from the point of view of the use of the plasma jet for the technological processes, but do not contain enough information for the analysis of the characteristics of the RF discharge. Research of [23, 24] showed that the plasma torch with the vortex stabilization has a very complex distribution of velocities. Tangential entrance of the gas into the discharge chamber leads to the formation of the vortex nucleus zone where there is a reverse current from the central zone of the plasma discharge. In the area of inductor the gas flow near the wall is directed to the exit of the plasma torch, in the axial region—in the opposite direction. Due to this fact there is a high recirculation of plasma in plasma torch with vortex stabilization, and it leads to decrease in the proportion of power supplied to the jet.

However, analysis of the experimental data of different authors shows that both in tangential and axial entering of the plasma-forming gas the backflow zone is observed in the inductor. The mechanism of gas-dynamic processes occurring in the induction plasma torch, based combustion model, was suggested by [25]. The area of active energy release in it—is a skin layer, where the heating and acceleration of the gas flow occurs in direction from the wall towards the axis. The lines of the gas current are refracted in the skin layer, and cold gas penetrates to the central region of the plasma torch through the side surface of the discharge. Productivity of pumping through the skin layer depends primarily on the power released at him. If the supplied gas does not have time to pass through the plasma, the plasmoid is a tube, and as a result of the formed "affluent" gas may be ejected from the area of inductor and the plasma torch is extinguished. If the gas is fed too little, then there is an internal recirculation of the gas. Then the appearance of local flows in the direction opposite to the jet stream perhaps. However, the experimental verification of the model showed that gas flow direction inside the plasmoid is absent. This means that neither the thermal mechanism nor the combustion mechanism [25], are not basic in this case.

Depending on the position of researchers the magnetic pressure in the RF discharge takes a decisive role in the dynamics of plasma flows [20], or it is ignored [26]. Chase [20, 21] pointed to the decisive role of the magnetic pressure in the induction plasma when the gas is accelerating. Conclusions of J. Chase are confirmed in [27], here the calculation results of the plasma jet velocity are obtained by solving the energy balance equations, the Maxwell equation, and the momentum equation. The calculation results are in good agreement with the experimentally measured values of the velocity of plasma jet. The measurements of air plasma velocity at the exit of the RF plasma torch are shown at Fig. 7.6. The experimental [28] data and the data of calculations [27] are in good agreement that plasma acceleration (in axial direction) occurs at the boundary of plasmoid and open space at the side of open space due to overpressure in the bunch (magnetic pinch effect). Cold gas, that is penetrating in a radial direction into a central discharge zone, leads to decrease in temperature on the axis of the plasma torch in the radial direction. Magnetic pressure gives a significant acceleration in the discharge of the heated gas toward the exit and the proportion of the power transmitted in the stream in the RF plasma torch will be higher than for vortex plasma torches. The mechanism of the gas acceleration by the electromagnetic forces, of course, exists also in the vortex plasma torches.

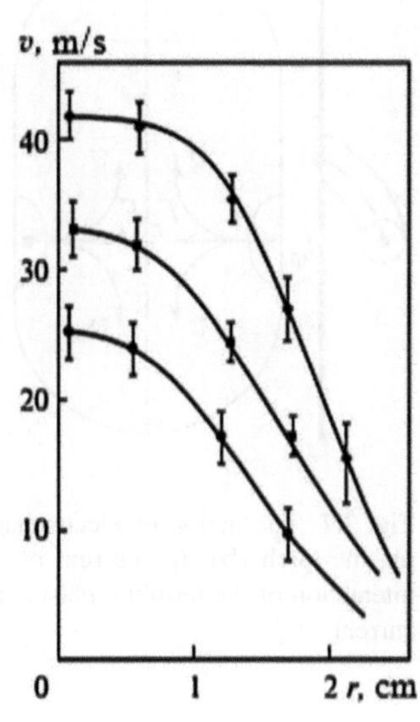

Fig. 7.6 Radial velocity distribution of the RF Air plasma torch (f = 5.28 MHz) at z = 20 cm from the cut at various discharge powers: *filled diamond*—P = 53 kW, *filled square*—P = 39 kW, *filled circle*—P = 25 kW, Q = 50 l/min

The gas acceleration in the RF plasma torch can be described as follows: RF discharge glows inside the inductor and the induced currents i_φ in the discharge interact with the magnetic field (for example, with a longitudinal component B_z of the magnetic field induction), a electromagnetic force F_r occurs as a result, compressing the plasma, and plasma is driven in the direction of the axis (Fig. 7.7a). In the future the mechanism of the thermal acceleration leads to the motion of plasma stream on the basis of the law of mass conservation. The amount of gas is drawn into the center of plasma torch in radial direction, and the same is supplied to the exit of plasma torch in the axial direction. The external "hot" layer of the gas penetrates into plasma in the radial direction, is reacted with plasma, is rapidly heated and accelerated (axially) as the penetration into discharge.

Thus, the electromagnetic force plays an important role in shaping of the plasma flow movement in the RF plasma torch. This force accelerates the plasma in the direction of the center axis of the plasma torch, leads to the emergence of front and aft vortex flows at the top and bottom (with sufficient force of inductor current) zones of plasma torch (Fig. 7.7b). The formation of such areas is known in the theory of induction heating as the effect of electromagnetic stirring.

However, the component of the electromagnetic force F_r is able not only to carry away the gas in the center of discharge, followed by heating and acceleration, but at the same time form an excessive magnetic (gasostatic) pressure in the discharge,

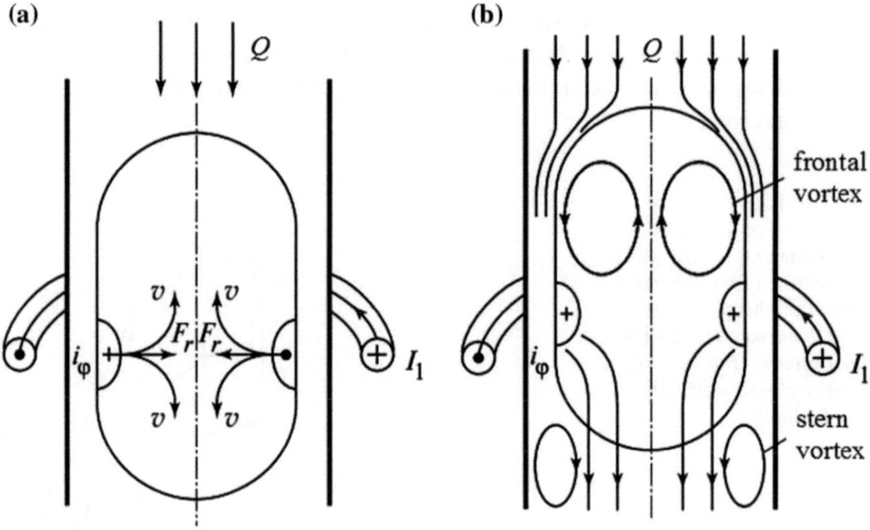

Fig. 7.7 The action of electromagnetic force (**a**) and qualitative picture of gas flow in the RF plasma torch (**b**): I_1—current of inductor; F_r—electromagnetic force that occurs as a result of interaction of the terminal plasma current with a longitudinal magnetic field; i_φ—the induced ring current

that has a maximum on the discharge axis. Thus the gradient of gas-static pressure $-\frac{\partial P}{\partial r}$ and component of electromagnetic force F_r are in opposite directions. There are just two mechanisms: the acceleration of plasma flow by the action of the electromagnetic force F_r and braking (preventing to the penetration of the gas flow into discharge) due to the effect of the gradient force gas-static pressure. Thus, the resulting force $F_r - \frac{\partial P}{\partial r}$ causes to acceleration of plasma stream in the radial direction that deliver a small amount of gas into discharge. Our most careful calculations based on a full simulation of the RF plasma torches show that the resulting force of the plasma acceleration $F_r - \frac{\partial P}{\partial r}$ is only (5–10) % of the electromagnetic force F_r. It allows enter into discharge in the radial direction to the external hot layer of gas ($T > 3000$ K) that is a small fraction of the total number of blown gas. Such a general mechanism is called as magnet-gas-static acceleration in the channel of plasma torch.

Therefore, assumption of some researchers about the equality of the electromagnetic force and the gradient of static pressure or disregard of them in solving the equations of plasma motion are erroneous. This undoubtedly leads to an erroneous determination of the resulting force and acceleration of plasma to an inaccurate calculation of the plasma flow velocity.

The answer to the question how much gas can penetrate into the discharge is found due to the interaction of two processes occurring simultaneously in the channel of plasma torch: acceleration and gas heating at a determination of discharge power. We see that for a given power the steady burning of plasma allows amount of gas to penetrate into plasma and this amount warms up to the high plasma temperature, that provides the conductivity and efficient allocation of energy in the discharge.

We assume that for some reason, more gas with some power than discharge can heat penetrates into the discharge. Then the external skin layer of discharge cools automatically and the temperature and electromagnetic force F_r acting on the plasma flow decrease. As a result the amount of gas entering to plasma reduces. Thus, the discharge automatically adjusts and maintains the steady burning in electromagnetic field of inductor and keeps the high temperature inside the channel of the RF plasma torch. This is possible due to the internal mechanism of negative feedback of thermodynamic and electromagnetic processes in the discharge. Due to the installation of such links it is possible to simulate these processes in the mathematical model and to ensure the stability of computational procedures for simulation of plasma processes (Sect. 6.3). Excessive amount of cold gas moves past near the plasma flow and forms a natural thermal protection layer along the wall of the plasma torch.

We see that the movement and heating of the plasma flow in the channel of plasma torch are the result of interaction of electromagnetic, thermodynamic and plasma processes in the RF discharge. The picture of its movement and heating can be based on the simulation of all these processes.

7.2 Electromagnetic Field in the RF Plasma Torches

The allocated power in the RF discharge and the degree of gas heating, as well as electromagnetic force acting on the plasma flow and plasma acceleration, have direct connection with the distribution of electromagnetic field; the calculation of the electromagnetic field inside the channel of plasma torch is also an important issue in the simulation of the RF plasma torches.

The RF plasma torch is a complex electromagnetic system. Plasma flow with different temperatures can be regarded as a conductive medium in the RF—plasma torch. It has a certain distribution of electrical conductivity along the length and radius of the channel of plasma torch and the magnetic field of inductor induces the terminal currents in plasma that are different to each other in amplitude and phase. The interaction of two fields generated by the current of inductor and plasma currents, it defines a single electromagnetic field inside the RF plasma torch.

The first works where the RF discharge was regarded as electrodynamic system with distributed parameters, should be considered as works by J. Thomson [29, 30]. Applying the solution of equations of electrodynamics for a long solenoid, he justified the magnetic induction nature of discharge without electrodes and defined the ignition threshold of discharge and certain conditions of its existence.

The number of works that are devoted to experimental measurements of the distribution of electric E and magnetic H fields and the current density j in the RF plasma torches is small. This is because of significant difficulties due to consequence of high temperatures to the sensors of measuring systems and due to plasma disturbances by the sensors at the change of its parameters. Therefore the temperature distribution and one of the parameters (E or H) are the most often determined. Other characteristics of discharge are calculated on the basis of these measurements using one of mathematical models. This is the approach that is used in [31–33] for study of the magnetic field topology in the RF plasma torches using a water-cooled magnetic probe. The scheme for measurement of the parameters of the RF plasma torch ($f = 2.6$ MHz, power 25 kW) are shown in Fig. 7.8, and the measurements of magnetic field in the central section of the inductor [32] are presented in Fig. 7.9. Measurements were carried out for two cases: in empty and loaded with plasma inductors at the same values of current in inductor $I_1 = 68$ A, the voltage at the inductor $U_1 = 5.7$ kV. Results of conductivity measurements and determination of electric field and current density in a cross section are shown in Fig. 7.10.

Unfortunately, the lack of information about the distribution of longitudinal magnetic field makes it impossible to obtain the integral parameters of the RF discharge. However, experimental data [31–33] allow making the estimate calculations and obtaining a qualitative picture of electromagnetic field inside the RF plasma torches.

Calculation of the electromagnetic field of the RF plasma torches is usually based on the solution of Maxwell equations

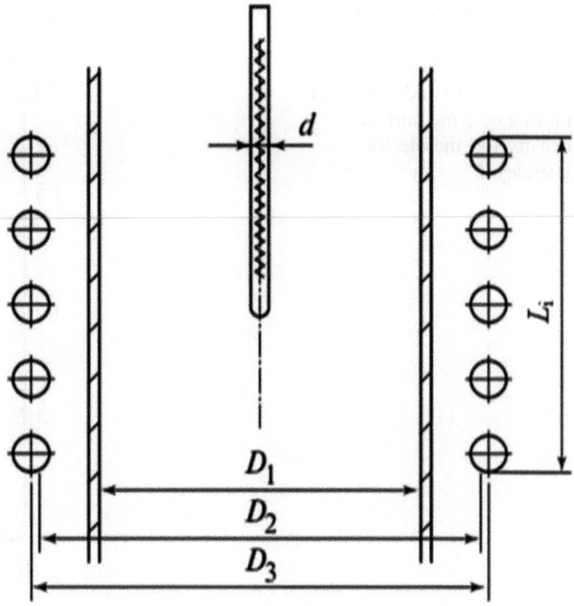

Fig. 7.8 Scheme of the field measurement using a magnetic probe [32]: D_1 = 155 mm—the internal diameter of discharge chamber, D_2 = 199 mm—the internal diameter of inductor, D_3 = 206 mm—diameter of the inductor on the coil axis, L_i = 134 mm—the length of inductor, d = 3 mm—diameter of the magnetic probe

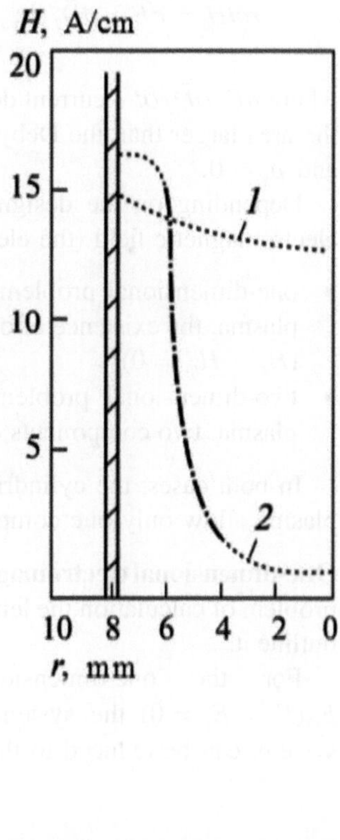

Fig. 7.9 Measurements result [32] of the magnetic intensity in unloaded inductor (*1*) and in the RF Argon plasma torch (*2*)

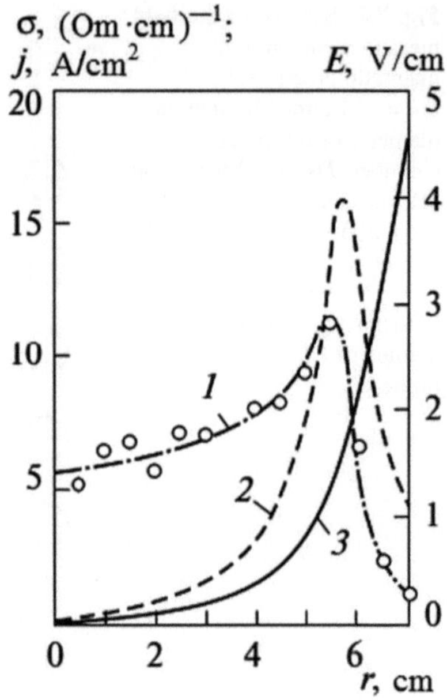

Fig. 7.10 Radial distribution of the electric field in the RF Argon plasma torch [32]: *1*—the electrical conductivity of plasma σ; *2* the current density *j*; *3* the electric intensity *E*

$$rot\vec{H} = \sigma\vec{E} + \partial\vec{D}/\partial t; \quad rot\vec{E} = -\frac{\partial\vec{B}}{\partial t}; \quad div\vec{B} = 0; \quad div\vec{E} = \rho_s, \tag{7.1}$$

where $\sigma\vec{E}$, $\partial\vec{D}/\partial t$—current density of conduction and of displacement in plasma; in the area larger than the Debye radius ($r > r_{\rm D}$) and quasi-neutral plasma $(Zn_i = n_e)$ and $\rho_s = 0$.

Depending on the design of inductors there are two types of models of the electromagnetic field (the electromagnetic problem) of the RF plasma torches:

- one-dimensional problem that is used to calculate the length of inductor with plasma: the existence of only one component of the magnetic field H_z is allowed $(H_r = H_\varphi = 0)$.
- two-dimensional problem, that can be used to calculate the inductor with plasma: two components of the magnetic field H_z, H_r, $(H_\varphi = 0)$ are considered.

In both cases, the cylindrical shape of the inductor and the ring currents in the plasma allow only one component of the electric field E_φ, $(E_z = E_r = 0)$.

One-dimensional electromagnetic problem Method of solving the one-dimensional problem of calculation the length of inductor with plasma is quite simple. We briefly outline it.

For the one-dimensional problem using H_z, $(H_r = H_\varphi = 0)$ and $E_\varphi(E_z = E_r = 0)$ the system of Eq. (7.1), written in the cylindrical coordinate system, can be reduced to the following form:

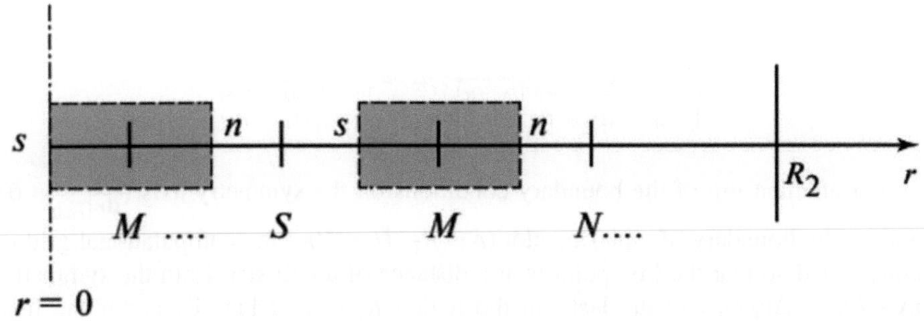

Fig. 7.11 The grid and control volumes for solving the one-dimensional electromagnetic problem of long inductor with plasma

$$\frac{1}{r}\frac{\mathrm{d}}{\mathrm{d}r}\left(\frac{r}{\sigma + j\omega\varepsilon_0}\frac{\mathrm{d}\dot{H}_z}{\mathrm{d}r}\right) - j\omega\mu_0\dot{H}_z = 0, \tag{7.2}$$

Here $\dot{H}_z \equiv \dot{H}$—the complex intensity of the magnetic field.

The solution of Eq. (7.2) is carried out by the method of control volume (Sect. 6. 1.1). Balance of field and the discrete analogue are obtained by integrating (7.2) of the control volume (Fig. 7.11).

$$\int\limits_{r_s}^{r_n}\frac{1}{r}\frac{\mathrm{d}}{\mathrm{d}r}\left(\frac{r}{\sigma + j\omega\varepsilon\varepsilon_0}\frac{\mathrm{d}\dot{H}}{\mathrm{d}r}\right)r\mathrm{d}r - j\omega\mu_0\int\limits_{r_s}^{r_n}\dot{H}r\mathrm{d}r = 0.$$

After integration we will have

$$\left(\frac{r}{\sigma + j\omega\varepsilon_0}\frac{\mathrm{d}\dot{H}}{\mathrm{d}r}\right)_n - \left(\frac{r}{\sigma + j\omega\varepsilon_0}\frac{\mathrm{d}\dot{H}}{\mathrm{d}r}\right)_s - 0,5j\omega\mu_0\dot{H}_M(r_n - r_s)(r_n + r_s) = 0,$$

Here the first derivative of the magnetic field at the edges of the discrete analogue is determined as follows:

$$\left.\frac{\mathrm{d}\dot{H}}{\mathrm{d}r}\right|_n = \frac{\dot{H}_N - \dot{H}_M}{r_N - r_M};\quad \left.\frac{\mathrm{d}\dot{H}}{\mathrm{d}r}\right|_s = \frac{\dot{H}_M - \dot{H}_S}{r_M - r_S}.$$

Consequently, the discrete analogue for Eq. (7.2) at internal points is:

$$a_M\dot{H}_M = a_N\dot{H}_N + a_S\dot{H}_S + b_M, \tag{7.3}$$

where

$$\begin{cases} a_N = \frac{r_n}{(\sigma_n + j\omega\varepsilon_0)(r_N - r_M)}; \; a_S = \frac{r_s}{(\sigma_s + j\omega\varepsilon_0)(r_M - r_S)} \\ a_M = a_N + a_S + 0.5j\omega\mu_0(r_n - r_s)(r_n + r_s) \end{cases}$$

For efficient use of the boundary conditions on the symmetry axis ($\left.\frac{d\dot{H}}{dr}\right|_{r=0} = 0$) and on the boundary of a plasma clot ($r = R_2$; $\dot{H} = H_{R_2}$) the computational grid is constructed so that the first point is at a distance of a half-step from the symmetry axis ($r_M = \Delta r/2$), and the last—in the region R_2 (Fig. 7.11). Then, for the first control volume in accordance with the boundary condition on the axis of symmetry we have the discrete analogue (7.3) with $a_s = 0$. For points at the distance R_2, $\dot{H} = H_{R_2}$.

Wherein

$$\begin{aligned} H_{R_2} = \frac{I_1 W}{\pi h}\sqrt{\frac{R_1}{R_2(z)}}\Bigg\{ & \operatorname{arctg}\frac{0.5h - z}{R_1 - R_2(z)} + \operatorname{arctg}\frac{0.5h + z}{R_1 - R_2(z)} \\ & - \left(1 - \sqrt{\frac{R_2(z)}{R_1}}\right)\left(\operatorname{arctg}\frac{0.5h - z}{R_1} + \operatorname{arctg}\frac{0.5h + z}{R_1}\right)\Bigg\}, \end{aligned} \tag{7.4}$$

where R_1—the radius of inductor; W—the number of turns in inductor; h—the length of inductor; z—coordinate of the plasma cross section, measured from the geometric center of inductor.

Two-dimensional electromagnetic problem The one-dimensional problem and the assumption of one-dimensional electromagnetic field (E_φ, H_z) are justified, if the edge effects in inductor are neglected. Physically, it is possible, if the ratio of length to diameter of inductor is much larger than unity and a current in the inductor is of relatively low frequency. However, for most real designs of the RF plasma torches the length of inductor is commensurate with its diameter ($h \approx D$), and the frequency of the current in the inductor can vary from a few to hundreds of megahertz. In this case, two-dimensional and edge effects play a significant role and consideration of two-dimensional electromagnetic field is necessary.

For two-dimensional electromagnetic field in the channel of the RF plasma torches, due to the symmetry of the ring current and inductor current in plasma (transfer current is equal to zero) we have the following relations for the components of electric and magnetic fields $E_z = E_r = 0$, $E_\varphi \neq 0$; $H_\varphi = 0$, $H_z \neq 0$, $H_r \neq 0$, and Maxwell equations can be written in the two-dimensional cylindrical system coordinate as

$$\left.\begin{aligned} & \tfrac{\partial H_r}{\partial z} - \tfrac{\partial H_z}{\partial r} = \sigma E_\varphi + \varepsilon_0 \tfrac{\partial E_\varphi}{\partial t}; \quad \tfrac{\partial H_z}{\partial z} + \tfrac{1}{r}\tfrac{\partial}{\partial r}(rH_r) = 0; \\ & \tfrac{\partial E_\varphi}{\partial z} = \mu_0 \tfrac{\partial H_r}{\partial t}; \quad \tfrac{1}{r}\tfrac{\partial}{\partial r}(rE_\varphi) = -\mu_0 \tfrac{\partial H_z}{\partial t}. \end{aligned}\right\} \tag{7.5}$$

There is an effective method for solving this system of equations using the magnetic vector potential $\vec{A}$, its advantage is the reduction in the number of equations to a single equation for $\vec{A}$ that simplifies greatly the solution of the problem.

7.2.1 Equation of the Vector Potential and Its Solution

General method for solving the two-dimensional electromagnetic problem is to find the distribution of the (magnetic) vector potential $\vec{A}$ ($\vec{B} = \mathrm{rot}\vec{A}$) [34, 35] that for symmetric ring current has only one component $A \equiv A_\varphi$, $A_z = A_r = 0$. Using the vector potential allows reduction of the number of equations that are necessary to solve to a minimum. Instead of solving the Maxwell equations (7.5) for the intensities of electric and magnetic fields E_φ, H_z, H_r, it is necessary to solve only one equation for the vector potential. For solving of general equations of the electromagnetic field for the vector potential $\vec{A}$, imagine that in each elementary region of the RF plasma torch, including possible areas of inductor, plasma, and whole space, there are the circular currents of conduction (current in plasma $\sigma\vec{E}$ and in inductor $\vec{j}_c$), and the displacement current $\partial\vec{D}/\partial t$, i.e., the total current density $\vec{j} = \sigma\vec{E} + \vec{j}_c + \partial\vec{D}/\partial t$; and it can be assumed that the current density in inductor $\vec{j}_c$ is determined as the ratio of the current in inductor $\vec{I}_c$ to the area S of the conductive section of inductor. For RF plasma torch the high-frequency current is substantially only on the internal side of inductor, since there is minimum inductance resistance in inductor, so a conductive sectional area of the internal side of the inductor should be considered. Presentation of total current density by the sum of current densities of conduction in inductor, in plasma, and displacement current allows obtaining the convenient equation for determination of the current density in any area of research, and not only inside the inductor or plasma flux. Thus, the total current density inside the plasma flux $\vec{j} = \sigma\vec{E} + \partial\vec{D}/\partial t$, in inductor $\vec{j} = \vec{j}_c + \partial\vec{D}/\partial t$, and in the other space, such as in the gap between the coils or between inductor and plasma.

Equation of the vector potential has the form [34]

$$\Delta\vec{A} - \mu_0\sigma\frac{\partial\vec{A}}{\partial t} - \varepsilon_0\mu_0\frac{\partial^2\vec{A}}{\partial t^2} = -\mu_0\vec{j}_c. \tag{7.6}$$

This equation, that describes the change of vector potential in space and time, can be used in the future to determine the dependence of the parameters of electromagnetic field from the current in inductor and in plasma. Here the term $\mu_0\sigma\frac{\partial\vec{A}}{\partial t}$ is due to the current of conduction in plasma, $\varepsilon_0\mu_0\frac{\partial^2\vec{A}}{\partial t^2}$—is due to the displacement current $\mu_0\vec{j}_c$—is due to the current in inductor. Thus, the equation of the vector potential (7.6) describes the change of the field in all the space, and not only inside the RF discharge.

Equation (7.6) in a cylindrical coordinate system based on only one component ($A_\varphi = A$) has the form

$$\frac{\partial^2 A}{\partial z^2} + \frac{1}{r}\frac{\partial}{\partial r}\left(r\frac{\partial A}{\partial r}\right) - \frac{A}{r^2} - \mu_0\sigma\frac{\partial A}{\partial t} - \varepsilon_0\mu_0\frac{\partial^2 A}{\partial t^2} = -\mu_0 j_c. \quad (7.7)$$

The solution of this equation is carried out by means of submission of the required variable A in the complex plane. By definition, the image of a sinusoidal vector potential $A = A_m \sin(\omega t + \psi_a)$ in the complex plane has the form:$A \to Ae^{j(\omega t+\psi_a)} = \dot{A}e^{j\omega t}$, $\dot{A} = Ae^{j\psi_a}$, where $j^2 = -1$; $\dot{A}$—the complex current value of the vector potential; A—the current value of the vector potential; ψ_a—the initial phase of the vector potential with for the current in inductor, the initial phase of the current in inductor is set equal to zero $\psi_i = 0$.

Equation (7.7) after the substitution of complex value of the vector potential becomes

$$\frac{\partial^2 \dot{A}}{\partial z^2} + \frac{1}{r}\frac{\partial}{\partial r}\left(r\frac{\partial \dot{A}}{\partial r}\right) - \left(\frac{1}{r^2} + j\mu_0\omega\sigma - \varepsilon_0\mu_0\omega^2\right)\dot{A} = -\mu_0 j_c. \quad (7.8)$$

For simplicity we denote $A \equiv \dot{A} \equiv \dot{A}_\varphi$

Vector potential Eq. (7.8) is obtained from Maxwell equations, so it like Maxwell equation expresses the law of conservation of the electromagnetic field in an infinitesimal volume. Therefore, for the implementation of this law of conservation of the field in the final control volume and in all the study area, we should integrate the Eq. (7.8) for each control volume (Fig. 7.12) and make up the balance of electromagnetic field for each control volume that will ensure the implementation of the law of conservation of electromagnetic field in the entire region. The result of

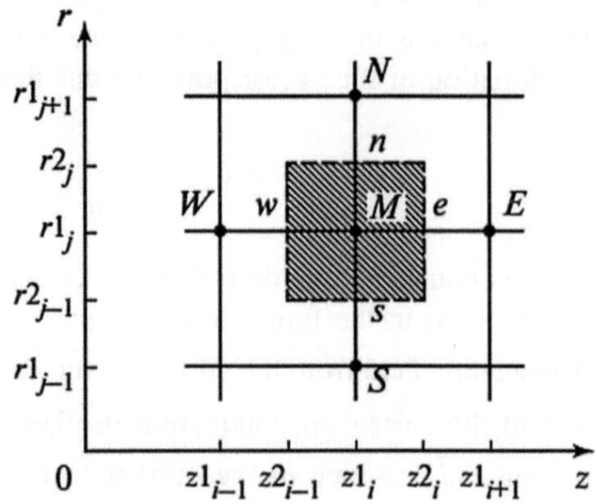

Fig. 7.12 The control volume and grid points for the solution of the equation of the vector potential

integration is the discrete analogue for equation of vector potential for the control volume. Integrate the vector potential Eq. (7.8) in two-dimensional control volume (Fig. 7.12).

$$\int\limits_{r_s}^{r_n}\int\limits_{z_w}^{z_e}\frac{\partial^2 A}{\partial z^2}\mathrm{d}z r\mathrm{d}r+\int\limits_{z_w}^{z_e}\int\limits_{r_s}^{r_n}\frac{1}{r}\frac{\partial}{\partial r}\left(r\frac{\partial A}{\partial r}\right)r\mathrm{d}r\mathrm{d}z$$

$$-\int\limits_{z_w}^{z_e}\int\limits_{r_s}^{r_n}\left(\frac{1}{r^2}+j\mu_0\omega\sigma-\varepsilon_0\mu_0\omega^2\right)Ar\mathrm{d}r\mathrm{d}z=-\mu_0\int\limits_{z_w}^{z_e}\int\limits_{r_s}^{r_n}j_c r\mathrm{d}r\mathrm{d}z.$$

At the same time, we note that under the integral of the first and second terms there is a return differential of the same order, so we get

$$\begin{gathered}\int\limits_{r_s}^{r_n}\left[\left(\frac{\partial A}{\partial z}\right)_e-\left(\frac{\partial A}{\partial z}\right)_w\right]r\mathrm{d}r+\int\limits_{z_w}^{z_e}\left[\left(r\frac{\partial A}{\partial r}\right)_n-\left(r\frac{\partial A}{\partial r}\right)_s\right]\mathrm{d}z\\-\int\limits_{z_w}^{z_e}\int\limits_{r_s}^{r_n}\left(\frac{1}{r}+j\mu_0\omega\sigma r-\varepsilon_0\mu_0\omega^2 r\right)A\mathrm{d}r\mathrm{d}z=-\mu_0\int\limits_{z_w}^{z_e}\int\limits_{r_s}^{r_n}j_c r\mathrm{d}r\mathrm{d}z.\end{gathered}\tag{7.9}$$

The third and fourth terms of this equation does not depend on the coordinates, so take them integrands constant in the central point M.

$$\begin{gathered}\int\limits_{z_w}^{z_e}\int\limits_{r_s}^{r_n}\left(\frac{1}{r}+j\mu_0\omega\sigma r-\varepsilon_0\mu_0\omega^2 r\right)A\mathrm{d}r\mathrm{d}z\\=\left(\frac{1}{r_M}+(j\sigma_M-\varepsilon_0)\mu_0\omega r_M\right)(z_e-z_w)(r_n-r_s)A_M;\end{gathered}\tag{7.10}$$

$$\mu_0\int\limits_{z_w}^{z_e}\int\limits_{r_s}^{r_n}j_c r\mathrm{d}r\mathrm{d}z=\mu_0 I_c r_M.$$

Further integration of the first and second members (7.9) should be made on each side of the control volume. The average values of the integrand in points e, w, n, s are accepted as an admission. The derivatives under the integral can be approximated as:

$$\left.\frac{\partial A}{\partial z}\right|_e=\frac{A_E-A_M}{z_E-z_M};\ \left.\frac{\partial A}{\partial z}\right|_w=\frac{A_M-A_W}{z_M-z_W};\ \left.\frac{\partial A}{\partial r}\right|_n=\frac{A_N-A_M}{r_N-r_M};\ \left.\frac{\partial A}{\partial r}\right|_s=\frac{A_M-A_S}{r_M-r_S}.\tag{7.11}$$

Substituting (7.10) (7.11) in (7.9) we get

$$\left[\frac{A_E - A_M}{z_E - z_M} + \frac{A_W - A_M}{z_M - z_W}\right]\frac{(r_n - r_s)(r_n + r_s)}{2} + \left[r_n\frac{A_N - A_M}{r_N - r_M} + r_s\frac{A_S - A_M}{r_M - r_S}\right](z_e - z_w)$$
$$-\left(\frac{1}{r_M} + (j\sigma_M - \varepsilon_0)\mu_0\omega r_M\right)(z_e - z_w)(r_n - r_s)A_M = -\mu_0 I_c r_M.$$

And the discrete analogue of Eq. (7.8) is

$$a_M A_M = a_E A_E + a_W A_W + a_N A_N + a_S A_S + b_M, \tag{7.12}$$

where

$$\begin{cases} a_E = \frac{(r_n - r_s)(r_n + r_s)}{2(z_E - z_M)} > 0; \quad a_W = \frac{(r_n - r_s)(r_n + r_s)}{2(z_M - z_W)} > 0; \\ a_N = \frac{r_n(z_e - z_w)}{r_N - r_M} > 0; \quad a_s = \frac{r_s(z_e - z_w)}{r_M - r_S} > 0; \\ a_M = a_E + a_W + a_N + a_S + S_p > 0; \\ S_p = \left(\frac{1}{r_M} + (j\sigma_M - \varepsilon_0\omega)\mu_0\omega r_M\right)(z_e - z_w)(r_n - r_s); \\ b_M = \mu_0 I_c r_M \text{ for the points on inductor;} \\ b_M = 0 \text{ for the points out of inductor.} \end{cases}$$

Solution of the system of discrete analogues (7.12) for all control volume gives the distribution of the magnetic vector potential $\vec{A}$ for the coordinates r, z. The resulting distribution $\vec{A}$, in turn, can be used to further define the parameters of the field in the RF plasma torch. Note, that for solving the system (7.12) it is necessary to define the boundary conditions for the vector potential.

7.2.2 *Boundary Conditions of the Equation of the Vector Potential*

The boundary conditions for solving the equation of the vector potential is generally defined in the initial section, or at the inlet of the plasma torch, on its wall, the axis of symmetry and at the exit. Consider ways to determine the vector potential at these borders

The zero boundary condition at infinity border The (magnetic) vector potential $\vec{A}$ has the direct communication with the electric field intensity $\vec{E}$, and at an infinite distance from the sources of the current value of the vector potential and electric field are equal to zero: $\vec{A} = \vec{E} = 0$ if $l = \infty$. For RF plasma torches due to the attenuation of the electromagnetic field far from inductor the approximate zero boundaries can be taken at a distance of a few diameters of inductor, and the step of the grid out of inductor can be made larger due to lack of currents of conduction in this area.

This way of defining the boundaries is simple but it is necessary to solve the equation of the vector potential for a much larger area, because the distribution of it should be determined, not only inside the inductor, but also outside it, although the parameters of the field of inductor are not of great interest in simulation of plasma processes inside the channel of plasma torch.

The boundary condition for the sum of the currents The second method of finding the vector potential at the boundary is based on the principle of superposition of the field as a sum of the external and internal currents in inductor and plasma. In this case, the domain for determination of the field is usually limited by a cylindrical channel of the plasma torch and the solution of the equation for the vector potential of (7.8) is carried out only for this area. The internal side of the wall of the plasma torch can be chosen as the external border. Determination of the value of vector potential for points on the boundary using the superposition principle, however, requires a large volume of computation, because the value of the vector potential in each such point depends on the cumulative effect of all the currents in the inductor and in plasma. The currents in plasma are obtained from the solution of Eq. (7.8) and change in the iterations. In this case, we have the solution of (7.8) with simultaneous approximation of its boundary condition.

We assume that the current in inductor is known and is constant for all the turns of inductor. The currents in plasma have different amplitudes and phases. The phase shift of the currents in plasma are different from the current in inductor, i.e., current in inductor can be presented as a real value, and the currents in plasma—the complex values. To further find the vector potential at the boundary of the plasma torch according to the principle of superposition of fields it is necessary to determine the field of loop with current.

The electromagnetic field of one circuit with the current The electromagnetic field outside the circuit with the current (where there is only the displacement current) is described by the magnetic vector potential

$$\Delta\vec{A} - \varepsilon_0\mu_0\frac{\partial^2\vec{A}}{\partial t^2} = 0. \tag{7.13}$$

The solution of this equation is:

$$\vec{A} = \frac{\mu_0}{4\pi}\int\limits_V \frac{\vec{j}(t)\mathrm{d}V}{r}, \tag{7.14}$$

Here the integral is taken over the volume where the current $\vec{i}$ flows.

At Fig. 7.13 the origin of coordinates 0 is placed in the center of circuit with current. The axis $0z$ is directed perpendicular to the plane of the loop. Here R_c—the

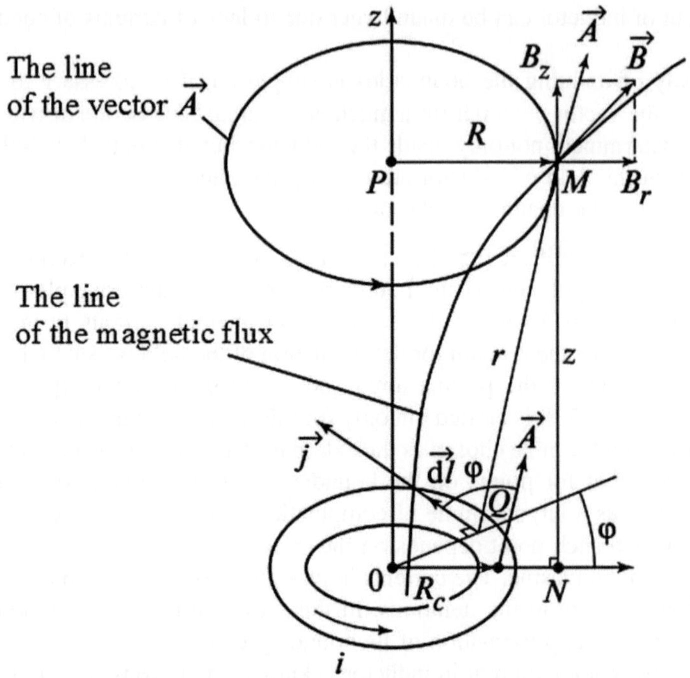

Fig. 7.13 To the definition of electromagnetic field with a loop with current

radius of the loop current; R—radius of the line for vector $\vec{A}$, with center at the point P; r—distance from the point M where the vector potential is determined, to the element of circuit $d\vec{l}(r = MQ)$; φ—angle of rotation $0Q$ for $0N$ or angle between the vector potential at the point M and the vector of element with a current circuit $d\vec{l}$; $\vec{j}$—the current density vector; elementary volume $dV = dSdl$. We write

$$\vec{A}_\varphi \equiv \vec{A} = \frac{\mu_0}{4\pi}\int\limits_V \frac{\vec{j}(t)dV}{r} = \frac{\mu_0}{4\pi}\int\limits_l\int\limits_S \vec{j}d\vec{S}\frac{d\vec{l}}{r} = \frac{\mu_0}{4\pi}\int\limits_l \frac{\vec{i}(t)d\vec{l}}{r} = \frac{\mu_0\vec{i}}{4\pi}\int\limits_l \frac{d\vec{l}}{r}.$$

The integration is performed around the loop with current. The magnetic vector potential has only one component A, i.e., the ring current $\vec{i}$ has only one component i_φ. Because of the symmetry to the axis $0z$ the lines of the vector potential with equal values must be the circumference, parallel to the loop with current. The projection of the vector $d\vec{l}$ to the vector $\vec{A}$ is equal to $\cos\varphi\, dl$, and the value of magnetic vector potential at the point M is calculated as [36]

$$A = \frac{\mu_0 i}{4\pi} \oint\limits_{l} \frac{\cos\varphi \, dl}{r} = \frac{\mu_0 i}{4\pi} \sqrt{\frac{R_c}{R}} F(k), \tag{7.15}$$

where $k = \sqrt{\frac{4RR_c}{z^2 + (R + R_c)^2}}$ is independent of l and r, $0 \le k \le 1$ The function $F(k)$ is defined as

$$F(k) = \frac{(2 - k^2)K(k) - 2E(k)}{k}, \tag{7.16}$$

Here $K(k)$ и $E(k)$—the full normal elliptic Legendre integrals of the I and the II types, the values of them can be found in the tables [37]

$$K(k) = \int\limits_0^{\pi/2} \frac{d\alpha}{\sqrt{1 - k^2 \sin^2 \alpha}}; \quad E(k) = \int\limits_0^{\pi/2} \sqrt{1 - k^2 \sin^2 \alpha} \, d\alpha. \tag{7.17}$$

Thus, the calculation of the vector potential A for the loop with current is reduced to finding the elliptic Legendre integrals $K(k)$ and $E(k)$ The variable k depends on the coordinates of the point where the vector potential (point M) is determined and on the geometric characteristics of the loop with current i. Note that if $k = 1$ (i.e., $z = 0$, $R = R_c$), $A = \infty$, that cannot be. Therefore, the Eq. (7.15) can be used to determine the parameters of the field only at a sufficient distance from the loop with current.

The boundary condition of the vector potential on the wall and on the entrance of the plasma torch This Eq. (7.15) for the calculation of the vector potential of the circuit with a current can be used to determine the boundary condition for the vector potential using the principle of superposition of the field. The value of the vector potential at the ith point on the boundary can be found as the sum of the vector potentials for all currents of inductor and currents of plasma using the following equation:

$$\dot{A}_i = \frac{\mu_0 I_c}{2\pi} \sqrt{\frac{R_c}{r_i}} \sum_{m=1}^{w_c} F(k_{im}) - j \frac{\mu_0 \omega}{2\pi \sqrt{r_i}} \sum_{j=1}^{M-1} \sum_{k=1}^{N-1} \sqrt{r_k} \sigma_{jk} \dot{A}_{jk} S_{jk} F(k_{ijk}), \tag{7.18}$$

where M, N—the number of grid points on the axis z and r, respectively; j, k—indices for axes r and z, respectively, for each of the ring currents in plasma; r_k—radius of jkth loop of the ring current in plasma; r_i—the distance to a point where a vector potential is determined; the complex density $\dot{j}_{jk}$ of the jkth ring current with plasma is determined by the vector potential $\dot{A}_{jk}$ using the equation $\dot{j}_{jk} = \sigma_{jk} \dot{E}_{jk} = -j\omega\sigma_{jk} \dot{A}_{jk}$, and the complex currents of plasma are found as $\dot{I}_{jk} = \dot{j}_{jk} S_{jk}$, here S_{jk}—sectional area of jkth circuit of the ring current.

Note that Eq. (7.18) can be used to determine the value of the magnetic vector potential on the wall and at the entrance to plasma torch, i.e., where there is no the current of conductivity.

The boundary condition of the vector potential on the axis of symmetry and at the exit of the plasma torch The values of the vector potential on the other boundaries are determined relatively simply. On the axis using the cylindrical conditions $A = 0$. At the exit of the plasma torch, where there is no current of conduction, the value of the vector potential can be found using two discussed ways or using a "soft" condition of determination the magnetic vector potential at the border $\frac{\partial A}{\partial z} = 0$. In this case, the discrete analogue for points on exit is

$$a_M A_M = a_W A_W + a_N A_N + a_S A_S, \tag{7.19}$$

where

$$\begin{cases} a_W = \frac{(r_n - r_s)(r_n + r_s)}{2(z_M - z_W)} > 0; \quad a_N = \frac{r_n (z_e - z_w)}{r_N - r_M} > 0; \quad a_S = \frac{r_s (z_e - z_w)}{r_M - r_S} > 0; \\ a_M = a_W + a_N + a_S + S_p > 0; \\ S_p = \left(\frac{1}{r_M} + (j\sigma_M - \varepsilon_0 \omega)\mu_0 \omega r_M\right)(z_e - z_w)(r_n - r_s). \end{cases}$$

Thus, a complete system of discrete analogues for internal and boundary points of the Eq. (7.8) is formulated. The solution of such a system can be made using the standard methods, such as the Gauss–Seidel method or using the method of alternating directions.

Note that in equation for calculating the value of the magnetic vector potential at the boundary (7.18) a vector potential at all internal points $\dot{A}_{jk}$ is included. However, $\dot{A}_{jk}$ is determined from the solution (7.8), i.e., the boundary condition is closed on the very solution. Thus, the solution of equation for the vector potential (7.8) should be performed using an iterative method with the approximation and clarification of the boundary condition (7.18). The iterative procedure for solving a discrete analogue of (7.12), (7.18), and (7.19) is

1. Set a starting field of the magnetic vector potential inside the channel of plasma torch. For the fast convergence the zero field can be taken as the initial condition.
2. The approximate values of the vector potential are determined at the boundary by the Eq. (7.18).
3. Solve the system of discrete analogues (7.12), (7.19) in order to define a new field of the vector potential.
4. Check the condition of the solutions convergence and the procedure is repeated, starting from position 2, if it is not running.

5. Calculate all electromagnetic quantities on the basis of the resulting field of the vector potential, and to solve other equations of plasma.

7.2.3 *Calculation of the Electromagnetic Field and Electric Characteristics of the RF Plasma Torches*

Determination of the inductance and current in the inductor The current in inductor I_1 is usually at the known voltage at the inductor U_1, and it is necessary to determine the inductance of inductor L and its inductive reactance $X_L = \omega L$. The inductance of a single-layer solenoid (empty inductor without plasma) is calculated by the well-known formula of Nagaoka [38]. According to this formula

$$L = K_L \frac{\mu_0 \pi D^2 W^2}{4h}. \tag{7.20}$$

The Nagaoka correction factor K_L is a function of the ratio of the inductor length to its diameter, $K_L = f(h/D)$, the value of this function is given in [38].

Correction for the inductance of the inductor for the winding pitch is calculated by equation

$$\Delta L = F \frac{\mu_0 W h}{2}. \tag{7.21}$$

The function $F(\Delta h / D_c)$ takes into account the ratio of the winding pitch Δh to the diameter of inductor coil D_c and the shape of the inductor coil (circular or rectangular). The value of this function is also shown in [38]. Author's calculations show that the correction for the step of winding for real inductors of the RF plasma torch is usually 2–7 % of the inductance, calculated by Eq. (7.20).

The total inductance

$$L = K_L \frac{\mu_0 \pi D^2 W^2}{4h} - F \frac{\mu_0 W h}{2}. \tag{7.22}$$

Consequently, the current in inductor can be calculated as

$$I_1 = \frac{U_1}{X_L} = \frac{U_1}{\omega L}. \tag{7.23}$$

To accurate determination of the total inductance of inductor with plasma it is necessary to know the total magnetic flux. The resulting value of the vector potential allows easily definition of it. Full magnetic flux is associated with the inductance of inductor with plasma according to equation $\Phi = LI_1$. Here I_1—the current in inductor with plasma. We write the following relation between the magnetic flux Φ and the magnetic vector potential $\vec{A}$ in inductor with radius R_c

$$\Phi = \int \vec{B}d\vec{s} = \int rot\vec{A}d\vec{s} = \oint \vec{A}d\vec{l} = 2\pi R_c A.$$

And the inductance of inductor with plasma

$$L = \frac{2\pi R_c A}{I_1}. \tag{7.24}$$

Definition of the electromagnetic field Knowing the distribution of the magnetic vector potential, we can determine the parameters of the electromagnetic field in the RF plasma torches, including the allocated power σE^2 is used to further solving of the equation of energy balance, and the electromagnetic forces F_z and F_r, required for solution of the momentum equations in plasma, as well as all the electrical parameters that are necessary for solving of the practical problems of coordination and adjustment of the generator and the RF plasma torches. Here are the necessary equations for the calculation of the field

1. The electric intensity

$$\dot{E}_\varphi = -j\omega\dot{A}. \tag{7.25}$$

On the axis due to conditions of the cylindrical axis symmetry $\dot{E}_\varphi = 0$.

2. The components of the magnetic field

$$\dot{H}_r = -\frac{1}{\mu_0}\frac{\partial\dot{A}}{\partial z};\ \dot{H}_z = \frac{1}{\mu_0}\frac{1}{r}\frac{\partial(r\dot{A})}{\partial r} = \frac{1}{\mu_0}\left[\frac{\dot{A}}{r} + \frac{\partial(\dot{A})}{\partial r}\right]. \tag{7.26}$$

From the condition of the cylindrical symmetry $\dot{H}_r = 0$. To determine the component of the magnetic field $\dot{H}_z$, on the axis magnetic vector potential can be presented near the axis of symmetry in the form of a cubic polynome $\dot{A}(r,z) = \dot{a}_0(z) + \dot{a}_1(z)r + \dot{a}_2(z)r^2 + \dot{a}_3(z)r^3$. $\dot{A}|_{r=0} = 0$ and $\dot{a}_0(z) = 0$. We have the following relation $\left.\frac{\partial\dot{A}}{\partial r}\right|_{r=0} = \dot{a}_1(z) = \left.\frac{\dot{A}}{r}\right|_{r=0}$. Substituting this relation into the Eq. (7.26), we obtain a new expression for determination of $\dot{H}_z$ on the axis

$$\dot{H}_z\Big|_{r=0} = \frac{2}{\mu_0}\left.\frac{\partial\dot{A}}{\partial r}\right|_{r=0}. \tag{7.27}$$

3. The density of the current of conduction in plasma

$$\dot{j}_\varphi = \sigma \dot{E}_\varphi = -j\omega\sigma\dot{A}. \tag{7.28}$$

4. Active power density of discharge in plasma

$$P = \sigma E_\varphi^2. \tag{7.29}$$

5. The components of the electromagnetic force in plasma

$$\begin{cases} F_r = Re\left[\mu_0 \sigma \dot{E}_\varphi \overset{*}{H}_z\right] = \omega\sigma Im\left[\dot{A}\mathrm{conj}\left(\frac{\dot{A}}{r} + \frac{\partial \dot{A}}{\partial r}\right)\right]; \\ F_z = Re\left[-\mu_0 \sigma \dot{E}_\varphi \overset{*}{H}_r\right] = \omega\sigma Im\left[\dot{A}\mathrm{conj}\left(\frac{\partial \dot{A}}{\partial z}\right)\right]. \end{cases} \tag{7.30}$$

Here *Re* and *Im*—real and imaginary parts of the complex value; "*" and conj —the notations of conjugate complex value.

6. Complete the active current in the plasma

$$\dot{I}_2 = \int_0^l \int_0^R \dot{j}_\varphi \mathrm{drdz} = -j\omega \int_0^l \int_0^R \sigma\dot{A}\mathrm{drdz}; \quad I_2 = \left|\dot{I}_2\right|, \tag{7.31}$$

where *l*, *R*—length and radius of channel of the RF plasma torch.

7. The phase shift of currents in plasma compared with current in inductor

$$\psi = \arg\left(\dot{j}_\varphi\right). \tag{7.32}$$

8. The total allocated power in plasma

$$P_2 = \int_V Re(\sigma \dot{E}_\varphi^2)\mathrm{dV} = 2\pi \int_0^l \int_0^R Re(\sigma \dot{E}_\varphi^2) r\mathrm{drdz}. \tag{7.33}$$

9. Active, reactive, and full equivalent resistance of plasma is

$$R_2 = \frac{P_2}{I_2^2}; \quad X_2 = \omega L_2 = \frac{2\pi \int_0^l \int_0^R Im(\sigma \dot{E}_\varphi^2) r\mathrm{drdz}}{I_2^2}; \quad Z_2 = \sqrt{R_2^2 + (\omega L_2)^2}. \tag{7.34}$$

10. The magnetic flux of the inductor with plasma

$$\Phi = \int \vec{B} d\vec{s} = \int \text{rot}\vec{A} d\vec{s} = \oint \vec{A} d\vec{l} = 2\pi r A. \tag{7.35}$$

These and other electrical parameters can be used to determine the optimal operation of generator and RF plasma torch.

Calculation results The following details of the calculation of electromagnetic field in the RF plasma torches are the result of the joint solution of the equation for the vector potential (7.8) and other equations from the model of equilibrium plasma: equation of the energy balance, the momentum equations, the continuity equation (Sect. 1.2), as well as equations for calculation of the composition and thermodynamic properties and transport coefficients of the plasma (Sect. 4.1–4.3). The following parameters and characteristics of electromagnetic field are identified

1. The inductance of inductor L, the inductive resistance $X_L = \omega L$ and the current of inductor I_1;
2. The distribution of vector potential $A(r, z)$;
3. The magnetic flux distribution $\Phi(r,z)$;
4. The distribution of intensity of the electric and magnetic fields $E(r,z)$, $H_z(r,z)$, $H_r(r,z)$;
5. The distribution of current density of conduction in plasma $j(r,z)$ and the phase shift of the plasma current $\psi(r, z)$;
6. The distribution of power density in plasma $P = \sigma E^2$;
7. The distribution of components of electromagnetic force in plasma $F_r(r, z)$, $F_z(r, z)$;
8. The total current in plasma I_2, the total power dissipated in plasma P_2, active R_2, reactive X_2 and full equivalent Z_2 plasma resistance that are necessary for determination of the agreed regime of operation for the RF plasma torch and generator.

The general scheme of the RF plasma torch with possible three streams of gases and the water cooling of the wall are shown in Fig. 7.14.

Initial data for the simulation of the RF plasma torches, the most commonly used in practice, are presented in Table 7.3

Because of the limited volume of the book we give only the results and analysis of electromagnetic field of the RF plasma torch (5 MHz); character of the field is similar at other plasma torches. Some characteristics obtained by their calculation, are shown in Table 7.4. Initial data for calculation the RF plasma torch with frequency of current in inductor 5 MHz (Table 7.3.) are the following: the voltage at the *4*th turn of inductor $U_1 = 4.4$ kW, that corresponds to the current $I_1 = 120$ A, two flows of plasma-forming and protective gas (Argon) with consumptions, respectively $Q_2 = 20$ l/min, $Q_3 = 40$ l/min; $Q_1 = 0$; $d_1 = 3$ mm; $D_1 = 5$ mm; $d_2 = 25$ mm; $D_2 = 30$ mm; $R_6 = 23$ mm; length of calculated area 120 mm.

The distributions of intensity of electric and magnetic fields in the channel of the RF plasma torch (5 MHz) are shown in Figs. 7.15, 7.16, 7.17, 7.18, 7.19, 7.20, 7.21, 7.22 and 7.23.

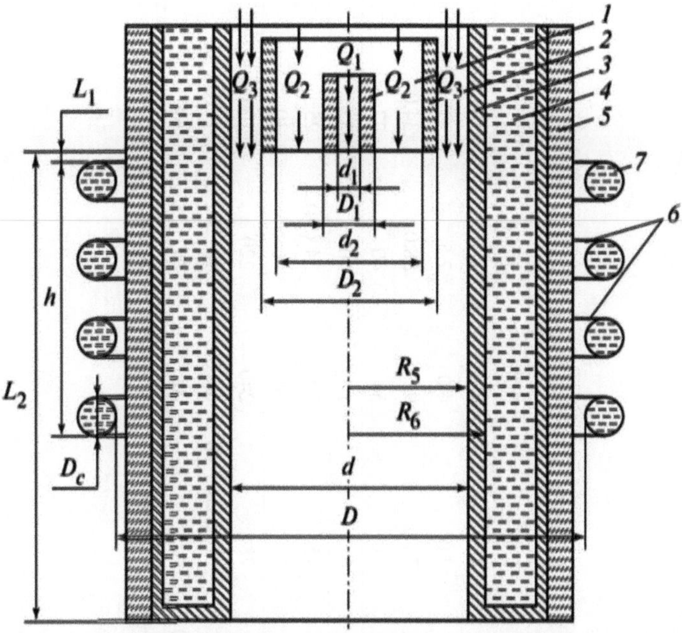

Fig. 7.14 Scheme of the RF plasma torch: *1* central tube for the possible entrance of chemicals, *2* tube for a plasma-forming flow, *3* a wall of the RF plasma torch, *4* water flow for cooling of the wall, *5* the external casing of RF plasma torch, *6* turns of inductor, *7* water flow for cooling of inductor, $d = 2R_5$ diameter of the plasma torch channel, d_1, D_1, d_2, D_2 internal and external diameters of the central channel and the plasma-forming flow, respectively, R_6 internal radius of a water-cooled channel of plasma torch, D the internal diameter of inductor, D_c diameter of the coil of inductor, h the length of inductor, L_1 the distance from the entrance of plasma torch to the first round of inductor, L_2 the length of plasma torch, Q_1, Q_2, Q_3 the consumption of plasma-forming and protective gas, respectively

The intensity of the electromagnetic field quickly decreases as the penetration of waves into the plasma, and the release of energy in the skin layer (Figs. 7.15, 7.16, 7.17, 7.18, 7.19, 7.20, 7.21, 7.22). In the gap between plasma and inductor, where there is no current of conduction, there is a slight enhancement of the magnetic field H_z (Fig. 7.23b) (compared with empty inductor) due to the interaction of magnetic fields of currents in the inductor and plasma. In this zone the value H_z, as in the case of empty inductor (Fig. 7.18), should be almost constant for r (Fig. 7.20), because the induced current of conduction is equal to zero, and the change due to the displacement current is insignificant. However, starting with the skin layer, where converting of electromagnetic energy into thermal one occurs actively, and the magnetic field decreases rapidly (Figs. 7.19, 7.20).

In the absence of plasma, an electromagnetic field is generated only by the current in inductor, and therefore in Figs. 7.15, 7.18 and 7.21 the field has a symmetrical shape for the central cross section of inductor.

Table 7.3 Initial data for the RF plasma torches

Parameter	Frequency in the inductor (MHz)							
	0.066	0.12	0.44	1.76	5.00	5.28	13.76	27.12
U_1 (kW)	1–3	1.2–2	8–9	6–7	4–4.5	6–6.5	4–6	3–3.5
W	11.0	3.0	4.0	5.0	4.0	4.0	2.5	3.0
D (mm)	130.0	200.0	220.0	96.0	78.6	92.0	58.0	34.0
d (mm)	100	140	165	60	41	54	40	27
D_c (mm)	10	20	22	10	6	10	10	5
h (mm)	260	100	160	100	45	70	34 (54)	25
L_1 (mm)	140	150	200	70	3	70	66	30
L_2 (mm)	600	500	600	300	78	240	220	100
$Q = Q_1 + Q_2 + Q_3$ (l/min)	10–20	10–20	6000–16,000	60–300	(8) + 20 + 40	60–80	10–20	7–10
Gas	Air	Air	Air + H_2S	Air (N_2)	Ar	Ar (N_2)	Ar	Ar
Position	Bottom–up	Bottom–up	Horizontal	Bottom–up	Up–bottom	Horizontal	Bottom–up	Bottom–up

U_1 voltage on inductor; W number of turns; Q total gas consumption

Table 7.4 The results of calculation of electromagnetic field in the RF plasma torches

Parameters	Frequency in the inductor (MHz)			
	0.44	1.76	5.00	5.28
K_L	0.615	0.697	0.563	0.626
L (10^{-7} H)	28.50	15.40	11.67	11.66
$X_L = \omega L$ (Ω)	7.878	17.03	36.38	38.68
I_1 (A)	1200	350	120	155
E_{max} (V/cm)	20.3	22.7	12.5	13.2
H_{zmax} (A/cm)	204.0	128.1	60.9	59.9
H_{rmax} (A/cm)	60.0	27.5	13.4	6.5
j_{max} (A/cm^2)	160	185	147	137
P_{max} (W/cm^3)	748	1019	808	769
P_{max}/j_{max} (V/cm)	4.67	5.51	5.50	5.61
F_{zmax} (10^{-3} N/cm^3)	0.15	1.75	1.90	0.30
F_{rmax} (10^{-3} N/cm^3)	−21.7	−13.1	−6.2	−5.4
I_2 (A)	5180	1690	400	580
P_2 (kW)	400–600	48.4	13.3–15.1	30
R_2 (Ω)	0.0150	0.0169	0.0834	0.0875

K_L the correction factor of Nagaoka in Eq. (7.20) for determination of inductance, E_{max} the maximum intensity of electric field on the wall of the RF plasma torch, H_{zmax}, H_{rmax} maximum components of the intensity of magnetic field, j_{max} the maximum current density of plasma, P_{max} maximum power density of plasma, F_{zmax}, F_{rmax} maximum components of the electromagnetic force in plasma, I_2 the total conductivity current of plasma, P_2 the total active power allocated in plasma, R_2 the total resistance of plasma

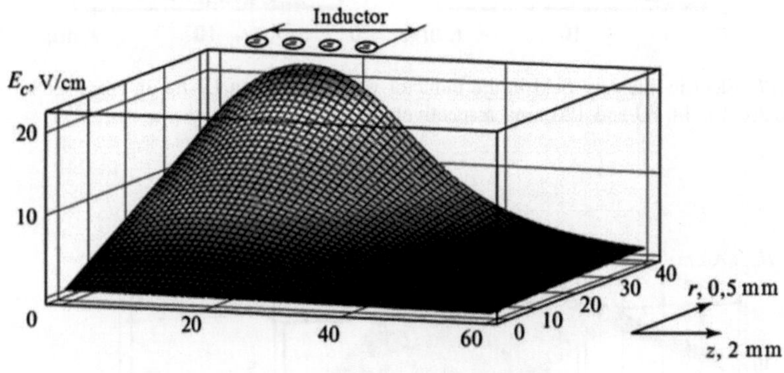

Fig. 7.15 Electric intensity field in the inductor without plasma

As expected, the intensity E and H_z have the peak and the intensity H_r—minimum in the central zone of inductor. Change of H_{zc} along the axis r inside inductor without plasma is slightly (Fig. 7.18) that is also confirmed using the results of calculations for empty inductor in accordance with the theory of induction heating. As bigger is the distance from inductor the value E and H_z reduce quickly, and at the end parts of the first and last turns of inductor H_r reaches maximum

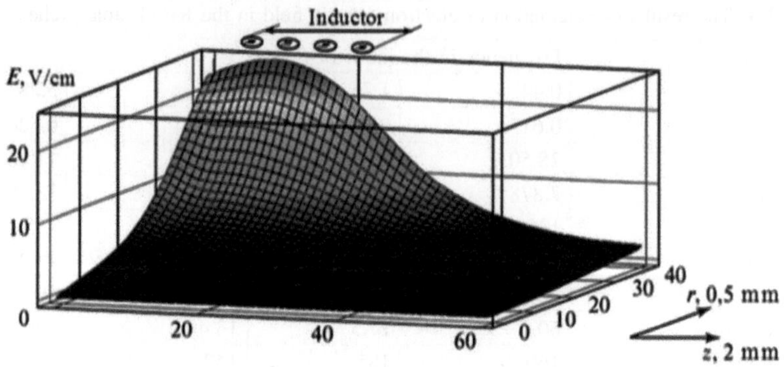

Fig. 7.16 Electric intensity field in the inductor with the plasma

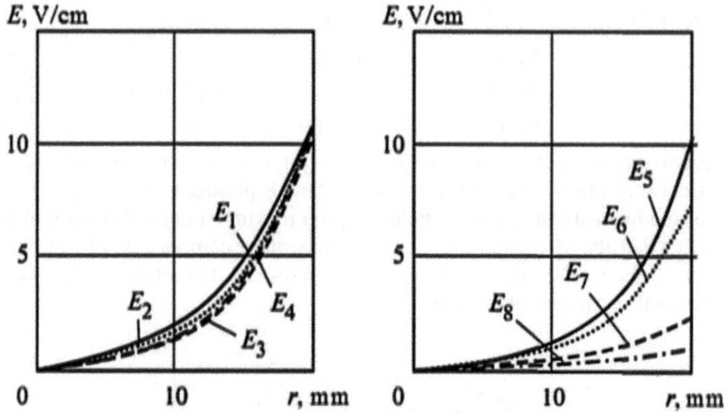

Fig. 7.17 Electric intensity field in the inductor with the plasma E_1–E_8, in cross-sections z = 4, 10, 16, 26, 36, 48, 80 and 120 mm, respectively

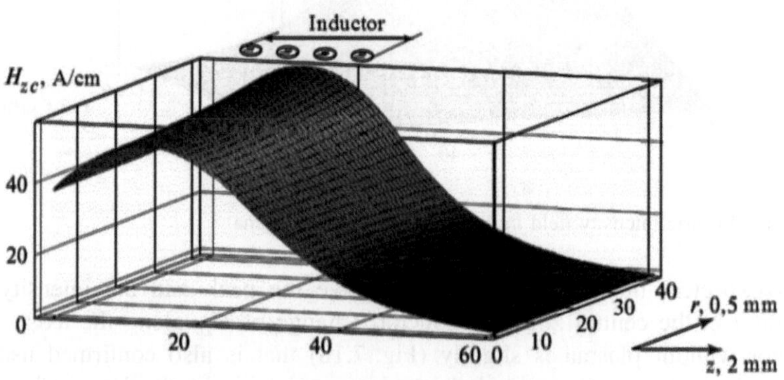

Fig. 7.18 Magnetic intensity component H_z of the inductor without plasma

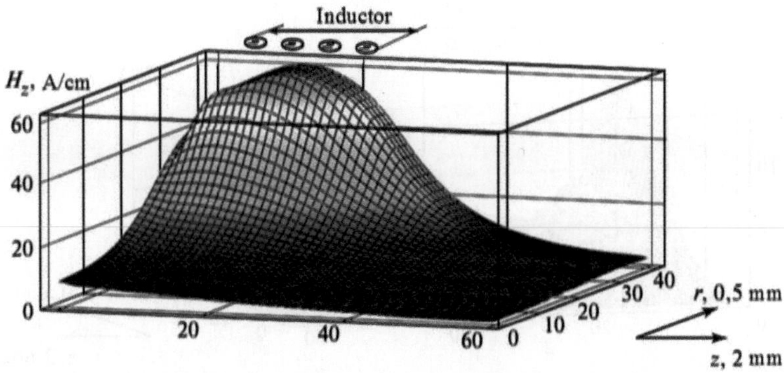

Fig. 7.19 Magnetic intensity component H_z of the inductor with the plasma

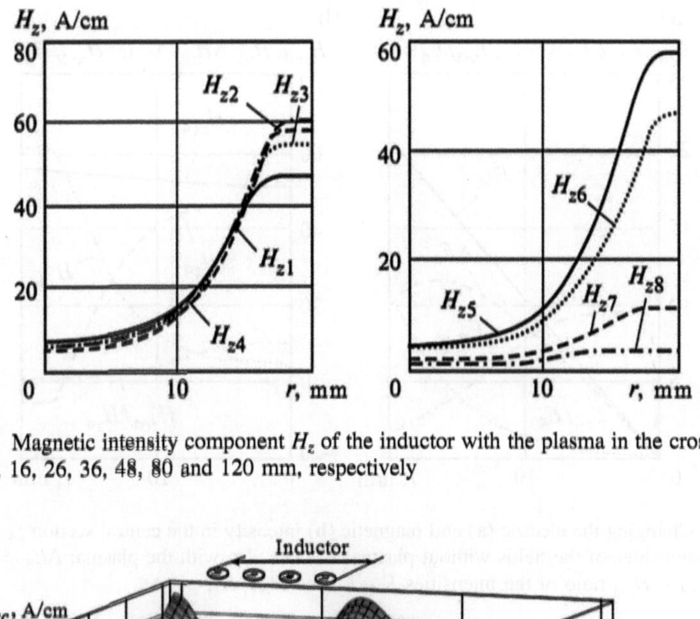

Fig. 7.20 Magnetic intensity component H_z of the inductor with the plasma in the cross-sections z = 4, 10, 16, 26, 36, 48, 80 and 120 mm, respectively

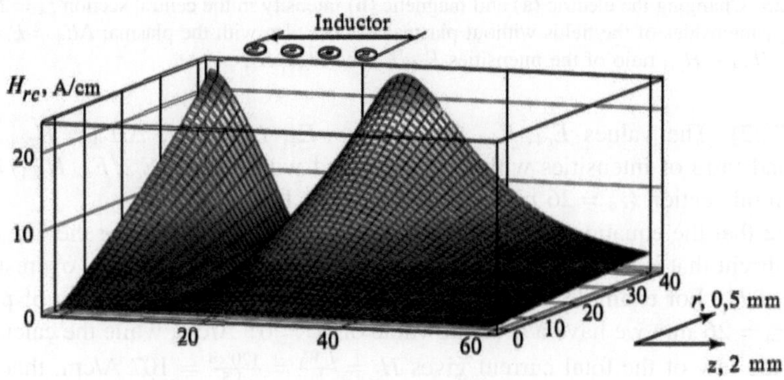

Fig. 7.21 Magnetic intensity component H_r of the inductor without plasma

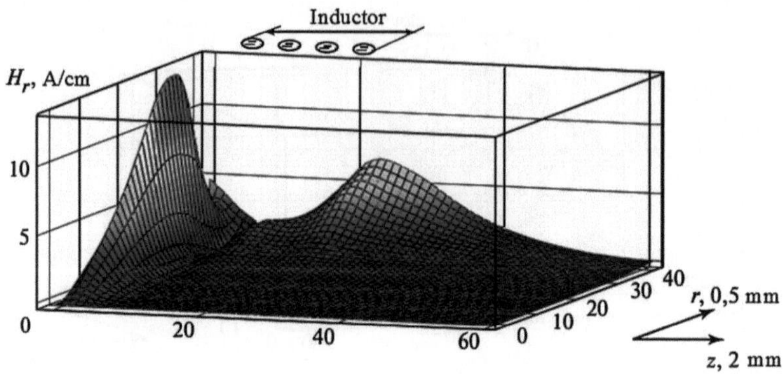

Fig. 7.22 Magnetic intensity component H_r of the inductor with the plasma

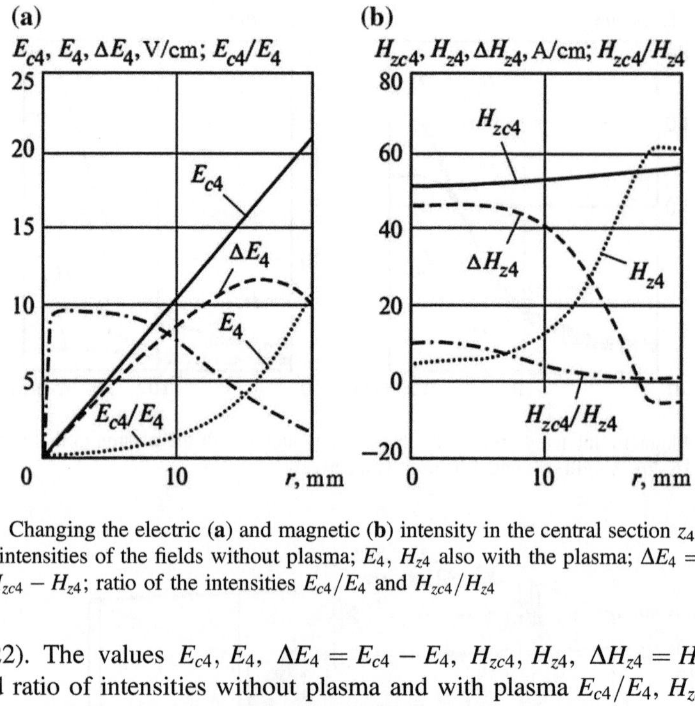

Fig. 7.23 Changing the electric (**a**) and magnetic (**b**) intensity in the central section z_4 = 26 mm: E_{c4}, H_{zc4} intensities of the fields without plasma; E_4, H_{z4} also with the plasma; $\Delta E_4 = E_{c4} - E_4$, $\Delta H_{z4} = H_{zc4} - H_{z4}$; ratio of the intensities E_{c4}/E_4 and H_{zc4}/H_{z4}

(Fig. 7.22). The values E_{c4}, E_4, $\Delta E_4 = E_{c4} - E_4$, H_{zc4}, H_{z4}, $\Delta H_{z4} = H_{zc4} - H_{z4}$ field and ratio of intensities without plasma and with plasma E_{c4}/E_4, H_{zc4}/H_{z4} in the central section (z_4 = 26 mm) are presented at Fig. 7.23.

Note that the equation for calculation of the magnetic field using the law of the total current that is often used in the theory of induction heating, gives overestimate value of H_z. For example, for a point on the wall in the central section of plasma torch z_4 = 26 mm we have a accurate value of H_z = 61 A/cm, while the calculation using the law of the total current gives $H_z = \frac{I_1 W_1}{h_1} = \frac{120\times4}{4.5} = 107$ A/cm, that gives the excess in 1.75 times. This difference is explained by the fact that the expression for calculating the magnetic field using the law of total current is valid only for an

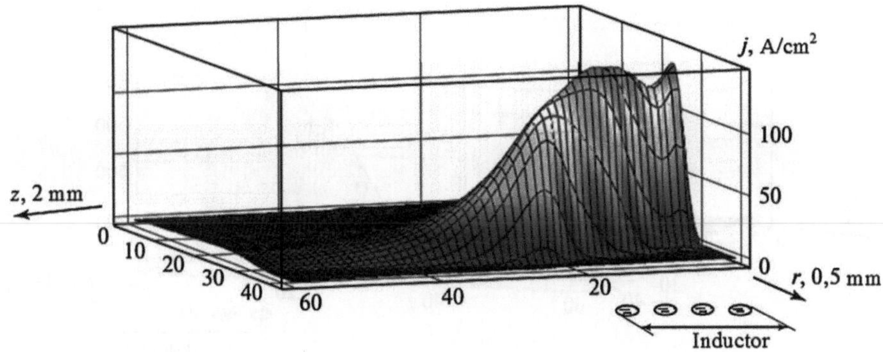

Fig. 7.24 The distribution of current density of plasma

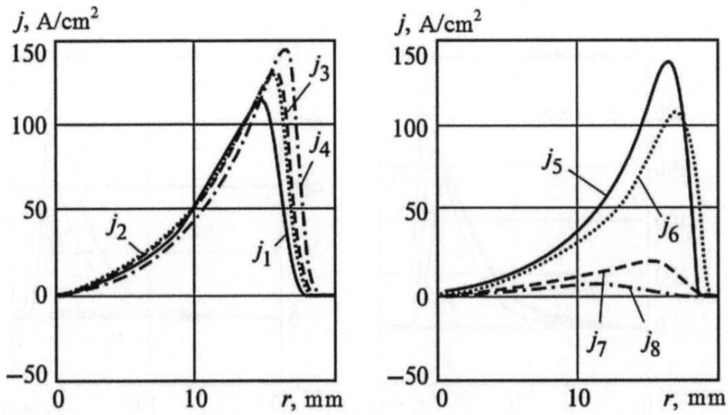

Fig. 7.25 Changes of current density of plasma j_1–j_8 in the sections z = 4, 10, 16, 26, 36, 48, 80 and 120 mm, respectively

infinitely empty long solenoid and cannot be used in the case of short inductor with plasma.

However in calculation of the boundary using Eq. (7.4) with R_2 = 16.5 mm (R_2 is determined using the temperature at boundary of Argon plasma that is equal to 7000 K) H_z = 62.4 A/cm, this value agrees with the exact calculations of author. Consequently, this expression can be used to solve one-dimensional electromagnetic problem.

The distribution of the current density $j(r, z)$ and allocated power density $P(r, z)$ in plasma are presented in Figs. 7.24, 7.25, 7.26 and 7.27. These data are input parameters for the solution of the equation of the energy balance. Note that the resulting distributions $j(r, z)$ and $P(r, z)$ are similar in shape, because there is a direct link between the allocated power and current density $P = \sigma E^2 = j^2/\sigma$. The thickness of the skin layer can also be determined using the obtained data, for the case Δ = 6–8 mm, and 70–90 % of plasma currents and allocates power are concentrated in this layer.

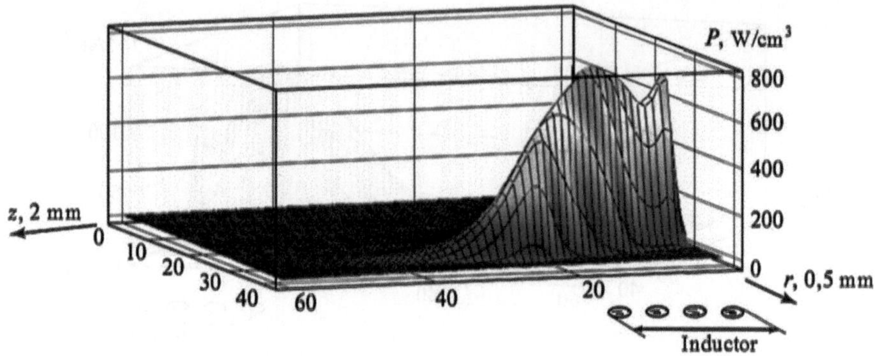

Fig. 7.26 The distribution of power density of discharge

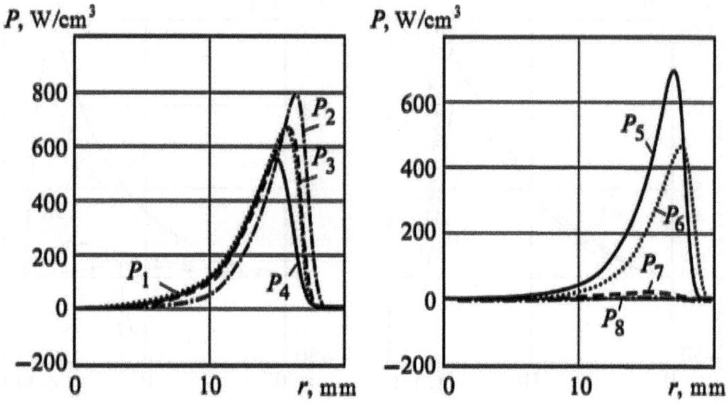

Fig. 7.27 Changes of power density in discharge sections P_1–P_8, $z = 4$, 10, 16, 26, 36, 48, 80, and 120 mm, respectively

The phase shifts of plasma currents, compared to the current in inductor, are shown in Fig. 7.28. Note that between the temperature and the phase shift of a certain connection: experimental results (Sect. 7.1) and simulation of the RF plasma torch show that the temperature in the channels of different the RF plasma torches are not significantly different, and the distribution of phase shifts of currents in plasma in these channels have approximately the same shape as in Fig. 7.28.

The knowledge of distribution is the electromagnetic force $F_r(r, z)$ and $F_z(r, z)$ (Figs. 7.29, 7.30) are another important factors in the calculation of electromagnetic field of the RF plasma torches, they are used to solve the equations of plasma motion and largely determine the dynamics of the plasma flow movement in the channel of plasma torch. The component $F_r(r, z)$ provides the movement of gas in plasma, its maximum is concentrated in the area of the skin layer in the center of inductor. The results of calculations show that the electromagnetic force in the RF plasma torch is much smaller than in an electric arc, it leads to the flow of plasma

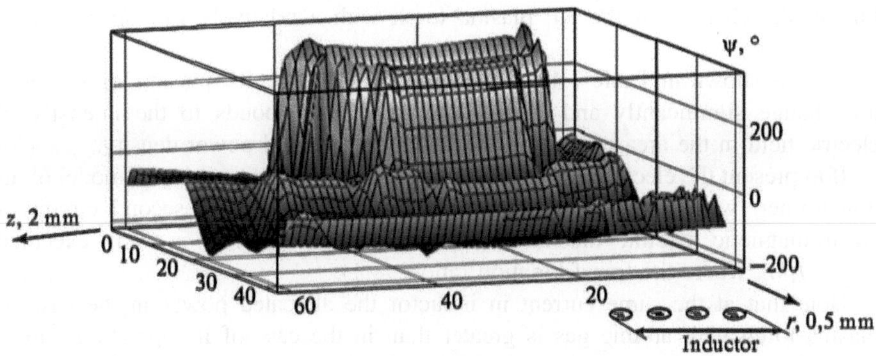

Fig. 7.28 The phase shifts of currents in the discharge in comparison with the current in inductor

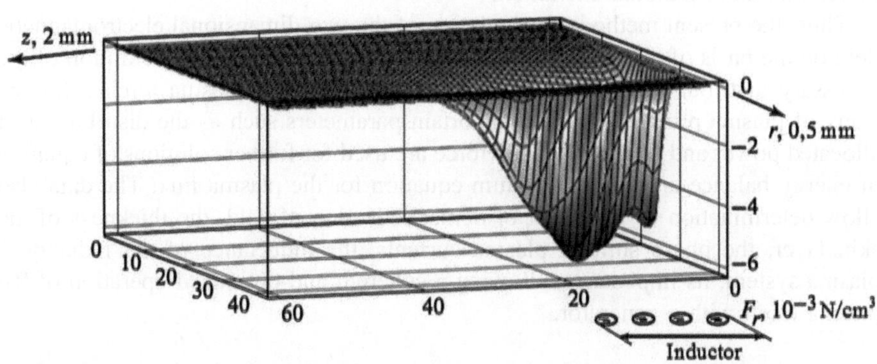

Fig. 7.29 The radial component of electromagnetic force F_r

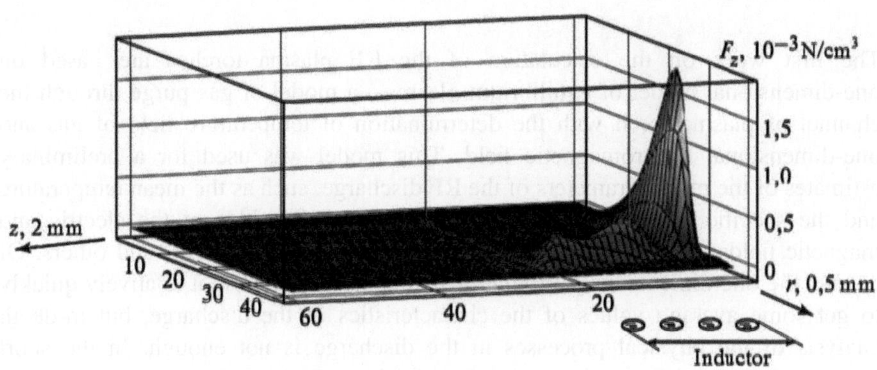

Fig. 7.30 The axial component of the electromagnetic force F_z

flux in the channel of the RF plasma torch with a relatively low speed (up to 100 m/s).

As it is shown in Table 7.4 the ratio P_{max}/j_{max} for the RF plasma torches does not change significantly and is 3–6 V/cm that corresponds to the intensity of electric field in the area of maximum current density and power density.

If to present the electromagnetic system of the RF plasma torch as a model of air transformer, where plasma is considered as a single turn of second circuit of electromagnetic circuit, the law of current transformation is well executed ($I_2 = kI_1 W$, where the transformation ratio $k \approx 1$).

Note that at the same current in inductor the allocated power in the case of plasma torch with atomic gas is greater than in the case of its operation with a molecular gas. For example, at operation of plasma torch with 0.44 MHz in Argon the allocated power P_2 = 600 kW, and when it is operating with Air P_2 = 400 kW. This is due to large energy content (enthalpy) of molecular gases because of their vibrational and rotational excitation.

Thus, the present method of calculation of the two-dimensional electromagnetic field on the basis of the equation of the vector potential allows the definition of the necessary distribution of electromagnetic field in the RF plasma torches for the study of plasma processes. These important parameters such as the distribution of allocated power and electromagnetic force are used for further solutions of equation of energy balance and the momentum equation for the plasma flux. The data also allow determination of the effect of demagnetization of field, the thickness of the skin layer, the phase shift of plasma currents, the inductance of the inductor—plasma system, its impedance, allowing a coherent and optimized operation of RF plasma torch with a generator.

7.3 Simulation of the Equilibrium Plasma in the RF Plasma Torches

The first work on the calculation of the RF plasma torches are based on one-dimensional model of equilibrium plasma—a model of gas purge through the channel of plasma torch with the determination of temperature field of gas and one-dimensional electromagnetic field. This model was used for a preliminary estimates of the main parameters of the RF discharge, such as the mean temperature and the specific power of discharge, the average intensities of the electric and magnetic fields, the total current and total power in the discharge, and others. Of course, the one-dimensional formulation of the problem makes it relatively quickly to get some average values of the characteristics of the discharge, but in-depth analysis of the physical processes in the discharge is not enough. In the short inductor the edge effects play a significant role, two-dimensional electromagnetic field and two-dimensional flow of plasma with reverse vortex of plasma flow are not uncommon, but the main feature of the RF plasma torch. The main characteristics of a discharge are defined on the basics of them—the field of the gas flow, heat flux, a complete picture of heat transfer in plasma, a single field of the discharge, the discharge power with the currents of inductor and plasma, and others. Therefore, the study of the basic physical processes of the RF discharge and calculation of parameters of the RF plasma torch should be carried out using a two-dimensional model of plasma based on the regime of operation of the plasma torch and its design features.

The model of thermal equilibrium plasma is formulated above (Sect. 1.2) and is used for study of dense plasma with a pressure close to atmospheric pressure. It is based on the assumption of the equality of all groups of temperature for the particles in plasma (Sect. 1.1), and the justice of Maxwell equations, the Boltzmann equation and the Saha equations (1.2), (1.4), (1.9). Quasi-neutral plasma flow extends over a distance much larger than the mean free path of particles, and is presented as a continuous medium with uniform transfer coefficients and thermodynamic properties depending on temperature and pressure. In the first approximation, the radiation around all the frequency range is considered to be optically thin, freely coming out of the discharge and is represented as a member of the radiation losses in the balance of power discharge.

Thus, the model of equilibrium plasma in the RF plasma torches include the following equation: equation of the balance energy (1.25), momentum equations (1.26), (1.27), continuity equation (1.29). Calculation of two-dimensional electromagnetic field in the RF discharge (1.30) is carried out using the equation for magnetic vector potential (7.8), and the distribution of electromagnetic field and electric parameters of system inductor-plasma can be determined by Eqs. (7.20–7.35). Furthermore, the equilibrium plasma model includes various equations for obtaining of composition, transfer coefficients, and thermodynamic properties of equilibrium plasma (Sect. 4.1–4.3). In many cases, the transfer coefficients and the thermodynamic properties of equilibrium plasma (particularly at atmospheric pressure) are known. These data are available in many reference books and manuals and are often specified as a discrete function of temperature and gas pressure. In the simulation procedure the determination of them, using the values of temperature and pressure, is reduced to the approximation. Methods for solving the problems of approximation are given in many books on the computational mathematics, and they will not be considered in this book.

7.3.1 Algorithms of the Mathematical Model

As it is shown in Sect. 6.1.2, the solutions of equation of the balance energy, the momentum equation, the continuity equation and other in the simulation of plasma processes it is convenient to use not one fixed grid, but several grids displaced from each other by a half pitch (Fig. 6.5), so that each equation can be solved with a separate grid. The Shifted Chess Grid should be constructed with maximum use of the boundary conditions for the solution of equations in the following way: the boundary points of each displaced grid should be placed directly on the border, where the absolute value of the necessary variable is known, and they are determined back down on half a step inside the computational area from those boundaries where the gradient of this variable is given.

Such a structure of shifted grids for simulation of the RF plasma torch using the specific of equations of equilibrium plasma model, boundary conditions, and their design features plasma torch are shown above (Sect. 6.1.3) in the development of the joint algorithm of solution the equation of the energy balance, the momentum equation, the continuity equation, etc., (Figs. 6.5a, 6.6–6.8). As it is shown above (Sect. 6.1.2), the structure of the shifted chess grids for simulation of the RF plasma torch and the Arc zone in the channel can be the same because of the similarity of many boundary conditions on the axis of symmetry, wall, and the exit of the RF plasma torch.

Discrete analogues of the equation of the balance energy, the momentum equation, the continuity equation for plasma flux are obtained for internal points (Sect. 6.1.1, 6.1.2), and their form based on internal feedback of physical processes are given earlier (Sect. 6.3.2, 6.3.3). Discrete analogue for equation of the magnetic vector potential and determination of the boundary conditions are given in Sect. 7.2. To form a complete system of discrete analogues and their matrix for determination of the coefficients of mathematical model of equilibrium plasma in the RF plasma torches we have to define the boundary conditions and the corresponding discrete analogues for the boundary points of the equation of the balance energy, the momentum equations, and the continuity equation.

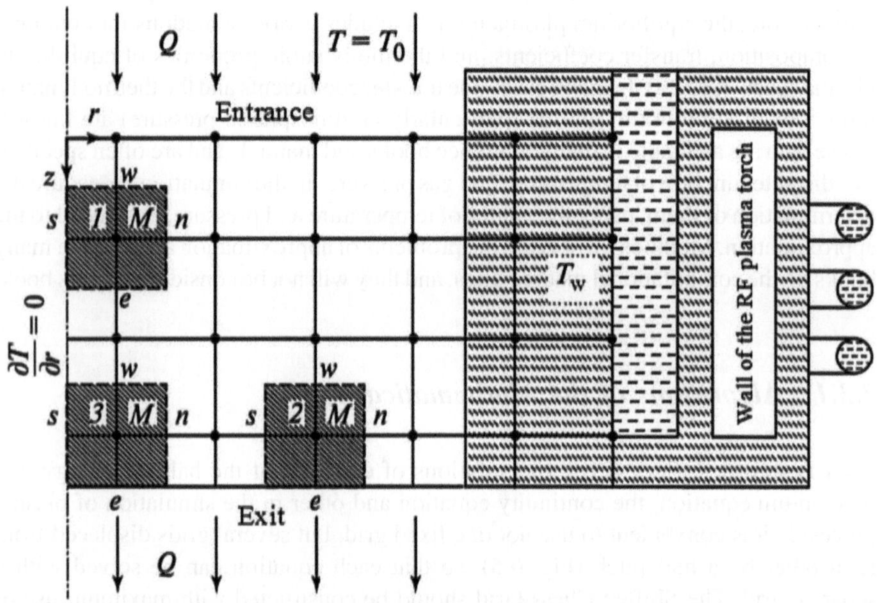

Fig. 7.31 Estimated grid and the control volumes for the boundary points at the decision of equation of the energy balance for the RF plasma torches

At the solving of various equations using a grid structure and the same boundary conditions for the discrete analogues for internal and boundary points are the same. Due to the displacement of the grid points the boundary points of grids in various cases can be not only on the border of calculated area, and for these points a special discrete analogues is required. Consider the boundary conditions for the solution of these equations and obtaining the corresponding discrete analogues in simulation of the RF plasma torches. In the future, these expressions could be used to simulate the arc zone in the channel (Sect. 8.2, 8.3).

1. **The boundary condition and the corresponding discrete analogues for equation of the balance energy**. As it is shown earlier (Sect. 6.1.1, 6.1.3, the discrete analogue of two-dimensional equation of balance energy for common internal point (Fig. 7.31) in a cylindrical coordinate system is

$$a_M T_M = a_E T_E + a_W T_W + a_N T_N + a_S T_S + b_M. \tag{7.36}$$

Hereafter, all the coefficients of the discrete analogue a_M, a_E, a_W, a_N, a_S, b_M are determined using the formula (6.61).

The computational grid is shown in Fig. 7.31, for the decision of equation of the energy balance there is an integral part of offset grid used in the simulation of the RF plasma torch (Fig. 6.5a). It encompasses not only the inside part of the plasma torch channel, but the area of the internal wall of the plasma torch. Thus, in the solution of the equation of energy balance the wall temperature of the plasma torch the degree of heating that is also an important parameter in the simulation and design of the RF plasma torches. Moreover, the boundary condition for the internal side of the water wall of the channel in plasma torch can be approximately determined using from the measured value of water temperature by thermocouple.

At the entrance of plasma torch (Fig. 7.31), the temperature of the cold gas is considered to be known $T = T_0 = 300$ K.

On the inner side of the water channel in the wall of plasma torch (Fig. 7.31) can be set as $T_W = 350$ K.

On the axis of plasma torch we have $\frac{\partial T}{\partial r} = 0$ because of the cylindrical symmetry. A discrete analogue for the points adjacent to the axis of symmetry is obtained by integrating the equation of the energy balance (6.39) for the control volume 1 (Fig. 7.31). The integration procedure is similar to the procedure of integration of generalized differential equation (Sect. 6.1.1). As $\frac{\partial T}{\partial r} = 0$ at $r = 0$, the heat flux due to heat conduction through the verge s of the control volume is equal to zero ($\lambda_s \frac{\partial T_s}{\partial r} = 0$). At the same time, at the axis of cylindrical symmetry $v_r = 0$ and convective heat flux through the verge s of the control volume is equal to zero ($\rho_s v_{rs} T_s = 0$). It follows that a discrete analogue of these points is (7.36) with a coefficient $a_S = 0$.

At the exit of plasma torch value of T is not specified, but is determined by decision of the equation of energy balance. However, the exit is often quite far from the inductor, and we can assume that all processes are established ($\frac{\partial T}{\partial z} \approx 0$), and the convective flow is much greater than the flow due to the thermal conductivity ($A(\mathrm{Pe}_e) = 0$). In this case, the approximate temperature value equal to its value at the nearest point upstream can be taken. We see that it corresponds to the real conditions, when the active physical processes in the simulation area (upstream), largely determine the value of the unknown quantities at the border (downstream), and not vice versa. It is natural to assume that for developed flux of plasma from the core of the inductor, the temperature at the exit of plasma torch depends mainly on the temperature of plasma flow in inductor (upstream), and the approximate boundary condition at the exit of the plasma torch little effect on the result of the decision of the equation of energy balance in the upstream, i.e., in the zone of inductor. A famous computational scheme (method) upstream (Sect. 6.2.1) in the solution of problems of gas dynamics is based on this real picture. A discrete analogue for the points adjacent to the exit is obtained by integrating the equation of balance energy in the control volume 2 (Fig. 7.31) with this assumption. As a result we have a discrete analogue (7.36) at $a_E = 0$. This form of the boundary conditions at the exit using a computational scheme "upstream" is highly effective, it gives the discrete analogue for points on this border and it will be used in the future.

The discrete analogue for a point located on the axis of symmetry at the exit of the plasma torch is obtained similarly (for the control volume 3 at Fig. 7.31) using two boundary conditions for the points on the axis of symmetry and for the points at the exit of the plasma torch, so the discrete analogue (7.36) for this point have coefficients $a_S = a_E = 0$.

2. **The boundary conditions and the corresponding discrete analogues of the momentum equations**. The discrete analogue of the two-dimensional momentum equation with for v_z (6.25) for common interior points in a cylindrical coordinate system, obtained above (Sect. 6.1.2) has the form (6.28)

$$a_M v_{zM} = a_E v_{zE} + a_W v_{zW} + a_N v_{zN} + a_S v_{zS} + b_{zM} + (p_w - p_e)\Delta S_r, \qquad (7.37)$$

where the coefficients a_M, a_E, a_W, a_N, a_S, b_{zM}, ΔS_r are determined using the Eq. (6.61).

At the entrance of the plasma torch (z = 0), the value v_z, m/s is determined using a given mass consumption of gas Q, l/min as $v_{z0} = \frac{Q \cdot 10^{-3}}{60\pi R^2}$, where R—the radius of the input channel of the plasma torch, m.

The discrete analogue for the points, adjacent to the axis of symmetry (the control volume 1 Fig. 7.32). Boundary conditions on the axis of the plasma torch $v_r = 0$ and $\left(\frac{\partial v_z}{\partial r} = 0\right)$ mean that the convective and viscous flows through the verge s of the control volume 1 equal to zero, i.e., for these points we have the discrete analogue (7.37) at $a_S = 0$.

The discrete analogue for the points, located at the exit of the plasma torch. Similarly to the boundary conditions for T, the boundary condition v_z at the exit of

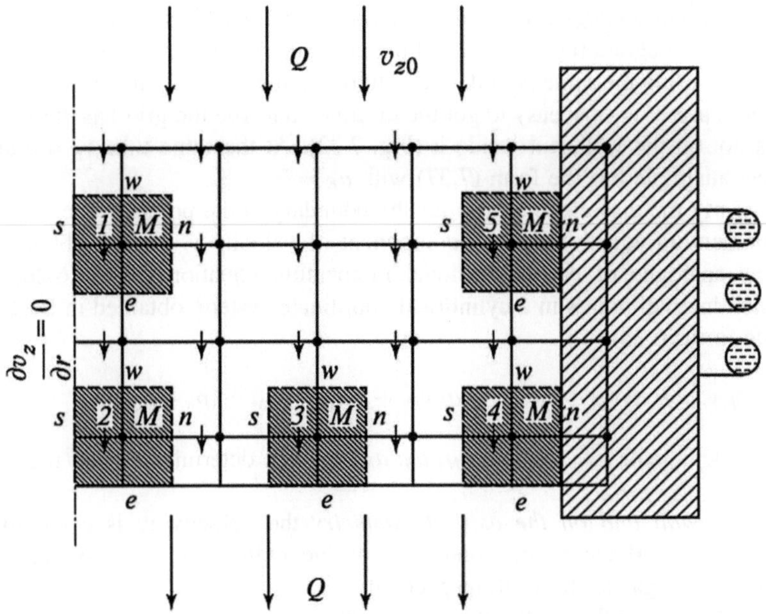

Fig. 7.32 The grid and the control volumes for the boundary points in the decision of the momentum equation for v_z in the RF plasma torches

the plasma torch is obtained with condition of a steady ($\frac{\partial v_z}{\partial z} = 0$) and the developed flows of the plasma stream. A discrete analogue for these points is obtained by integrating the momentum Eq. (6.25) for v_z at the control volume 3 (Fig. 7.32). As $\frac{\partial v_z}{\partial z} = 0$ at the exit, and viscous forces $f_{\mu z} \sim \mu \frac{\partial v_z}{\partial z}$, there is a loss of momentum (the viscous flow) on the axis z is equal to zero. The established character of the gas flow at the exit allows to receive the speed $v_{ze} = v_{zM}$. This assumption about the boundary condition allows in obtaining the discrete analogue for these points that does not include information about the plasma velocity at the exit of the plasma torch. It has the same form as (7.37) at $a_E = 0$.

The discrete analogue for a point at the exit and on the axis of symmetry (control volume 2, Fig. 7.32). For two boundary conditions on the symmetry axis and at the exit it can be obtained that, for a point located on the axis of symmetry at the exit of plasma torch, the discrete analogue (7.37) is using $a_E = a_s = 0$.

The discrete analogue for the points adjacent to the wall of plasma torch, is obtained integrating the momentum equation (6.25) for control volume 5 (Fig. 7.32) using the boundary conditions for the points on the wall of plasma torch: $v_z = v_r = 0$. Gas moves without "slipping", clinging to the wall. If we take the structure of the grid for determination of v_z such that the point on the right boundary is shifted by half a step inside the walls, you can simply set the speed at this boundary (Fig. 7.32) $v_z = 0$, and the discrete analogue for points of the control volume 5 will have the form (7.37).

The discrete analogue to a point adjacent to the wall and to the exit, it is also obtained by integration of the momentum equation (6.25) for the control volume 4 (Fig. 7.32) using the same boundary conditions at the exit and on the wall. As in the previous case, it is very easy to get the discrete analogue the grid has the boundary points not on the wall, but inside it (Fig. 7.32). At the same time $v_z = 0$ and the discrete analogue has the form (7.37) with $a_E = 0$.

Similarly the discrete analogue for the boundary nodal points in the decision of the momentum equation for the component of velocity v_r can be obtained. The discrete analogue of two-dimensional momentum equation for v_r (6.26) at the common internal points in a cylindrical coordinate system, obtained in Sect. 6.1.2, has the form (6.28)

$$a_M v_{rM} = a_E v_{rE} + a_W v_{rW} + a_N v_{rN} + a_S v_{rS} + b_{rM} + (p_s - p_n)\Delta S_z. \tag{7.38}$$

The coefficients a_M, a_E, a_W, a_N, a_S, a_M, b_{rM} are determined using the formula (6.61).

On the wall and on the axis of symmetry the velocity v_r is equal to zero: $v_r|_{r=0} = v_r|_{r=R} = 0$ using the conditions of the cylindrical axial symmetry and sticking of the gas to the wall, respectively.

The discrete analogue for the points adjacent to the entrance is obtained by integrating of the momentum equation (6.26) for the control volume 1 (Fig. 7.33).

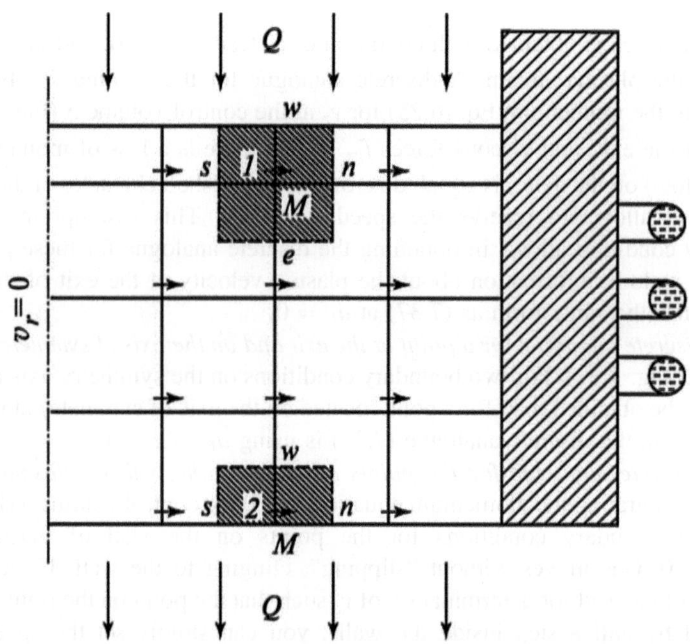

Fig. 7.33 The grid and the control volume of the boundary points in the solution of the momentum equation for v_r in the RF plasma torches

In integrating using the boundary conditions at the entrance: $v_r = 0$ (for axial gas supply) and $\frac{\partial v_r}{\partial z} = 0$, the convective and viscous flows through the verge w of the control volume 1 are equal to zero, i.e., the discrete analogue for these points has the form (7.38) with $a_W = 0$.

The discrete analogue for the points at the exit is obtained by integrating the momentum equation (6.26) for half of the control volume 2 (Fig. 7.33). Here the boundary condition at the exit $\frac{\partial v_r}{\partial z} = 0$ as a result of the steady flow of plasma, in the end we have the discrete analogue in the form (7.38) with $a_E = 0$.

3. **The boundary conditions and the corresponding discrete analogues of the continuity equation for p.** The pressure at the exit of the plasma torch is considered to be known and for most of the RF plasma torches with an open channel is equal to atmospheric pressure. Therefore, the boundary points of the grid for determination of pressure are directly on the exit of the plasma torch (Fig. 7.34). At the same time, the values of the inlet pressure on the wall and on the axis of symmetry are usually unknown, but the values of the velocity components on these boundaries are known. Therefore, the grid for determination of pressure compared with the grid for definition of v_z and v_r is shifted inward from the boundaries on a half-step, that makes it possible to determine accurately the velocity components v_z and v_r on the verges of the control volume (Fig. 7.34).

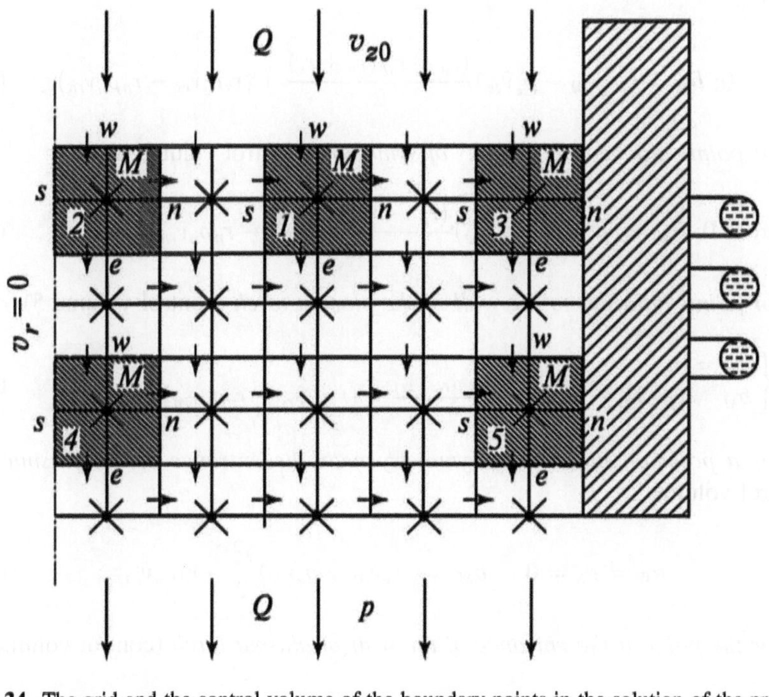

Fig. 7.34 The grid and the control volume of the boundary points in the solution of the pressure p in the RF plasma torches

Due to this structure of the shifted grid the special discrete analogue for pressure using the continuity equation and through the components of the velocity and v_r using the above methodology (Sect. 6.1.2) is obtained. General discrete analogue for the continuity equation (6.38) is

$$a_M p_M = a_E p_E + a_W p_W + a_N p_N + a_S p_S + b_M, \tag{7.39}$$

where $b_M = (\rho_w \hat{v}_{zw} - \rho_e \hat{v}_{ze})\frac{(r_n - r_s)(r_n + r_s)}{2} + (r_s \rho_s \hat{v}_{rs} - r_n \rho_n \hat{v}_{rn})(z_e - z_w)$; The coefficients a_M, a_E, a_W, a_N, a_S are determined by (6.36–6.38).

The main advantage of the shifted grid is obtaining of the discrete analogue for points, adjacent to the exit, the axis of symmetry and to the wall, i.e., at the borders where the pressure is unknown. In Fig. 7.34 five types of control volumes for the boundary points for solving of the continuity equation and obtaining of the discrete analogue for pressure can be identified. As it is shown earlier (Sect. 6.1.2), the discrete analogue for these points is obtained by integrating of the continuity equation for the control volume. Using the boundary conditions the continuity equation for v_z and v_r at the entrance ($z = 0; v_z = v_{z0}$), on the wall ($r = R$; $v_r = v_{rn}$ for the plasma torch with a porous wall the velocity at the wall is not equal zero) and at the axis of symmetry ($r = 0$; $v_r = 0$) we have to discrete analogue for the boundary points of continuity equation in a general form (7.39) with the following coefficients:

For the points, located at the entrance of the plasma torch (control volume 1 in Fig. 7.34.

$$a_W = 0;\ b_M = (\rho_w v_{z0} - \rho_e \hat{v}_{ze})\frac{(r_n - r_s)(r_n + r_s)}{2} + (r_s \rho_s \hat{v}_{rs} - r_n \rho_n \hat{v}_{rn}) z_e; \tag{7.40}$$

For points adjacent to the axis of symmetry (control volume 4)

$$a_S = 0;\ b_M = (\rho_w \hat{v}_{zw} - \rho_e \hat{v}_{ze})\frac{(r_n - r_s)(r_n + r_s)}{2} - r_n \rho_n \hat{v}_{rn}(z_e - z_w); \tag{7.41}$$

For points adjacent to the wall of the plasma torch (control volume 5)

$$\begin{cases} a_N = 0; \\ b_M = (\rho_w \hat{v}_{zw} - \rho_e \hat{v}_{ze})\frac{(r_n - r_s)(r_n + r_s)}{2} + (r_s \rho_s \hat{v}_{rs} - r_n \rho_n v_{rn})(z_e - z_w); \end{cases} \tag{7.42}$$

For a point on the axis of symmetry near the entrance of the plasma torch (control volume 2)

$$a_W = a_S = 0; \quad b_M = (\rho_w v_{z0} - \rho_e \hat{v}_{ze})\frac{r_n^2}{2} - r_n \rho_n \hat{v}_{rn} z_e; \tag{7.43}$$

For the point at the entrance of the wall of plasma torch (control volume 3)

$$\begin{cases} a_W = a_N = 0; \\ b_M = (\rho_w v_{z0} - \rho_e \hat{v}_{ze}) \dfrac{(r_n - r_s)(r_n + r_s)}{2} + (r_s \rho_s \hat{v}_{rs} - r_n \rho_n v_{rn}) z_e. \end{cases} \quad (7.44)$$

Thus, using the method of control volume and the structure of shifted grids the complete systems of discrete analogue of mathematical model of equilibrium plasma in the RF plasma torches were prepared. The obtained discrete analogues represent a closed system of discrete equations, that include internal and external communication of various physical processes and obtained from the boundary conditions, the regime of operation and the features of construction of the RF plasma torches, so they can the most accurately simulate the physical processes and be a reliable tool for the study of the characteristics of plasma.

7.3.2 Analysis of the Simulation Results

Construction of the grids for calculations As it is shown above (Fig. 6.5a), for simulation of the RF plasma torches four shifted grids are used: [*z*1, *r*1]—for determination of the vector potential *A*, [*z*1, *r*2]—for defining of the velocity v_z and temperature *T*, [*z*2, *r*1]—for obtaining of the speed v_r, and [*z*2, *r*2]—for the determination of the pressure *p*. Depending on the control volume the transfer coefficients and thermodynamic properties can be found by approximating the value of temperature on the different grids. The coordinates of the grid [*z*1, *r*1] are given, and the coordinates of the grid [*z*2, *r*2] can be determined in the middle of the grid coordinates [*z*1, *r*1], respectively. Grids for simulation are built taking into account the peculiarities of the temperature field and the flow of the RF plasma. Constructing the grids it is also necessary to take into account the geometric characteristics of the plasma torch, zone of active release of energy and current flow, the zone of active interaction between the field and plasma, and others. All active processes in the RF discharge occur in the area of inductor and in the skin layer, so the grid steps in these areas should not be large. Grids are built in the zones, and where sharp temperature and speed gradients are expected and a smaller step of the grid should be chosen, and vice versa. For example, in the area between the skin layer and the wall of the plasma torch it should take the steps Δr = 0.3–0.5 mm compared with the average Δr = 1–2.5 mm in the remaining zones. In some cases, for further input of chemicals and processed particles into the discharge the entrance of plasma torch can be located close to the inductor. The sharp temperature and velocity gradients in this zone are expected. Therefore it is necessary to take a step Δz = 0.1–0.5 mm in this area, compared with Δz = 2–4 mm in the zone of inductor and Δz = 5–8 mm in the zone of the jet of plasma torch. Thus, for example, at the geometric characteristics of the channel of the RF plasma torch (5 MHz) (Table 7.3; Fig. 7.14) by the number of grid points at the *z*-axis is obtained and it should be equal to 40–50 and at the axis *r* = 30–40, and the number of unknown variables of the equation will be 1200–2000.

The selection of the step of the grid and the determination of the coefficients of internal feedback for various equations are carried out taking into account the region of stability (Figs. 6.17, 6.18, 6.19), that are obtained as a result of numerous experiments on simulation of plasma processes.
Note that using of a uniform grid (Δz = const, Δr = const) the process of solving will be more stable, but in this case the number of nodes and respectively the number of discrete equations will be considerably large.

Analysis of the simulation results The RF plasma torch (5 MHz) and similar plasma torches (Table 7.3) with three streams are used in practice for spectral analysis and plasma decomposition of chemical compounds. They work in pure argon in a mixture of Ar and N_2 and in other molecular gases. Some of the results and analysis of researches of electromagnetic fields in the RF plasma torch (5 MHz)

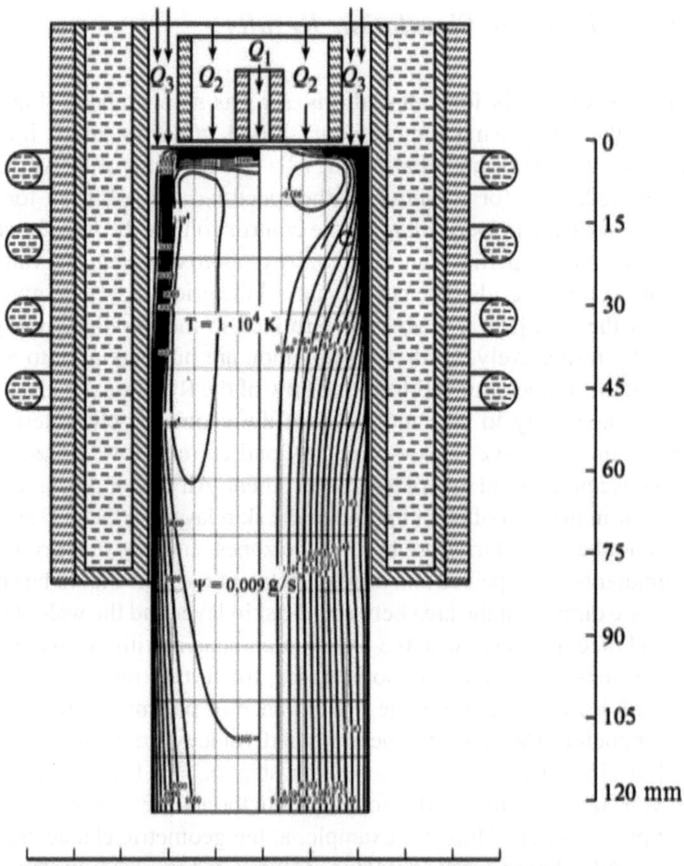

Fig. 7.35 The temperature and flow fields of Argon in the RF plasma torch (5 MHz): ψ—current function; (Q_1 = 0, Q_2 = 20 l/min, Q_3 = 40 l/min)

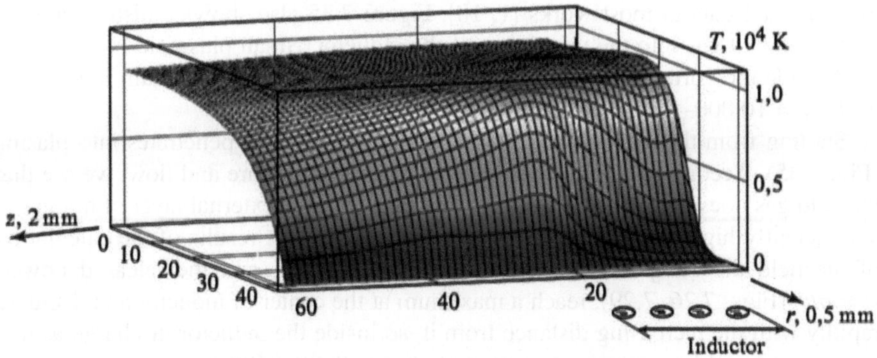

Fig. 7.36 The temperature of Argon in the RF plasma torch (5 MHz)

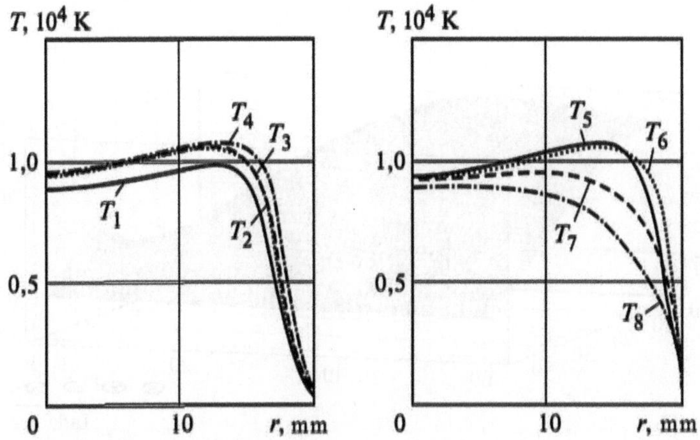

Fig. 7.37 The temperatures of Argon T_1–T_8 in the sections z = 4, 10, 16, 26, 36, 48, 80, and 120 mm, respectively

is shown above (Sect. 7.2.3), we present here the rest of the simulation results of an equilibrium plasma [39, 40]. For comparison with data of the field calculation (Sect. 7.2.3) and in simulation of nonequilibrium plasma (Sect. 7.5.2) the results are shown in the same cross-sections: $z_1 = 4$ mm, $z_2 = 10$ mm, $z_3 = 16$ mm, $z_4 = 26$ mm, $z_5 = 36$ mm, $z_6 = 48$ mm, $z_7 = 80$ mm, $z_8 = 120$ mm.

Figures 7.35, 7.36 and 7.37 shows the fields of temperature and velocity of argon flow of plasma in the RF plasma torch (5 MHz). Comparing the data on the temperature (Fig. 7.35), and released power (Figs. 7.26, 7.27), we see that the maximum plasma temperature is not on the axis of the plasma torch, and a zone close to the region of maximum power allocation, and is T_{max} = 9500–10,500 K. The temperature of argon jet at the exit of plasma torch reaches 8000–9000 K. These temperatures of argon plasma in the RF plasma torches are confirmed by experimental data in most works [1, 18]. Figure 7.35 also shows a distinct area of vortex flow of gas into the upper part of the inductor. Small plasmoid is constantly held inside the vortex, and the plasma temperature reaches a maximum in the zone of T_{max} = 10,000–10,500 K.

Starting from the second round of inductor, gas actively penetrates into plasma (Fig. 7.35). According to the data of the fields of temperature and flow, we see that the cold gas does not penetrate into the discharge but the external layer of hot gas at a sufficiently high temperature T = 5000–8000 K. By the results of the calculation of the field the component of electromagnetic force F_r and the released power, $P = \sigma E^2$ (Figs. 7.26, 7.29), reach a maximum at the center of inductor and decrease rapidly with the increasing distance from it, so inside the inductor discharge as it is compressed and has an average temperature T = 9000–9500 K.

At the exit of plasma torch, where there is practically no energy release and electromagnetic force, there is expansion of the flow with decreasing temperature (Fig. 7.35) to T = 8000–9000 K due to diffusion and thermal conductivity.

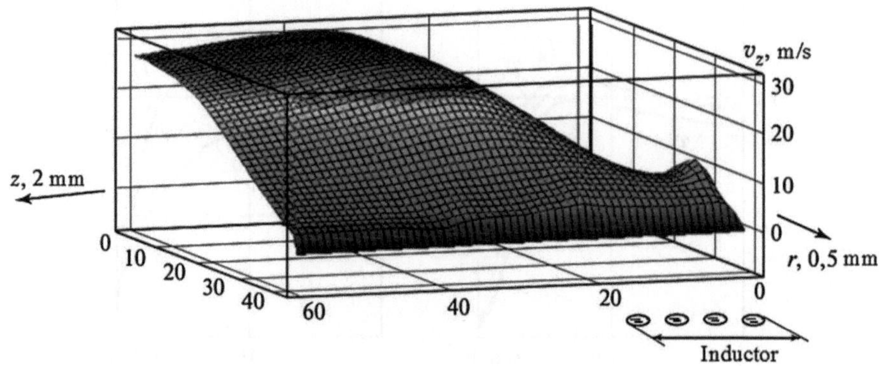

Fig. 7.38 The field of the velocity components v_z of Argon in the RF plasma torch (5 MHz)

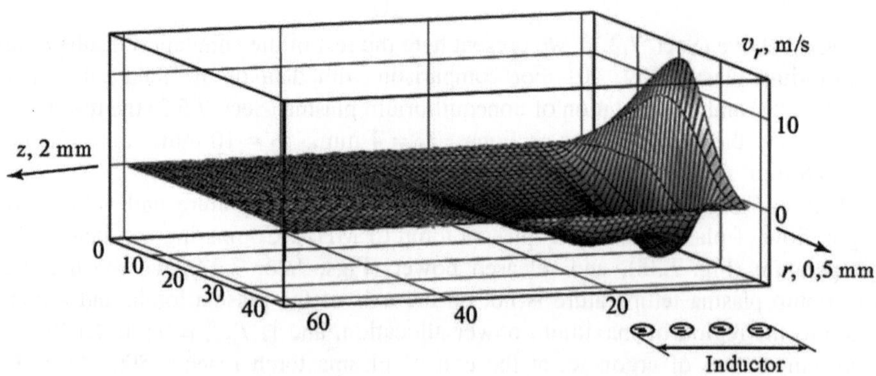

Fig. 7.39 The field of the velocity components v_r of Argon in the RF plasma torch (5 MHz)

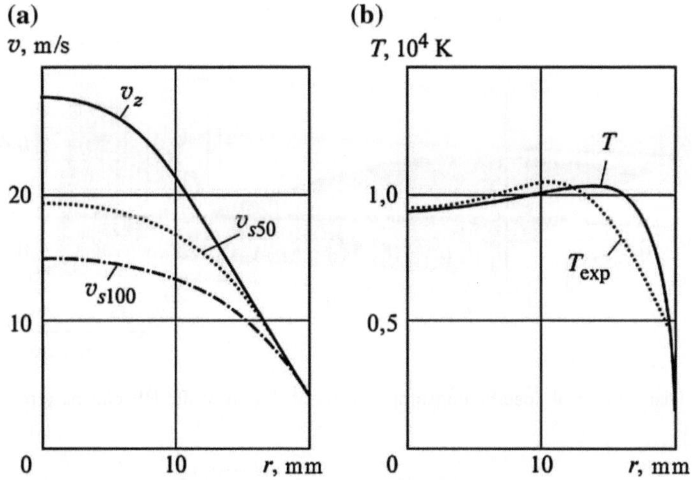

Fig. 7.40 The experimental and calculated data of Argon in the RF plasma torch (5 MHz): **a** the distribution of the measured particle velocity silicon v_{s50} (d_s = 50 mm), v_{s100} (d_s = 100 mm) [41] and the estimated plasma velocity v_z, at z = 120 mm; **b** distribution measured T_{exp} [39, 42] and the estimated T plasma temperature in the section z = 54 mm

Expansion flow in some cases leads to overheating of the walls of the plasma torch, so in practice the introduction of shielding gas is carried out by using the plasma torch exit gas stream oncoming.

Distribution components speed v_z and v_r of the plasma flux are shown in Figs. 7.38 and 7.39. The speed of plasma reaches a maximum at the exit of the plasma torch and is v_{zmax} = 32 m/s and a maximum speed of reverse plasma flow $v_{zmin} = -10$ m/s and is equal to the velocity component v_r in this zone.

The results of measuring the velocity of the silicon particles with diameter of 50–100 μm in the jet of plasma torch (z = 120 mm) and the temperature of plasma in cross section z = 54 mm, and a comparison with the simulation result in these sections are shown at Fig. 7.40. The measurement results are taken from [39, 41, 42], experiments were carried out at the same conditions for simulation of the RF plasma torch (5 MHz). To measure the velocity of the particles using the laser method of registration, and to measure the temperature of plasma the method of the absolute intensity with the integral Abel transformation were used.

Processing data of other regimes at the constant voltage on inductor shows that varying amount of plasma-forming and protective gas is 20–100 l/min does not significantly change the temperature field and the plasma flow. This confirms our earlier conclusion (Sect. 7.1) that movement of the plasma flow depends mainly on the electromagnetic force and power density in the discharge.

For analysis of the stream of plasma flow the distribution of specific gas consumption ρv_z and ρv_r. are presented in Figs. 7.41 and 7.42. We see that gas not « blows » into the discharge, excessive amounts of cold gas flowing out of the

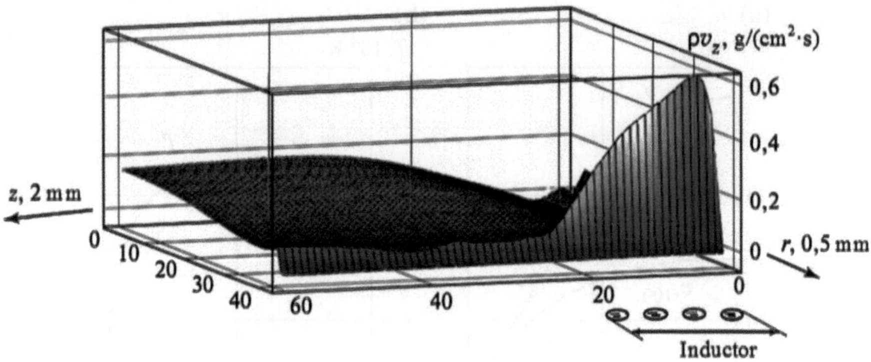

Fig. 7.41 Distribution of specific consumption ρv_z of Argon in the RF plasma torch (5 MHz)

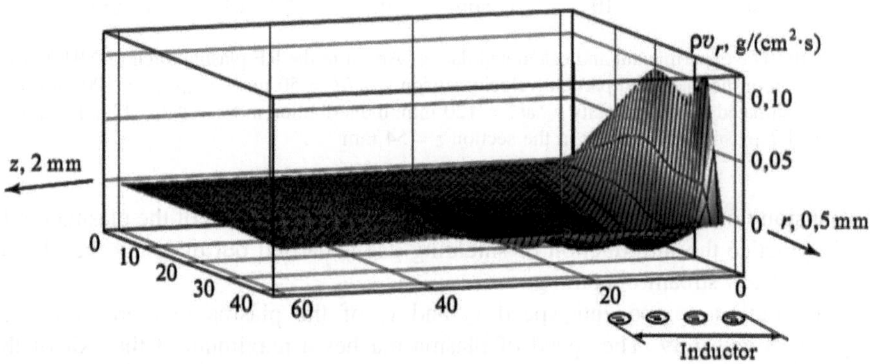

Fig. 7.42 Distribution of the specific gas consumption ρv_r of Argon in the RF plasma torch (5 MHz)

plasma core (in the zone near the wall), forming a natural gas thermal protection layer.

As expected, the electromagnetic force F_r (Fig. 7.29) compresses the plasma flow, thus creating the maximum pressure in the center of inductor (Fig. 7.43). At the entrance a small pressure is created by the counter flow of plasma-forming gas flow and reverse vortex of plasma. The pressure at the entrance zone of the protective gas is enough high to provide a large gas consumption. Since the second spiral inductor, active penetration of gas into plasma by an electromagnetic force leads to the formation parietal under pressure zone (Figs. 7.35, 7.43). This explains the number of cases detected in the parietal vortex flow (aft vortex), exciting additional mass of gas from the exit of the plasma torch in the event of inadequate supply of plasma gas

The resulting pressure field in the channel of plasma torch allows determination of the static pressure gradient in the discharge (Fig. 7.44). As it was previously noted (Sect. 7.1) the movement of gas in the RF discharge takes place under the

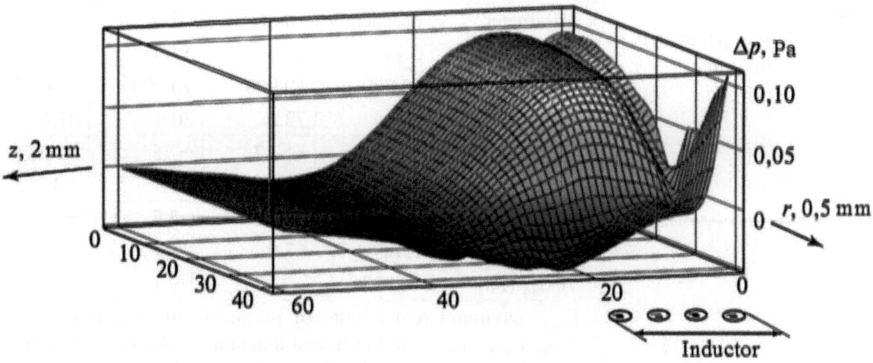

Fig. 7.43 Field of the static pressure $\Delta p = p - p_{out}$ inside the RF plasma torch (5 MHz)

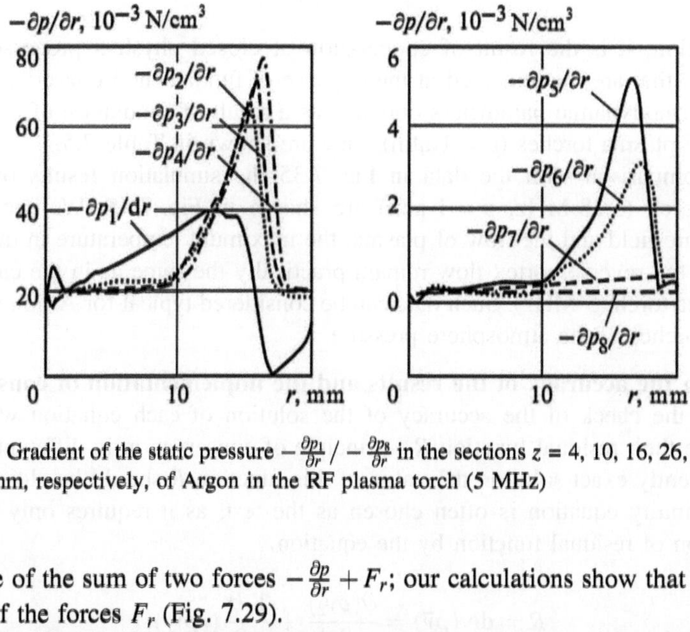

Fig. 7.44 Gradient of the static pressure $-\frac{\partial p_1}{\partial r} / -\frac{\partial p_8}{\partial r}$ in the sections z = 4, 10, 16, 26, 36, 48, 80, and 120 mm, respectively, of Argon in the RF plasma torch (5 MHz)

influence of the sum of two forces $-\frac{\partial p}{\partial r} + F_r$; our calculations show that it is only 3–8 % of the forces F_r (Fig. 7.29).

We have seen that for the calculation of plasma the gradient of static pressure cannot only be taken into account or to equate it to the electromagnetic force F_r. Created field of static pressure in plasma acts as a negative feedback loop that prevents to the excessive penetration of cold gas into plasma, and provides the ability to hold the high temperature of discharge, creating the favorable conditions for the flow of current, the release of energy and maintenance of discharge.

Thus, the plasma parameters depend on close cooperation between the various physical processes inside the discharge. The discharge in the plasma torches as it is self-regulating and "chooses" the best variant of its sustainable implementation with minimal energy consumption. Some researchers, such as *M.* Steinbeck, based on the analysis of the thermal heating in the Arc call it as a principle of minimum energy

Table 7.5 Some thermodynamic parameters of the RF plasma torches

Parameter	Frequency in inductor (MHz)			
	0.44	1.76	5.00	5.28
T_{max} (K)	10,800	10,900	10,700	10,600
v_{zmax} (m/s)	128.4	72.8	30.9	27.6
v_{zmin} (m/s)	−177.5	−87.7	−7.5	−11.8
v_{rmax} (m/s)	40.6	43.2	13.3	1.35
v_{rmin} (m/s)	−40.8	−9.98	−7.7	−4.5
Δp_{max} (Pa)	156.9	67.0	21.9	15.8
Δp_{min} (Pa)	−212.7	−27.7	−9.1	−7.2

T_{max} maximum temperature of plasma in the discharge, v_{zmax}, v_{zmin}, v_{rmax}, v_{rmin} minimum and maximum values of the velocity components, respectively, Δp_{max}, Δp_{min} difference of maximum and minimum pressures in the channel compared with the pressure at the exit of the plasma torch

consumption, it is the result of cooperation of closed physical processes in the discharge, that are implemented at the expense of fundamental conservation laws. Some thermodynamic parameters obtained as a result of simulation of other types of the RF plasma torches (p = 1 atm), they are shown in Table 7.5.

For comparison with the data in Fig. 7.35 the simulation results of the RF plasma torch (5.28 MHz, p = 1 atm) are shown in Fig. 7.45. We see that the temperature field and the flow of plasma, the maximum temperature in the area of inductor, the reverse vortex flow remain practically the same as in the case of the RF plasma torch (5 MHz). Such data can be considered typical for Argon in the RF plasma torches of the atmosphere pressure.

Checking the accuracy of the results and the implementation of conservation laws To the check of the accuracy of the solution of each equation we can be determine their residual function R—principle of conservation in differential form. At sufficiently exact solution the value of this function R should be close to zero. The continuity equation is often chosen as the test, as it requires only a simple calculation of residual function by the equation

$$R = \mathrm{div}(\rho\vec{v}) = \frac{\partial(\rho v_z)}{\partial z} + \frac{1}{r}\frac{\partial}{\partial r}(r\rho v_r). \tag{7.45}$$

In order to establish the accuracy of the simulation of different processes the compliance of the conservation laws in the equations in integral form for each variable can also be checked. For example, knowing a predetermined amount of gas Q it can be verified the law of mass conservation in integral form

$$\psi = \frac{Q\rho}{2\pi} = \int_0^R \rho v_z r\mathrm{d}r. \tag{7.46}$$

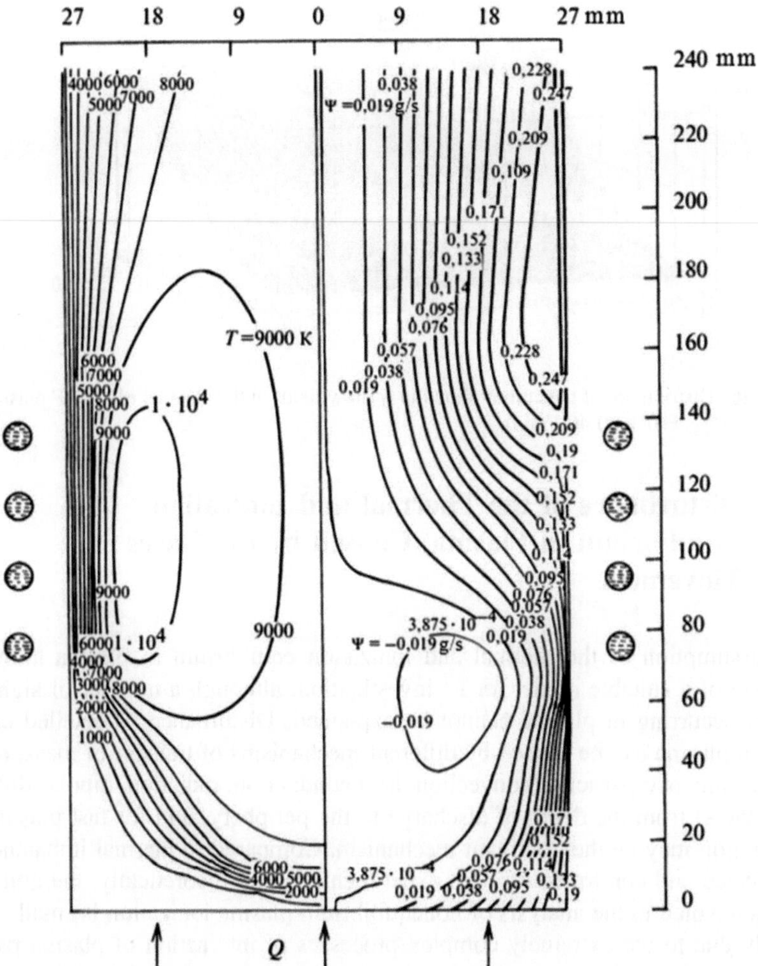

Fig. 7.45 The temperature and flow fields of Argon in the RF plasma torch (5.28 MHz) ($Q = 60$ l/min, $U_1 = 6$ kV, $I_1 = 155$ A, $P_2 = 30$ kW)

The right side of (7.46) is the function of current on the wall of the plasma torch ($R = R_5$) is determined from the resulting velocity and temperature fields. Since the field of temperature and speed, in fact, are the result of the combined action of all closed physical processes of mathematical model, and the observance of equality (7.46) indicates the proper and correct simulation results. Figure 7.46 shows the distribution of the current function in three dimensions. The obtained value of the current function at the wall of plasma torch is equal $\psi_{Ar}(R_5) = 0.219$ g/s and corresponds to the total amount of gas at the entrance to the plasma torch $Q_{Ar} = 60$ l/min, that indicates the accurate result of our solutions.

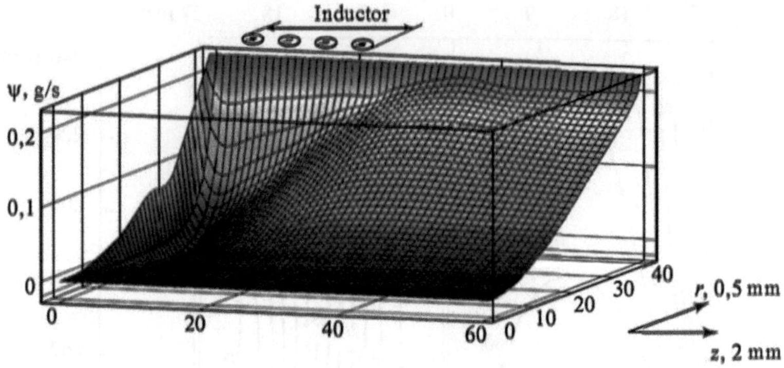

Fig. 7.46 Distribution of the current function ψ (flow field) in the channel of the RF plasma torch (5 MHz) (Q_{Ar} = 0 + 20 40 l/min)

7.4 Disturbance of the Thermal and Ionization Equilibrium in Plasma, Caused by the Gases Movement

The assumption of the thermal and ionization equilibrium in plasma allows the creation of a suitable model for its investigation, although a number of significant events occurring in plasma cannot be explained. Disturbance of detailed equilibrium in plasma can be caused by different mechanisms of transfer of mass, energy, momentum, and particles (convection, heat conduction, radiation, bipolar diffusion, and others) from the depth of discharge to the periphery, and for fast plasma flow convection may be the dominant mechanism. Compared to thermal imbalance that is detected and confirmed by both experimentally and theoretically, the number of works devoted to the analysis of nonequilibrium plasma ionization is small. This is mainly due to the extremely complex processes of interaction of plasma particles (energy exchange between particles, momentum exchange, and transformation of one type of particles into particles of another sort, radiative processes, the presence of external fields, etc.). In such a situation it is important to identify the main factor that disorders thermal and ionization equilibrium of plasma and to determine its impact and the degree of disturbances of these equilibriums. Such analysis can be done using a relatively simple model of plasma.

One of the first studies about the thermal equilibrium disturbances caused by the movement of gas through the two-temperature model is [43]. The task of heating it in a half-heated plasma space is considered for analyze of the thermal equilibrium of gas: cold gas with an initial temperature T_{a0} penetrates into the half-plasma space with electron temperature T_e at the given gas consumption ρv_z and pressure p. It is believed that the plasma is in a state of partial local thermal equilibrium (PLTE), i.e., Maxwell (1.2), Boltzmann (1.4) and Saha (1.9) equations are valid. The one-dimensional equation of the energy balance was decided for atom–ion component in the form of

$$\rho v_z c_p \frac{dT_a}{dz} = b(T_e - T_a), \tag{7.47}$$

Here $b = \frac{3}{2}k\frac{2m_a}{m_a}n_e\sqrt{\frac{8kT_e}{\pi m_e}}(n_a Q_{ea} + n_i Q_{ei})$—coefficient of energy exchange between electrons and atom–ion components.

The model does not take into account the energy flux due to thermal conduction, especially in the initial section where the gas is actively heated and high thermal gradient $\frac{dT_a}{dz}$ and hence a considerable flow of heat by conduction $\lambda_{ai}\frac{dT_a}{dz}$ should be expected. In the case of small consumptions ρv_z, corresponding to the real conditions of the RF plasma torches, it can lead to inaccurate results. This statement of the problem makes it possible to eliminate the need of the decision of the momentum equation, the continuity equation and Maxwell equations and to restrict with the decision of the simplified equation of energy balance for atom–ion component in order to determine the degree of the gas heating.

On the basis of solving the problem of gas heating in half-space of plasma the first results of the analysis of disturbances of thermal ionization equilibrium of the moving gas were obtained in [44]. It turned out that the disturbance of the ionization equilibrium of plasma can significantly influence to the speed of heating of atomic–ion components, i.e., there is a definite connection between the thermal and ionization processes. Our task is to determine this connection.

In the nonequilibrium model of heating gas in a half-heated space of plasma includes the following equation [44]:

Equation of the energy balance for atom–ion component

$$\frac{d(\rho v_z T_a)}{dz} = \frac{d}{dz}\left(\frac{2m_a\lambda_{ai}}{5k}\frac{dT_a}{dz}\right) + \frac{3}{2}k\frac{2m_e}{m_a}n_e\sqrt{\frac{8kT_e}{\pi m_e}}(n_a Q_{ea} + n_i Q_{ei})(T_e - T_a); \tag{7.48}$$

Equation of the ionization equilibrium in one-dimensional form

$$\begin{aligned} \frac{d}{dz}(\rho v_z n_e) = {} & \frac{d}{dz}\left(\rho D_{amb}\frac{dn_e}{dz}\right) + \rho k_r n_e n_a\left(\frac{n_e^{*2}}{n_a^*} - \frac{n_e^2}{n_a}\right) \\ & + \left(n_e v_z - D_{amb}\frac{dn_e}{dz}\right)\frac{d\rho}{dz}. \end{aligned} \tag{7.49}$$

Here n_e^*, n_a^*—the concentration of electrons and atoms in the two-temperature ionization equilibrium, calculated on the basis of the law of mass action by the Saha (4.1), or Potapov equation (4.4), and Dalton law of partial pressures (4.6). Plasma is considered to be quasi-neutral ($n_i \approx n_e$), singly and doubly ionized, atomic, and the charged particles are the result of two opposite processes—impact (stepped) ionization and triple (impact-radiative) recombination.

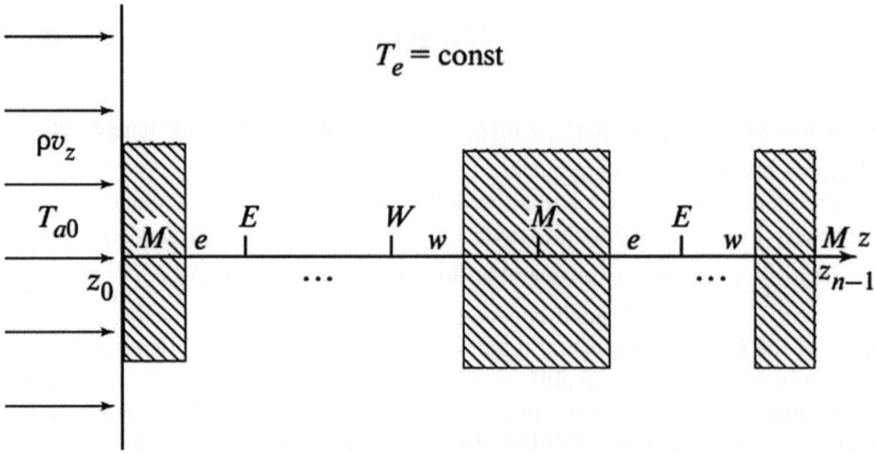

Fig. 7.47 The grid and control volumes for solving of one-dimensional equation of the balance energy and equation of ionization equilibrium in the problem of gas heating

These data of the effective sections Q_{ea}, Q_{ei} and the ways of defining the transfer coefficients and thermodynamic properties of two-temperature plasma are given in Sect. 2.2, 4.1–4.4.

Total discrete analogue for solving the equation of the energy balance for atomic and ionic components and equation of ionization equilibrium (Fig. 7.47) can be obtained by solving the one-dimensional GDE (Sect. 6.1.1), taking into account the linearization of the source function (6.55):

$$a_M\Phi_M = a_E\Phi_E + a_W\Phi_W + b_M, \tag{7.50}$$

where

$$\begin{cases} a_E = D_e A(P_e) \geq 0; \quad a_W = D_w A(P_w) + F_w \geq 0 \\ a_M = a_E + a_W - S_p(z_e - z_w) \geq 0; \quad b_M = S_c(z_e - z_w); \end{cases}$$

for equation of the energy balance

$$\Phi \equiv T_a; \Gamma = \frac{2m_a\lambda_{ai}}{5k},$$

$$S = \frac{3}{2}k\frac{2m_e}{m_a}n_e\sqrt{\frac{8kT_e}{\pi m_e}}(n_a Q_{ea} + n_i Q_{ei})(T_e - T_a) = S_c + S_p T_a;$$

for equation of ionization equilibrium

$$\Phi \equiv n_e; \Gamma = \rho D_{amb};$$

$$S = \rho\left[k_r n_e n_a\left(\frac{n_e^{*2}}{n_a^*} - \frac{n_e^2}{n_a}\right) - n_e\frac{dv_z}{dz}\right] - D_{amb}\frac{dn_e}{dz}\frac{d\rho}{dz} = S_c + S_p n_e.$$

Sustaining the stability of computational procedures In a joint decision of the two Eqs. (7.48) and (7.49) for the discrete analogue (7.50) the circuit with internal feedback equation of the energy balance for atom–ion components can be installed by the method described in Sects. 6.3.2, 6.3.3. In particular, the linearization of the source function for equation of the energy balance can be carried out by the algorithm (6.56), i.e., according to its absolute value. However, at the linearization of the source function the equation of ionization equilibrium it is necessary to consider the following factors: the temperature difference of electron and atomic–ion components varies as the heating of atoms and ions in the range of $\Delta T = T_e - T_a$ = 5000–20,000 K, so the electron density in this area also varies in a wide range of several orders of magnitude. In this case, rapid convergence is important to determine the initial concentration of electrons that correspond to the range of possible changes in the electron and ion atomic components temperature. The most convenient way to get the initial electron density from the initial field of temperature of electron and atom–ion components is in ionization equilibrium (4.1) or (4.4).

This method is convenient to define the initial field of the electron density, possesses, however, with a serious disadvantage. In the first iteration, we have near-zero source function in equation of ionization equilibrium $n_e^* = n_e$; $n_a^* = n_a$; $S \approx 0$, so that the internal feedback coefficient for the first iteration in equation of ionization equilibrium is very small ($S_p \approx 0$) and at the beginning the computational process will be carried out virtually using the open circuit without internal feedback. Practice has shown that this is enough to cause a divergence of solutions of equation of ionization equilibrium. So a special way of linearization should be found for the source function in equation of ionization equilibrium and the factor of internal feedback should be set.

The effectiveness of internal feedback loop in the stability of solutions of differential equations is not violated if the source function for equation of ionization equilibrium (7.49) is linearized as

$$S_p = \beta\rho k_r n_e^2; \quad S_c = S - S_p n_e. \tag{7.51}$$

Here, the coefficient of internal feedback $\beta \geq 1$, and is subject for possible regulation. An important result of this method of installation of internal feedback loop is nonzero coefficient S_p even for the first iteration, when the source function for equation of ionization equilibrium $S \approx 0$, i.e., internal feedback loop is always established and continues to operate effectively. It should be noted that with the (7.51) it was possible to ensure the stability of the computational procedure at the joint decision of equation of the energy balance (7.48) and equation of ionization equilibrium (7.49).

Boundary condition and the discrete analogue for the points on the boundary are obtained by integrating the corresponding equation for half of the control volume (Fig. 7.47). In the integration it should take into account the following boundary conditions: $z = z_0$, $T_a = T_{a0}$, but the concentration of electrons in $z = z_0$ is unknown; for steady-state process in the end ($z = z_{n-1}$), $\left.\frac{dT_a}{dz}\right|_{z_{n-1}} = \left.\frac{dn_e}{dz}\right|_{z_{n-1}} = 0$.

In the case when the values of variables on the boundary are unknown, it can be assumed that the value of Φ at the surface is completely determined by an internal process. It is justified for the processes that are actively occurring not at the border but inside the study area. This boundary condition is completely determined by the internal process (like a computer scheme "upstream" in Sect. 6.2.1). In view of this boundary condition the discrete analogue for equation of ionization equilibrium at the starting point ($z = z_0 = 0$, Fig. 7.47) has the form

$$a_M \Phi_M = a_E \Phi_E + b_M, \tag{7.52}$$

where

$$a_E = D_e A(P_e) \geq 0;\ a_M = a_E - S_p z_e \geq 0;\ b_M = S_c z_e$$

For the last point ($z = z_{n-1} = 0$, Fig. 7.47) the discrete analogue of equation of the balance energy and equation of ionization equilibrium is

$$a_M \Phi_M = a_W \Phi_W + b_M, \tag{7.53}$$

Here

$$a_W = D_w A(P_w) + F_w \geq 0;\ a_M = a_W - S_p(z_M - z_w) \geq 0;\ b_M = S_c(z_M - z_w).$$

The decision procedure of equation of balance energy and equation of ionization equilibrium is as follows.

1. Take the initial temperature field of atomic and ionic components $T_a = T_{a0}$ and using (4.1) or (4.4) determine the initial field of the electron density n_{e0}.
2. Calculate the coefficients of discrete analoueg of equation of ionization equilibrium (7.49).
3. Solve the system of the discrete analogue for equation of ionization equilibrium (7.50) and a new field of electron density n_e is obtained.
4. Calculate the coefficients of the discrete analogue (7.48).
5. Solve the system of the discrete analogue for equation of the balance energy (7.50) and a new temperature field T_a is obtained.
6. Check on convergence and repeat the procedure from position 2, until the full convergence.

Analysis of the obtained results Analysis of disturbances of thermal and ionization equilibrium caused by the movement of gases is carried out at the example of Argon plasma in two stages in order to further possible comparing of the results:

(1) Analysis of disturbance of thermal equilibrium with maintaining of the ionization equilibrium. In this case, the equation of balance energy for only atom–ion component is solved (7.48), the concentration of particles is determined by (4.4).
(2) Analysis of disturbance of thermal equilibrium using disturbances of ionization equilibrium. The energy balance equation of the atom–ion component (7.48) can be solved with equation of ionization equilibrium (7.49), and the equilibrium values n_e^*, n_a^* are defined using (4.4).

Data on the specific consumption of Argon were chosen for the conditions in RF and Arc plasma torches close to real: $\rho v_z = 0.05; 0.1; 0.2; 0.3; 0.5$ g/(cm^2 c). The values of cold gas temperature are set $T_{a0} = 300 - 9000$ K, the electron temperature $T_e = 9000 - 20{,}000$ K and the pressure $p = 10^5$ Pa.

The results of calculations for the ionization equilibrium are shown in Fig. 7.48, and its disturbance is in Fig. 7.49. The comparison of these figures shows that the

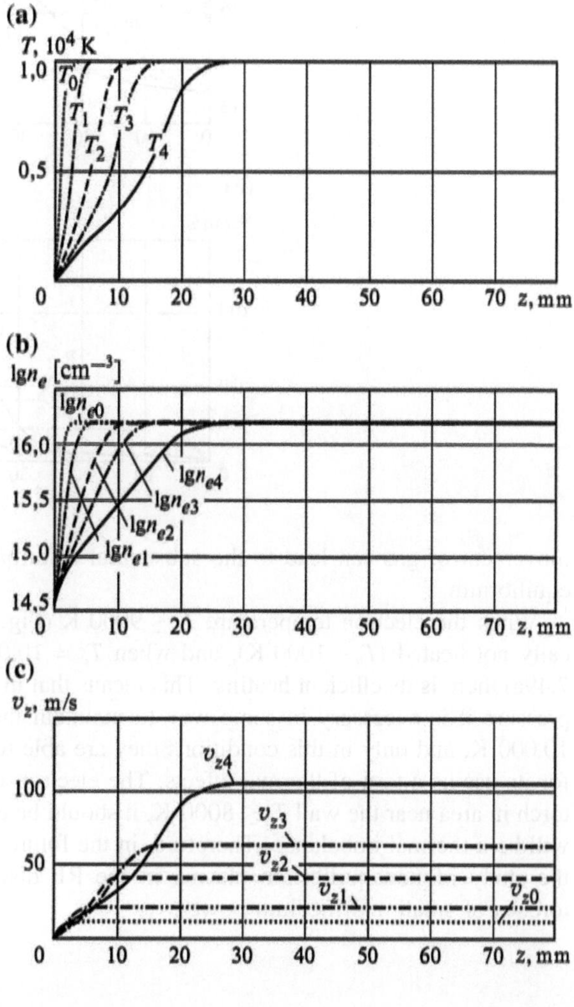

Fig. 7.48 The temperature for the atom–ion component (**a**), electron density (**b**) and the gas velocity (**c**) at the ionization equilibrium and T_e = 10,000 K: *doted line* $\rho v_z = 0.05\,\text{g/(cm}^2\text{s)}$, *dash line with double doted line* $\rho v_z = 0{,}1\,\text{g/(cm}^2\text{s)}$, *dash line* $\rho v_z = 0{,}2\,\text{g/(cm}^2\text{s)}$, *dash line with single doted line* $\rho v_z = 0{,}3\,\text{g/(cm}^2\text{s)}$, *solid line* $\rho v_z = 0{,}5\,\text{g/(cm}^2\text{s)}$

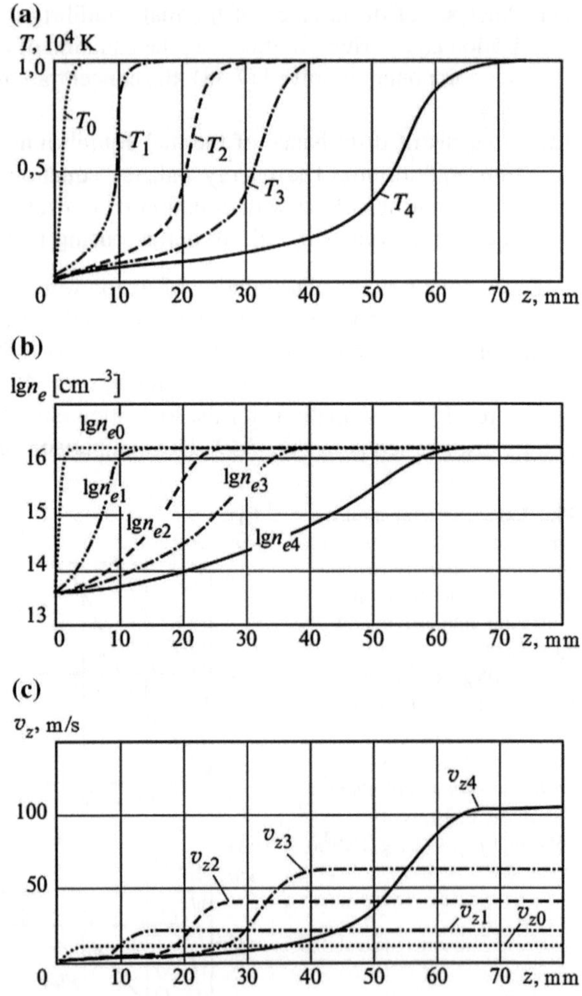

Fig. 7.49 The temperature for the atom–ion component (**a**), the electron density (**b**) and the gas velocity (**c**) at the disturbance of ionization equilibrium and T_e = 10,000 K: notation is the same as in Fig. 7.48

movement of gas can lead to the substantial disturbance of the plasma ionization equilibrium.

When the electron temperature $T_e \leq 9000$ K (Fig. 7.50), the cold gas is practically not heated ($T_a < 1000$ K), and when T_e = 10,000 K and above (Figs. 7.48, 7.49a) there is its efficient heating. This means that in Argon plasma at atmospheric pressure it is necessary in some way to maintain the electron temperature above 10,000 K, and only in this conditions they are able to effectively heat the cold gas for its use as a tool of thermal effects. The electron temperature in the RF plasma torch in area near the wall $T_e \leq 8000$ K, it should be expected that in this area there will be a thermal imbalance. Therefore, in the future it is better to use a model for the study of nonequilibrium plasma in the RF discharge, especially for plasma torches of small and medium power.

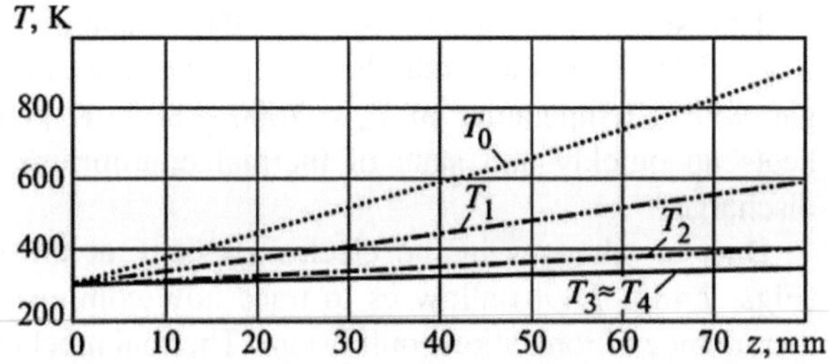

Fig. 7.50 The change of the temperature of atom–ion components along the z-axis in disturbance of ionization equilibrium and $T_e = 9000$ K, $T_{a0} = 300$ K: notations are the same as in Fig. 7.48

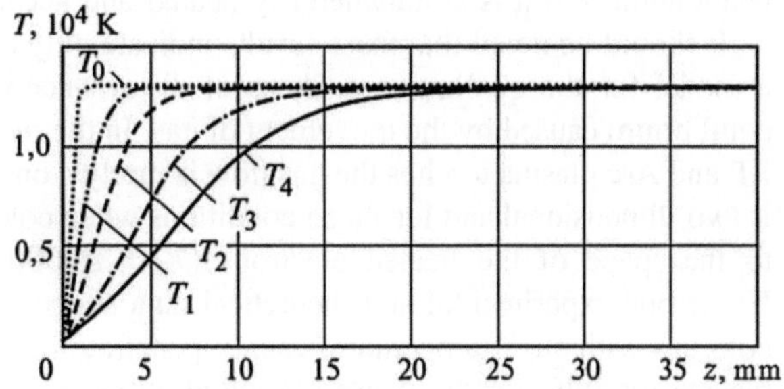

Fig. 7.51 The change of the temperature of atom–ion component along the z-axis in disturbance of ionization equilibrium and $T_e = 13{,}000$ K, $T_{a0} = 300$ K: notations are the same as in Fig. 7.48

When the electron temperature $T_e \geq 13{,}000$ K (Fig. 7.51) the atom–ion components are heated up quickly, that leads to a leveling of their temperature and electron temperature, so for most of Arc plasma torches with medium and large powers at atmospheric pressure ($T_e \geq 13{,}000$ K) there is a thermal equilibrium, and for study of these plasma torches it is sufficient to use the model of equilibrium plasma.

Gas with temperature $T_a = 2500 - 3000$ K begins effectively heating. The data for the cases when the gas with such a temperature can be heated even at $T_e = 9000$ K are shown in Fig. 7.52. From this it follows that the thermal efficiency of various plasma processes can be substantially increased if the plasma-forming gas is preheated to the indicated temperature. For other gases, at a certain temperature of electrons and pressure the temperature of the gas can be determined using this model, in this conditions the atom–ion components are rapidly heated and maintain the state of thermal equilibrium of discharge.

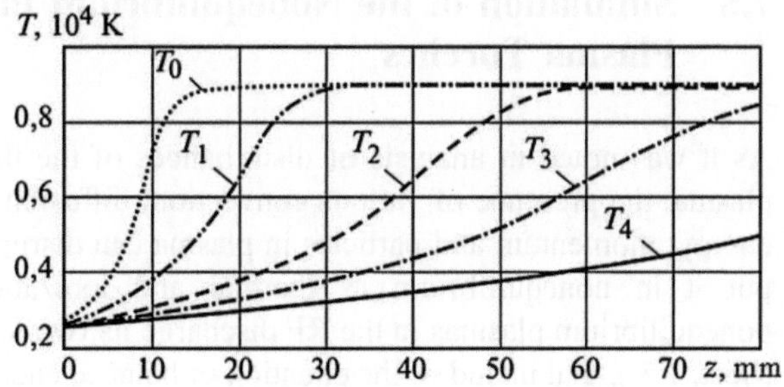

Fig. 7.52 The temperature of atom–ion components in disturbance of ionization equilibrium and $T_e = 9000$ K, $T_{a0} = 2500$ K: notations are the same as in Fig. 7.48

The simulation results for the RF plasma torch at atmospheric pressure (Figs. 7.35, 7.45) show that the cold gas does not penetrate into plasma and the hot gas with a temperature of $T_a \approx 7000 - 8000$ K penetrates. In this case, the gas heats up quickly and state of thermal equilibrium is established in the core of discharge.

Data on changes in the electron density at the gas penetration in two cases (Figs. 7.48b, 7.49b) allow us to trace how composition of plasma changes in the disturbance of ionization equilibrium. Thermal mechanism of plasma acceleration is clearly revealed from the results of the calculation of plasma velocity—gas at the penetration into it is simultaneously heated and accelerated (Figs. 7.48c, 7.49c).

It should be noted that these results indicate only the convenience in use of such a model for the qualitative analysis of disturbance of the thermal and ionization equilibrium caused by the movement of gas. In the real conditions of working of the RF and Arc plasma torches the gas flow is rarely a one-dimensional, in most cases it is two-dimensional and idealized conditions when cold gas ($T_a \approx 300$ K) penetrates to the space of the heated electrons ($T_e > 10{,}000$ K), is virtually nonexistent. Numerous experimental and theoretical data suggest that in natural conditions the cold gas with such temperature cannot penetrate into plasma, and if to feed it into plasma forcibly, for the dense plasma at atmospheric pressure, the temperature of electrons in such a cold channel will be not higher than several thousand degrees of Kelvin.

The electron temperature and specific consumption of gas do not remain constant as the gas penetration into the discharge. For example, for the RF plasma torch at atmospheric pressure the temperature of electrons in the internal layer of discharge $T_e = 7000$–8000 K, and in central and medium layers $T_e = 9000$–$11{,}000$. As it is described above, the calculation results depend strongly on the consumption of gas ρv_z and temperature of electrons T_e, so the field of gas heating in the discharges of real plasma torches may be different.

Thus, the reduced one-dimensional model allows quickly and qualitatively determination of the heating degree of heavy components of gas and analyze the disturbance of thermal and ionization equilibrium in plasma, but for the study of real processes in the RF discharges the full two-dimensional model of nonequilibrium plasma should be used.

7.5 Simulation of the Nonequilibrium Plasma in the RF Plasma Torches

As it was noted in analysis of disturbances of the thermal ionization equilibrium plasma, the presence of various convection, diffusion and radiative fluxes of mass, energy, momentum and particles in plasma can disrupt its detailed equilibrium and put it in nonequilibrium as thermal and ionization state. For the study of nonequilibrium plasmas in the RF discharge its two-temperature model can be used (Sect. 1.3.), that includes: the equation of balance energy for electron gas (6.43), the equation of balance energy for atom–ion component (6.44), the momentum equation of gas (6.25), (6.26) the continuity equation of gas (6.33), the equation if ionization equilibrium (6.47), electromagnetic Maxwell equation (7.1) transformed into the equation of the vector potential (7.8), the system of boundary conditions, system of equations and for determination of composition, thermodynamic properties and transport coefficients of two-temperature plasma (Sect. 4.1–4.4).

7.5.1 Discrete Analogues of the Mathematical Model

The equations of balance energy of the electron gas and atom–ion component, the momentum equations, the continuity equation for all the gas, that are included in the model of nonequilibrium plasma can be written in the general form of GDE (6.2) with the corresponding coefficients (Table 6.1). Formation of the discrete analogues of these equations (Sect. 6.1.2, 6.1.3) is obtained on the basis of an algorithm of solving the GDE (Sect. 6.1.1), taking into account the construction of internal feedback loop (Sect. 6.3.2, 6.3.3).

Because of the similarity in structure, the equations of balance energy of the atom–ion component, the electron gas, and the equation of ionization equilibrium will be solved on the same grid [$z1$, $r2$] (Fig. 6.5a), as for solving the equation of balance energy for the model of equilibrium plasma. Structure of the shifted grid for solving of the momentum equationы and the continuity equation is presented above (Sect. 6.1.2, Figs. 6.5, *a*, 6.6–6.8). The boundary conditions and the corresponding discrete analogue of the momentum equations and the continuity equation were obtained above (Sect. 7.3.1). Boundary conditions and the discrete analogue of equation of balance energy for atom–ion gas (6.44) for internal and boundary points have the same form as the equation of the balance energy in the model of equilibrium plasma (6.39).

For the maximum use of boundary conditions the equation for the vector potential (7.8) is solved using the grid [$z1$, $r1$] (Fig. 6.5a), its discrete analogue and the boundary condition, the equations for calculation of the field and the electrical parameters of the RF plasma torches are given above (Sect. 7.2.3).

Thus, for the complete formation of the systems of discrete analogues and matrix of coefficients in two-temperature model of nonequilibrium plasma of the RF plasma torches it is necessary to define the boundary conditions and the corresponding discrete analogues for the boundary points of the equation of balance energy of the electron gas and the equation of ionization equilibrium.

1. *On the axis of plasma torch* using the condition of cylindrical symmetry we have

$$\frac{\partial T_e}{\partial r} = \frac{\partial n_e}{\partial r} = 0.$$

Note that at the solving the equation of energy balance in the model of equilibrium plasma (6.39) is taken into account such a boundary condition as on the axis of plasma torch. Therefore, if the equation of energy balance for electron gas and the equation of ionization equilibrium are solved using the same grid [$z1$, $r2$], as for the equation of energy balance using the model equilibrium plasma (Fig. 6.5a), the discrete analogues in equation of energy balance for electron gas and the equation of ionization equilibrium for points adjacent to the axis of symmetry, have the same form as for the equation of balance energy in model of equilibrium plasma (Sect. 7.3.1).

2. Definition of the boundary conditions for the electron gas *on the wall of the plasma torch* is more complex, as the total recording of the boundary conditions for electrons requires consideration of transport phenomena in the regions near the wall of plasma torch, taking into account the interaction of electrons with material on the surface of the wall. This problem is quite complex and difficult to achieve because of lack of information about the processes of interaction of electrons with the material of the wall in plasma torch in terms of the external field and the processes occurring in plasma. Therefore, the boundary condition at the wall of plasma torch is usually defined using simple models of the interaction of electrons with cooled walls of plasma torch [43, 45–52]. In most cases, these are the one-dimensional balance equations for energy, current and particles of plasma in the boundary layer of gas.

Model of the collisionless layer In the model of collisionless layer it is assumed that the electrons with a kinetic energy $m_e v_e^2/2$ at the distance from the wall that are equal to the mean free path l for particles do not have the collisions with other particles. Because of the high mobility electrons overcome this layer faster than the ions and charge the wall negatively. As a result is a potential drop φ_R occurs in the layer with thickness of the order of Debye radius r_D—in the layer of charge separation, that prevents the flow of electrons and accelerates the flow of ions, and also leads to the equalization of ion and electron currents to the wall. Ononly those electrons can access to the wall of plasma torch whose kinetic energy is greater or equal to the potential energy of barrier $\frac{mv_e^2}{2} \geq e\varphi_R$. In this case, the current density of electrons j_e, ions j_i and energy flux q_e to the wall can be determined as

$$j_e = -e \int\limits_{\sqrt{\frac{2e\varphi_R}{m_e}}}^{\infty} v_e f_{M,e}(v_e) dv_e = -\frac{1}{4} e n_e \bar{v}_e \exp\left(-\frac{e\varphi_R}{kT_e}\right); \tag{7.54}$$

$$j_i = e \int\limits_0^{\infty} v_i f_{M,i}(v_i) \mathrm{d}v_i = \frac{1}{4} e n_i \bar{v}_i; \tag{7.55}$$

$$q_e = - \int\limits_{\sqrt{\frac{2e\varphi_R}{m_e}}}^{\infty} \frac{m_e v_e^2}{2} v_e f_{M,e}(v_e) \mathrm{d}v_e = -\frac{j_e}{e}(e\varphi_R + 2kT_e), \tag{7.56}$$

where n_e, n_i—the concentration of electrons and ions at the boundary of the collision layer; $f_{M,e}$, $f_{M,i}$—the Maxwell distribution functions (1.2) for electron and ion velocity; $\bar{v}_e = \sqrt{\frac{8kT_e}{\pi m_e}}$, $\bar{v}_i = \sqrt{\frac{8kT_i}{\pi m_i}} = \sqrt{\frac{8kT_a}{\pi m_a}}$—the average thermal velocity of electrons and ions; m_e, m_i, m_a—the mass of electron, ion and atom, respectively; e—the charge of electron.

Potential φ_R is determined by the condition of the balance of current densities to the nonconductive wall $j_e + j_i = 0$; taking into account (7.54) and (7.55) we have

$$e\varphi_R = kT_e \ln\left(\frac{\bar{v}_e}{\bar{v}_i}\right) = \frac{kT_e}{2} \ln\left(\frac{T_e m_a}{T_a m_e}\right). \tag{7.57}$$

Conditions for the electron density and temperature (n_e, T_e) on the border between collision and collisionless areas can be determined from the balance of flows of particles and energy for the condition of the quasi-neutrality on this border. Due to the small radius of Debye these conditions at this boundary are valid to the wall of plasma torch: $n_e = n_{eR}, T_e = T_{eR}$.

It is assumed that the flow of electrons moving to the wall comes from the area of the collisions due to the bipolar diffusion of charged particles, so from the condition of the balance of particles through this boundary we obtain the following boundary condition for the electron density

$$-\left(D_{ambR}\frac{\partial n_e}{\partial r}\right)_R = \frac{n_{eR}\bar{v}_i}{4} = n_{eR}\sqrt{\frac{kT_{aR}}{2\pi m_a}}. \tag{7.58}$$

In determination of the electron temperature at the wall it is assumed that the cooling of electron gas occurs due to diffusion of high-energy electrons that can overcome the potential barrier in the layer of charge separation. In this case, the total energy flux density in view of transfer of the ionization energy that is released at the recombination at the wall, will be

$$q_e = -\frac{j_{eR}}{e}(e\varphi_R + 2kT_{eR} + E_i). \tag{7.59}$$

At the same time, this energy from the collision area is supplied with the diffusion flux of electrons and due to the thermal conductivity, while

$$q_e = -\lambda_e\frac{\partial T_e}{\partial r} - \left(\frac{5}{2}kT_e + E_i\right)\frac{j_{eR}}{e}. \tag{7.60}$$

The energy balance on the border of the collision and collisionless regions (7.59) (7.60) is

$$-\left(\lambda_e \frac{\partial T_e}{\partial r}\right)_R = \left(\frac{kT_{eR}}{2} - e\varphi_R\right)\frac{j_{eR}}{e}.$$

Taking into account (7.55), (7.57) and condition of the balance in current densities at the boundary of two areas we obtain the boundary condition for the electron temperature in the form

$$-\left(\lambda_e \frac{\partial T_e}{\partial r}\right)_R = \frac{kT_{eR}}{2}\left[\ln\left(\frac{T_{eR} m_a}{T_{aR} m_e}\right) - 1\right] n_{eR}\sqrt{\frac{kT_{aR}}{2\pi m_a}}. \quad (7.61)$$

Relation (7.61) is consistent with the data of [45], where the heat flux to the wall using its catalytic properties was calculated:

$$-\left(\lambda_e \frac{\partial T_e}{\partial r}\right)_R = \frac{1-k}{1-\beta}\frac{n_e \bar{v}_i}{4}\left[\left(\frac{1-\beta}{2} - (1-\beta)\ln\frac{\bar{v}_i}{\bar{v}_e}\right)kT_e + \frac{3}{2}\beta kT_e\right].$$

Here k, β—the coefficients of reflection from the wall for electrons and heavy particles. As it is shown in calculations [45], the change in reflection coefficients from 0 to 0.99 does not affect the integral characteristics and axial values T_{e0}, n_{e0}, on the axis, but close to the wall, it may have a significant impact.

Model of elastic collisions In the conditions of strong energy exchange between atom and ion components and electron gas it can be offered that all the energy of electric field near the wall is spent on heating of the heavy plasma components in elastic collisions with electrons. The energy balance near the wall of the plasma torch can be written as

$$\sigma_e E^2\big|_{r=R} = b(T_e - T_a)\big|_{r=R}, \quad (7.62)$$

Here $b = \frac{3}{2}k\frac{2m_e}{m_a}n_e\sqrt{\frac{8kT_e}{\pi m_e}}(n_a Q_{ea} + n_i Q_{ei})$—the coefficient of energy exchange of electrons with atoms and ions (Sect. 1.3).

Since the electric intensity E on the wall of plasma torch and the temperature of the heavy particles $T_a = T_{aR}$ are known, then from (7.62) we can obtain the equations for the boundary temperature

$$T_{eR} = T_{aR} + \frac{\sigma_e E^2}{b}\bigg|_{r=R} = T_{aR} + \frac{e^2 m_a E^2}{3k\left[\sqrt{\frac{8kT_{eR}}{\pi m_e}}(n_a Q_{ea} + n_i Q_{ei})\right]^2}. \quad (7.63)$$

Calculation of the power transmitted to the heavy particles by electrons (Fig. 4.27) through elastic collisions showed that at low electron temperature near the wall of plasma torch ($T_e < 7000$ K), they almost cannot transfer the energy to atoms and ions due to the low concentrations of electrons and effective interaction cross section. In this case, the assumption about the heating of the heavy components of plasma due to electron impact in accordance with (7.62) is no longer justified. In particular, in the region near the wall of the plasma torch (electrons in this region have relatively little energy) atoms and ions are heated not due to elastic collisions with electrons, but as a result of other mechanisms of heat transfer such as conduction and convection.

Model of adiabatic wall The simplest condition for determination of temperature and concentration of electrons on the wall is the assumption of adiabatic walls of plasma torch for the electron gas

$$\left(\frac{\partial T_e}{\partial r}\right)_R = \left(\frac{\partial n_e}{\partial r}\right)_R = 0, \tag{7.64}$$

i.e., energy that electrons have when they reach the wall is the same that they have in reflection from it (there is no conductive heat flux for electrons). The presence of collisionless shielding layer with a negative potential on the wall makes this condition more reasonable.

As a result of numerical computations using these models it is found that differences T_{eR} near the cold walls of the channel of electric Arc are quite large. However, this does not affect to the characteristics of plasma distribution over the cross-section of the channel. The deviations of calculation results for the different models are already at $r/R < 0.9$ and less than 3 %. It is noted in [53, 54]. At the heating of the walls of the channel or decreasing the arc current the differences in the values T_{eR}, found using different models reduce. This is due to the fact that for environments such as plasma, the boundary values depend on the dynamics of the processes in the internal areas, but not vice versa, so a small error in the determination of the boundary conditions can not have a significantly influence to the development of processes in plasma. Therefore, in a two-temperature plasma model, the simpler boundary condition can be used for determination of T_{eR} on the wall of plasma torch.

3. In the case where the entrance of plasma torch ($z = 0$) is at the long distance from the discharge, the boundary condition for the electron gas *at the entrance* of plasma torch can be determined using a simple model (7.64) or take the equilibrium value $T_e = T_a$, $n_e = n_{e0}$ (n_{e0} is found by the Saha equation (4.1)).

From the condition of steady state at the exit of plasma torch the boundary condition for the electron gas in this area will be:

$$\frac{\partial T_e}{\partial z} = \frac{\partial n_e}{\partial z} = 0. \tag{7.65}$$

Note that the decision of equation of balance energy for atom–ion gas (6.44) and the equation of balance energy for equilibrium plasma (6.39) such as the boundary condition at the exit of the plasma torch is taken into account. If the equation of balance energy of the electron gas and the equation of ionization equilibrium is resolved on the same grid [z1, r2] as the equation of balance energy of equilibrium plasma (Fig. 6.5a), then the discrete analogue of the equation of balance energy of electron gas and the equation of ionization equilibrium for the points adjacent to the exit, have the same form that the equation of balance energy of equilibrium plasma (Sect. 7.3.1).

7.5.2 Comparative Analysis of the Simulation Results

For comparison with the results of simulation of equilibrium plasma (Sect. 7.3.2) perform the simulation of nonequilibrium plasma in the RF plasma torch (5 MHz) with the same regime of operation (with two streams of argon gas $Q_1 = 0$, $Q_2 = 20$ l/min, $Q_3 = 40$ l/min) [55, 56]. The obtained results are also listed for sections $z = 4$, 10, 16, 26, 36, 48, 80 and 120 mm.

Figure 7.53 presents the distribution of the electron and atom–ion temperature in these sections. Inside the discharge ($r < 15$ mm) there is equality of temperatures for electrons and heavy particles ($T_e = T_a$). Divergence of temperatures takes place only in the external layer ($r > 15$ mm) of the discharge. In the layer near the wall the temperature difference can be up to $\Delta T = T_e - T_a = 5000$–$6000$ K, that indicates the existence of a zone of thermal nonequilibrium in the RF plasma torch at atmospheric pressure.

Comparison of the results also shows that the temperature of electrons and heavy particles inside the discharge (Fig. 7.53) does not differ from the temperature obtained by the model of equilibrium plasma (Figs. 7.36, 7.37). High temperature ($T_e > 9500$ K) and electron density ($n_e > 10^{16}$ cm^{-3}) in the core layer of the plasma can effectively transfer energy to atoms and ions through elastic collisions and maintain plasma in thermal equilibrium.

Figures 7.54 and 7.55 shows the distribution of the densities for electrons and atoms in argon plasma in the different sections, and in Fig. 7.56—the distribution of frequencies at the elastic collisions ν_{eai} of electrons with atoms and ions. These data allow us to determine the rate of energy exchange between electrons and heavy particles and analyze the heating of gas in discharge.

In the near-wall zone $n_e < 10^{15}$ cm^{-3} (Fig. 7.54) and the frequency of collisions $\nu_{eai} < 10^{10}$ s^{-1} (Fig. 7.56),—the electrons cannot effectively heat the atoms and ions in elastic collisions (Fig. 7.57), plasma is in the state of thermal nonequilibrium (Fig. 7.53).

Balance of power in the channel of plasma torch allows determination of the ratio of energy exchange in the discharge. Figure 7.57 shows the distribution of the total specific power density $P = \sigma E^2$, received by electrons from the electromagnetic field. The electrons through the elastic collisions give the power P_{ai} to atoms and ions (Fig. 7.58), and some of this power comes free from plasma by radiation U_{rad} (Fig. 7.59).

Accounting of nonequilibrium ionization in this model allows us to find the part of the power $P_I = \mathrm{div}\left(E_i n_e - \frac{5}{2} k T_e n_a\right)\vec{v}$ spent on uncompensated ionization that occurs mainly in the areas of active ionization (Fig. 7.60), flow of the current

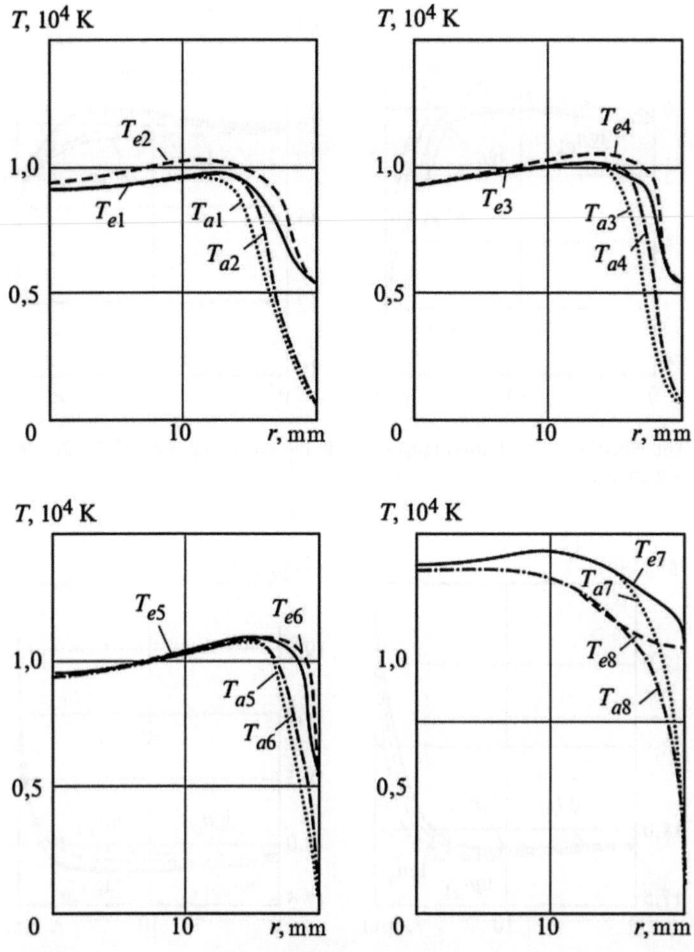

Fig. 7.53 The electron T_e–T_{e8} and atom–ion T_{a1}–T_{a8} temperatures in the cross-sections z = 4, 10, 16, 26, 36, 48, 80, and 120 mm, respectively

(Figs. 7.24, 7.25) and the allocation of capacity (Figs. 7.26, 7.27, 7.57) in the skin layer of plasma.

The distribution of the specific heat capacity of the electron gas heating is shown in Figs. 7.61 and 7.62. The balance of power in the central section of inductor.

An important characteristic of the state of ionization equilibrium in plasma is shown in Fig. 7.63 using the source function of equation of ionization equilibrium $S_{n_e} = k_i n_e n_a - k_r n_e^2 n_i$—the resulting source of production (birth) and loss (dead) of charged particles in the reactions of ionization and triple recombination.

As it was expected, the area of ionization disequilibrium in plasma is in the field of active current and maximum energy release. In the remained part of the channel in the RF plasma torch $S_{n_e} \simeq 0$ and plasma is in ionization equilibrium. The charged

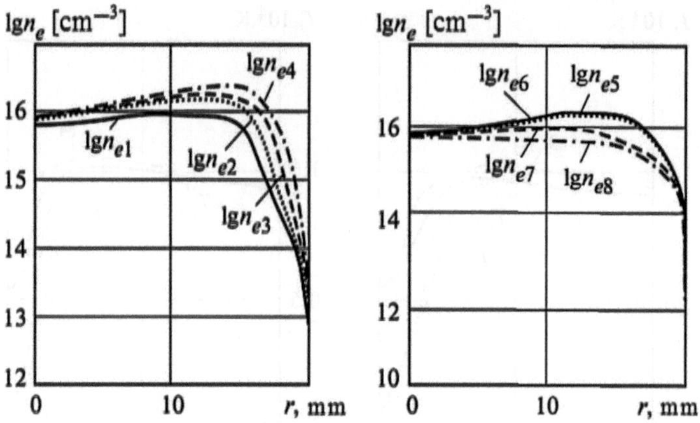

Fig. 7.54 The electron concentration lgn_{e1}–lgn_{e8} in the sections z = 4, 10, 16, 26, 36, 48, 80, and 120 mm, respectively

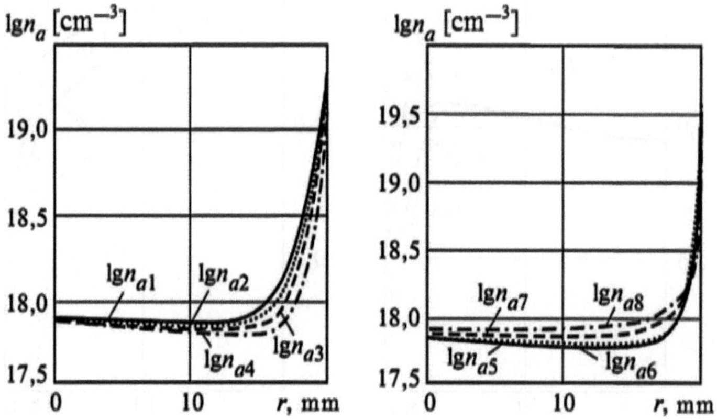

Fig. 7.55 The atom concentration lgn_{a1}–lgn_{a8} in the sections z = 4, 10, 16, 26, 36, 48, 80, and 120 mm, respectively

particles appear mainly in the area of current flow and allocation of power in plasma, but due to convection and diffusion, they are carried to other parts of the discharge, where they recombine.

Thus, using the two-temperature model a complete picture of gas heating in the RF discharge is obtained. The area of thermal (area near the wall) and ionization (area of flow current and energy release) nonequilibrium plasma are clearly defined, as well as the area of the active heating of the gas in the channel of the RF plasma torches.

Basically, the RF discharge at atmospheric pressure is thermal equilibrium, except for the area near the wall where the gap of the electron temperature and the temperature of heavy components may reach several thousand degrees of Kelvin.

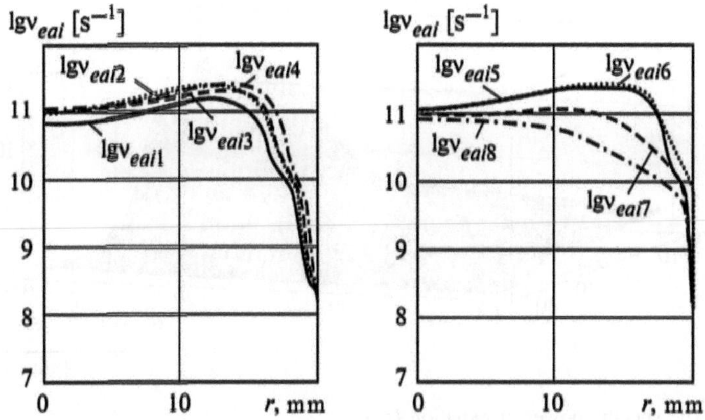

Fig. 7.56 The frequency of elastic collisions of electrons with atoms and ions $\lg\nu_{eai1}$–$\lg\nu_{eai8}$ in the sections z = 4, 10, 16, 26, 36, 48, 80, and 120 mm, respectively

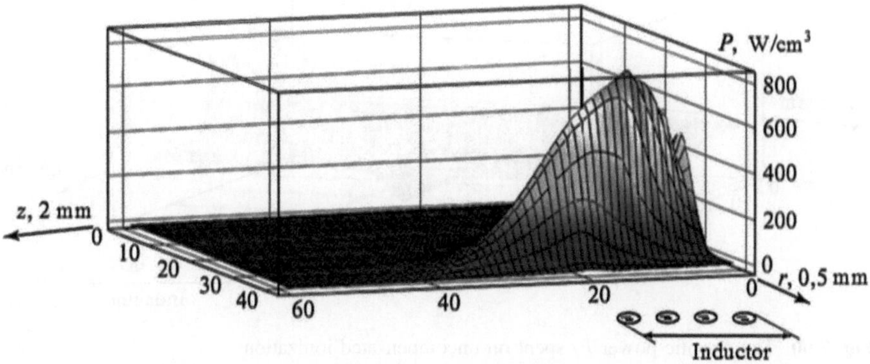

Fig. 7.57 The specific power $P = \sigma_e E^2$ obtained by electrons from electromagnetic field

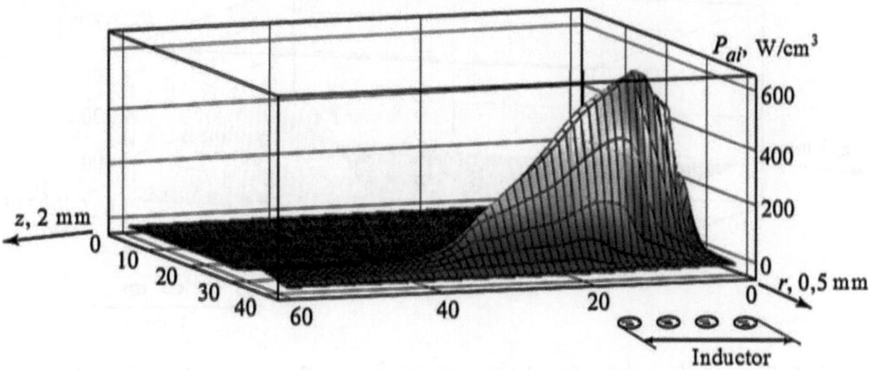

Fig. 7.58 The distribution of specific heat power of the heavy components from the electron gas $P_{ai} = b(T_e - T_a)$

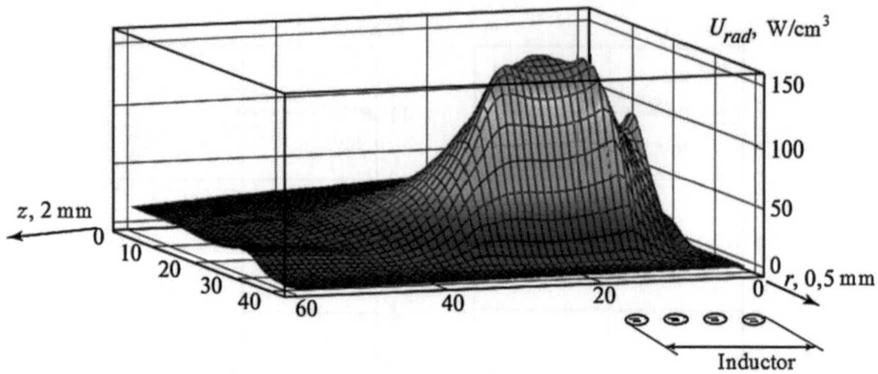

Fig. 7.59 The specific power of radiation U_{rad}

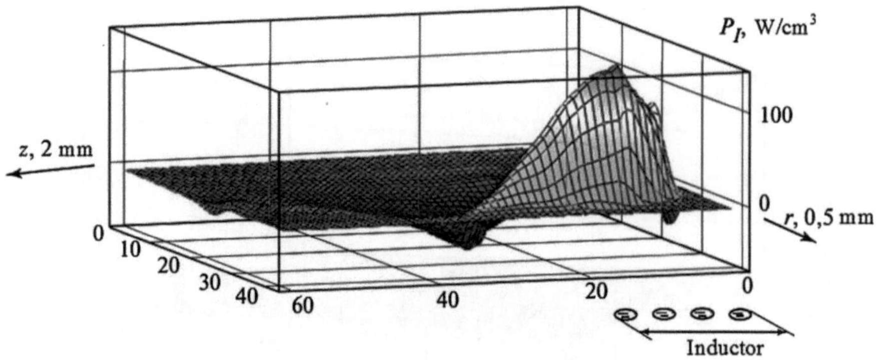

Fig. 7.60 The specific power P_I spent on uncompensated ionization

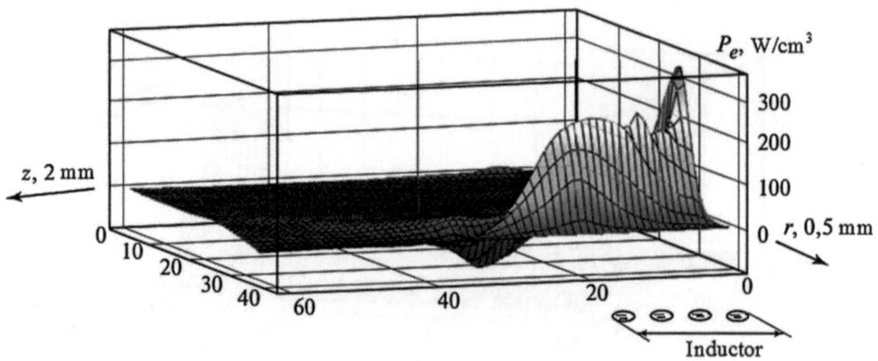

Fig. 7.61 The specific power of the heating of electron gas $P_{\text{e}} = \sigma E^2 - P_{\text{ai}} - U_{\text{rad}} - P_I$

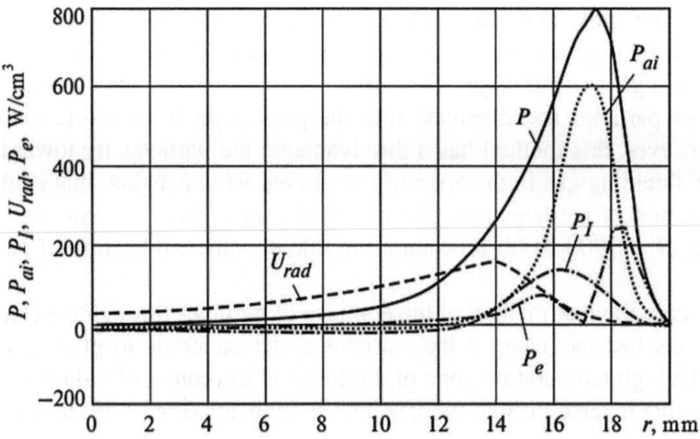

Fig. 7.62 The specific power in the central section of inductor $z_4 = 26$ mm

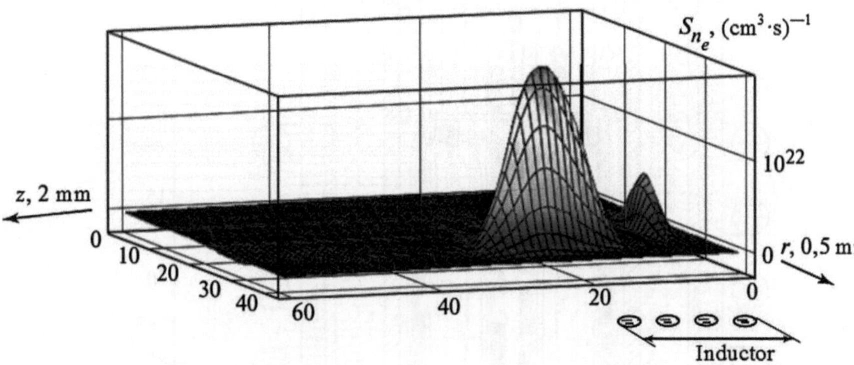

Fig. 7.63 The source of production and loss of charged particles S_{n_e}

However, the temperature and the electron density in this zone are low, and the thermal nonequilibrium is almost no effect on the discharge parameters. This means that we can use the approximate models to determine the boundary conditions of the electron gas on the wall of plasma torch. The coincidence of the simulation results with the use of two models (one- and two-temperature) in the study of the RF discharge at atmospheric pressure makes it possible in the future to solve the technological problems to use a simple model of equilibrium plasma.

The RF plasma torch with a cold channel As it was noted above the RF discharge due to the technological properties, such as purity of plasma that is not contaminated by materials of electrodes, the relatively large spatial stream, a virtually unlimited lifetime in any gaseous medium is widely used in plasma chemistry and processing of material.

One of the most common methods of introducing the powder material and chemicals for the processing into the channel of the RF plasma torch is the input passing through a central gas flow (Fig. 7.64). There is also a method of the entrance of particles and chemical into the plasma jet from a side of the plasma torch. However, this method has a disadvantage: the particles fly towards the cold stream of shielding gas in the zone of the plasma jet expansion, that itself prevents the penetration of particles into plasma. At the same time, the low temperature of plasma jet (T = 7000–8000 K) cannot provide the efficient heating of particles and materials.

In this case the first method of the entrance of particles through the central tube it is preferable, because using it the particles easier penetrate to plasma and rapidly fall into the high temperature zone of discharge at the center of inductor. In order to overcome this reverse flow of the gas flow in inductor (Fig. 7.35) it is necessary to

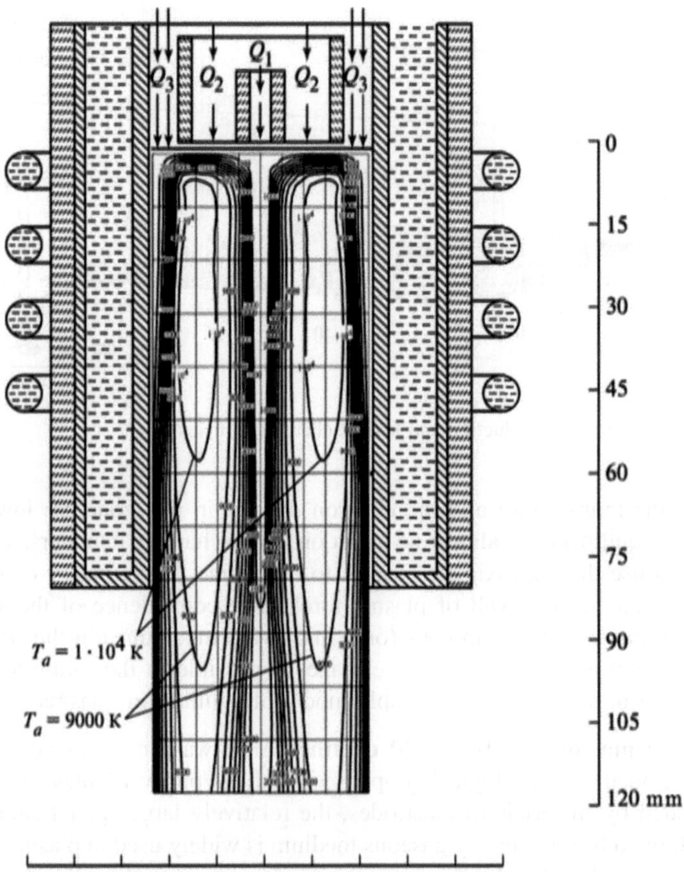

Fig. 7.64 The atom–ion temperature in the RF plasma torch (5 MHz) with a cold channel

give a certain velocity to particles, i.e., feed a corresponding amount of transport gas through the central channel. In the RF plasma torches for chemical analysis and decomposition of harmful substances such a channel is also used to enter the chemical reagents into the plasma core (Fig. 7.64).

The transport gas together with particles or chemicals, as their penetration into plasma is heated and simultaneously cools plasma, thereby forming a cold channel in the center of plasma torch. Heating rate and efficiency of chemical reactions and processing in the RF discharge is directly dependent on the temperature, gas flow rate in a channel and discharge parameters. Therefore, the definition parameters of the discharge in the event of such a channel are an important practical task simulation of the RF plasma torches.

For comparison with the results obtained above, simulation of the RF plasma torch (5 MHz) was carried out with the same initial data (Table 7.3), but with additional stream of transport argon gas (Q_1 = 8 l/min) supplied to the entrance by the central tube with diameter of 3 mm [57, 58]. As in the case of the RF plasma torch (5 MHz) with two flows, we present results in the same cross-sections: z = 4, 10, 16, 26, 36, 48, 80, and 120 mm.

The atom–ion temperature field T_a is shown in Fig. 7.64, and the distribution of T_e and T_a in the same sections is shown in Fig. 7.65. At these pictures we can see a gap of the temperatures T_e and T_a, and formation of cold channel in the center of RF plasma torch. In addition to the near-wall zone there is another zone of thermal nonequilibrium—the zone of cold channel. Here, there is also some discrepancy between the temperatures of electron and atom–ion gases. Disturbance of thermal equilibrium, caused by the movement of gases, was obtained on the basis of one-dimensional problem of gas heating in the half-space of plasma (Sect. 7.4). However, unlike to this case in cold channel the electron temperature is not specified, but it is determined by solving the equation of the energy balance for the electron gas.

From these results we can say that the electron temperature in a cold channel is not higher than 3000 K, and the temperature of the atom–ion gas is lower. Cold channel is located on the axis of symmetry, the electrons, atoms and ions inside a channel are heated mostly due to mixing with the other layers of more heated plasma. On the axis of discharge electrons also cannot get the energy from electric field due to zero intensity of this field.

The data presented in Figs. 7.64 and 7.65 also allow the determination of the zone of thermal equilibrium in discharge as compared with the data of the RF plasma torch (5 MHz) with two streams (Figs. 7.35, 7.53) it is smaller and is biased towards the wall of the plasma torch. When the cold channel appears, the plasma flow as it is displaced to the wall, and a picture of plasma flow is stored mainly in the core of discharge, including the reverse vortex plasma flow (Fig. 7.66) in the upper zone of inductor. However, due to the displacement of plasma to the wall the coupling coefficient of inductor with plasma, current density and the allocated power increase. As a result, we obtain the total allocated power in plasma P_2 = 14.6 kW, the total current in plasma I_2 = 407 A (compared with

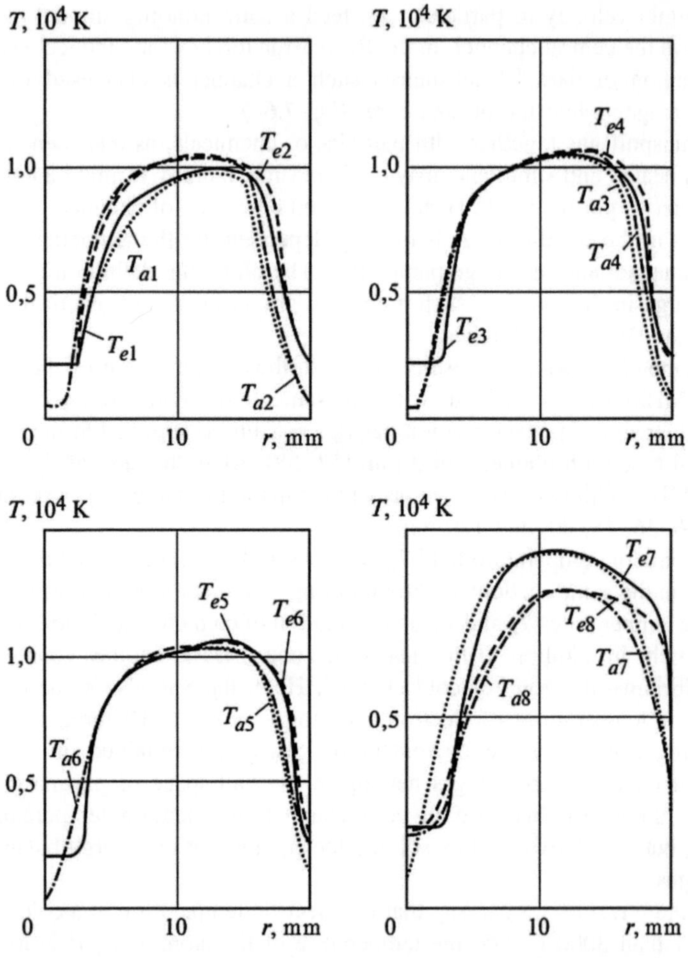

Fig. 7.65 The electron T_{e1}–T_{e8} and atom–ion T_{a1}–T_{a8} temperatures in the sections z = 4, 10, 16, 26, 36, 48, 80, and 120 mm, respectively of the RF plasma torch (5 MHz) with a cold channel

P_2 = 13.3 kW and I_2 = 400 A in the case of two gas streams at the same voltages of inductor U_1 = 4.4 kV and a current in inductor I_1 = 120 A).

The field of the gas flow is shown in Fig. 7.66, and the axial gas flow velocity v_z—in Fig. 7.67. The distribution of the specific consumption of the gas flow is shown in Fig. 7.68. The big values of specific consumption (ρv_z = 3 g/(cm^2 s)) in the center of plasma torch should be noted. The above analysis (Sect. 7.4) shows that even the high-temperature electrons (T_e = 10,000–11,000 K) cannot provide a heating of such amount of cold gas and the discharge stay cold.

From the comparison of the obtained results it follows that there are large differences between the parameters and areas of the flow (Figs. 7.35, 7.66), speeds (Figs. 7.38, 7.67) and specific consumptions (Figs. 7.41, 7.68) of plasma flows

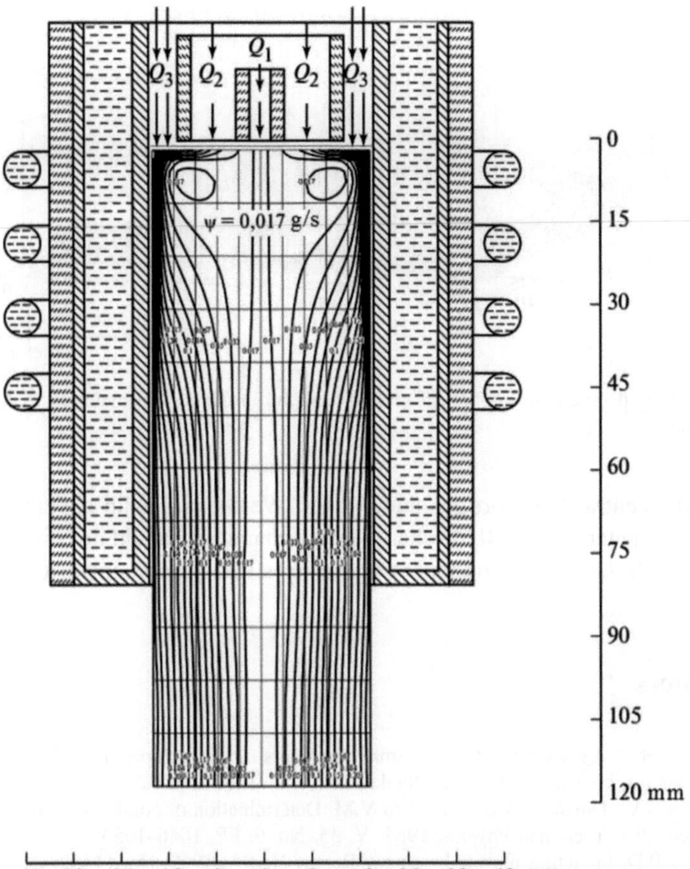

Fig. 7.66 The field of gas flow in the RF plasma torch (5 MHz) with a cold channel

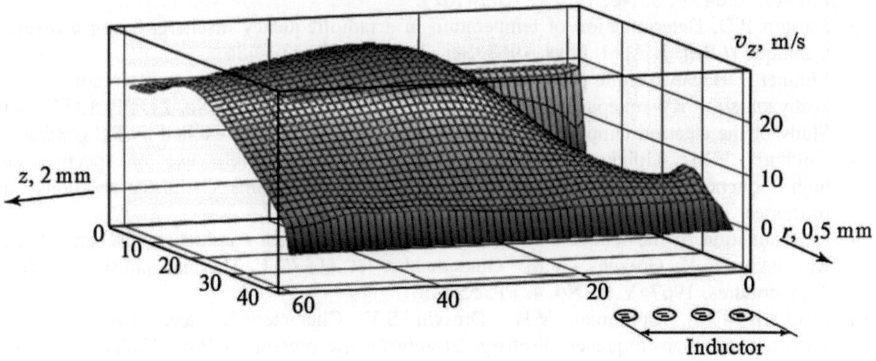

Fig. 7.67 The axial gas flow velocity v_z in the RF plasma torch (5 MHz) with the cold channel

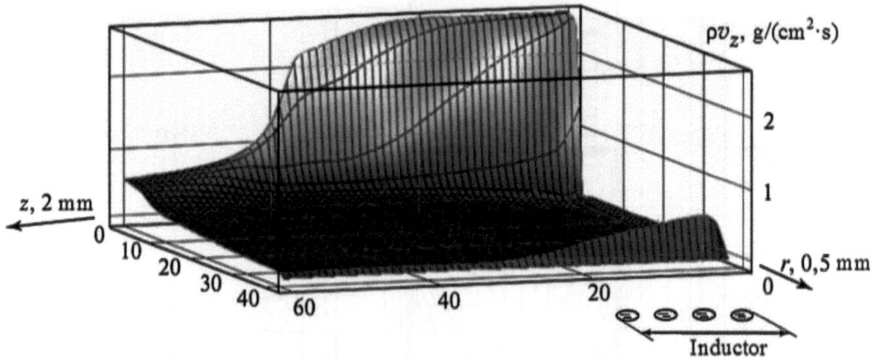

Fig. 7.68 The specific consumption of plasma flow ρv_z in the RF plasma torch (5 MHz) with a cold channel

without the central transport gas and with it. We see that with advent of the cold channel parameters of the RF discharge rarely change, it should be considered using RF plasma torches for the treatment of particles and plasma chemistry.

References

1. Dresvin S.V. Low-temperature plasma generators // Encyclopedia of Low-Temperature Plasma / Ed. by V.E. Fortov. M.: Nauka, 2000. V. II. PP 280–328.
2. Dresvin S.V., Donskoy A.B., Goldfarb V.M. Determination of conductivity of RF discharge in Argon // J. Technical Physics. 1965. V. 35. No. 9. PP. 1646–1653.
3. Dundas P.D. Induction plasma heating // Report NASA. 1969. No. 11487.
4. Vermeulen R.S., Lee Bowden, Vieram P.N. Regulation of the moving state of the plasma using RF electromagnetic fields // Rocket technical and Astronautics. 1967. V. 5. No. 12. PP. 251–260.
5. Dymschits B.M., Koretsky J.P. Experimental study of inductive discharge // J. Technical Physics. 1964. V. 5. No. 19. PP. 1677–1679.
6. Jonston P.D. Determination of temperature in a radiofrequency discharge using a reversal technique // Brit. J. Appl. Phys. 1968. Ser. 2. No. 1. PP. 479–484.
7. Molinet F. Исследование распределения электронной температуры в аргоновой плазме, возбуждаемой ВЧ-генератором // C.r. Acad. Sc. 1966. Vol. 262. No. 21. PP. 1377–1380. Study of the electron temperature distribution in an argon plasma excited by RF generator
8. Goldfarb V.M., Goikhman V.H. Characteristics and possible use of spectroscopic high-frequency discharge at atmospheric pressure // J. Plasma chemistry reactions and processes. 1968. V. 8. No. 2. PP. 193–196.
9. Determination of the temperature in the steady high-frequency induction discharge / R.E. Rovinskii, V.A. Gruzdev, V.M. Gutenmakher et al. // J. Thermal physics of High Temperatures. 1967. V. 5. No. 4. PP. 557–561.
10. Goldfarb V.M., Goikhman V.H., Dresvin S.V. Characteristics and possible use of spectroscopic high-frequency discharge at atmospheric pressure // Proc. Colloq. Spectrosc. Hungary. 1967. PP. 751–760.
11. Eckert H.U., Kelly F.L., Olsen H.N. Spectroscopic observation on induction-coupled plasma flames in air and argon // J. Appl. Phys. 1968. Vol. 39. No. 3. PP. 1846–1852.

12. A study of RF plasma flame of Argon burner / V.M. Goldfarb, Donskoy A.V., Dresvin S.V. et al. // J. Thermal physics of High Temperatures. 1967. V. 5. No. 4. PP. 549–555.
13. Apsit A.R., Goikhman V.H. Preparation and study of pulsed RF inductive discharge at atmospheric pressure // J. Technical Physics. 1970. V. 5. No. 7. PP. 1551–1560.
14. Dresvin S.V., Klubnikin V.S. The study of nonequilibrium in a stream of Argon plasma RF discharge at atmospheric pressure // J. Thermal physics of High Temperatures. 1971. V. 9. No. 3. PP. 475–480.
15. Biberman L.M., Vorob'ev V.S., Yakubov I.T. Low-temperature plasma with nonequilibrium ionization // Successes of physical sciences. 1979. V. 128. No. 2. pp. 233–271.
16. Velikhov E.P., Kovalev A.S., Rakhimov A.T. Physical phenomena in gas-discharge plasma. M .: Nauka, 1987.
17. Goldfarb V.M., Ilyina E.V. Non-equilibrium low-temperature plasma sources. The populations of the levels and spectral diagnostics // Proceedings of the Academy of Sciences of the USSR. 1975. No. 3. Vol. 1. pp. 28–38.
18. Goikhman V.H., Goldfarb V.M. RF thermal discharge. // J. Plasma chemical reactions and processes / Ed. L.S. Polak. M.: Science. 1977. PP. 232–278.
19. Reboux J. Ingeniers et techniciens. 1963. No. 166. P. 109.
20. Chase J.D. Magnetic Pinch Effect in the Thermal RFI Plasma // J. Appl. Phys. 1969. Vol. 40. No. 1. PP. 318–325.
21. Chase J.D. Theoretical and Experimental Investigation of Pressure and Flow in Induction Plasmas // J. Appl. Phys. 1971. Vol. 42. No. 12. PP. 4870–4879.
22. Klubnikin V.S. Thermal and gas-dynamic characteristics of the RF discharge in a stream of Argon // J. Thermal physics of High Temperatures. 1975. V. 13. No. 3. PP. 473–474.
23. Sorokin L.M. Theory electric Arc in a forced heat exchange. M.: Nauka. 1977.
24. Sorokin L.M. // J. Physics and chemistry of material treatments. 1980. PP. 32–34.
25. Raiser Y.P. // Successes of physical sciences. 1969. V. 99. No.. 4. P. 545.
26. Soshnikov V.N., Trekhov E.S. On the theory of RF vortex high pressure discharge // J. Thermal physics of High Temperatures. 1966. V. 4. No. 2. P. 166.
27. Dresvin S.V. // Abstracts of the VII Conf. by generators of Low-temperature plasma. Alma-Ata 1977.
28. Dresvin S.V., El-Mikati H. // J. Thermal physics of High Temperatures. 1977. No. 2.
29. Thomson J.J. // Philos. Mag. 1926. Vol. 2. P. 674.
30. Thomson J.J. // Philos. Mag. 1927. Vol. 4. P. 1128.
31. Fomenko A.A., Trekhov E.S. Questions of Low-temperature plasma physics. Minsk: Science and Technology, 1970. PP. 195–198.
32. Eckert H.U. Measurement of the Magnetic Field: Distribution in a Thermal Induction Plasma // J. Appl. Phys. 1971. Vol. 42. No.. 8. P. 3108–3113.
33. Eckert H.U. Dual Magnetic Probe Systems for Phase Measurement in Thermal Induction Plasma // J. Appl. Phys. 1972. Vol. 43. No. 6. P. 2707–2713.
34. Nguyen Quoc S. 2D-Electromagnetic field calculation on the RF plasma torch // Proc. of the Inter. Sem. on Heating by Internal Sources. Padua. Italy, 2001. P. 609–615.
35. Mostaghimi J., Boulos M. Two-Dimensional electromagnetic field effects in induction plasma modeling // J. Plasma Chem. Plasma Process. 1989. Vol. 9. P. 25.
36. L.R. Neiman, K.S. Demirchyan. Theoretical Foundations of Electrical Engineering. L.: Energoizdat, 1981. V. 2.
37. Korn G., Korn T. Handbook of mathematics for scientists and engineers. M.: Science. 1984.
38. Kalantarov P.L., Zeitlin L.A. Calculation of inductances: Handbook. L.: Energoatomisdat 1986.
39. RF multiflux plasmatron study. Experimental measurement and modeling / D. Morvan, J. Erin, S. Magnaval et al. // Proc. of the XII Inter. Sym. on Plasma Chemistry — Minneapolis. USA, 1995. PP. 17430–1748.
40. Modelling of the RF multiflux plasmatron / D. Morvan, J. Erin, S. Magnaval et al. // Proc. of the IV European Conference on Thermal plasma Processes. Athens. Greece, 1996. PP.712–717.

41. In-flight measurement of particle size and velocity in a radio frequency plasma torch / E. Franke, F. Krayem, D. Morvan et al. // Proc. of the V European Conference on Thermal plasma Processes. St. Petersburg, 1998. PP. 347–353.
42. Experimental investigation of impurity evaporation from powders injected in an RF plasma torch // D. Morvan, J. Amouroux, S. Magnaval et al. // Proc. of the IV European Conference on Thermal plasma Processes. Athens. Greece, 1996. PP. 713–720.
43. Physics and technology of Low-temperature plasma / Ed. by S.V. Dresvin. M.: Atomizdat. 1972.
44. Dresvin S., Amouroux J., Nguen Quoc Shi. Analisis of deviation from thermal and ionization equilibrium in an Argon plasma flow // J. High temperature Material Processes. 1997. Vol. 1. No. 3. PP. 369–381.
45. Clarke C.J., Inkropera F.P. Thermochemical nonequilibrium in a stabilized Argon Arc plasma // J. Rocket and technical complex. 1972. V. 10. No. 1. PP. 19–21.
46. The calculation of the initial section of the channel characteristics of the electric Arc on the basis of two-temperature plasma model / V.M. Lelevkin, E.P. Pakhomov, V.F. Semenov, V.S. Engelsht // J. Thermal physics of High Temperatures. 1986. V. 24. No. 3. PP. 587–593.
47. Mathematical modeling of electric Arc / Ed. V.S. Engelsht. Frunze: Ilim. 1983.
48. Asinovsky E.I., Pakhomov E.P. An analysis of the temperature field in a cylindrically symmetric pole of electric Arc // J. Thermal physics of High Temperatures. 1968. V. 6. No. 2. PP. 333–336.
49. Litvinov I.I., Lyumkis E.D., Philipov S.S. Nonequilibrium model of a strongly radiating electric discharge in an inert gas // Preprint No. 135. M. 1976. P. 72.
50. Gogosov V.V., Shelchkova I.N. The conclusion of the boundary conditions for concentration, velocity and temperature components of a partially ionized plasma taking into account the capacity of the parietal falls // USSR Academy of Sciences. Ser. Fluid Mechanics. 1974. No. 5. PP. 76–88.
51. Gogosov V.V. Shelchkova I.N. Examples of simple boundary conditions for the electron and ion densities and temperatures of electrons and heavy particles // USSR Academy of Sciences. Ser. Fluid Mechanics. 1976. No. 6. PP. 169–172.
52. Nazarenko I.P., Panevin I.G. Calculation canal stabilized Arcs based on radiative transfer and nonequilibrium plasma // Electric Arc theory in conditions of forced heat. Nauka. Novosibirsk. 1977. PP. 61–67.
53. Alievsky M.Y., V.M. Zhdanov. The transport equations for a many-component non-isothermal plasma // J. Appl. Phys. 1963. No. 5. pp. 11–17
54. Dresvin S.V. Fundamentals of theory and design of RF plasma torches. L.: Energoatomizdat 1991.
55. Modelling of the RF multiflux plasmatron / D. Morvan, J. Amouroux, S. Magnaval et al. // J. High Tem. Mat. Proc. 1998. Vol. 1. No. 3. P. 78–85.
56. Caculation of RF plasma torch parameters by means of nonequilibrium model of Ar Plasma / S.V. Dresvin, S. Nguyen-Kuok, D. Ivanov, J. Amouroux // Proc. of the VI European Conference on Thermal plasma Processes. Strasbourg, 2000. P. 257–267.
57. Calculation of temperature and flow in RF plasma torch. Influence of cold channel produced by axial gas injection / S. Nguyen-Kuok, S. Dresvin, J. Amouroux, D. Morvan // Proc. of the V European Conference on Thermal plasma Processes. St. Petersburg, 1998. P. 169–175.
58. Modeling of coupled motion, heat and mass transfer, electromagnetic and atomic hydrogen in a Ar — H2 thermal RF plasma / Ph. Mandin, D. Morvan, J. Amouroux et al. // Proc. of the XIV Inter. Sym. on Plasma Chemistry. Prague, 1999. Vol. 1. P. 415–420.

Chapter 8
The Arc Plasma Torches

8.1 The Features of the Arc Plasma Torches

Arc plasma torches are one of the most common types of discharge with medium and high pressures and are widely used in engineering and technology. The differences of Arc plasma torches from other types of discharge are the high density ($j = 10^2$–10^7 A/cm^2) and low cathode voltage drop. Thus, if the cathode voltage drop in the glow discharge is usually $U_k \geq 100$ V, in Arc discharge, it is only $U_k = 10$–40 V. This is due to the fact that effective mechanisms of thermo-emission of electrons from the cathode of the Arc provides a current close to the total current of the Arc, and it is not necessary to increase it due to high U_k.

The created devices for controlling the gas flow in the electric Arc—the Arc plasma torches allow to obtain the plasma flow with temperature $T = (10$–$50) \times 10^3$ K—and the velocity of the flow $v = 10$–10^4 m/s in inert and molecular gases at the atmospheric pressure. The large number of types and constructions of electric Arc plasma torches is developed that are intended for different purposes. They are used for cutting, welding and melting of metals, covering with heat-resistant coatings, heating of the gas in the aerodynamic tubes and chemical reactors, heating of the gas in the direct reduction of metals from ores, the intensification of blast and open-hearth processes. The Arc plasma torches are also used for thermal drilling of hard rocks, in spectrometric study of properties of the gas, to obtain a stable standard of brightness, as the jet thrusters for spacecraft and others.

The Arc plasma torches can be divided into two great classes, fundamentally different from each other. First class—the plasma torches with electric Arc of the direct action, where the electric Arc lights and closed on technological object—metal (Fig. 8.1). The second class—Arc plasma torch, generating a plasma jet, where the electric Arc is in the channel of the plasma torch (cathode and anode inside the channel) and technological effect—is achieved indirectly through exposure of the Arc to the plasma jet (Fig. 8.2).

S. Nguyen-Kuok, *Theory of Low-Temperature Plasma Physics*,
Springer Series on Atomic, Optical, and Plasma Physics 95,
DOI 10.1007/978-3-319-43721-7_8

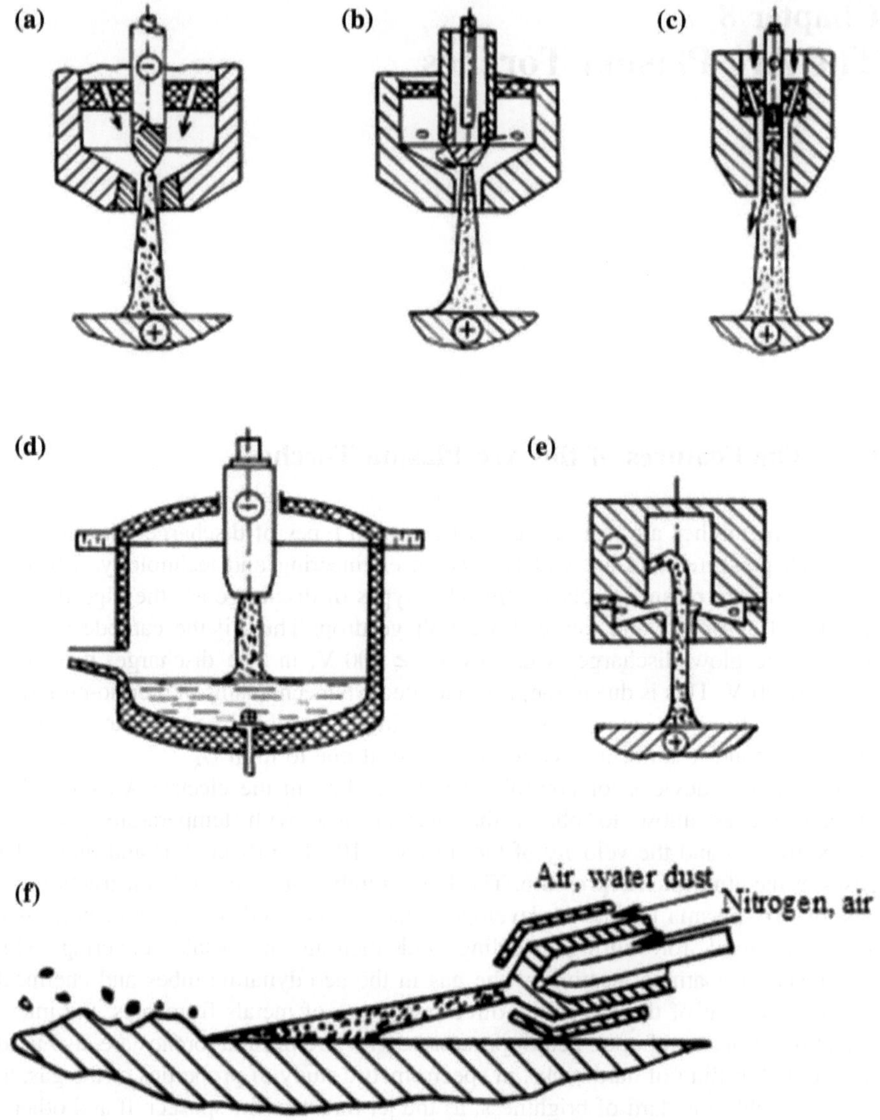

Fig. 8.1 The Arc plasma torches with the direct action [1]: **a**, **b**, **e**, **f** with the compressed Arc in the channel (cutting, spraying and planing); **c**, **d** with open Arc (welding, melting)

By the way of exposure to the Arc and its stabilization in the space the plasma torches of the first class are divided into (1) plasma torches with the Arc stabilized by the gas flow in a space (welding and melting of metals); (2) plasma torches with the Arc, compressed in the channel and stabilized by the gas flow (plasma Arc cutting of metals).

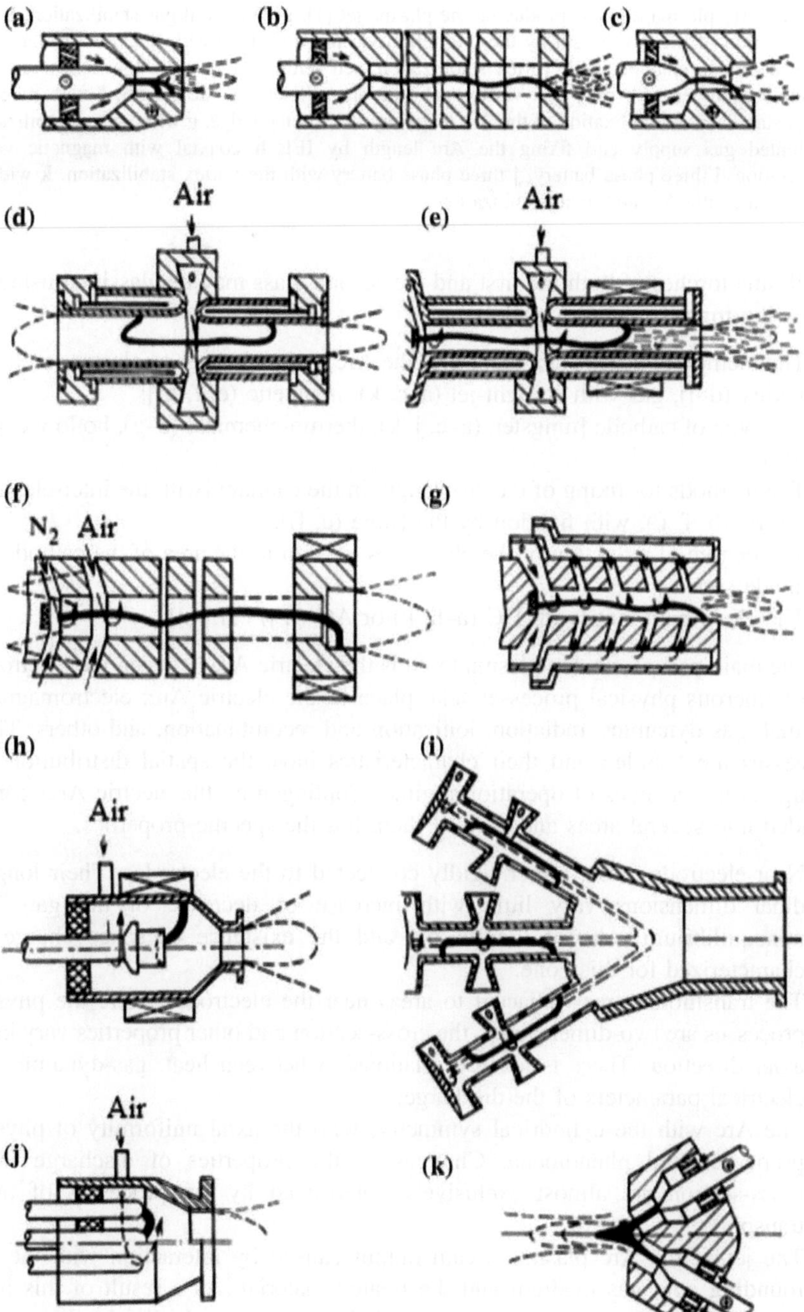
(a)
(b)
(c)
(d)
Air
(e)
Air
(f)
N2 Air
(g)
(h)
Air
(i)
(j)
Air
(k)

◀ **Fig. 8.2** Arc plasma torches, producing the plasma jet [1]: **a** with axial gas stabilization, **b** with axial fixation of the Arc length by the inter-electrode insert (IEI), **c** with fixation Arc length by ledge, **d** double chamber with vortex stabilization, self-setting Arc length, the hollow cathode, **e** double chamber with magnetic vortex stabilization, self-setting Arc length, **f** with magnetic vortex stabilization and fixation of the Arc length by IEI and the ledge, **g** with vortex stabilization, distributed gas supply and fixing the Arc length by IEI, **h** coaxial with magnetic vortex stabilization, **i** three phase battery, **j** three phase battery with the vortex stabilization, **k** with the imposition of the Arc and axial stabilization

Plasma torches of both the first and the second class may be classified using the following main features (Fig. 8.2):

1. The method of space stabilization of the Arc in the channel of plasma torch [gas vortex (d–j), gas with straight-jet (a–c, k), magnetic (e, f, h)];
2. The type of cathode [tungsten (a–c, j, k), thermo-chemical (e–g), hollow copper (d, i)];
3. The methods for fixing of the Arc length in the channel [with the inter-electrode inserts (b, f, g), with fixation by the ledge (c, f)].
4. The degree of reduction of the electrodes erosion in the area of the cathode and anode spots.
5. The electric current [with DC (a–h, k) or AC (i, j) current].

The main part of the Arc plasma torch is the electric Arc between the electrodes. The numerous physical processes take place in the electric Arc: electromagnetic, thermal, gas dynamics, radiation, ionization and recombination, and others. These processes are coupled and their characteristics have the spatial distribution and change at the changes of operation regime. Contingently, the electric Arcs can be divided into several areas and each of them has the specific properties.

1. Near-electrode areas, rather rigidly connected to the electrodes. Their longitudinal dimensions vary little with increase or decrease of the gap. The nonequilibrium state of the plasma and the existence of space charge are characterized for this zone.
2. The transitional areas adjacent to areas near the electrodes. Here the physical processes are two-dimensional, the cross-section and other properties vary in the axial direction. There is a close relationship between heat, gas-dynamic, and electrical parameters of the discharge.
3. The Arc with the cylindrical symmetry, with the axial uniformity of physical properties and phenomena. Changes of the properties of discharge in its cross-section are almost exclusively determined by the processes of radial transport.
4. The jet of the Arc plasma stream output caused by interaction with the surrounding cold gas medium and the treated material. As a result of this interaction, the temperature and velocity of the Arc jet gradually decrease to the temperature and velocity of the environment.

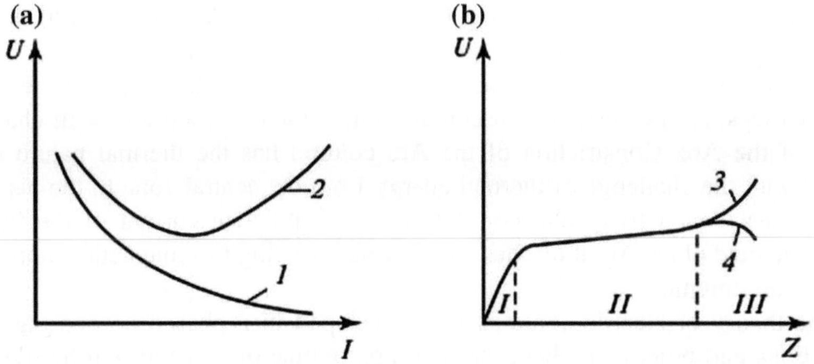

Fig. 8.3 Current–voltage characteristics of the electric Arc (**a**) and the characteristic distributions of electric potential along the length of the Arc (**b**): *1* decreasing characteristic, *2* characteristic with the growing region, *3* the distribution with the positive decrease of anode, *4* the same with negative decrease, *I* near-cathode region, *II* the Arc column, *III* near-anode region

The current–voltage characteristics of electric Arc in many cases are decreasing. However, for the Arc in the channel where the growth of the diameter of the plasma column is limited by the walls at the increase of current, the growing area can appear at the current–voltage characteristic (Fig. 8.3a). The choice of a method for stabilizing of the discharge, of the power supply, the efficiency of the Arc installation and other properties of the Arc are determined using the current–voltage characteristic.

Near-electrode processes of the electric Arc are emerged as an independent branch of plasma physics. In addition to clarification of the mechanisms of electric, heat and mass transfer, important practical problem of the increase of the service life for the electrode was solved here. And the perspectives for simulation of electric discharges from electrode to electrode are emerged only recently. Conditions in the near-electrode regions are signally different from the conditions in plasma column, and three zones can be distinguished at the study of the processes in electric Arc: the Arc column, near-cathode, and near-anode regions (Fig. 8.3b).

The electric current in the near-cathode region is created mainly due to the motion of electrons emitted from the cathode surface and due to the ions coming from the Arc column. In the near-anode region, the electric current mainly occurs at the moving of electrons from the Arc column to anode. The length of the near-electrode zones is very small and is in the order of a few mean free paths of the particles. Consequently, the intensity of electric field in the near-electrodes areas should be very large. For the Arc at the atmospheric pressure it is estimated at approximately $\sim 10^4$–10^6 V/cm near the cathode surface.

At the Arc column, the intensity of electric field is largely dependent on the diameter of the discharge channel, current, type and regime of the gas flow and other conditions and has a much smaller value compared with the value at the electrodes $\sim$(5–20 V/cm). Some larger value of the intensity of electric field can be obtained by holding of the Arc column using the additional magnetic field. The Arc column is constricted and a strong constriction of the Arc is observed near the electrode; and the diameter of the Arc near the cathode is typically less than near the anode. The physical processes, leading to the narrowing of the Arc near the surface of electrodes, are associated with condition near the electrode and with characteristics of the Arc. Constriction of the Arc column has the thermal nature and is linked with the challenge of thermal energy from the central zone to the periphery due to heat conductivity and convection. At a sufficient current of the intrinsic magnetic field of the Arc it reaches a large value, leading to a magnetic compression of the Arc column.

The theory of electric columns was developed in a phased accounting of the conditions and processes where the used approximation is valid. Further development of the theory is mainly due to various external factors that are taken into account: the longitudinal and transverse blowing of the Arc discharge, interaction of the Arc with electric and magnetic fields, taking into account the impact of the near-electrode processes and walls of the channel, and others.

The series of monographs, summaries and reviews such as the famous book of Finkelberg and Mekker [2] are devoted to the theory of electric Arcs. It provides a detailed qualitative description of physical processes in the electric Arc, the basic theoretical models; the methods of diagnostic and experimental data are presented. The monograph of Granovsky [3] electrodynamics of gas is described in details. The second part of this book published by a group of authors [4] includes the information about the classification of the types of discharges and currents. Written by experimentalists and theoreticians, it includes data on virtually all types of gas discharges and their classification. An important review of the analytical methods and results of researches of the Arc discharges is given in the book edited by Polak [5]. In addition to the problems of the plasma chemistry the new results in the theory of electric Arcs and RF-discharges in plasma are described in it, they stimulate the researches, the theoretical description and numerical modeling of processes in electric Arc. The physical ideas in the theory of open free Arcs, their movement and stability are described in the monographs of Anderson [6] and Novikov [7].

The papers [8–10], theory and experimental description of the electric Arcs edited by Zhukov M. and the monographs [11, 12] are devoted to the theory and experimental description of the electric Arcs in the conditions of forced heat exchange. The authors, in addition to the presentation of original results, summed up the researches in this area. Stated in monographs, [11, 12] made it possible to identify the directions for development of the theory of electric Arc plasma generators. The described problems of the transfer of radiation can be found in [13, 14]. The work of Alexandrov and Ruhadze [15] is devoted to the theory of high-current pulsed Arc discharge. The criteria approach to the description of generators low-temperature plasma is considered in the works [8, 16, 17]. The communication of electric Arc with other types of discharges and a single theoretical approach to the description of a wide range of processes in gas discharges are considered by Raiser [18].

The theory of thermal plasma of the cylindrical Arc column and extended electric Arcs is well-developed now, where the pattern of flow and heating of the gas by the electric Arc is presented on the basis of qualitative, channel, integrated, self-similar, and numerical methods for the solving, and the characteristics of plasma are determined depending on the external adjustable parameters.

The basis for most of the theoretical studies of the electric Arcs is assumption of the existence of local thermodynamic equilibrium (LTE) of plasma. However, the application of the equilibrium plasma model for description of the processes in electric Arc is not always justified. The experiments [19, 20] show that, in the Arc plasma, there are spatial areas where the equality of temperatures of electrons and heavy particles (atoms, ions, and molecules) is violated, the Maxwell velocity distribution and Boltzmann distribution over the excited levels are not installed, and the condition of ionization equilibrium is not fulfilled. Near the electrodes, in the initial section of the Arc and in the peripheral regions of the discharge, the nonequilibrium of plasma is the most significant, and the taking into account of deviations from LTE becomes necessary.

In connection with the development of a new branch of industry—plasma chemistry is often used for modeling of nonequilibrium plasma and modeling of plasma and chemical reactions in a channel of the plasma torch. It is often necessary to determine the velocity of the chemical reactions in a predetermined plasma flow, high energy flows, the degree of heating of the chemical components, the overall efficiency of the plasma-chemical process, etc.

Another application of the model of nonequilibrium plasma is connected with the investigation of electrode processes in the Arcs. The relevance of researches of such processes is related not only with the checking of physical hypotheses, with the expansion of the circle of tasks, but also with determination of the ways for improve of the efficiency of the practical use of Arc plasma torches. It is known that the investigations of near-electrode processes using the model of equilibrium plasma cannot give a satisfactory result. Because of the complexity and the mutual influence of physical processes and the difficulties of their mathematical description, taking into account nonequilibrium, to a first approximation, it can be performed using the model of two-temperature plasma [21]. The mathematical formulation of this approach for the initial area of the Arc using the approximation of boundary layer proposed at first in [22]. In [23], two-temperature approximation for the three-fluid plasma model is proposed, but in this paper the goal is actually stated. In [24], the method for calculation of the initial part of the electric Arc is presented; it is based on the deviation of the plasma composition from the equilibrium at the assumption that the temperatures of all the components of plasma are equal. Attempt of calculation of the initial part of the Arc on the basis of two-temperature plasma model is carried out in [25, 26], here a relatively simple system of equations using the assumption of ionization equilibrium in plasma is used. The development of this approach is reflected in the work [27], wherein the framework of assumption of two-temperature layer the calculation of initial part of the Arc in different hydrodynamic regimes of the flow of the plasma-forming gas is made. Attempt to take into account the thermo-chemical nonequilibrium of plasma takes place in [28–31], here the analysis of two-temperature plasma flow in the transition region of the channel and asymptotic Arc is carried out.

Further development of the theoretical method for research of the Arc is associated with the use of models of equilibrium and nonequilibrium plasma for description of the dynamics of gas heating in the Arc, for the study of phenomena in the near-electrode, near the wall areas and interaction of the plasma flows with the disperse material. Theoretical methods should also be used to study the curved Arc, Arc of alternating voltage and high-voltage discharge, for description of the kinetic processes in the moving nonequilibrium plasma, especially molecular, turbulence and interaction of the kinetic processes with the turbulence, for solving of the unsteady problem and problem with violation of the axial symmetry [32].

8.2 The Arc Plasma Torches for Cutting and Spraying

Arc Plasma Torches for Cutting of Metals Using of the compressed plasma Arc for cutting of the refractory metals and alloys is one of the broad areas of the practical application of low-temperature plasma. The electric Arc, compressed in the channel of the plasma torch and stabilized by the gas flow, has the strong thermal and gas-dynamic effect on the anode (detail). The abilities of the compressed electric Arc for intensively melting of the metal anode (detail) and actively blow out the melt using a high-speed gas-dynamic flow are in the basis of plasma Arc cutting of metals. Currently the plasma cutting is one of the most effective methods of thermal cutting of metal. Its quality of the cut is only slightly inferior to the laser cutting, but exceeds the latest in economic indexes and broader opportunities, especially relating to the maximum thickness for the cutting of metals.

The installation for plasma cutting Maxigraph DD/3.5/4.0/4.5/5.0/5.5 mm of the company KOIKE (Japan) is shown in Fig. 8.4 has the following characteristics: the maximum number of burners—6; the maximum velocity of cutting for parts of steel

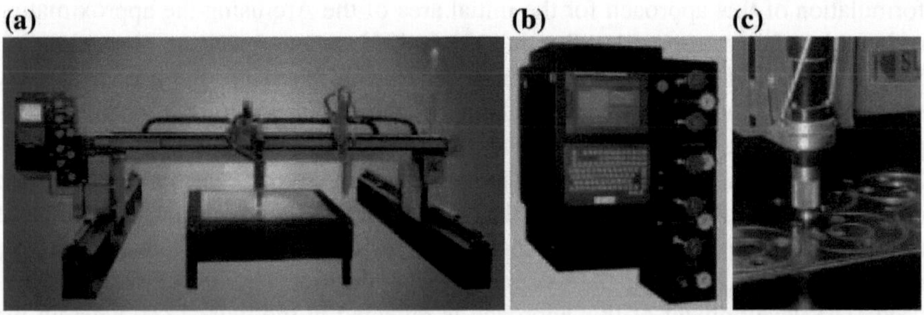

Fig. 8.4 The installation for plasma cutting Maxigraph (**a**) with device of the numerical control (**b**), and plasma burner (**c**) of the company KOIKE (Japan)

St3 with the thickness 10 mm–6000 mm/min; the effective cutting width—2.6, 3, 1, 3.6, 4.1 and 4.6 mm; effective cutting length—the length of the rail minus 1.75 m; the library of standard shapes for cutting—60; the system of numerical control (CNC)—Hybrid 400/D–180i FS; the format of the control system EIA (US); external interface RS–232C; plasma cutting system—max-200/HPR 130/260, Super 200/400.

The characteristics of installations for plasma cutting of metals from the series APR of the company "Power Engineering" are shown in Table 8.1.

High velocity of plasma and high thermal flow in anode spot of the electric Arc is achieved by the forced compression of electric Arc in the channel of small diameter (d = 1–5 mm) and its stabilization it in space using straight-jet or vortex gas flow (air, nitrogen, oxygen, etc.). Below as an example there are the specifications of one of the most effective plasma torches for cut RPM-6, produced serially with the installation of OPR-6. The construction of such plasma torch (plasma cut) and the scheme of process are shown in Fig. 8.5.

There are some technical features of the use of compressed Arc for cutting of metals. They include, first of all, the correct choice of power, current, the diameter of nozzle and conformity of their thickness and velocity of movement of the cut detail. The parameters of developed regimes of plasma Arc cutting of stainless steel X10CrNiTi18-10 (Euro analogous) using plasma torch RPM-6 are shown in Table 8.2. There are the values of optimum cutting velocity and the width of resulting cut. Increase of the thickness of cut metal makes it necessary to increase the current of the cutting Arc. However, current and cutting velocity is limited by resistance of the nozzle in the plasma torch.

Big values of the dynamic pressure $\rho v^2/2$ and the velocity v of the gas should be noted. The electric Arc for cutting the velocity of the flow plays a very important role in the process of heat exchange, and at the destruction of the layer of the liquid metal. At velocities of the flow of several kilometers per second, obviously, not only the active melt takes place, but also mechanical action of the jet on the molten metal. The compressed Arc simultaneously melts and blows the molten metal from the surface of the cut. The effect of pressure p on the regime of plasma cutting and its strong dependence on the current gas flow and the diameter of the nozzle determine significant differences in the dynamic pressures in plasma torches of various constructions. This, in turn, explains the difference in technical parameters of plasma cuts of many companies. Successfully selected configuration and the diameter of the nozzle, the proper organization of the gas vortex (there is no recommendations in this area) can significantly increase the velocity and the dynamic pressure of the Arc and, therefore, at the same current I to provide the great velocity and better quality of cutting.

Table 8.1 Characteristics of installations for plasma cutting of series APR

Characteristic	APR-40	APR-60	APR-90	APR-91	APR-140	APR-150
Supply voltage, V	220	380	380	380	380	380
Consumption power, kW	4	16	20	14	25	29
Cutting current, A	25 (35)	40 (60)	60 (90)	60 (100)	70 (130)	50 (100, 150)
Duty cycle, %	100 (60)	70 (60)	70 (60)	80 (70)	70 (60)	100 (90, 85)
The air pressure, MPa	4–5	5–6	5–6	5–6	5–6	5–6
Air consumption, l/min	100	170	180	180	210	210
Thickness of cut material, mm						
Maximum	8	18	30	35	45	50
High-quality cut	6	10–12	20–25	25–30	30–35	35–40
Mass, kg	28	60	90	100	125	170
Dimensions (length × width × height), mm	320 × 400 × 625	535 × 320 × 625	535 × 320 × 625	550 × 420 × 820	550 × 420 × 820	640 × 420 × 820

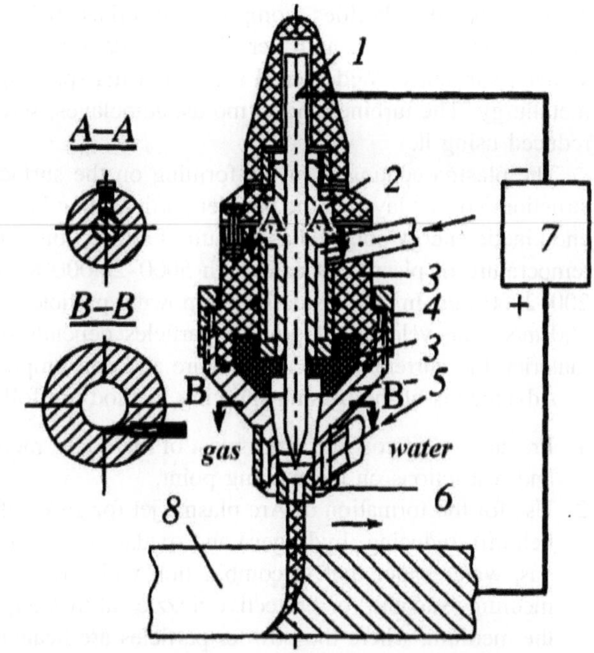

Fig. 8.5 General view of the plasma torch RPM-6 and scheme of the process of metal cutting using the compressed Arc: *1* cathode, *2* sleeve, *3* insulators, *4* union nut, *5* housing, *6* nozzle, *7* power supply, *8* anode-detail

Table 8.2 Recommended parameters of regimes of plasma torch operating RPM-6 (nitrogen) in the plasma Arc cutting of stainless steel of different thickness [1]

Thickness of cut metal, mm	Diameter of channel, mm	Current of cutting Arc, A	Cutting velocity, mm/s	Width of cut, mm	
				Up	Bottom
20	2.8	320	1700	5.5	2.8
40	4.0	500	900	6.5	3.5
60	4.8	610	540	7.5	4.5
80	5.2	650	420	8.5	7.0
100	5.6	710	300	10.5	8.5
120	6.0	740	250	11.5	9.5

Arc Plasma Torches and Technology of Plasma Spray of the Coatings Arc plasma torches for plasma spraying are designed for the covering with the corrosion-resistant, wear-resistant, sealing, friction, insulation, and other special coatings on the surfaces of details. Coating often consists of refractory oxides, carbides, nitrides, borides, composite materials. It is applied by spraying of wire and powder materials, and there is no oxidation of the surface of detail and the sprayed substance. Widespread use of plasma spraying is in heat engineering and metallurgy. The turbine blades, molds, autoclaves, screws, bearings and others are reduced using it.

The plasma coating is in the forming on the surface of details (products, constructions) of the layer of the powder particles that have a certain margin of thermal and kinetic energy obtained as a result of interaction with the jet of Arc plasma. The temperature of plasma jet can reach 5000–20,000 K, and the velocity of the flow 200–2000 m/s. In the plasma jet, the powder particles melt and get the velocity 50–800 m/s. The velocity of powder particles depends on their size, density of the material, the current of the Arc, nature and consumption of plasma-forming gas.

Advantages of the plasma spraying method are follows.

1. Preparation of coatings from most of materials, melting without decomposition, and restrictions on the melting point.
2. Use for the formation of Arc plasma jet for gases of various types: inert (argon, helium), reducing (hydrogen) and oxidant (air, nitrogen) and ammonia, natural gas, water vapor that in combination with the use of cameras with protective medium (vacuum) or protective nozzles allows adjustment of characteristics of the medium, where the powder particles are heated and move.
3. Flexible regulation of electric and gas regimes of plasma torches operation, including the process of coating of materials that allows controlling of electrical properties of sputtered particles and the conditions for formation of the coating.

DC Arc plasma torches, operating in inert, neutral and oxygen containing mediums, are the most widely used in industry. By way for supplying of the plasma-forming gas there are plasma torches with tangential and axial feeds of gas. The plasma torches of indirect action are widely used for plasma spraying of coverings of powders.

Today in the world, there is a large number of constructions of Arc plasma torches for spraying, working on direct current, and that have the different configurations of the discharge channel, the nozzle, anode and cathode, the method of input of powder mixture, construction, materials and methods of construction of electrodes, the principles of the Arc stabilization. The linear (axial) scheme of plasma torches with traverse of plasma-forming gas are most widely used (Fig. 8.6). Electric Arc discharge burns between the cylindrical anode and the cathode with smooth or stepped channel, and in the latter case, the expansion of the

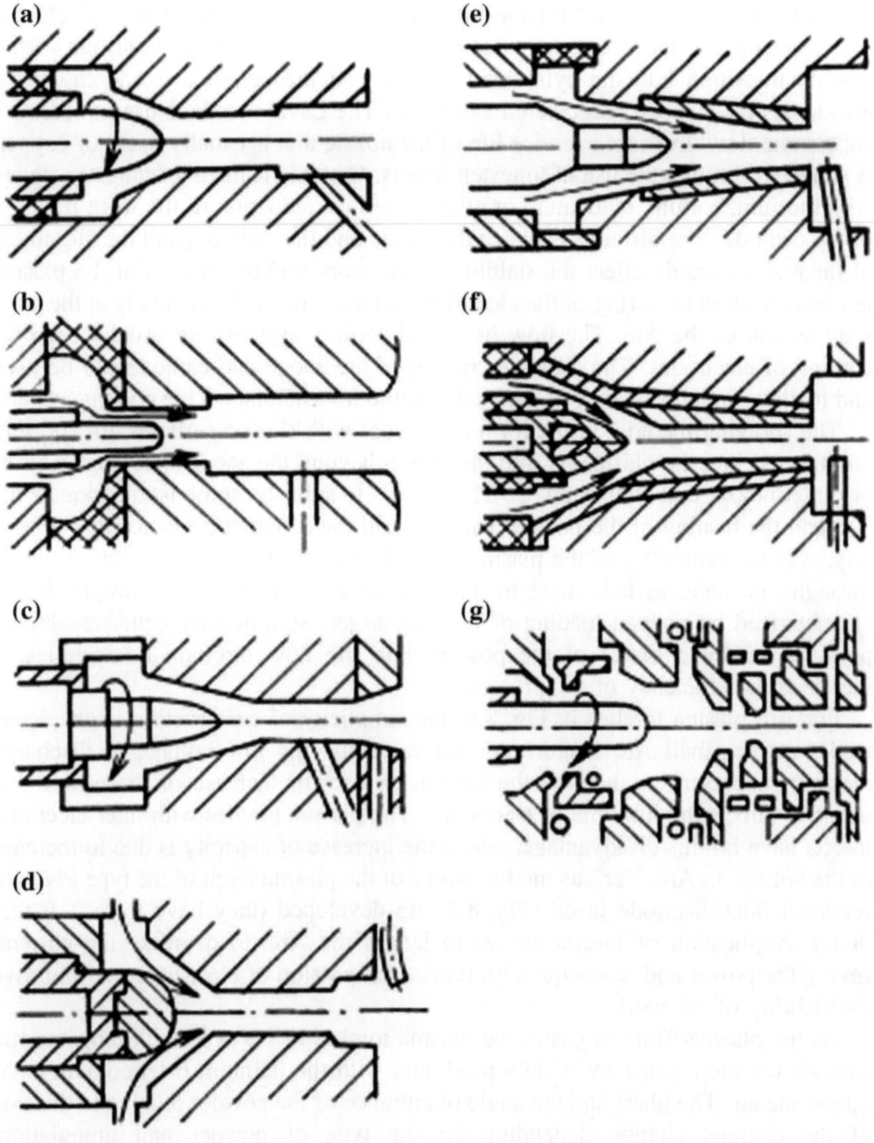

Fig. 8.6 Schematic view of the channels of Arc plasma torches for spraying of different constructions [33]

channel at the output is used for a gas-dynamic fixation of anode spots of discharge. The front part of the nozzle is usually made in the form of the confuser with a smooth transition into the cylindrical channel. At the output in some cases, the nozzle has the stepped and tapered extension. The Laval nozzle is used to create a supersonic flow. Increased service life of the nozzle that is usually made of copper, is achieved through the use of tungsten inserts. Cathode is made of tungsten doped with thorium, barium, lanthanum or other materials, reduction of the work function of the cathode. The distance between the anode and the cathode, and the alignment of them significantly affect the stability of the work and parameters of the plasma jet, and the strict centering of the electrodes is recommended, especially at the axial stabilization of the Arc. The flow or closed cooling systems are usually used for cooling of electrodes. The separate cooling of the anode and cathode can be used and it allows a quick change of electrodes without switching off the cooling system.

The spraying material in the form of powder with size of particles 40–100 μm, are injected into the plasma jet into the region beyond the anode spot (Fig. 8.6a–c) or outlet nozzle exit (Fig. 8.6d–g). This allows to split the combustion zone of the Arc and the heating of the powder and to avoid their mutual influence that greatly improves the reliability of the plasma torch. Entrance of the powder into the flow through one hole, as it is done in most of the plasma torches, is simple, but is characterized by uneven loading of the plasma jet. Significantly better results are provided at the entrance of the powder into the flow through a few holes, it increases the efficiency of the process.

For Arc plasma torches in Fig. 8.6, the simplicity of construction is characterized, but the small Arc length and, consequently, the low voltage of discharge makes it necessary to increase the current of Arc for increase of the power that negative affects the life time of electrodes. The plasma torches with inter-electrode inserts have no this disadvantage, where the increase of capacity is due to increase of the voltage in Arc. Various modifications of the plasma torch of the type PN with sectional inter-electrode insert (Fig. 8.7) are developed (they have from 2–6 sections). Application of inserts allows to length the Arc, to decrease the current, saving the power and, consequently, reduce the erosion of electrodes and improve the stability of the work.

As the plasma-forming gas in the plasma torches of PN-type with tungsten rod cathode the inert gases are widely used, and with the hafnium or zirconium cathodes—the air. The place and the angle of entrance of the powder relative to the axis of the channel change depending on the type of powder and granulation. Performance of the coating of aluminum oxide particles with diameter of 50 μm is 10 kg/h.

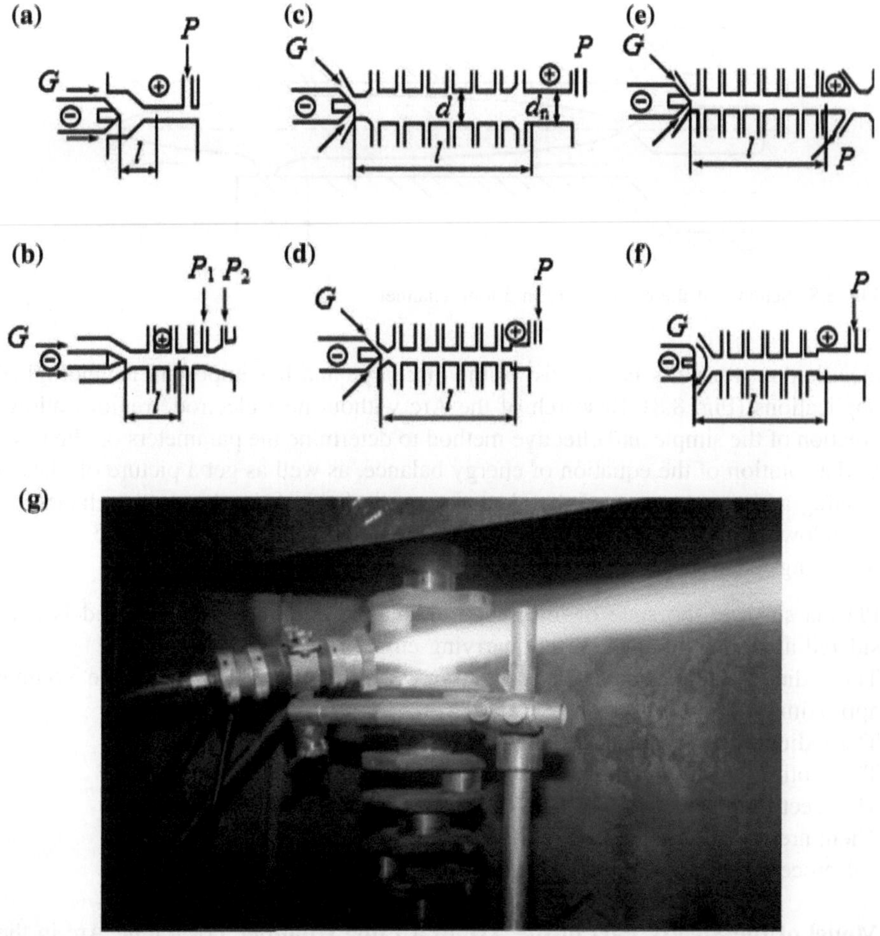

Fig. 8.7 Schematic view of the Arc plasma torches of PN-type [33]: **a** PN-1; **b** PN-2; **c** PN-6; **d** PN-14 m, PN-15, PN-16; **e** PN-20, PN-21, PN-22; **f** PN-23; **g** plasma spraying installation: I = 100–250 A, U = 80–150 V, P = 10–30 kW, Q_{Ar} = 18–75 l/min, Q_{H_2O} = 10–12 l/min, p_{H_2O} = 6 atm, the efficiently of powder (Ni–Cr–B–Si) with diameter 40 μm up to 5 kg/h; G plasma gas supply, P entrance of the powder, d_n diameter of the nozzle, d, l diameter and length of the Arc channel

8.3 Simulation of the Arc Plasma Torches in the Channel

8.3.1 Models of the Arc Plasma Torches in the Channel

As it was noted, a feature of Arc plasma torches for cutting and spraying is maintain of steady burning of plasma flow between the cathode and the anode in a cylindrical channel, the length of them substantially exceeds its diameter. From all areas of the

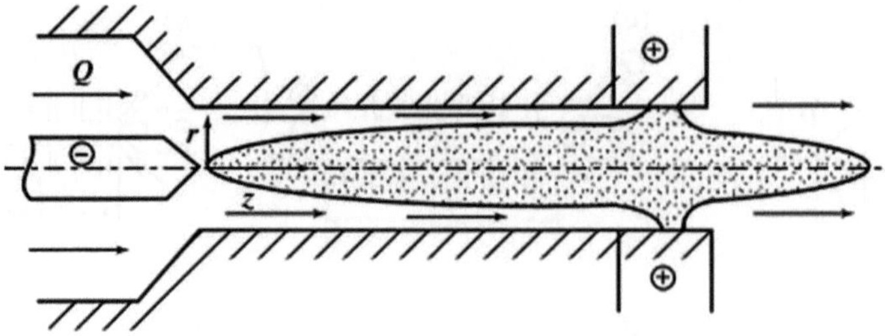

Fig. 8.8 Scheme of the electric Arc in a long channel

electric Arc it usually takes most of the discharge and has important technological applications (Fig. 8.8). Research of the Arc without near-electrode regions allows creation of the simple and effective method to determine the parameters on the basis of the solution of the equation of energy balance, as well as get a picture of plasma moving and the structure of the electromagnetic field of the Arc in the channel.

Below we consider the model of the channel for the study of the Arc with the following features:

Plasma stays in the state of local thermodynamic equilibrium (LTE) and is considered as the continuous current-carrying environments;
The radiation of plasma is optically thin, and may be described in the volume approximation;
The radius of the channel is much smaller than its length;
The motion of gas flow in the Arc channel is one-dimensional;
The electromagnetic field in the Arc is one-dimensional;
There are no external magnetic fields;
All processes in the Arc channel are stationary.

Model of the Steady Part of the Arc in a Long Channel For a long Arc in the channel with radius much smaller than its length $d \ll 1$ the flow can be considered one-dimensional ($v_r = v_\varphi = 0$). Heating of the gas occurs very fast, and constant along the channel temperature field is set in the Arc, so there is a large established part $\left(\frac{\partial T}{\partial z} = 0\right)$, that takes up almost all the space of the Arc, except for the initial part. At the steady part it can be assumed that the balance of convective flows is equal to zero and the withdrawal of energy from the Arc is made by thermal conductivity in the radial direction. In this case, the calculated model of steady-state part of the long Arc includes a one-dimensional equation of energy balance for obtaining of temperature and one-dimensional Maxwell's equations for determination of electromagnetic field.

The one-dimensional equation of energy balance has the form:

$$\sigma E^2 - U_{\text{rad}} = -\frac{1}{r}\frac{\text{d}}{\text{d}r}\left(r\lambda\frac{\text{d}T}{\text{d}r}\right) \tag{8.1}$$

The one-dimensional intensity of electric field ($E_r = E_\varphi = 0;\ E = E_z$), current density and specific power are calculated as

$$E = \frac{I}{2\pi\int\limits_0^R \sigma r \text{d}r};\quad j_z = \sigma E;\ P = \sigma E^2 \tag{8.2}$$

The one-dimensional intensity of magnetic field ($H_z = H_r = 0;\ H = H_\varphi$):

$$H_\varphi = \frac{E}{r}\int\limits_0^r r\sigma \text{d}r \tag{8.3}$$

The electric and magnetic fields are one-dimensional and the only one component of electromagnetic force, that appears as a result of interaction of the Arc current with its own magnetic field ($F_z = F_\varphi = 0$):

$$F_r = -\mu_0 j_z H_\varphi = -\mu_0\frac{\sigma E^2}{r}\int\limits_0^r r\sigma \text{d}r \tag{8.4}$$

The decision of the equation of energy balance (8.1) is carried out using the method of the control volume (Sect. 6.1.1). The balance of powers and the discrete analogue are obtained by integrating of the equation of balance energy in the control volume (Fig. 8.9).

$$\int\limits_{r_s}^{r_n}\frac{1}{r}\frac{\text{d}}{\text{d}r}\left(r\lambda\frac{\text{d}T}{\text{d}r}\right)r\text{d}r + \int\limits_{r_s}^{r_n}\left(\sigma E^2 - U_{\text{rad}}\right)r\text{d}r = 0 \tag{8.5}$$

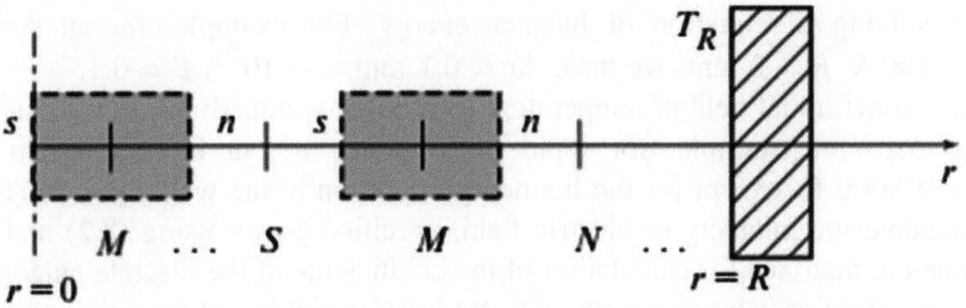

Fig. 8.9 Grid and the control volumes for solving of the equation of energy balance for the steady part of the Arc in a long channel

After integrating we will have

$$\left(r\lambda\frac{\mathrm{d}T}{\mathrm{d}r}\right)_n-\left(r\lambda\frac{\mathrm{d}T}{\mathrm{d}r}\right)_s+0.5\left(\sigma E^2-U_{\mathrm{rad}}\right)_M(r_n-r_s)(r_n+r_s)=0 \qquad (*)$$

Here, the derivative of temperature in the radial direction on the verge of the control volume *n, s* can be approximated as:

$$\left(\frac{\mathrm{d}T}{\mathrm{d}r}\right)_n=\frac{T_N-T_M}{r_N-r_M};\ \left(\frac{\mathrm{d}T}{\mathrm{d}r}\right)_s=\frac{T_M-T_S}{r_M-r_S} \qquad (**)$$

Substituting Eqs. (**) in (*), we obtain the discrete analogue of the equation of energy balance for the Arc in the channel at steady part for the internal points

$$a_M T_M = a_N T_N + a_S T_S + b_M$$

where

$$\begin{cases} a_N=\frac{r_n\lambda_n}{r_N-r_M}>0; \quad a_S=\frac{r_s\lambda_s}{r_M-r_S}>0; \\ a_M=a_N+a_S-0.5S_{pM}(r_n-r_s)(r_n+r_s)>0; \\ b_M=0.5S_{cM}(r_n-r_s)(r_n+r_s); \quad S_M=\left(\sigma E^2-U_{\mathrm{rad}}\right)_M. \end{cases} \qquad (8.6)$$

Coefficients S_{pM}, S_{cM} are determined by (6.56). For efficient use of the boundary conditions on the axis of symmetry $\left(\frac{\mathrm{d}T}{\mathrm{d}r}\Big|_{r=0}=0\right)$ and on the wall ($T=T_R$, where T_R—the temperature on the wall) construct the computational grid so that the first point is located on half a step from the axis of symmetry ($r_M=\Delta r/2$), and the latter —on the wall (Fig. 8.9). Taking into account the boundary conditions on the axis of symmetry for the first control volume the flow due to heat conduction through the verge *s* is equal to zero. Consequently, discrete analogue of the equation of energy balance for the first point has the form (8.6) at $a_s=0$. For the latest point on the wall, we have $T=T_R$.

We give the *algorithm of solving.*

1. Build the calculated grid lines (Fig. 8.9) with constant or variable step Δr, and take the values of convergence criteria ε and internal feedback coefficient β for the solving of equation of balance energy. For example, for an Arc with $I = 100$ A, $R = 2$ mm, we have $\Delta r = 0.1$ mm, $\varepsilon = 10^{-3}$, $\beta = 0.1$.
2. Take some initial field of temperature (providing conductivity) in the channel of the Arc, for example, for rapid convergence it can be a constant value $T = 10{,}000$ K, except for the boundary condition at the wall $T_R = 500$ K.
3. Calculate the intensity of electric field, specified power using (8.2) and determine the matrixes for calculation of the coefficients of the discrete analogue for the equation of balance energy (8.6) for interior and boundary points (Fig. 8.9).

4. Solve the system of the discrete analogue for the equation of balance energy (8.6) using iterative or direct sweep method in order to find a new field temperature.
5. Check the condition of convergence, if it is not fulfilled, repeat the decision from the 3 before the full convergence.
6. Determine the results of calculation of the field using the Eqs. (8.2–8.4).

The results of calculation and experimental data for the steady part of the Arc in the channels with diameter $d = 4$; 10 mm are shown in Fig. 8.10. Although not taking into account of the removal of energy by the convective flow in this model some inflated gas temperature is obtained in the area near the wall, but in an Arc there is a good coincidence of the calculation result with the experimental data that suggests the applicability of this model for calculation of the steady-state area of the Arc in the channel.

In addition, this model can be successfully used for determination of temperature in the zone of conductivity of the initial section (after the near-cathode region) of the Arc and as an initial approximation at the full simulation of the Arc in the channel.

Model of the Long Arc in the Channel The model of steady part of the Arc (8.1) allows to quickly get the basic parameters of the Arc in the channel, but does not allow the determination of the parameters of the initial part, in the flow in the area near the wall where the heat removal from the depths of the Arc is not only by conductive, but also the convective mechanisms.

For a long channel of Arc with axial gas supply (Fig. 8.8) the radial and tangential velocities are assumed equal to zero ($v_r = v_\varphi = 0$). Specific consumption through the cross-section of the channel of the Arc is given or is determined by taking into account a given gas consumption G, kg/s, or Q, m^3/s: $\rho v_z(r) = \frac{G}{\pi R^2} = \frac{\rho Q}{\pi R^2}$. Note that the specific consumption $\rho v_z(r)$ may not remain constant over the section, the Arc always operates as a magnetic pump, and only a certain amount of gas enters into the discharge for heating and the remaining part (80–90 %) is moving along the wall of the channel forming the most natural thermal protection layer. Thus, the channel can be divided into two zones with different specific consumption: the discharge and the zone near the wall. The exact distribution $\rho v_z(r)$ can be obtained by the calculation of the field of the gas flow and solving the momentum and continuity equations of gas.

The assumption of one-dimensional motion of gas in a long narrow channel allows determination of the axial velocity component of gas from the condition of the specific consumption through the cross-section.

In contrast to the model of steady part of the Arc, the equation of balance energy for the model of the long Arc in the channel is:

$$\rho c_p v_z \frac{\partial T}{\partial z} = \frac{\partial}{\partial z}\left(\lambda \frac{\partial T}{\partial z}\right) + \frac{1}{r}\frac{\partial}{\partial r}\left(r\lambda \frac{\partial T}{\partial r}\right) + \sigma E^2 - U_{\text{rad}} \tag{8.7}$$

This model includes the equations for determination of the transfer coefficients and thermodynamic functions of approximation functions of equilibrium plasma or

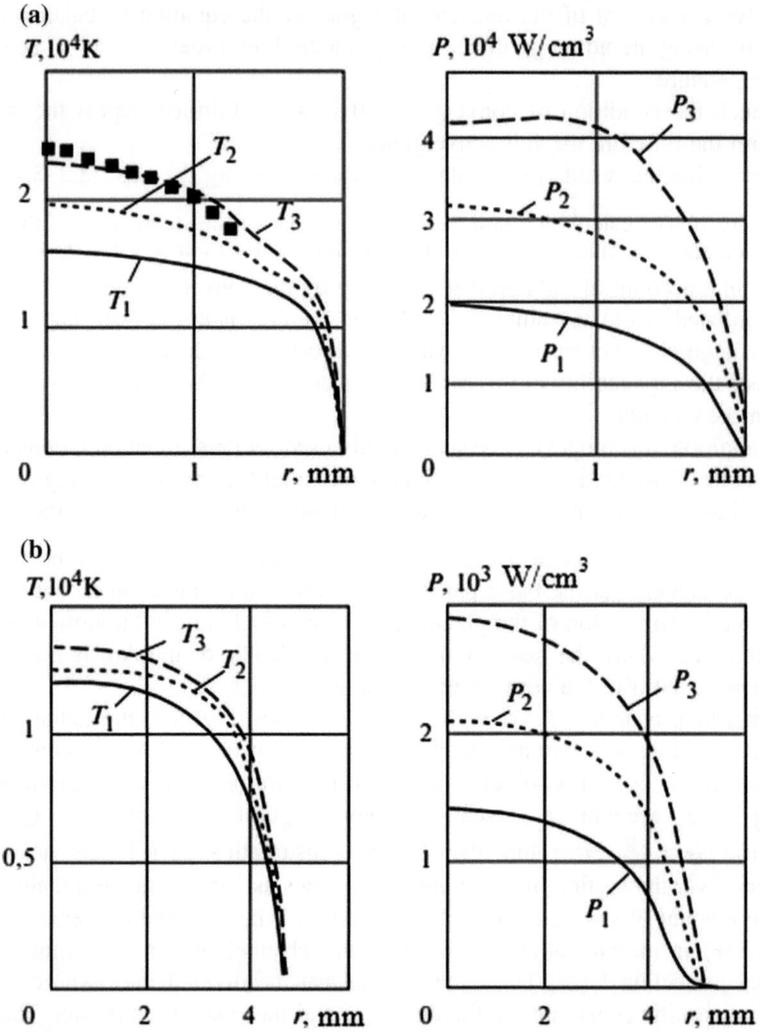

Fig. 8.10 Temperature and specified power of Ar-Arc in the channel $d = 4$ mm (**a**), and $d = 10$ mm (**b**): *lines* calculated data, T_1, P_1—$I = 100$ A, T_2, P_2—$I = 150$ A, T_3, P_3—$I = 200$ A; points—the experiments for $d = 4$ mm, $I = 200$ A [34]

approximation equations of their temperature dependences: $\sigma(T)$, $\lambda(T)$, $c_p(T)$, $\rho(T)$, $U_{rad}(T)$. Calculation of the parameters of the electromagnetic field is carried out according to Eqs. (8.2–8.4).

Such a model of the Arc in the channel is used in many theoretical studies, for example [9–11].

The boundary condition: in initial (zero) section ($z = 0$), a one-dimensional model of energy and current balance can be used (8.1), (8.2) for determination of the temperature field in the zone of conduction at the zero section, but it is necessary to set or find the radius of the zone of conduction R_k in this section. Radius R_k can be obtained by assuming that the current flowing through the section of the Arc I, and the resulting distribution of the intensity of the field along the length of the Arc has a decreasing character (due to thermal expansion of the Arc in the channel). Out of the zone of conduction ($r > R_k$) the temperature of the cold gas is known: $T = 300$ K.

It is believed that at the wall ($r = R$) the gas is in thermal equilibrium with plasma, the temperature of the wall (taking into account heating of the wall from the plasma flow due to radiation and heat conductivity) is $T = 300$–1000 K.

On the axis, we have $\frac{\partial T}{\partial r} = 0$ from the condition of the cylindrical axial symmetry.

At the exit (at the end of a long cylindrical channel), where all active processes are considered steady, a simple boundary condition $\frac{\partial T}{\partial z} = 0$ can be taken, and the temperature is equal to the temperature upstream [application of the scheme «upstream» (Sect. 6.2.1)].

The decision of the equation of balance energy (8.7), together with the calculation of the intensity of electric field for the Arc in the channel is carried out using the method of control volume for solving of general differential equation (Sect. 6.1.1). First, we transform the equation of balance energy (8.7) in the general differential equation for the control volume. Because of the smallness of the control volume it can be assumed that the specific heat of plasma is constant ($c_p = \text{const}$) inside each of the control volume. The law of mass conservation for the one-dimensional plasma flow in the channel has the form $\frac{\partial(\rho v_z)}{\partial z} = 0$, and $\rho v_z = \text{const}$ along the z axis. Thus, the equation of energy balance (8.7) can be written as:

$$\frac{\partial(\rho v_z T)}{\partial z} = \frac{\partial}{\partial z}\left(\frac{\lambda}{c_p}\frac{\partial T}{\partial z}\right) + \frac{1}{r}\frac{\partial}{\partial r}\left(r\frac{\lambda}{c_p}\frac{\partial T}{\partial r}\right) + \frac{1}{c_p}\left(\sigma E^2 - U_{\text{rad}}\right) \tag{8.8}$$

The procedure for integrating of the equation of balance energy (8.8) is similar to the procedure for the integration of general differential Eq. (6.2) (Sect. 6.1.1), and $\Phi = T$; $\Gamma = \frac{\lambda}{c_p}$; $S = \frac{1}{c_p}\left(\sigma E^2 - U_{\text{rad}}\right) = S_c + S_p T$, $\rho v_r = 0$. The discrete analogue of the equation of energy balance (8.8) for internal points (Fig. 6.5a) is:

$$a_M T_M = a_E T_E + a_W T_W + a_N T_N + a_S T_S + b_M$$

where

$$\begin{cases} a_E = D_e A(\mathrm{Pe}_e) = D_e A(|\mathrm{Pe}_e|) + \max(-F_e, 0) \geq 0; \\ a_W = D_w A(\mathrm{Pe}_w) + F_w = D_w A(|\mathrm{Pe}_w|) + \max(F_w, 0) \geq 0; \\ a_N = D_n A(\mathrm{Pe}_n) = D_n \geq 0; \quad a_S = D_s A(\mathrm{Pe}_s) + F_s = D_s \geq 0; \\ a_M = a_E + a_W + a_N + a_S - S_{pM}\Delta V > 0; \quad b_M = S_{cM}\Delta V; \\ \Delta V = (z_e - z_w)\frac{(r_n - r_s)(r_n + r_s)}{2}. \end{cases} \tag{8.9}$$

Function $A(\mathrm{Pe})$ that characterize the computational scheme (Sect. 6.1.1) is $A(\mathrm{Pe}) = \max\left[0, (1 - 0.1\mathrm{Pe})^5\right]$. The coefficients S_{pM}, S_{cM} are determined using (6.56).

$$S_{pM} = -\left|\frac{S_M}{\beta_T T_M}\right|; \quad S_{cM} = S_M - S_{pM} T_M \tag{8.10}$$

Here $S_M = \left[\frac{1}{c_p}\left(\sigma E^2 - U_{\mathrm{rad}}\right)\right]_M$; β_T—coefficient of internal feedback (Sect. 6.3.2).

For simulation of the long Arc in the channel, we use the similarity of boundary conditions of Arc and RF-discharge in the channel and we use the higher developed (Sect. 6.1.2) structure of the shifted chess grids (Fig. 6.5a). And the discrete analogue for the boundary points are obtained in the same way and have the same form as in the case of RF-plasma torch (Sect. 7.3.1).

Algorithm for solving of the equation of balance energy for this model is similar to the above algorithm for solving of the model of established part of the Arc. Note some features for modeling of the long Arc in the channel:

1. First, for the solution the initial profile of the temperature field in the channel of the Arc should be specified. The intensity of electric field is calculated using the Eq. (8.2) and the specified power σE^2 are sensitive for the temperature distribution, and the stability of solution of the problem depends on a predetermined initial temperature field in the Arc channel and on the selection of the coefficient of internal feedback β_T. The one-dimensional model of the steady part of the Arc can be used again for obtaining of the initial field of the temperature [(8.1), (8.2)]. This method usually gives the initial field of temperature sufficiently close to the solution of the equation of balance energy (8.7), because it satisfies to the one-dimensional balance of energy and of the Arc current. Using the obtained initial temperature field can ensure a smooth change of the solution and fast convergence of the problem.
2. The coefficient of internal feedback β_T is determined based on the following reasoning: if the solution is carried out using a coarse mesh or using the initial temperature field that is significantly different from the result of the final decision the lower value of the coefficient β_T to provide more effect of feedback should be taken into account. For example, for argon Arc in the channel with diameter of 6 mm and a length of 140 mm with the current of 150 A in the

selection of the grid with the steps $\Delta z = \Delta r = 0.05$–$0.1$ mm at the initial part of the Arc and $\Delta z \equiv 0.1$–5 mm, $\Delta r = 0.1$–0.2 mm in the other zone of the Arc for sustainable solution the coefficient $\beta_T = 0.1$–0.5 is sufficient. However, for the same Arc in the selection of the grid with steps $\Delta z = \Delta r = 0.2$–$0.5$ mm at the initial part of the Arc and $\Delta z = 0.5$–10 mm, $\Delta r = 0.2$–0.5 mm in the other zone of the Arc the value of coefficient $\beta_T < 0.01$ for obtaining of the sustainable solutions. The number of iterations is also different: for the same conditions the lower values of β_T are able to provide the greater stability of the computational process, but obtaining of the convergence of solutions will be slower, i.e., the number of iterations will be greater, respectively.

8.3.2 Analysis of the Simulation Results

The results of modeling of different Arcs are different to each other in the technological purpose, structure, and the regime of operation. The input data of several types of Arcs in the channel are presented in Table 8.3. They are selected in accordance with the characteristic data of the serial industrial models of the Arc plasma torches RPM-6 [21], PV-25 [35], PN-6, PRD-9 [36]. These and other types of Arcs in the channel in practice are often used for cutting, plasma spraying, spheronization of the particles, as well as for plasma-chemical reactions and treatment of the surfaces of materials. Most of these technological processes occurs out of the nozzle of the plasma torch, in the plasma jet, the definition of the Arc parameters, including temperature and velocity at the exit of plasma torches, is of great practical importance.

Table 8.3 The input data, parameters of the regime and the technological purpose of different Arcs in the channel

Diameter of the channel d, mm	Arc current I, A	Consumption and type of gas Q, l/min	Length of the Arc l, mm	Technological purpose
3	50–200	7.5–56, air, N_2, Ar	50	Cutting
4	50–250	7.5–56, air, N_2, Ar	50	Cutting
6	50–300	7.5, 15, 25, 46, Ar	140	Spray, spheroidization
10	50–300	7.5, 15, 25, 46, Ar	140	Spray, spheroidization, treatment of the surfaces, implementation of plasma-chemical reactions
20	80–160	11, 22, 45, Ar	45	Plasma-chemical reactions, spheroidization, spray, treatment of the surfaces

The Arc for the cutting of metal is usually characterized by the small diameter of the channel ($0.4 \le d \le 4$ mm), high-current compressed flow of plasma with a high velocity of the flow (up to several kilometers per second), high temperature and the specific power of the Joule heating. These conditions are necessary for the effective destruction of the metal and at the same time making of the smallest seams and the high speed of cutting.

Arcs for the plasma spraying and for the plasma-chemical reactions and, in contrast, have a larger diameter of the channel ($d > 5$ mm) for creation of the plasma flow with a large cross-section, and favorable conditions for its interaction with material. The processed is often entered into the plasma jet through the exit of the channel of the plasma torch, so it does not disturb the flow of plasma and does not pollute the electrodes (Figs. 8.6, 8.7).

The Intensity of Electric Field in the Arcs The intensity of electric field characterizes the local allocation of energy along the length of the Arc. In the electric Arcs with the longitudinal stabilization, the intensity of electric field has the maximum value close to the input electrode and is reduced to a certain limit along the length of the Arc (Figs. 8.11, 8.12, 8.13). This describes the output of the Arc to the established zone, where the parameters of plasma remains practically constant along the length of the Arc. In general, a large number of parameters affects to the intensity of electric field, however, there are two main ones: the Arc current I and the diameter of the channel d. The data on the changes of the intensity of electric field depending on these variables are presented in Figs. 8.11, 8.12 and 8.13.

For comparison of the results obtained in Fig. 8.14 the experimental data [36] are presented, that have similar conditions with the calculations. These data correspond to the electric Arc that is burning in a cylindrical Arc channel with

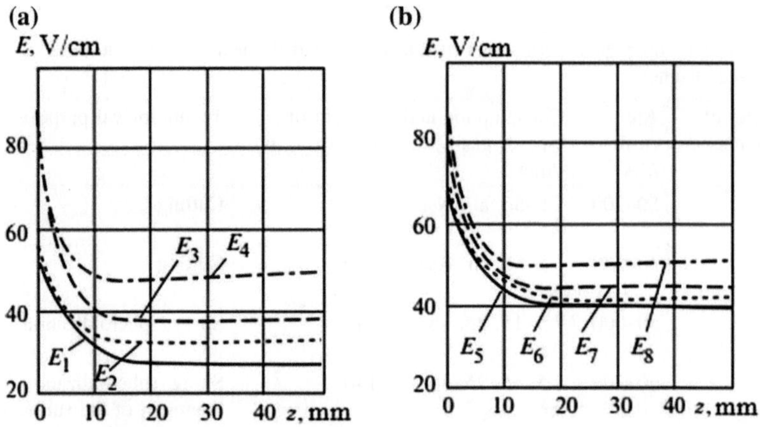

Fig. 8.11 The intensity of electric field along the length of the Arc in the channel $d = 3$ mm (**a**) and $d = 4$ mm (**b**): E_1, E_5 intensities of electric field at $I = 100$ A; E_2, E_6 the same at $I = 150$ A; E_3, E_7 the same at $I = 200$ A; E_4, E_8 the same at $I = 300$ A; gas consumption $Q_{Ar} = 37.5$ l/min

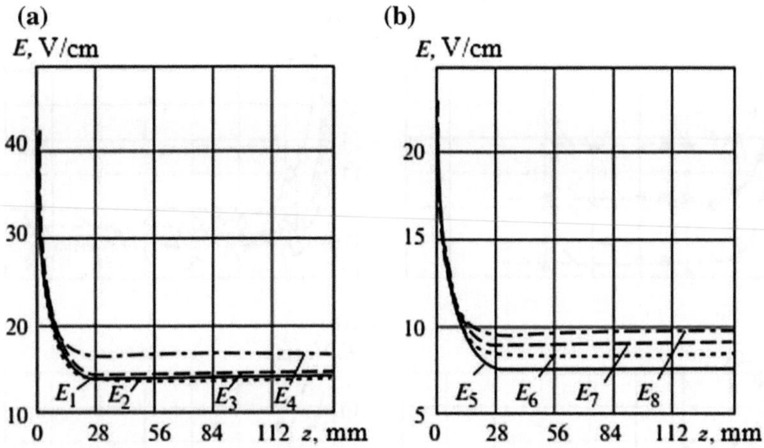

Fig. 8.12 The intensity of the electric field along the length of the Arc in the channel with $d = 6$ mm (**a**) and $d = 10$ mm (**b**): notation is the same as in Fig. 8.11

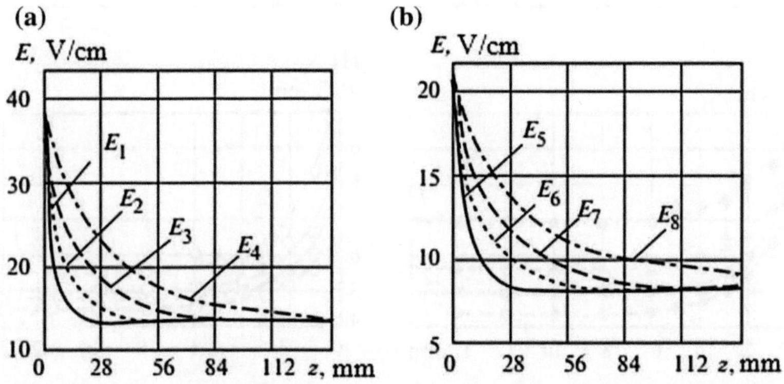

Fig. 8.13 The intensity of electric field along the length of the Arc in the channel $d = 6$ mm (**a**) and $d = 10$ mm (**b**) at $I = 150$ A: E_1, E_5 intensities of electric field at $Q_{Ar} = 7.5$ l/min; E_2, E_6 the same at $Q_{Ar} = 15$ l/min; E_3, E_7 the same at $Q_{Ar} = 25$ l/min; E_4, E_8 the same at $Q_{Ar} = 46$ l/min

sectioned inter-electrode insert with axial feed of argon. The intensity of electric field was determined by differentiating of the potential distributions along the length of the Arc, measured by the sections of the Arc channel relative to the cathode. The difference between the theoretical (Figs. 8.12, 8.13) and the experimental data (Fig. 8.14) is 5–15 %.

A constant value of the intensity of electric field in the steady zone (Figs. 8.11, 8.12, 8.13, 8.14) shows the linear dependence of the potential along the length of the Arc, and confirms that the parameters of the Arc in the channel remains

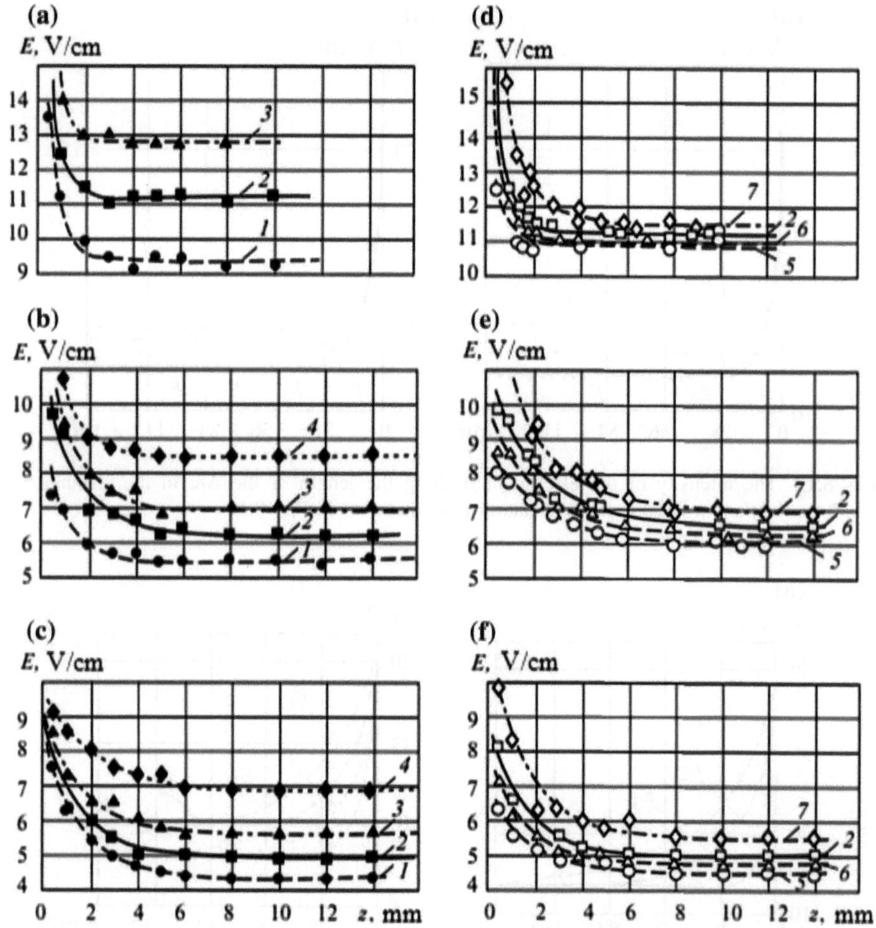

Fig. 8.14 The experimental data of the intensity of the electric field along the length of the Arc [36] at a flow rate of Argon of 25 l/min (**a–c**) and a current of 150 A (**d–f**): **a**, **d** $d = 6$ mm; **b**, **e** $d = 10$ mm; **c**, **f** $d = 12$ mm; *1* $I = 100$ A; *2* $I = 150$ A; *3* $I = 200$ A; *4* $I = 300$ A; *5* $Q_{Ar} = 9.4$ l/min; *6* $Q_{Ar} = 15$ l/min; *7* $Q_{Ar} = 46$ l/min

practically constant along its length. These results justify the use of the above described model for calculation of the Arc at the steady zone.

For the Arc in the channel the intensity of electric field at the steady area mainly depends on the current of the Arc and to a lesser degree depends on the gas consumption, the similar result was obtained in [36]. The data on the intensity of the electric field at the steady zone in dependence on the Arc current channels of various diameters are presented in Fig. 8.15. The results of calculation differ from the results of experiments, not more than 10 % that corresponds to the error of experimental methods for determination of *E*.

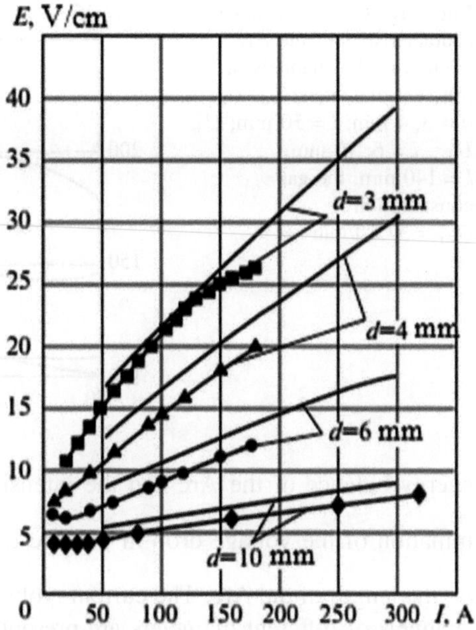

Fig. 8.15 The intensity of electric field on the steady part of the Arc: the gas consumption $Q_{Ar} = 37.5$ l/min. *lines without points* calculations; *lines with points* experimental data [37–40]

Note that the value of the intensity of electric field in the initial section depends on the set temperature distribution in the initial section, the radius of the conduction zone and the current of the Arc. Using the one-dimensional model (8.1), (8.2), the temperature field and the radius of the conductive zone in the initial section taking into account the falling characteristics of the intensity of electric field and the temperature along the length of the Arc channel (with the expansion of the Arc) can be determined. This technique allows the simple definition the boundary condition in the initial section of the Arc and to eliminate some of the errors associated with the assumption of one-dimensional motion of gas and fields. It was used for selection of the temperature field in the first cross-sectional area of the cathode.

The Current–Voltage Characteristics of the Arc The most important characteristic is the drop of the voltage in the Arc, depending on the current of the Arc. Taking into account the voltage drop in the electrode areas of these features allow to define a coherent operation of the arc plasma torch with a power source and set the range of stable operation of the generator in the regulation of its parameters. The lack of data on the local parameters of the Arc leads to the necessary of the use of published data on the approximate current–voltage characteristics of different Arcs depending on the current of the Arc, the diameter of the channel and the consumption of the gas [8, 36, 41].

Simulation of the Arc allows obtaining of the local distributions of the parameters and to establish their relationship with the external parameters of the plasma torch using the current–voltage characteristic. For the long Arc in the channel voltage drop in the Arc can be greater than in the near-electrode regions. In the model of the long Arc in the channel the simultaneous solving of equation of the

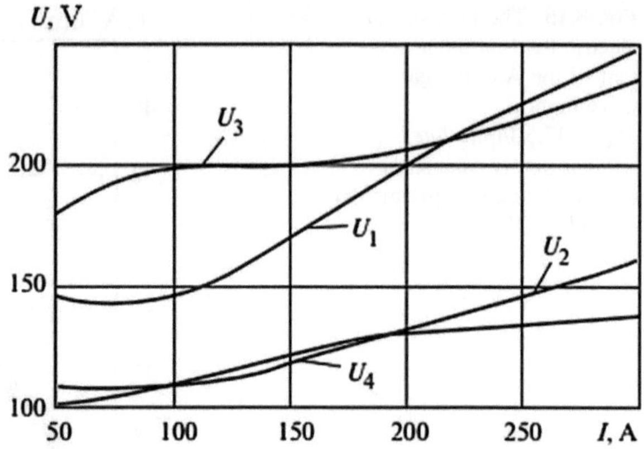

Fig. 8.16 Current–voltage characteristics of the Arc column in the channels of different diameters: U_1, U_2—$d = 3$; 4 mm, $l = 50$ mm; U_3, U_4—$d = 6$; 10 mm, $l = 140$ mm; the gas consumption $Q_{Ar} = 37.5$ l/min

energy balance of the Arc and the intensity of the electric field allows the determination of the voltage drop in the Arc using the equation $U = \int_0^l E(z)\mathrm{d}z$—where l—the length of the Arc. The current–voltage characteristics of the Arc produced in channels of different diameters are presented in Fig. 8.16.

These obtained results indicate that for the Arcs in the channel of the small diameter ($d < 4$ mm) at the currents $I < 100$ A the current–voltage characteristic can have a small decreasing zone. In the other area of the currents ($I > 100$ A), and for the Arcs of large diameter ($d > 4$ mm) the current–voltage characteristic has the increasing form (Fig. 8.16). This can be explained using the nonlinear relationship between the voltage and the current of the Arc:

$$U = \frac{I}{2\pi} \int_0^l \frac{\mathrm{d}z}{\int_0^R \sigma(T) r \mathrm{d}r}.$$

In the zone of low current and at its increase the Arc is rapidly expanding due to the sharp increase in the size of the conductive zone, leading to a possible reduction of the voltage drop in the Arc. However, from a certain value of the current of the Arc due to the presence of the wall of plasma torch, as well as a streamlined flow of gas and power losses due to the radiation and heat conduction in the radial direction, the growth of the conductive area of the Arc is constrained. So, the current–voltage characteristic of the Arc has a growing character.

The Current Density and Specified Power in the Arc Column For practical, the most important are the parameters of the Arc in the steady zone, because the technological process of the Arc are implemented using these parameters. Thus, the thermal effect of the Arc to the treated materials is determined by the heat flux and momentum, i.e., by temperature and velocity of the flow of plasma at the exit. The results of calculation of the main parameters at the exits of different channels of the

Arcs: distribution of current densities and specified powers are presented in Fig. 8.17. Total power from the current of the Arc is calculated based on the received local parameters and presented in Fig. 8.18. After reviewing the formula P = UI, we have a good agreement with the data of Fig. 8.18, indicating that the main Arc voltage drop in its long column and accuracy of simulation results.

The Fields of Temperatures and Velocities of Plasma Very pressed Arcs can have the best thermal and dynamic effects on the materials. So, for the cutting of metals due to the narrow channel of the Arc the value of current density reaches 2000–6000 A/cm^2, and the value of specific power 40–300 kW/cm^3 (Fig. 8.17a, b). Such values of these parameters provide the high temperature and the velocity of the gas flow at the exit of the Arc channel. The temperature of plasma at the exit of the channel of the Arc for the cutting of metals is (20–40) × 10^3 K, and the velocity of plasma can reach 1300–6000 m/s, that creates very favorable conditions for the cutting of metal (Fig. 8.19a, b).

At the increase in diameter of the Arc channel to 6 mm and 10 mm, there is a sharp decrease of the current density (400–1800 A/cm^2) and a specific power of the Arc (3–28 kW/cm^3) (Fig. 8.17c, d), that leads to a rare reduction of the temperature (12–24) × 10^3 K and the velocity of plasma flow (60–800 m/s) at the exit of the channel (Fig. 8.19b, c). However, a large spatial extent and high-enthalpy of plasma flow with such characteristics at the exit of the channel create more favorable conditions for such technological processes such as spheroidization of refractory particles, spraying and treatment of the surfaces, as well as the suitability of the environment for the implementation of various plasma-chemical reactions.

The temperature and velocity of plasma at the exit of the channel depending on the Arc currents and the diameter of channel are presented in Fig. 8.19, and the changes of temperature at the axis along the length of the Arc are shown in Fig. 8.20. The maximum temperature of the Arc is achieved on the axis in the initial section and at the move away from it rapidly decreases to the value of temperature at the steady zone of the Arc. From these data, it is also possible to determine the length of the initial, transient and steady zones of the Arc. Comparing the data of the temperature field and the specified powers of two models (Figs. 8.10, 8.17, 8.19) and comparing them with results of some experiments, we see a good their match. The fields of temperature and specified power obtained using the model of steady zone of the Arc are more flat, and they are wider at a lower maximum values at the axis compared with the fields obtained using the model of the long Arc in the channel. This not so big difference is explained by the relatively small role of convection and thermal conductivity in the axial direction in the energy balance of the electric Arc.

Thus, on the basis of the model of equilibrium plasma of the Arc in the channel model the main parameters of a number of electric Arcs used in practice are obtained: they are used for cutting, spraying, in plasma chemistry and for spheroidization of particles. Using these data we can determine the appropriate flows of the heat at the exit, the power losses due to radiation and heat conductivity to the wall of plasma torch, the power carried away by the flow of plasma to the jet, the efficiency of the installation and effectiveness of the technological process.

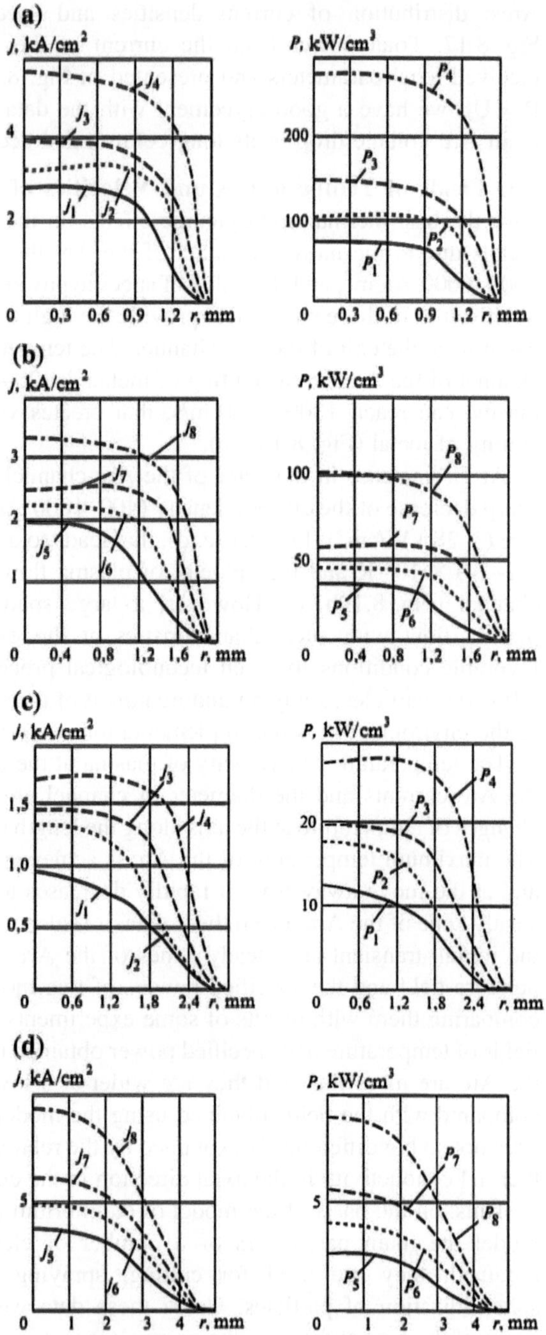

Fig. 8.17 Distribution of current density and specified power along the radius of the Arc at the steady zone of the Arc in the channels of different diameters: **a** $d = 3$ mm; **b** $d = 4$ mm; **c** $d = 6$ mm; **d** $d = 10$ mm; j_1, j_5, P_1, P_5 the current densities and specific powers at $I = 100$ A; j_2, j_6, P_2, P_6 the same at $I = 150$ A; j_3, j_7, P_3, P_7 the same at $I = 200$ A; j_4, j_8, P_4, P_8 the same at $I = 300$ A; gas consumption $Q_{Ar} = 37.5$ l/min

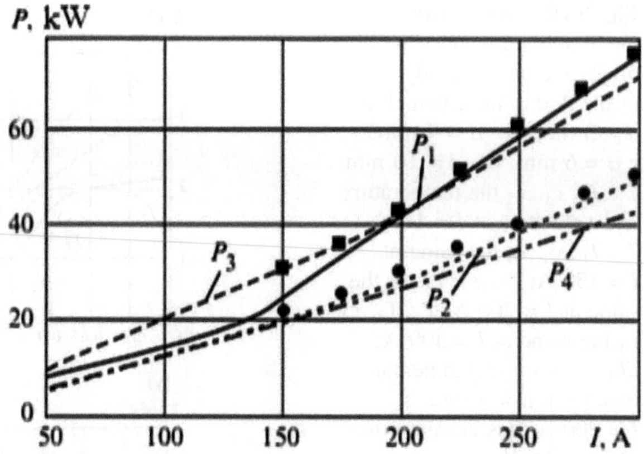

Fig. 8.18 Dependence of the total power from the current of the Arc in the channel with diameters $d = 3$; 4 mm ($l = 50$ mm, P_1 and P_2) and $d = 6$; 10 mm ($l = 140$ mm, P_3 and P_4): points—the experimental data [35]: *filled square* $d = 3$ mm; *filled circle* $d = 4$ mm; gas consumption $Q_{Ar} = 37.5$ l/min

Using these values the connection can also establish between such local parameters of the electric Arc as the temperature field and velocity field, intensity of the electric field and current density with external parameters of the Arc, such as the current and the diameter of the Arc in the channel, consumption of the plasma-forming gas, the length, and current–voltage characteristic of the Arc and others. The data about these links are extremely important in the choice of a coherent and stable operation of the Arc with the power supply, as well as for solving of the problem of direct control of the electric Arc.

8.4 Two-Dimensional Electromagnetic Problem for the Arc Plasma Torches

Calculation of the electromagnetic field in the Arc plasma torches is often carried out using the one-dimensional model with assumption that in the Arc there is only one component of the intensity of electric field E_z ($E_r = E_\varphi = 0$) and one component of the intensity of magnetic field H_φ ($H_z = H_r = 0$). Physically, this is possible only for the Arc with the cross-section constant along the length of the flow of current, or in the long compressed Arc in the channel.

In many cases, this assumption of one-dimensional electric field is not entirely justified, especially for the free-stabilized Arcs, uncompressed in the channel, for short and curved Arcs, for the Arcs with high power and active expansion of the conductive zone. It cannot be applied even for Arcs in the channel at the calculation of the initial and transition zones and the zone of attachment of the Arc to electrodes.

In particular, the change of the conductive zone of the electric Arc can lead to a second component of the intensity of electric field E_r, especially in the zones where the Arc is actively growing or shrinking. For example, in the zones of the cathode and anode anchor in the initial and transition zones of the Arc, very often the radial component of the intensity of electric field can be comparable with axial ($E_r > E_z$)

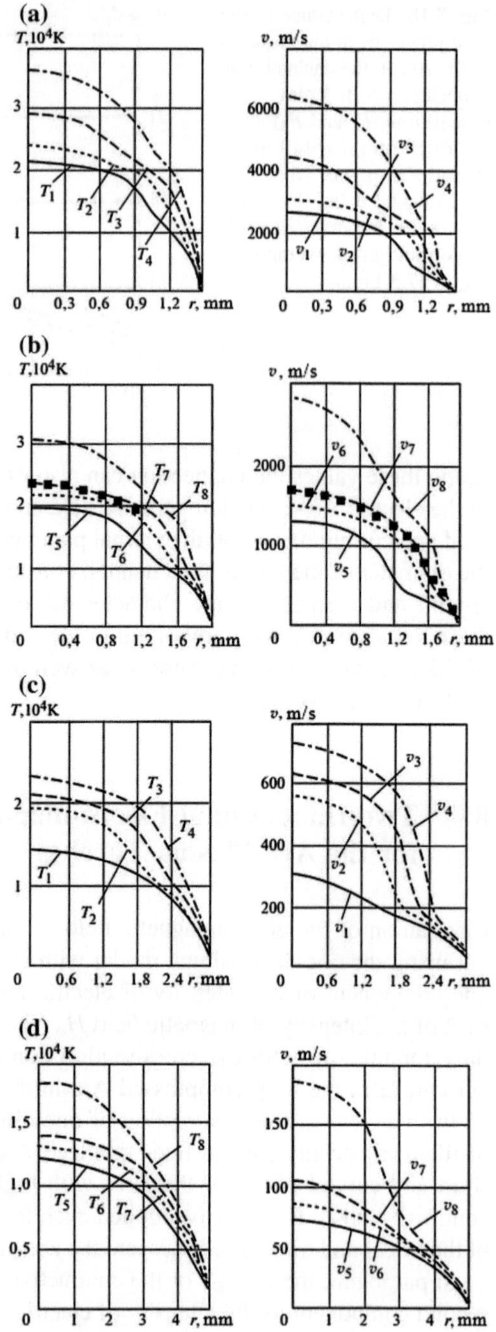

Fig. 8.19 The distribution of the temperature and velocities of plasma at the steady zone of the Arc in the channel: **a**—$d = 3$ mm; **b**—$d = 4$ mm; **c** $d = 6$ mm; **d**—$d = 10$ mm; T_1, T_5, v_1, v_5 the temperature and velocities at $I = 100$ A; T_2, T_6, v_2, v_6 the same at $I = 150$ A; T_3, T_7, v_3, v_7 the same at $I = 200$ A; T_4, T_8, v_4, v_8 the same at $I = 300$ A; *filled square* experimental data [34], $d = 4$ mm, $I = 200$ A; gas consumption $Q_{Ar} = 37.5$ l/min

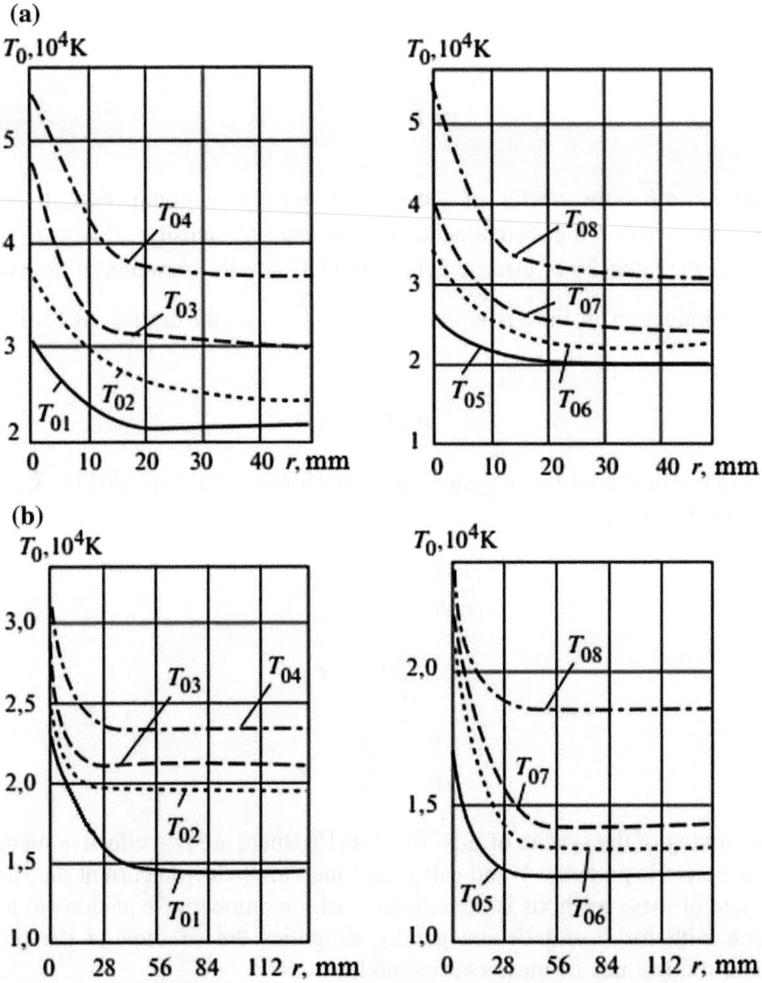

Fig. 8.20 Axial temperature distribution along the length of the plasma channel: **a**—$d = 3$ mm (T_{01}, T_{02}, T_{03}, T_{04}), $d = 4$ mm (T_{05}, T_{06}, T_{07}, T_{08}); **b**—$d = 6$ mm (T_{01}, T_{02}, T_{03}, T_{04}), $d = 10$ mm (T_{05}, T_{06}, T_{07}, T_{08}); T_{01}, T_{05}—$I = 100$ A; T_{02}, T_{06}—$I = 150$ A; T_{03}, T_{07}—$I = 200$ A; T_{04}, T_{08}—$I = 300$ A; gas consumption $Q_{Ar} = 37.5$ l/min

in its value. Moreover, in these zones, the most active processes of the electric Arc very often occur and their characteristics are largely determined by these processes. All of this suggests the need for the calculation of a two-dimensional electromagnetic field at the simulation of electric Arcs.

Two-dimensional electromagnetic problem presupposes the existence of two components of electric Arc E_z, E_r ($E_\varphi = 0$) and one component of the magnetic field H_φ ($H_z = H_r = 0$). For its productions we can use the Maxwell's system of equations for description of the field in the form of the following two equations:

$$\text{rot}\vec{H} = \vec{j}; \tag{8.11}$$

$$\text{rot}\,\vec{E} = -\frac{\partial \vec{B}}{\partial t}, \tag{8.11a}$$

Here $\vec{j} = \sigma \vec{E}$—the vector of the current density of conduction in the Arc, $j_z = \sigma E_z$, $j_r = \sigma E_r$—the components of the current density; $E = \sqrt{E_r^2 + E_z^2}$—module of the intensity of electric field. For a DC Arc the variation of the vector of magnetic induction in time is equal to zero, $\frac{\partial \vec{B}}{\partial t} = 0$ and the Eq. (8.11a) has the form:

$$\text{rot}\,\vec{E} = 0. \tag{8.12}$$

In a cylindrical coordinate system, the two-dimensional Eqs. (8.11) (8.12) have the following form:

$$\frac{1}{r}\frac{\partial}{\partial r}(rH_\varphi) = \sigma E_z; \tag{8.13}$$

$$-\frac{\partial H_\varphi}{\partial z} = \sigma E_r; \tag{8.14}$$

$$\frac{\partial E_r}{\partial z} - \frac{\partial E_z}{\partial r} = 0. \tag{8.15}$$

For solving of the system of Eqs. (8.13–8.15), there are two effective methods—using the electric potential V and using the function of electric current Φ. The main advantage of these methods is the reduction of the number of equations to a single equation with for V and Φ that greatly simplifies the solution of the problem. Consider the features of these two methods.

8.4.1 Solution Using the Electric Potential

To reduce the number of Eqs. (8.13–8.15) and bringing them to a single equation the electric potential V can be used, that is connected with the intensity of electric field: $\vec{E} = -\text{grad}V$. This is possible because for every scalar potential V from the Eq. (8.12) the equality $\text{rot}(\text{grad}V) = 0$ is always satisfied. Consequently, the components of the intensity of electric and magnetic fields can be expressed using V:

$$E_z = -\frac{\partial V}{\partial z}; \quad E_r = -\frac{\partial V}{\partial r}; \quad E = \sqrt{\left(\frac{\partial V}{\partial z}\right)^2 + \left(\frac{\partial V}{\partial r}\right)^2}; \tag{8.16}$$

$$H_\varphi = -\frac{1}{r}\int_0^r r\sigma\frac{\partial V}{\partial z}\mathrm{d}r. \tag{8.16a}$$

And the components of current density, expressed using V have the form:

$$j_z = -\sigma\frac{\partial V}{\partial z}; \quad j_r = -\sigma\frac{\partial V}{\partial r}; \quad j = \sqrt{\left(\sigma\frac{\partial V}{\partial z}\right)^2 + \left(\sigma\frac{\partial V}{\partial r}\right)^2}. \tag{8.17}$$

On the other hand, for any continuous vector $\vec{H}$ we always have $\mathrm{div}(\mathrm{rot}\vec{H}) = 0$, therefore, from the Eq. (8.11) a well-known law of continuity in the current form $\mathrm{div}\vec{j} = 0$ can be obtained. This equation is written in the two-dimensional cylindrical coordinate system:

$$\frac{\partial j_z}{\partial z} + \frac{1}{r}\frac{\partial}{\partial r}(rj_r) = 0. \tag{8.18}$$

Substituting (8.17) into (8.18), we obtain a single equation describing the field using the electric potential V:

$$\frac{\partial}{\partial z}\left(\sigma\frac{\partial V}{\partial z}\right) + \frac{1}{r}\frac{\partial}{\partial r}\left(r\sigma\frac{\partial V}{\partial r}\right) = 0. \tag{8.19}$$

The solution of (8.19) for the electrical potential is carried out using the method of control volume (Sect. 6.1.1). The differential Eq. (8.19) expresses the law of the field conservation in electric Arc for the infinitely small volume, so, as it is shown in (Sect. 6.1.1) to carry out the law of conservation of the field in the finite control volume Eq. (8.19) should be integrated over the control volume (Fig. 8.21).

$$\int_{r_s}^{r_n}\int_{z_w}^{z_e}\frac{\partial}{\partial z}\left(\sigma\frac{\partial V}{\partial z}\right)\mathrm{d}z\mathrm{d}r + \int_{z_w}^{z_e}\int_{r_s}^{r_n}\frac{1}{r}\frac{\partial}{\partial r}\left(r\sigma\frac{\partial V}{\partial r}\right)r\mathrm{d}r\mathrm{d}z = 0. \tag{8.20}$$

Under the integral of each member of Eq. (8.20) there is the first derivative of the same order, so we can accurately produce the integration of (8.20) at least once, and in the end we get:

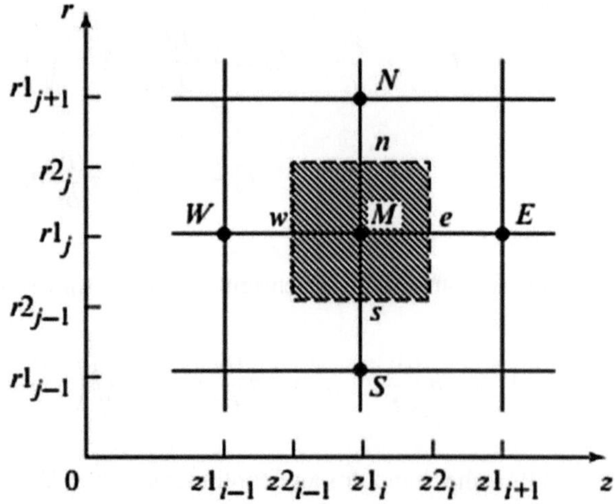

Fig. 8.21 Control volume and location of the grid points [$z1$, $r1$] (Fig. 8.22) for the solution of two-dimensional problem of electromagnetic Arc

$$\int_{r_s}^{r_n}\left(\sigma\frac{\partial V}{\partial z}\right)_e r\mathrm{d}r - \int_{r_s}^{r_n}\left(\sigma\frac{\partial V}{\partial z}\right)_w r\mathrm{d}r + \int_{z_w}^{z_e}\left(r\sigma\frac{\partial V}{\partial r}\right)_n \mathrm{d}z - \int_{z_w}^{z_e}\left(r\sigma\frac{\partial V}{\partial r}\right)_s \mathrm{d}z = 0. \tag{8.21}$$

Note that the integration is carried out along each verge of the control volume (Fig. 8.21). For the further integration of (8.21) assume that integrands take its averages in the points *e, w, n, s*, and the first derivatives are approximated using the finite differences:

$$\begin{cases} \left(\sigma\dfrac{\partial V}{\partial z}\right)_e = \sigma_e\dfrac{V_E - V_M}{z_E - z_M}; \left(\sigma\dfrac{\partial V}{\partial z}\right)_w = \sigma_w\dfrac{V_M - V_W}{z_M - z_W}; \\ \left(\sigma\dfrac{\partial V}{\partial r}\right)_n = \sigma_n\dfrac{V_N - V_M}{r_N - r_M}; \left(\sigma\dfrac{\partial V}{\partial r}\right)_s = \sigma_s\dfrac{V_M - V_S}{r_M - r_S}. \end{cases} \tag{8.22}$$

Substituting Eqs. (8.22) into (8.21), we obtain the discrete analogue for Eq. (8.19) for the interior points:

$$a_M V_M = a_E V_E + a_W V_W + a_N V_N + a_S V_S,$$

where

$$\begin{cases} a_E = \frac{\sigma_e(r_n - r_s)(r_n + r_s)}{2(z_W - z_M)} > 0; \quad a_W = \frac{\sigma_w(r_n - r_s)(r_n + r_s)}{2(z_M - z_W)} > 0; \\ a_N = \frac{\sigma_n r_n(z_e - z_w)}{r_N - r_M} > 0; \quad a_S = \frac{\sigma_s r_s(z_e - z_w)}{r_M - r_S} > 0; \\ a_M = a_E + a_W + a_N + a_S > 0. \end{cases} \tag{8.23}$$

Note that the coefficients of the discrete analogue (8.23) are positive and satisfy to the criterion of stability (6.49).

Example of the mesh for solving the two-dimensional electromagnetic problem of open Arc is shown in Fig. 8.22.

The Boundary Condition of the Equation for Electric Potential We believe that the electric potential of anode is zero ($V_a = 0$). Then the potential of all other points of the electric Arc is the relative potential compared with anode.

On the axis of the symmetry we have that the radial component of the current density is equal to zero using the condition of the cylindrical axial symmetry. Taking into account (8.17) we have: $\frac{\partial V}{\partial r}\Big|_{r=0} = 0$. At the integration of the equation for electric potential (8.19) in the half of the control volume 1 (Fig. 8.22), this boundary condition leads to the fact that the coefficient of the discrete analogue (8.23) $a_S = 0$ (taking into account the boundary condition for V on the axis, the Eq. (8.19) can be solved using the displaced grid [$z1$, $r2$] (Fig. 6.5b)]. The discrete analogue for the points on the axis of symmetry is:

$$a_M V_M = a_E V_E + a_W V_W + a_N V_N,$$

where

$$\begin{cases} a_E = \frac{\sigma_e r_n^2}{2(z_E - z_M)} > 0; \quad a_W = \frac{\sigma_w r_n^2}{2(z_M - z_W)} > 0; \quad a_N = \frac{\sigma_n r_n (z_e - z_w)}{r_N - r_M} > 0; \\ a_M = a_E + a_W + a_N > 0. \end{cases} \tag{8.24}$$

On the wall ($r = R$) for the Arc in the channel or on the free boundary of the open Arc that is located well away from the Arc it can be put that $\frac{\partial V}{\partial r}\Big|_{r=R} = 0$ or $V\big|_{r=\infty} = 0$ for open Arc.

At the cathode, the potential V should be negative and can be defined using the predetermined value of the Arc current I. If we take into account that the constant and one-dimensional intensity of the electric field E_z at the cathode can be determined using the predetermined Arc current, then using (8.16) for V we have the following boundary condition on the cathode:

$$\frac{\partial V}{\partial z}\Big|_{z=0} = -E_z\big|_{z=0} = \frac{I}{2\pi \int_0^{R_c} r\sigma dr}, \tag{8.25}$$

where R_c—the radius of the cathode spot.

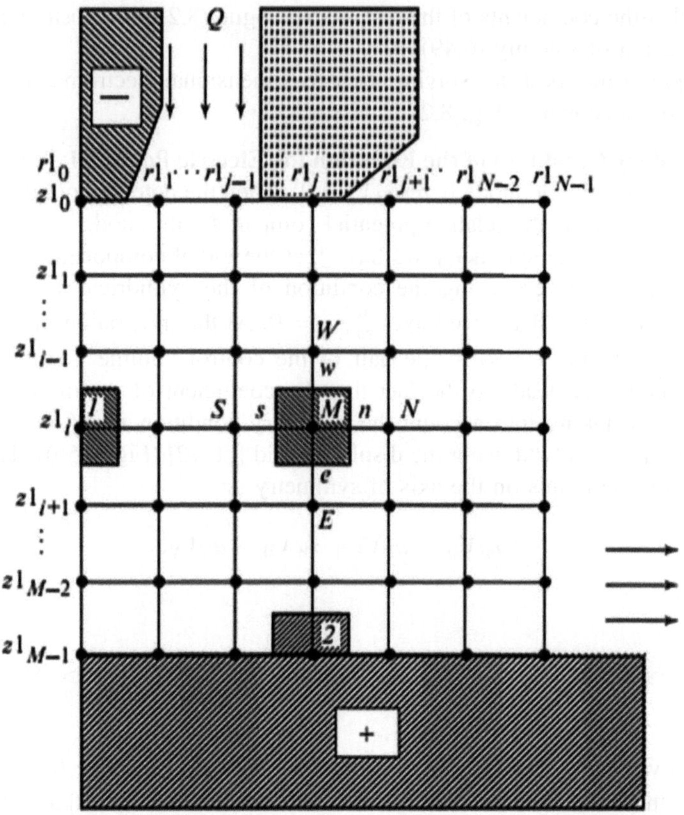

Fig. 8.22 The grid and the control volume for solving the two-dimensional electromagnetic problem for the open Arc

8.4.2 Solution Using the Function of the Electric Current

At the solving of the two-dimensional problem using the electrical potential the boundary condition for the gradient of electric potential at the cathode (8.25) is not very convenient for obtaining of the appropriate discrete analogue. Usually, the approximation of the gradient V leads to an additional term in the discrete analogue for points on the cathode that is not desirable. On the other hand, the approach approximation of the gradient $\frac{\partial V}{\partial z}$ in a narrow zone of the cathode spot with a high temperature gradient can lead to an inaccurate definition of V on the cathode border and affect to the result of calculation of the field in the Arc zone. In this regard, consider the second method for solving of the two-dimensional system of electromagnetic Eqs. (8.13–8.15) using the function of electric current Φ, in this case the boundary conditions is easily implemented and does not affect to the accuracy of the solution.

The function of an electric current express the law of total current for the cross-section of the Arc and can be used for the reduction of the number of equations that are necessary for solving of the two-dimensional electromagnetic problem of the electric Arc. The function Φ is also an excellent feature for the analysis of the field of the current flow in the electric Arc and one of the main parameters in the study and design of Arc plasma torches. In a cylindrical coordinate system, the function of the electric current Φ can be expressed using the intensity of the magnetic field H_φ:

$$\Phi = rH_\varphi. \tag{8.26}$$

Accordingly, the component of the intensity of the magnetic field can be determined using the function Φ:

$$H_\varphi = \Phi/r. \tag{8.27}$$

Substituting (8.27) into (8.13) and into (8.14), we obtain new equations for calculation of the components of the current density, expressed using the function Φ:

$$j_z = \sigma E_z = \frac{1}{r}\frac{\partial \Phi}{\partial r}; \quad j_r = \sigma E_r = -\frac{1}{r}\frac{\partial \Phi}{\partial z}. \tag{8.28}$$

In this case it is possible to express the components of the electric field E_z and E_r using the function Φ:

$$E_z = \frac{1}{\sigma r}\frac{\partial \Phi}{\partial r}; \quad E_r = -\frac{1}{\sigma r}\frac{\partial \Phi}{\partial z}. \tag{8.29}$$

Substituting (8.29) into (8.15), we obtain a new equation for description of the field using the function of the electric current Φ, that is symmetric for the derivatives for z and r.

$$\frac{\partial}{\partial z}\left(\frac{1}{\sigma r}\frac{\partial \Phi}{\partial z}\right) + \frac{\partial}{\partial r}\left(\frac{1}{\sigma r}\frac{\partial \Phi}{\partial r}\right) = 0 \tag{8.30}$$

Thus, using function of electric current Φ we can also reduce the system of field Eqs. (8.13–8.15) to a single equation.

Discrete Analogue for the Function of the Electric Current The numerical solution of the Eq. (8.30) can also be produced using the method of the control volume (Sect. 6.1.1). For maximize effective of the boundary conditions the Eq. (8.30) should be solved using the grid $[z1, r1]$ (Fig. 8.22), border points of it are located directly on the cathode ($z = 0$), at the axis of symmetry ($r = 0$), on the wall or at the free boundary ($r = R$).

To implement the law of the conservation of the field in the finite control volume it is necessary to integrate Eq. (8.30) over the control volume (Fig. 8.21).

$$\int_{r_s}^{r_n}\int_{z_w}^{z_e}\frac{\partial}{\partial z}\left(\frac{1}{\sigma r}\frac{\partial \Phi}{\partial z}\right)\mathrm{d}z r\mathrm{d}r+\int_{z_w}^{z_e}\int_{r_s}^{r_n}\frac{\partial}{\partial r}\left(\frac{1}{\sigma r}\frac{\partial \Phi}{\partial r}\right)r\mathrm{d}r\mathrm{d}z=0.$$

We obtain the following equation after the first integrating:

$$\int_{r_s}^{r_n}\left(\frac{1}{\sigma}\frac{\partial \Phi}{\partial z}\right)_e\mathrm{d}r-\int_{r_s}^{r_n}\left(\frac{1}{\sigma}\frac{\partial \Phi}{\partial z}\right)_w\mathrm{d}r+\int_{z_w}^{z_e}\left(\frac{1}{\sigma r}\frac{\partial \Phi}{\partial r}\right)_n\frac{r_n+r_s}{2}\mathrm{d}z$$
$$-\int_{z_w}^{z_e}\left(\frac{1}{\sigma r}\frac{\partial \Phi}{\partial r}\right)_s\frac{r_n+r_s}{2}\mathrm{d}z=0.$$

For the further solving the following finite-difference equations of approximation are used:

$$\left.\frac{\partial \Phi}{\partial z}\right|_e=\frac{\Phi_E-\Phi_M}{z_E-z_M};\quad \left.\frac{\partial \Phi}{\partial z}\right|_w=\frac{\Phi_M-\Phi_W}{z_M-z_W};\quad \left.\frac{\partial \Phi}{\partial r}\right|_n=\frac{\Phi_N-\Phi_M}{r_N-r_M};\quad \left.\frac{\partial \Phi}{\partial r}\right|_s=\frac{\Phi_M-\Phi_S}{r_M-r_S}.$$

Substituting these formulas into the last equation, we get the discrete analogue for Φ in the form:

$$a_M\Phi_M=a_E\Phi_E+a_W\Phi_W+a_N\Phi_N+a_S\Phi_S,$$

where

$$\begin{cases} a_E=\frac{r_n-r_s}{\sigma_e(z_E-z_M)}>0;\quad a_W=\frac{r_n-r_s}{\sigma_w(z_M-z_W)}>0;\\ a_N=\frac{(r_n+r_s)(z_e-z_w)}{2\sigma_n r_n(r_N-r_M)}>0;\quad a_S=\frac{(r_n+r_s)(z_e-z_w)}{2\sigma_s r_s(r_M-r_S)}>0;\\ a_M=a_E+a_W+a_N+a_S>0. \end{cases} \tag{8.31}$$

All the coefficients of the discrete analogue of (8.31) are positive and satisfy the stability criterion (6.49).

The Boundary Conditions of the Equation for the Functions of Electric Current On the axis of the plasma torch $\Phi|_{r=0}=0$ (8.26).

For the cathode the existence of only one component of the electric and magnetic fields $E_z(r)=\mathrm{const}$ and H_φ, we have from (8.28):

$$\Phi(r)|_{z=0} = \int_0^r \sigma E_z r\mathrm{d}r = \frac{I\int_0^r \sigma r\mathrm{d}r}{2\pi \int_0^{R_c} \sigma r\mathrm{d}r},$$

Here, R_c—the radius of the conductive zone of the Arc on the cathode (the radius of the cathode spot).

On the wall of the Arc channel or at the free boundary of open Arc ($r = R$), where there is no the conduction current, and from (8.28) we have $\Phi(R,z) = \int_0^R \sigma E_z r\mathrm{d}r = \frac{I}{2\pi} = \mathrm{const}$. This function of the electric current should be obtained out of the conductive zone of the Arc, and it is an excellent criterion for testing of the accuracy of field calculation.

We believe that at the anode there is only one component of the intensity of the electric field $E_z(r) = \mathrm{const}$, $E_r = 0$. From (8.29) we obtain $\frac{\partial \Phi}{\partial z}\Big|_{z=L} = 0$. The discrete analogue for the points on the anode is obtained by integrating Eq. (8.30) for the half of the control volume 2 (Fig. 8.22).

Taking into account the boundary conditions at the anode, the discrete analogue for these points is:

$$a_M \Phi_M = a_W \Phi_W + a_N \Phi_N + a_S \Phi_S,$$

where

$$\begin{cases} a_W = \frac{r_n - r_s}{\sigma_w (z_M - z_W)} > 0; \quad a_N = \frac{(r_n + r_s)(z_M - z_w)}{2\sigma_n r_n (r_N - r_M)} > 0; \\ a_S = \frac{(r_n + r_s)(z_M - z_w)}{2\sigma_s r_s (r_M - r_S)} > 0; \quad a_M = a_W + a_N + a_S > 0. \end{cases} \tag{8.32}$$

Equations for Calculation of the Two-Dimensional Field of the Electric Arc
After solving of the equation for the electric potential V (8.19) or the function of electric current Φ (8.30), the obtained field of V or Φ can be used for determination of the parameters of the field of the electric Arc and their distribution using the following equations:

1. The components of the intensity of the electric field are determined using Eqs. (8.16) or (8.29).
2. The intensity of the magnetic field is determined using Eqs. (8.16a) or (8.27).
3. The components of the current density are determined using Eqs. (8.17) or (8.28).
4. Specific allocated power of the Arc is determined as:

$$P(r,z) = \sigma E^2 = \sigma\left[\left(\frac{\partial V}{\partial z}\right)^2 + \left(\frac{\partial V}{\partial r}\right)^2\right] = \frac{1}{\sigma r^2}\left[\left(\frac{\partial \Phi}{\partial r}\right)^2 + \left(\frac{\partial \Phi}{\partial z}\right)^2\right]. \quad (8.33)$$

5. The components of the electromagnetic force are determined as:

$$\left.\begin{aligned} F_z(r,z) &= \mu_0 \sigma E_r H_\varphi = \mu_0 \tfrac{\sigma}{r}\tfrac{\partial V}{\partial r}\int_0^r r\sigma \tfrac{\partial V}{\partial z}\,\mathrm{d}r = -\mu_0 \tfrac{\Phi}{r^2}\tfrac{\partial \Phi}{\partial z}; \\ F_r(r,z) &= -\mu_0 \sigma E_z H_\varphi = -\mu_0 \tfrac{\sigma}{r}\tfrac{\partial V}{\partial z}\int_0^r r\sigma \tfrac{\partial V}{\partial z}\,\mathrm{d}r = -\mu_0 \tfrac{\Phi}{r^2}\tfrac{\partial \Phi}{\partial r}. \end{aligned}\right\} \quad (8.34)$$

6. The total power allocated in the Arc is determined as:

$$\begin{aligned} P &= 2\pi\int_0^l\int_0^R \sigma E^2 r\mathrm{d}r\mathrm{d}z = 2\pi\int_0^l\int_0^R \sigma\left[\left(\frac{\partial V}{\partial z}\right)^2 + \left(\frac{\partial V}{\partial r}\right)^2\right] r\mathrm{d}r\mathrm{d}z \\ &= 2\pi\int_0^l\int_0^R \frac{1}{\sigma r^2}\left[\left(\frac{\partial \Phi}{\partial r}\right)^2 + \left(\frac{\partial \Phi}{\partial z}\right)^2\right] r\mathrm{d}r\mathrm{d}z, \end{aligned} \quad (8.35)$$

Here l and R—length and radius of the Arc.

7. Strictly speaking, from the formulation of the two-dimensional problem of electromagnetic Arc the voltage drop at the Arc has the distribution along the radius of the Arc. However, as in the one-dimensional problem we take the total value of the voltage drop as a voltage drop on the axis of the Arc. It is defined as:

$$U = \int_0^l E_z(0,z)\mathrm{d}z = \int_0^l \left(\frac{1}{\sigma r}\frac{\partial \Phi}{\partial r}\right)\bigg|_{r=0} \mathrm{d}z. \quad (8.36)$$

By solving Eqs. (8.13–8.15) for the electric potential V the absolute value in the center of the cathode spot is a voltage drop in the Arc (the electrical potential at the anode is taken equal to 0).

8. On the axis of the Arc plasma torch from the condition of the cylindrical axial symmetry we have the following relations: $E_r = 0$, $j_r = \sigma E_r = 0$, $\frac{\partial \Phi}{\partial z} = \frac{\partial \Phi}{\partial r} = H_\varphi = 0$, $\Phi = 0$, $F_z = F_r = 0$.
9. From Eq. (8.15) we have $\frac{\partial E_z}{\partial r} = \frac{\partial E_r}{\partial z}$. Out of the conductive zone of the Arc we have: $E_r = E_z = 0$, $\Phi = \frac{I}{2\pi}$, $F_z = F_r = j_z = j_r = P = 0$.

Thus, using the electric potential V or the function of the electric current Φ we are able to reduce the number of equations to be solved to one equation for V or Φ. As a result of the mathematical decision appropriate discrete analogue is obtained in the form (8.23) or (8.31). In each case the boundary conditions and the corresponding discrete analogues for points at the borders are considered. We can also obtain the necessary algorithms for determination of the two-dimensional electromagnetic field of the electric Arc. As an example, we use the proposed methodology for calculating of two-dimensional electromagnetic field of open Arc.

8.4.3 Calculation of the Electromagnetic Field of the Arc Plasma Torches

The electromagnetic field of the Arc plasma torches depends on many factors, but it largely depends on the current, diameter and the length of the channel of discharge, type and regime of the gas flow and other conditions. For the calculation of electromagnetic field of the Arc it is necessary to know the temperature field. The temperature field is determined from the solving of the equation of the balance energy that in turn can be solved if the electromagnetic field, the velocity field of the Arc and the pressure field the composition of the plasma are known for the determination of the properties of plasma, etc. Thus, the calculation of the electromagnetic field of the Arc should be carried out with the calculation of the temperature field, velocity, pressure, composition, transport coefficients and thermodynamic functions of plasma, etc., i.e., the calculation of the electromagnetic field is the part of a complex the task of mathematical modeling of plasma processes in electric Arc. The found results of the calculation of the fields, so are correct, because they satisfy not only the Maxwell's equations, but also all the laws of conservation: of energy, momentum, mass and others.

At this stage, we believe that the temperature field of the Arc is already set correctly (it was actually found from the joint result of simulation of all the physical processes in electric Arc), and determine the corresponding intensities of electromagnetic field. Here, according to the above described method of determining the intensity of the two-dimensional electromagnetic field of the electric Arc using the function of the electric current Φ we analyze the results of calculation of the field of open Arc for welding using the current I = 90 A and 160 A, the plasma-forming gas is Argon, the gas consumption—Q_{Ar} = 20 l/min, the diameter of the nozzle d_n = 14 mm, the length of the Arc l = 18 mm. Open Arc between vertically arranged cathode and the anode has a cylindrical axial symmetry. The scheme of such Arc, and obtained field of a function of the electric current are shown in Fig. 8.23.

From the obtained field of the electric current (Fig. 8.23), it can be seen that at the initial and transition areas the free open Arc is actively expanding. A similar pattern (narrowing of the electric Arc) is observed at the anode, that in some case is determined by the form of attachment of the Arc to the anode (contracted or diffusion). In the zone where there is uniform distribution of the function of the electric current along the radius, we have the zero value of the current density of the Arc according to the Eq. (8.28). On this basis, the radius of the zone of the current flow can be determined using the cross-sections of the electric Arc. In this case, for the current I = 90 A this area of the Arc is bounded with the function of electric

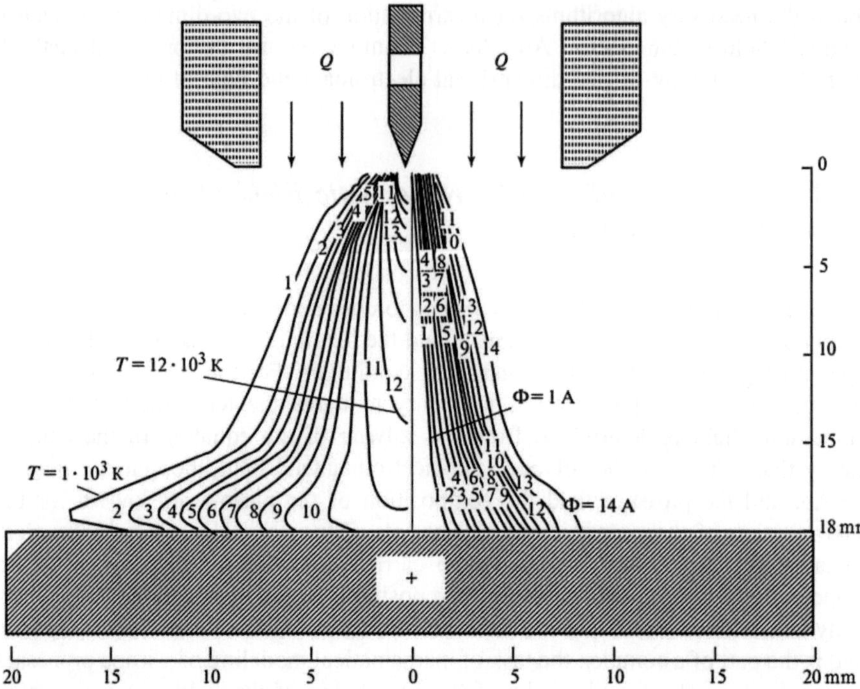

Fig. 8.23 The fields of temperature and the function of electric current of the welding Arc (Q_{Ar} = 20 l/min, I = 90 A)

current Φ = 14.3 A (Fig. 8.23). If we compare the field of the current flow with a field of the Arc temperature (Fig. 8.23), then for Argon plasma the boundary of the current flow roughly coincides with the line of the isotherm T = 7000 K, at this temperature the electrical conductivity $\sigma \sim$ 3–4 Ω^{-1} cm^{-1}. These conclusions are also confirmed in [8, 32].

The maximum value of the intensity of the electric field E_z = 22.9 and 23.64 V/cm is achieved in the initial section of the Arc at the regimes of operation with the currents I = 90 and 160 A, respectively. At the cathode the conductive zone is minimal, but the value of the intensity of the electric field is maximal. With the expansion of the Arc from the cathode to the anode the maximal value on the axis reduces, but the zone of the electric field is extended (Fig. 8.28) for keeping unchanged current in the cross-section of the Arc. The maximal value of the radial component of the electric field (E_r = 7.2 V/cm) is achieved in the zone of active extension of the Arc current near the cathode (Fig. 8.25).

From the comparison of the obtained distributions E_z and E_r (Figs. 8.24, 8.25) it can be seen that E_r can be up to $\sim$ 30 % of E_z and more in the initial and transitional areas of the free open Arc. This means that the components of the current density j_r and the electromagnetic force F_z and will also be significant in comparison with the basic components j_z and F_r. So, the pattern of current flow

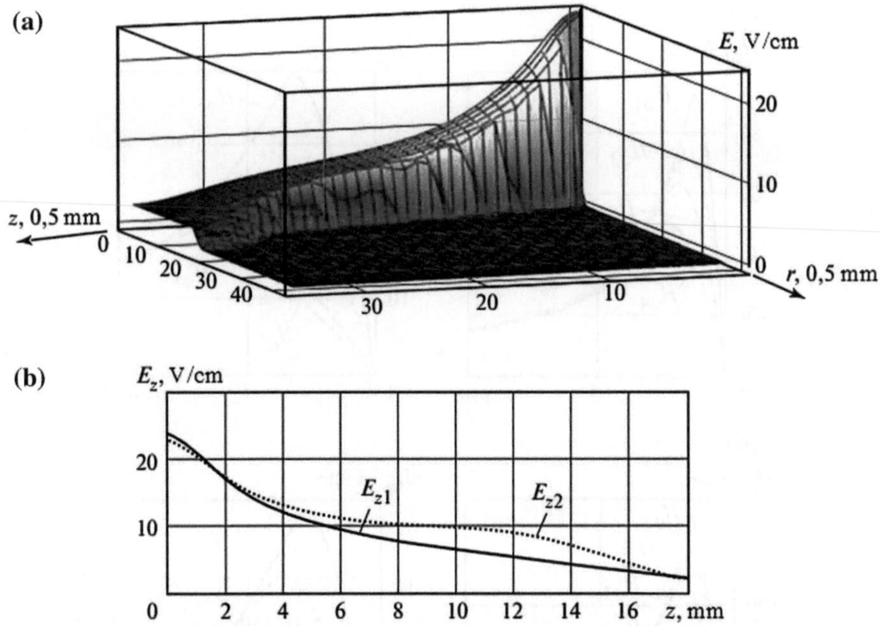

Fig. 8.24 The distribution of the intensity of the electric field in Argon welding Arc at $I = 90$ A (**a**) and its changes on the axis E_{z1}, E_{z2} (**b**): E_{z1} the intensity of the electric field at $I = 90$ A, E_{z2} the same at $I = 160$ A

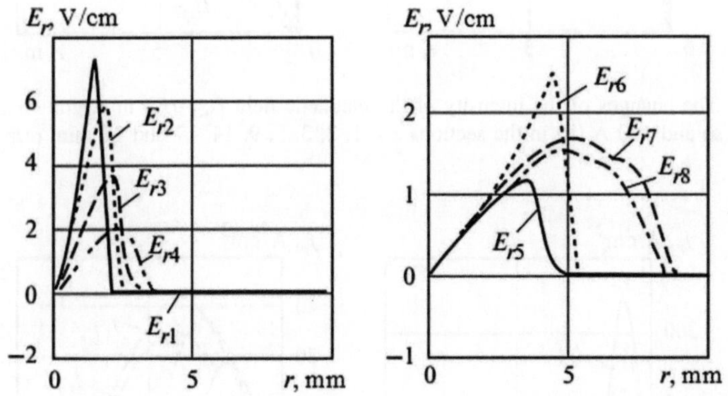

Fig. 8.25 The changes of the radial components of the intensity of the electric field in Argon welding Arc at $I = 90$ A in the sections $z = 1, 2, 3, 5, 9, 14, 17$ and 18 mm, respectively

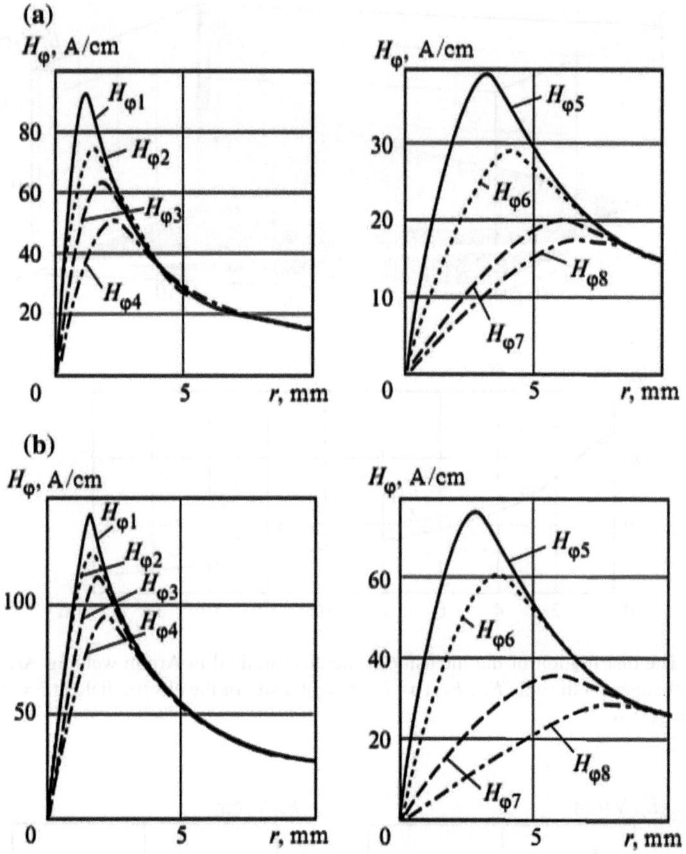

Fig. 8.26 The changes of the intensity of the magnetic field $H_{\varphi 1}$–$H_{\varphi 8}$ in Argon welding Arc at $I = 90$ A (**a**) and 160 A (**b**) in the sections z = 1, 2, 3, 5, 9, 14, 17 and 18 mm, respectively

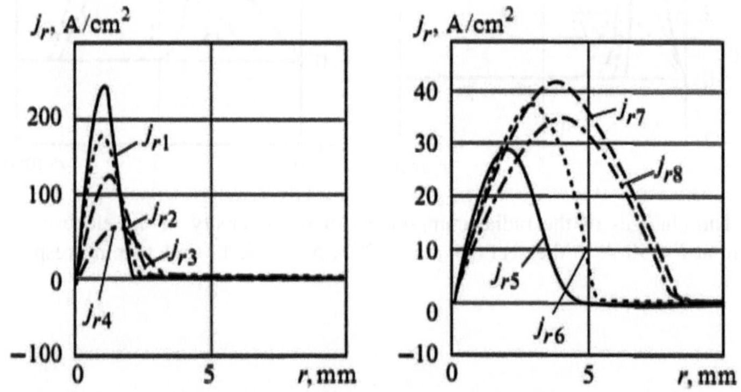

Fig. 8.27 The changes of the radial components of the current density j_{r1}–j_{r8} in Argon welding Arc at $I = 90$ A in the sections z = 1, 2, 3, 5, 9, 14, 17 and 18 mm, respectively

(Fig. 8.23), the released energy (Fig. 8.29), and the distribution of electromagnetic force (8.30), that directly affect to the velocity of heating and acceleration of plasma in the open free Arc will be significantly distorted if we neglect the radial

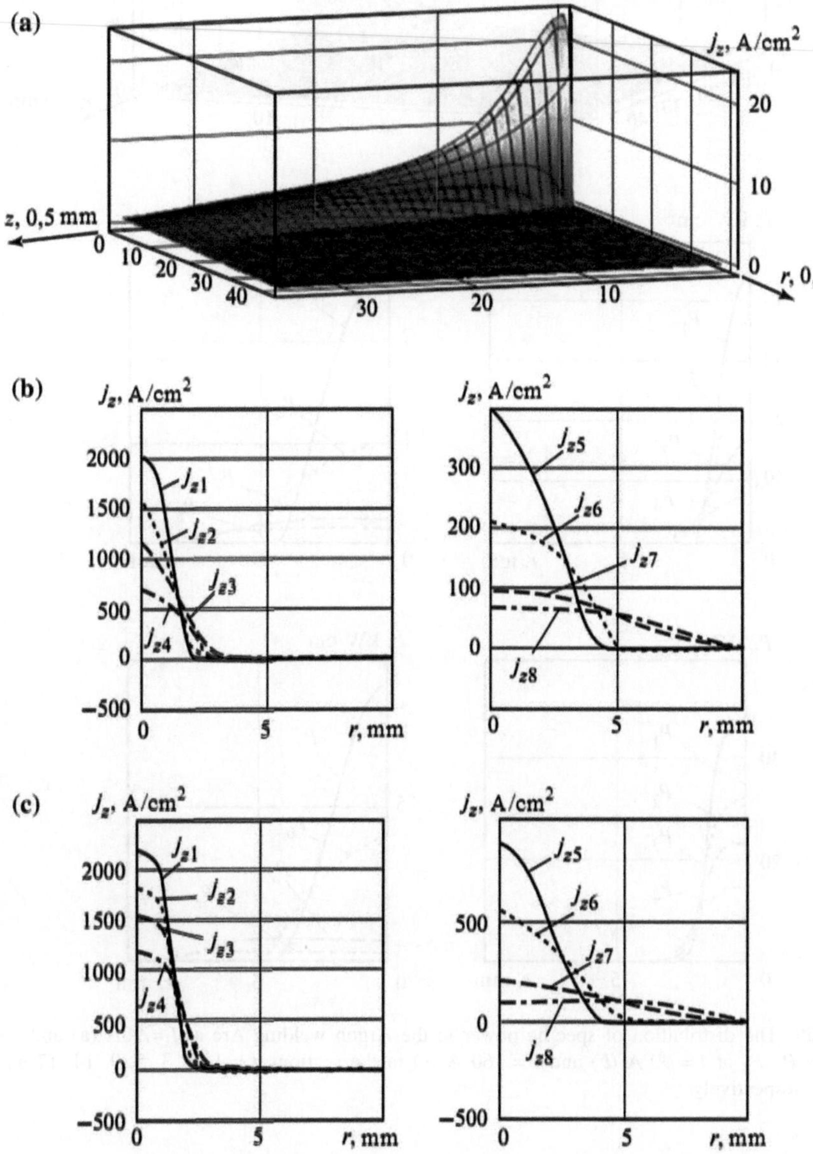

Fig. 8.28 The distribution of the axial component of the current density in Argon welding Arc I = 90 A (**a**) and the changes of j_{z1}–j_{z8} at I = 90 A (**b**) and 160 A (**c**) in the sections z = 1, 2, 3, 5, 9, 14, 17 and 18 mm, respectively

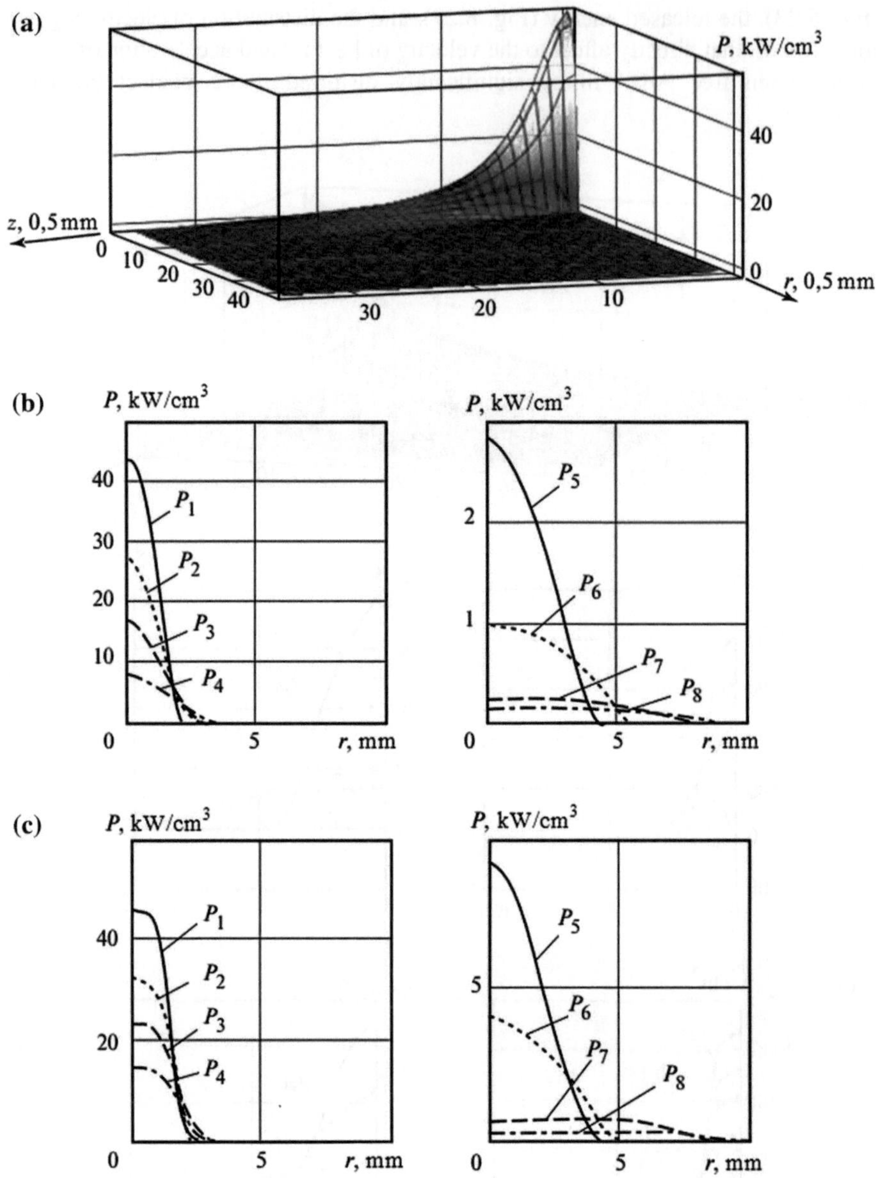

Fig. 8.29 The distribution of specific power in the Argon welding Arc at $I = 90\text{A}$ (**a**) and the changes P_1–P_8 at $I = 90$ A (***b***) and $I = 160$ A (**c**) in the sections $z = 1, 2, 3, 5, 9, 14, 17$ and 18 mm, respectively

component of the intensity of the electric field that proves the necessity of a two-dimensional setting of the problem of the field calculation in electric Arc.

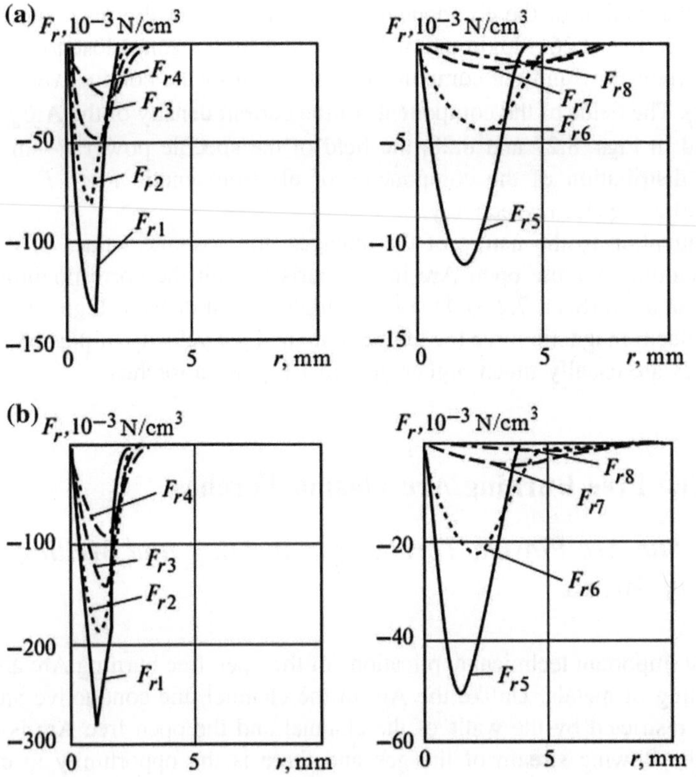

Fig. 8.30 The distribution of radial components of the electromagnetic force F_{r1}–F_{r8} in the Argon welding Arc at I = 90 A (**a**) and I = 160 A (**b**) in the sections z = 1, 2, 3, 5, 9, 14, 17 and 18 mm, respectively

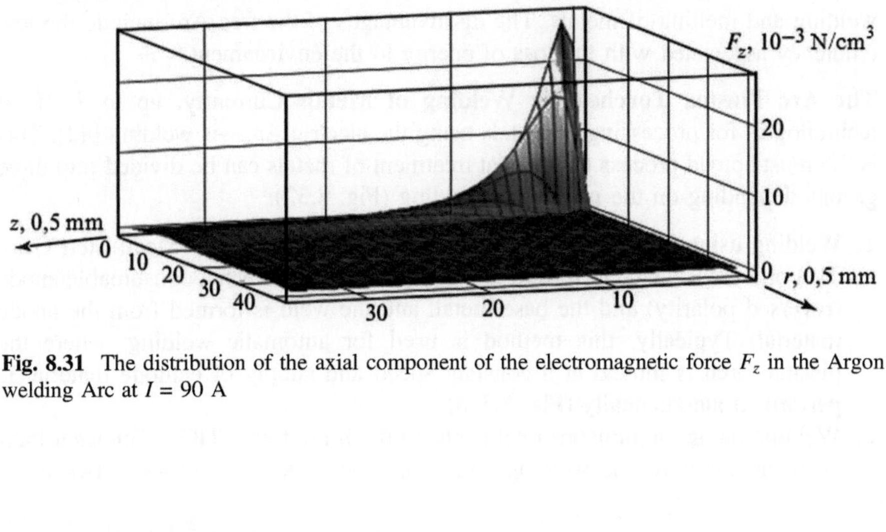

Fig. 8.31 The distribution of the axial component of the electromagnetic force F_z in the Argon welding Arc at I = 90 A

The distribution of the components of the intensity of the magnetic field H_φ is presented in Fig. 8.26. Maximum of H_φ is achieved in a zone close to the boundary of the current flow and the component reduces out of the conductive zone by law $H_\varphi = \frac{I}{2\pi r}$. The fields of the components of the current density of the Arc j_r and j_z are presented in Figs. 8.27 and 8.28, the field of the specific power P—in Fig. 8.29, and the distribution of the components of electromagnetic force F_r and F_z—in Figs. 8.30 and 8.31, respectively.

Pay attention to the nature of the changes, the absolute values of the ratio of these quantities for the open Arc in comparison with the corresponding data for RF-plasma torch (Sect. 7.2.3). Due to the high current density, high specific power and the electromagnetic force the temperature and the velocity of plasma flow in the open Arcs are usually much higher than in RF-plasma torches.

8.5 The Free Burning Arc Plasma Torches

8.5.1 The Arc Plasma Torches for Welding and Melting of Metals

The most important technical applications of the open free burning Arc are welding and melting of metals. Unlike the Arc in the channel, the conductive area of it is severely restricted by the walls of the channel and the open free Arc is stabilized due to the flowing stream of the gas and there is the opportunity to expand its conductive area that leads to relatively low current density and released power. The temperature and velocity of plasma in the free Arc are typically lower than the temperature and velocity of the Arc in the channel and are $T \approx (10\text{–}20) \times 10^3$ K and several tens of meters per second [8, 9]. Lack of the wall of the channel greatly simplifies the construction of the plasma torch and creates favorable conditions for welding and melting of metals. The disadvantages of the free Arc include the low efficiency associated with the loss of energy to the environment.

The Arc Plasma Torches for Welding of Metals Currently, up to 80 % of technologies for processing of metals using the electric Arc—is welding [42]. This is the most spread process of the heat treatment of metals can be divided into three groups depending on the method of welding (Fig. 8.32):

1. Welding using the consumable electrode in inert gas (MIG—Metal Inert Gas). This process is characterized in that the Arc burns between a consumable anode (reversed polarity) and the base metal, and the weld is formed from the anode material. Typically, this method is used for automatic welding, where the plasma torch is moved at a constant speed and supply of cathode material is performed automatically (Fig. 8.32a).
2. Welding using the nonconsumable electrode in inert gas (TIG—Tungsten Inert Gas). In this way, the welding Arc is similar to Arc for cutting between the

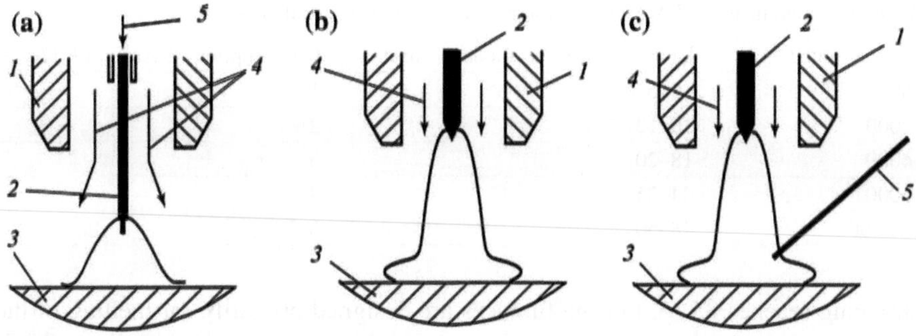

Fig. 8.32 Schemes of the Arc welding methods: **a** using the consumable electrode, **b** using the nonconsumable electrode, **c** the same using the metal additive; *1* nozzle, *2* electrode, *3* welded products, *4* plasma–forming gas, *5* metal additive

nonconsumable (tungsten) cathode and anode of the welded metal. The Arc in this regime significantly differs from the Arc for cutting. This difference is due to the fact that in the welding of metals the melting of the metal is only required and spattering of it is unacceptable, so the welding Arc has lower speeds (30–150 m/s) and lower temperatures $T \approx (10\text{–}15) \times 10^3$ K. At the welding using the nonconsumable electrode (Fig. 8.32b, c) weld seam is formed from the molten metal (with a small thickness of the metal) or from the metal additives, that is fed into the welding area (with a large thickness of the welded metal). This welding method can be applied for welding of metals that cannot be welded by the other means (titanium, aluminum).

3. Welding of metals using the submerged consumable electrode.

Similar to the process of cutting of metals the efficiency of welding is largely determined by the speed of the process. At the same time the requirements for the regime of the Arc are tougher because of the necessity of forming the weld of specified quality that is determined not only by its geometry, but also by the chemical composition of the weld metal. In all developed countries in the recent decades the number of equipment and devices for plasma welding has been created.

The Arc Plasma Torches for Melting of Metals Melting Arc plasma torches are plasma torches with the Arc of direct action. One of their spots is out of the molten metal. The heat released in the anode spot and to a lesser extent, the radiation energy of the Arc is also useful.

Currently, there are the industrial examples of the plasma torches operating at the currents lower than 10 kA. Some data on plasma torches for melting of metals are given in Table 8.4 (the cathode is made of thoriated or lantaniumed tungsten). The DC Arc plasma torches of several thousand of amperes are used for the melting. The voltage of the Arc depends on the composition, consumption of the gas and its length in the range of the voltage is 30–150 V and higher. Very often, for increase the power of the furnace for melting of the metal in one bath several plasma torches operates in parallel. Efficiency of plasma torches during the

Table 8.4 Parameters of Arc plasma torches for melting of metals [1]

Arc current *I*, A	Diameter of the cathode d_k, mm	Consumption of Argon *G*, g/s
800	5–6	1
2000	10–12	2
4000	18–20	3
5000	24–25	4
10,000	48–50	6

surfacing reaches 85 %. Plasma furnaces are designed primarily for melting of the steels, but can also be used for melting non-ferrous metals and their alloys. Their main advantage over conventional Arc steel-melting furnaces is the absence of graphite electrodes that eliminates carbonization of metal and allows smelting of low carbon steels.

Another important advantage of these furnaces is the opportunity to seal the furnace space. For graphite electrodes it was not possible to create a sealed input device into the furnace, in contrast with it the sealing of metal water-cooled housings of plasma torches have no the great difficulties. This fact allows you to maintain any predetermined atmosphere in the furnace space in the process of melting (neutral, oxidizing or reducing), and change it for a given program. Last not least important feature of the plasma furnaces is a significant reduction in pollution of the environment, and improving of the working health. Plasma furnaces almost operate silently, unlike the Arc ones. On the other hand, the cost of melting is higher than for the Arc, due to the consumption of Argon and large power consumption due to the greater number of water-cooled units and higher temperatures for the furnace lining. Therefore, the use of plasma furnaces is most advisable in the smelting of high-strength steels and alloys hen the reduction of the consumption of alloying elements by reducing of the burn has the determining value.

Currently, plasma furnaces are being built that in the form of furnace space are like the Arc steel-melting furnaces (Fig. 8.33). At the using of a single plasma torch strengthen it is stayed vertically in the roof (Fig. 8.33a). Several plasma torches can be positioned vertically or obliquely in the side walls (Fig. 8.33b).

As an example, the main there are parameters of industrial plasma furnaces with a capacity of 5, 10 and 30 t that are in operation at plants in Russia and Germany. All furnaces are equipped with a DC Arc plasma torches, and the melting space is sealed so that the overpressure is hold on in it during the melting process.

The furnace with capacity of 5 t has one plasma torch, introduced vertically through the roof. Power of the plasma torch is 3500 kW. Energy consumption for melting is about 700 kWh/t. Argon consumption is 200–250 l/min. The lifetime is over 70 melting and the hearth—150–160 heats. The furnace with capacity of 10 t is equipped with three Arc plasma torches, entered through the side walls of the furnace. Plasma torches are mounted on special mechanisms that allow axial and angular move them. Each plasma torch is powered by a separate power supply thyristor with a wide control range. Operating current of the plasma torches is 6 kA at the voltage of 200–600 V. The real power introduced into the furnace during the melting process 7–

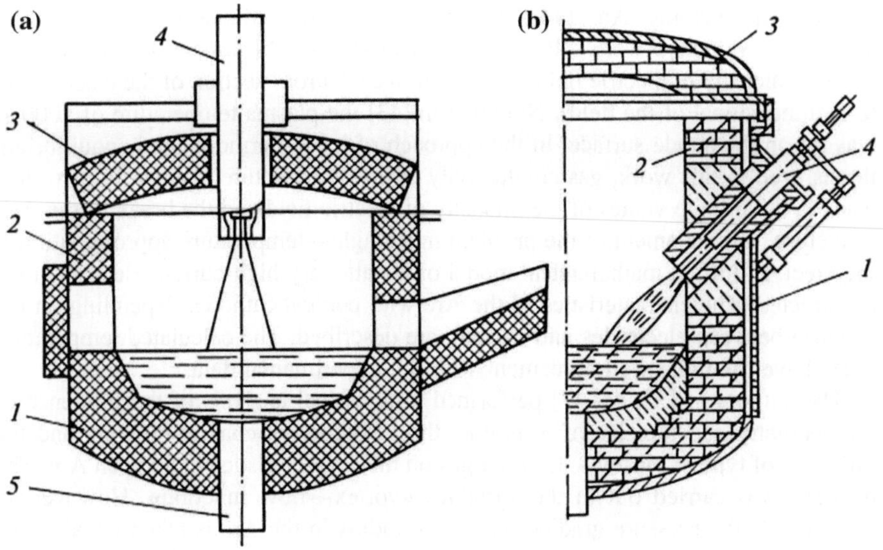

Fig. 8.33 Scheme of the plasma furnace with ceramic crucible [1]: **a** a plasma torch in the roof; **b** the plasma torches mounted in the walls: *1* housing, *2* lining, *3* an arch, *4* plasma torch, *5* bottom electrode

8.5 MW, the melting speed—7–10 t/h, the energy consumption—650 kWh/t, the consumption of argon—400–500 l/min.

The largest plasma furnace is a 30 t furnace of Russian OKB—1556. Four plasma torches are introduced obliquely through the side walls using the hydraulic actuators they can be moved along the axis and change the angle of the incline. Power of DC plasma torches is carried by three thyristor converters, so that both three plasma torch operates simultaneously. At the operating, any of the plasma torches can be turned off, and the other can be turned on to speed up the melting process. Maximum current of the power supply is 10 kA at the voltage of 150–660 V, introduced during the melting capacity of 12–15 MW, the specific energy consumption for melting is 625 kWh/t, the consumption of Argon—1000 l/min. It would be expected that with the advent of more resistant refractory (ZrO_2-based) technical and economic parameters of plasma furnaces the characteristics of the furnaces will improve at a faster rate than for conventional Arcs.

8.5.2 Algorithms of the Mathematical Models

One of the works devoted to the theoretical study of the opened electric Arcs is a work of Lovke J., where the results of the of electric Arcs calculation at the low currents are presented, taking into account the natural convection [43] and the high-current welding Arc [44]. Magnet-gas-dynamic equations with variables «pressure—velocity» were solved by the method of establishment. It was assumed that the intensity of electric field is constant in each cross-section of the electric Arc (one-dimensional of the field). Note that in [43] the plasma temperature of 3000 K was set on the anode surface. In the approach of local thermodynamic equilibrium that is used in this work, gas conductivity at this temperature is equal to zero, that leads to the infinite values of the intensity of electric field and the break of the Arc. Therefore, this statement of the problem in a single—temperature approximation is incorrect. In [30], a mathematical model of a stationary high-current electric Arc is constructed. The characteristics of the Arc with conical cathode, depending on the distance between electrodes and current, are described. The calculated temperature fields have the quite well agreement with the experimental data.

The authors of [32, 45–47] performed a numerical analysis of the influence of the geometry and the size of electrode, the velocities of coaxial blowing and the influence of type of plasma—forming gas on the characteristics of an open Arc. The decision was carried out in the variables «vortex—flow function». However, in these works the pressure gradient along the radius in the Arc is taken equal to the radial electromagnetic force that leads to the overestimate field of the pressure in the Arc and to elimination of the role of acceleration in plasma by the electromagnetic force. In the studied electric devices used in these works, cathode is a rod electrode with a conical or a hemispherical shape of the end and a composite cylindrical with a flat end and a coaxial blowing. The anode in all these cases is an infinite flat plate. The physical variables «pressure—velocity» and on the basis of two-temperature plasma model in [31] analyzed the Argon open free Arc for welding of metals is carried out. The field of electron and ion-atom temperatures are calculated as well as the velocity field of plasma, all the electrical parameters of the welding Arc current with $I = 75$; 100 A with the argon consumption $Q = 10$–20 l/min.

As it is noted in Sects. 1.2 and 1.3, equilibrium and nonequilibrium plasma models are widely used for simulation of plasma processes in the electric Arc. Here, on the basis of these models the simulation of open free Arc and the comparative analysis of simulation results using the examples of the Arcs for welding and melting of metals will be carried out. In contrast to the simplified model of the Arc in the channel (Sect. 8.3.1) the used models take into account the two-dimensional electromagnetic field of the Arc (Sect. 8.4) and the motion of gas, i.e., full two-dimensional field momentum and continuity equations of gas included in the decision process. Note that the technique outlined here can also be successfully used for the study of the Arc in the channel, especially to its initial, transitional zones and in the zone of the Arc attachment taking into account the boundary conditions at the wall.

The full two-dimensional model of equilibrium plasma in Arc plasma torches in a cylindrical coordinate system includes the following equation: equation of the balance energy (1.25), the momentum Eq. (1.26, 1.27), the continuity equation of all the gas flow (1.29), the Maxwell's Eqs. (1.31), that can be solved using the equation for the electric potential V (8.19) or a function of electric current Φ (8.30). In Sect. 8.4, the algorithms for solving of these equations and their boundary conditions corresponding to the discrete analogue for internal and boundary points are and equations for calculation of the parameters of the field of the electric Arc are presented. For the two-temperature model, instead of the equation of the energy balance for all the gas (1.25) two separate equations of the energy balance for electrons (1.34) and heavy components (1.39) are used. In the case of ionization equilibrium the equation of ionization equilibrium is included into the model (1.42). The general algorithms for solving of these equations are given in Sect. 6.1. The issue of sustainability and organization of the computational procedures are considered in Sects. 6.2 and 6.3. In Sect. 6.3.3, presented the discrete analogues for various equations related to the construction of additional internal feedback loop.

These models of plasma also include the equations for determination of the composition, the transport coefficients and thermodynamic functions of plasma Sects. 4.1, 4.2, 4.3 and 4.4. In many cases, the coefficients of the transfer and the thermodynamic functions of equilibrium plasma (especially at atmospheric pressure) are known and are often specified as a table depending on the temperature and pressure of the gas. The determination of them using the set value of the temperature and the pressure are reduced to normal approximation problem, for example, using the linear or cubic spline.

For the complete formation of two-dimensional systems of discrete analogues for equilibrium and nonequilibrium models of the electric Arc we can determine only the boundary conditions and the corresponding discrete analogues of the boundary points of the equation of the balance energy in the model of equilibrium plasma, the momentum equation, the continuity equation for all the gas, the equation of the balance energy for electron gas and heavy components in the model of nonequilibrium plasma and equation of ionization equilibrium.

Boundary Conditions *At the cathode,* the temperature and concentration of plasma should be determined from the experiment or the calculation of near-cathode processes. However, the complexity of calculation of the near-electrode processes and the complexity of experiments limits these approaches. In this case, the spot of cathode can be excluded from the general review of the Arc and use the simple approximate models for determination of the plasma temperature in the zero section. Note that at the cathode the plasma velocity is relatively small, convection does not play a significant role in the balance of power, and the heat extraction is mainly carried out due to the thermal conductivity in radial direction. It is also assumed that movement of the gas in the near-electrode regions does not violate the conditions of the ionization equilibrium, that the concentration of particles can be determined using the Saha equation (4.1). Thus, we have, respectively, the following approximate models for determination of the boundary conditions on the cathode surface:

(a) One-dimensional model of equilibrium plasma

$$\sigma E^2 - U_{\mathrm{rad}} = -\frac{1}{r}\frac{\mathrm{d}}{\mathrm{d}r}\left(r\lambda\frac{\mathrm{d}T}{\mathrm{d}r}\right); \quad E = \frac{I}{2\pi\int\limits_0^{R_c}\sigma r\mathrm{d}r}; \tag{8.37}$$

(b) One-dimensional model of nonequilibrium plasma

$$\begin{aligned} &\sigma E^2 - U_{\mathrm{rad}} - b(T_e - T) = -\frac{1}{r}\frac{\mathrm{d}}{\mathrm{d}r}\left(r\lambda_e\frac{\mathrm{d}T_e}{\mathrm{d}r}\right); \\ &b(T_e - T) = -\frac{1}{r}\frac{\mathrm{d}}{\mathrm{d}r}\left(r\lambda_{ai}\frac{\mathrm{d}T}{\mathrm{d}r}\right); \quad E = \frac{I}{2\pi\int\limits_0^{R_c}\sigma r\mathrm{d}r}, \end{aligned} \tag{8.38}$$

where R_c—the radius of the cathode spot is equal to the radius of the optical emission of the gas in the zero section at the temperatures of T_R, T_{eR} T_{aR}; T_R, T_{eR} T_{aR}—the gas temperature in equilibrium plasma (8.37), electrons and atoms (8.38) at the border of optical radiation of the conductive cathode zone. For example, for argon T_R = 6500–7000 K, T_{eR} = 7000–7500 K, T_{aR} = 300–500 K. The other notations are entered similar to the case of equilibrium and nonequilibrium models (Sects. 1.2, 1.3).

Approximate models (8.37) and (8.38) can be used for determination of the temperature in the conductive area in the initial section of the Arc. The main advantage of such models is the guarantee of implementation of the energy balances and the electromagnetic field in the zero section.

Outside the conductive zone temperature of the cold gas is equal to T = 300 K. At the cathode, the plasma velocity is zero, and outside the conductive zone is considered that the gas is uniformly applied on the inner channel of the Arc nozzle and the velocity is determined using the predetermined gas consumption G, kg/s or Q, m^3/s using the following equation:

$$v_z = \frac{G}{\pi\rho\left(R^2 - R_k^2\right)} = \frac{Q}{\pi\left(R^2 - R_k^2\right)}; \quad v_r = 0, \tag{8.39}$$

where R_k and R—radiuses of the cathode and the nozzle of the Arc, respectively.

On the axis of symmetry from the condition of the cylindrical axial symmetry we have:

$$\frac{\partial T}{\partial r} = \frac{\partial T_e}{\partial r} = \frac{\partial n_e}{\partial r} = \frac{\partial v_z}{\partial r} = v_r = 0 \tag{8.40}$$

At the anode, the gas velocity is zero. For determination of the temperature at the anode models (8.37) and (8.38) as the analogy can be used with the set of anode spot radius R_a instead of the cathode spot R_c. Unlike the cathode the anode is usually downstream and the near-electrode processes on the anode are much less active (in comparing with cathode) occur, and therefore to determine the boundary conditions at the anode a simple model of steady-state process at the exit can be used, i.e., the target variable gradients are equal to zero:

$$\frac{\partial T}{\partial z} = \frac{\partial T_e}{\partial z} = \frac{\partial n_e}{\partial z} = v_z = v_r = 0 \tag{8.41}$$

To validate these models various boundary conditions were calculated using (8.37), (8.38) and (8.41) for the open free Arc with currents of 90 and 160 A. The discrepancy between the temperature at the anode (downstream) are not more than 10 %, and the resulting temperature field in the Arc (upstream) practically do not differ from each other.

To simulate the open free Arc border of *the free boundary* should be chosen at the sufficient distance from the Arc. If the anode is a flat surface and the distance between the electrodes is small, the plasma flow of the Arc, striking with anode can spread along long sides. In this case, the zone of interaction between the Arc and the environment can occupy a large space, here the temperature and velocity of the plasma flow gradually reduced until the temperature and the velocity of the environment. Therefore, the selected area of calculation should be sufficient for simulation of plasma processes in the Arc and the establishment of the zone of interaction between the Arc and environment. For the majority of electric free Arcs such a free boundary can be selected in the region of three–five radius of the Arc.

As in the case of the boundary conditions at the exit (downstream) of the channel of RF-plasma torch (Sect. 7.3.1), on the free boundary all the processes can be considered established, the changes of the temperature and the velocity at the normal to the boundary are non-essential, i.e.:

$$\frac{\partial T}{\partial r} = \frac{\partial T_e}{\partial r} = \frac{\partial n_e}{\partial r} = \frac{\partial v_z}{\partial r} = \frac{\partial v_r}{\partial r} = 0 \tag{8.42}$$

For the free Arc the pressure on the free surface is known for the Arc at the atmospheric pressure at this boundary p = 1 atm.

At some borders of the free Arc not all the unknown variables can be predetermined, for example the gas pressure at the cathode, at the anode and at the axis of symmetry are unknown. At these boundaries only the values of the velocity components or their normal gradients are known. Therefore, the determination of the pressure on these boundaries can be carried out using the Shifted Chess Grid and the techniques, described above in Sect. 6.1.2. On the other hand, the points for determination of the target variable are desirable to be placed directly on the borders, where their values are known. This is done for obtaining of the maximum use of this information on the boundary conditions.

Using the considered boundary conditions the following structure of Shifted Chess Grids for modeling of plasma processes in the open free Arc is proposed (Figs. 6.5b, 8.34, 8.35)

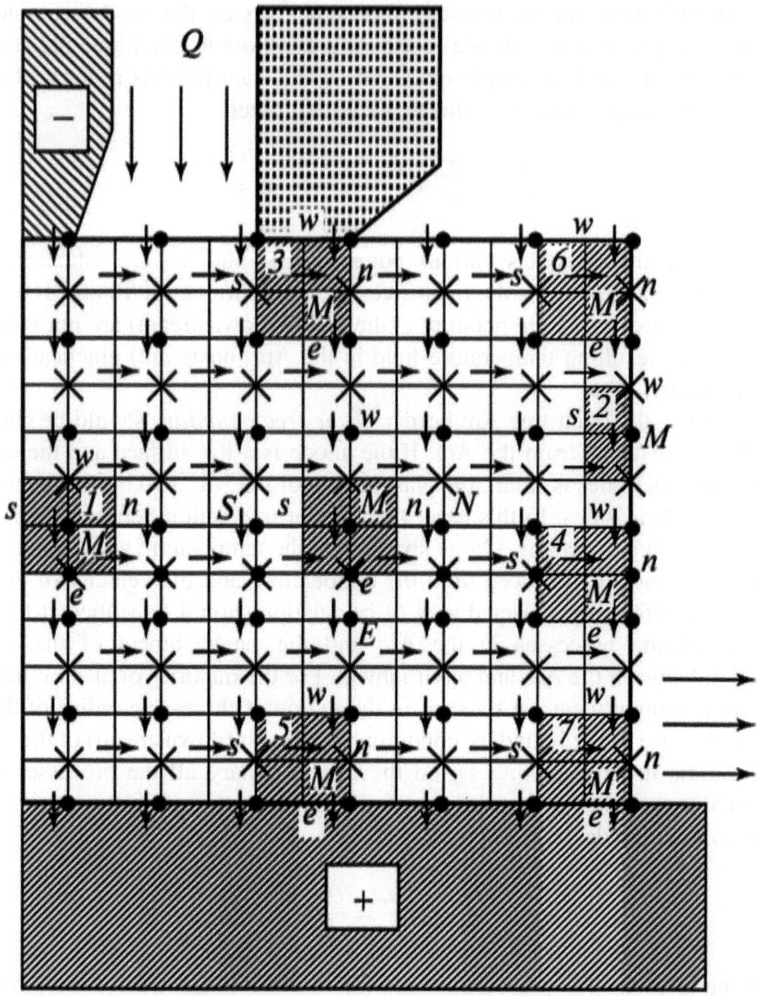

Fig. 8.34 The shifted chess grid and the control volumes for obtaining of the discrete analogues of the equation of the energy balance, equation of ionization equilibrium and the momentum equation for open free Arc: *filled circle* points for determination of the temperature and density, *down arrow* the same for the component of velocity v_z, *right arrow* the same for component of velocity v_r, *cross line* the same for the pressure p

Electromagnetic equation for the function of electric current Φ is solved on the grid [$z1$, $r1$], its origin of coordinates begins on the axis of symmetry and at the cathode surface. Grid takes the area between the cathode and the anode, the boundary points are on both electrodes and on the axis of symmetry;

Equation of the balance energy and equation of ionization equilibrium are resolved on the grid [$z1$, $r2$], shifted compared with the grid [$z1$, $r1$] for a half step

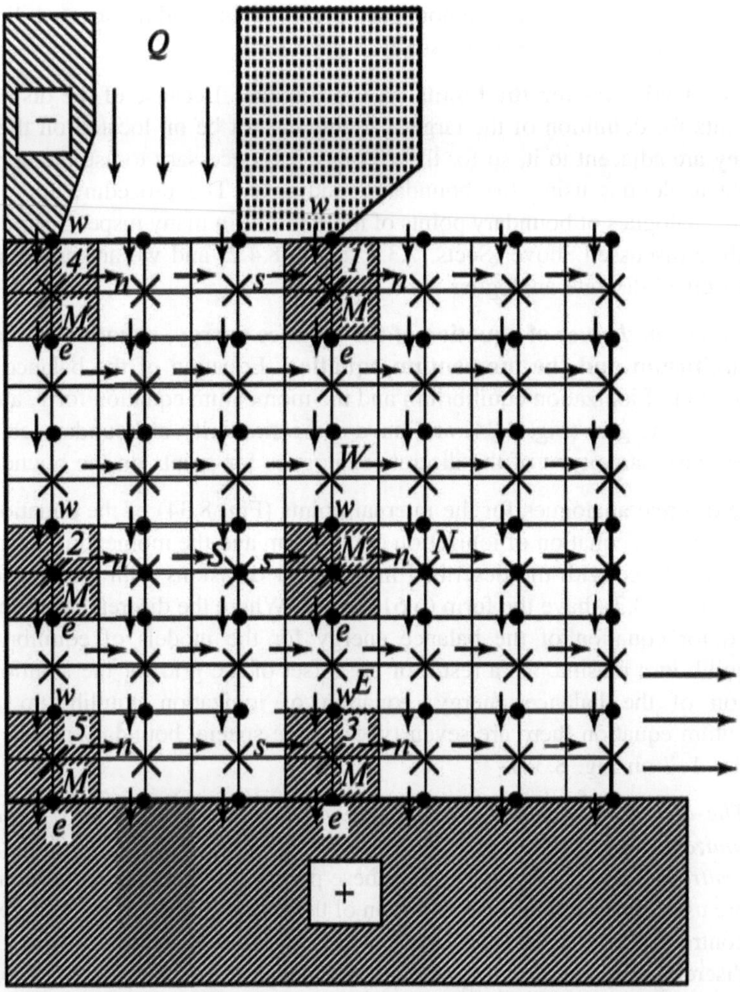

Fig. 8.35 The shifted chess grid and the control volumes for obtaining of the discrete analogues of the continuity equation for the pressure of the open free Arc: the notations are the same as in Fig. 8.34

to the side of the free boundary. The boundary points of this grid are placed directly on the cathode, the anode and on the free boundary;

The momentum equation of plasma for v_z is solved using the grid [$z1$, $r2$];

The momentum equation of plasma for v_r is solved on the grid [$z2$, $r1$], shifted compared with the net [$z1$, $r1$] for a half step toward the anode, the boundary points are located on the axis of symmetry and on the free boundary;

The continuity equation is solved and the pressure field is defined using the mesh [$z2$, $r2$], shifted toward the free surface and the anode, respectively, for a half step relative to the grid [$z1$, $r1$]. The boundary points of this grid are located directly on the free surface, the pressure is usually known here.

Discrete Analogues for the Points on the Borders Because of the displacement the points for definition of the target variable cannot be on located on the border, but they are adjacent to it, so for these points it is necessary to use the appropriate discrete analogues using the boundary conditions. The procedure for obtaining discrete analogues of boundary points of the free Arc in many respects similar to the procedure discussed above (Sects. 7.3.1, 8.4.1, 8.4.2) and we are presenting here only received discrete analogues.

1. ***Discrete analogues* of equation of the balance energy, equation of ionization equilibrium and the momentum equation.** Equation of the balance energy, equation of ionization equilibrium and the momentum equation for v_z are solved at the same grid (Fig. 8.34) and have some similarity in boundary conditions, they have same type of the discrete analogues for points on the boundary.

The discrete analogues for the internal points (Fig. 8.34) of the equation of the balance energy, equation of ionization equilibrium and the momentum equation for v_z, in accordance with the described in Sect. 6.1 decisions with internal feedback (Sect. 6.3.2, 6.3.3), have the form (6.61) (6.34). Where the discrete analogue (6.61) is used for equation of the balance energy for the models of equilibrium and nonequilibrium plasma. As a result of the offset of the grid for the solution of the equation of the balance energy, equation of ionization equilibrium and the momentum equation there are seven types of the special boundary points (control volumes 1–7 on Fig. 8.34).

(a) *The discrete analogues of the equation of the balance energy, equation of ionization equilibrium and the momentum equation for v_z for points of the control volume* 1 (Fig. 8.34). For these points, the boundary conditions (8.40) are used. Therefore, the contribution of the flows through the left verge s of the control volume 1 into the total balance is equal to zero. It follows that the discrete analogue of the equation of the balance energy, equation of ionization equilibrium and the momentum equation for v_z for these points has the same form (6.61) and (6.34) with $A_S = 0$.

(b) *The discrete analogues of the equation of the balance energy, equation of ionization equilibrium and the momentum equation for points of the control volume* 2 *or* 4. At the obtaining the discrete analogues of the equation of the balance energy, equation of ionization equilibrium and the momentum equation for v_z for the points on the free border we use the approximate boundary condition (8.42), and a predetermined pressure on this boundary p. Integration of the equation of the energy balance, equation of ionization equilibrium and the momentum equation for v_z is only in the half of the control volume 2, and the momentum equation for v_r—in the control volume 4. Taking into account the boundary condition on the free boundary the discrete analogues for these points have the same form (6.61), (6.34), at $A_N = 0$.

(c) *The discrete analogues of the momentum equation for* v_r *for points of the control volume* 3 *or* 5. At the entrance and at the anode there are the similar boundary conditions $v_r = \frac{\partial v_r}{\partial z} = 0$. Therefore, for the points at the entrance the contribution of flows into the momentum balances through the upper verge w of the control volume 3 (Fig. 8.34) is equal to zero. Accordingly, for these points the discrete analogue has the form (6.34) at $A_W = 0$. Similarly, for the points adjacent to the anode, $A_E = 0$.

(d) *The discrete analogues of the momentum equation for* v_r *for points of the control volume* 6 *or* 7. For these points there are two types of boundary conditions for the points adjacent to the free boundary ($\frac{\partial v_r}{\partial r} = 0$) and to the points adjacent to the entrance or the anode ($v_r = \frac{\partial v_r}{\partial z} = 0$). Therefore, the contributions of the flows through the verges w and n for the control volume 6 and flows through the verges e and n for the control volume 7 are not taking into account at the obtaining of the discrete analogue. And for these points the discrete analogue has the form (6.34) at $A_W = A_N = 0$ for points of the control volume 6 and at $A_E = A_N = 0$ for the point of the control volume 7.

2. **The discrete analogues of the continuity equation for the pressure p.**

According to the decision of the continuity equation for the pressure (Sect. 6.1.2) its discrete analogue for the internal points (Fig. 8.35) has the form (6.38). Using the shifted chess grid for the solution of the continuity equation for the open free Arc, there are five types of special boundary points (the control volume 1–5 Fig. 8. 35), their discrete analogue is different from the discrete analogue for the internal points because of the boundary conditions.

(a) *The discrete analogues of the continuity equation for points of the control volume* 1. The feature of the points of the control volume 1 is the given boundary condition for the velocity v_z on the up verge w. In accordance with the general procedure for the obtaining of the discrete analogue of the continuity equation (Sect. 6.1.2) the discrete analogue for these points has the same form (6.38) with the following coefficients:

$$a_W = 0;$$

$$b_M = (\rho_w v_{zw} - \rho_e \hat{v}_{ze}) \frac{(r_n - r_s)(r_n + r_s)}{2} + (r_s \rho_s \hat{v}_{rs} - r_n \rho_n \hat{v}_{rn})(z_e - z_w).$$

From these expressions it is seen that using the shifted grid it is not necessary to determine the unknown pressure at the point W, and well-known velocity at the entrance v_{zw} is included into the source coefficient b_M instead of $\hat{v}_{zw}$.

(b) *The discrete analogues of the continuity equation for points of the control volume* 2. The boundary condition for points of the control volume 2 is the zero radial component of the velocity on the axis of symmetry, $v_{rs} = 0$.

Therefore, the discrete analogue for these points has the same form (6.38) with the following coefficients:

$$a_S = 0; \quad b_M = (\rho_w \hat{v}_{zw} - \rho_e \hat{v}_{ze})\frac{r_n^2}{2} - r_n \rho_n \hat{v}_{rn}(z_e - z_w).$$

(c) *The discrete analogues of the continuity equation for points of the control volume* 3. The boundary condition at the anode $v_{ze} = 0$. The discrete analogue for these points has the same form (6.38) and:

$$a_E = 0; \quad b_M = \rho_w \hat{v}_{zw} \frac{(r_n - r_s)(r_n + r_s)}{2} + (r_s \rho_s \hat{v}_{rs} - r_n \rho_n \hat{v}_{rn})(z_e - z_w).$$

d) *The discrete analogues of the continuity equation for points of the control volume* 4 *and* 5. For the points 4 and 5 there are two boundary conditions: on the axis of symmetry ($v_{rs} = 0$) and at the cathode ($v_{zw} = 0$ or it is set) or at the anode ($v_{ze} = 0$). So, the discrete analogue for these points has the form (6.38) with the following coefficients:

For points of the control volume 4

$$a_W = a_S = 0; \quad b_M = (\rho_w v_{zw} - \rho_e \hat{v}_{ze})\frac{r_n^2}{2} - r_n \rho_n \hat{v}_{rn}(z_e - z_w);$$

For points of the control volume 5

$$a_E = a_S = 0; \quad b_M = \rho_w \hat{v}_{zw} \frac{r_n^2}{2} - r_n \rho_n \hat{v}_{rn}(z_e - z_w).$$

Thus, taking into account the features of the shifted grids and the boundary conditions we have all the discrete analogues of equation of the balance energy, equation of ionization equilibrium, the momentum equation and the continuity equation for the open free Arc. For internal and boundary points, they have the total view, respectively (6.61), (6.34) and (6.38) with the change of the coefficients for the boundary points.

Using the similar techniques, the discrete analogues of different equations for points on the boundary of the electric Arc in the channel and others can be easily obtained. The shifted grids presented in Fig. 6.5 also can be successfully used.

8.5.3 Comparative Analysis of the Simulation Results

The results of simulation of typical plasma processes of open free Arc for welding and melting of metals are presented here.

The open welding Arc with the operating range of currents $I = 50–300$ A and a length $l = 5–18$ mm have the radius of nozzle $R = 5–7$ mm, the radius of the cathode spot $R_c = 0.3–0.6$ mm, the radius of the conductive zone in the zero section is selected for the radius of the radiation $R_1 = 1.3–1.6$ mm, respectively, the radius of the modeling zone $R_2 = 30$ mm, the radius of result presentation area $R_3 = 20$ mm. The consumption of Argon $Q_{Ar} = 10–40$ l/min. At the simulation of the welding Arc with a nominal current $I = 90$ A and 160 A the main obtained results are: the temperature field, the field of the plasma flow, the field of velocities, the field of the static pressure, distribution of densities of particle, etc. The results of the calculation of electromagnetic field of these Arcs and analyze of them were presented in Sect. 8.4.3.

The boundary conditions for the temperature at the cathode and at the anode are obtained using the one-dimensional model (8.37) and (8.38) at the predetermined radiuses of the conductive zones at the cathode and anode. The boundary condition for the electromagnetic field is defined like in Sects. 8.4.1 and 8.4.2. The shifted grids, boundary conditions for the velocity, pressure and corresponding discrete analogues are presented in Sect. 8.5.2. The initial temperature in the Arc is equal to 10,000 K. The initial components of the velocity, the function of electric current and the electric potential are equal to zero, the atmospheric pressure is taken as the initial one.

1. **The results of the simulation of the Arc on the basis of equilibrium model.** The obtained fields of temperature and the gas flow in welding Arc with different regimes of the operation are presented in Figs. 8.36, 8.37 and 8.38. From the figures it is seen that the open free Arc has a conical shape with a maximum temperature in the zero section: at the regime with $I = 90$ A, $Q_{Ar} = 20$ l/min, $T = 18{,}500$ K; at the regime with $I = 160$ A, $Q_{Ar} = 20$ l/min, $T = 21{,}500$ K. The total specific power in Arc at these two regimes, respectively, $P = 1.45$ kW and 3.2 kW. The voltage drop in the Arc $U = 16.1$ and 20.3 V, respectively.

The distribution of the current function (gas-dynamic) $\psi = \int_0^r \rho v_z r dr$ gives us the picture of the plasma flow in the open free Arc (Figs. 8.36a, 8.37a, 8.38). Thus, the main part of the plasma-forming gas does not penetrate into plasma, and wraps around it, forming an external stabilizing gas flow.

The used in the model concept of a free boundary allows the determination of the degree of interaction between the plasma flow and the environment. The fields of the gas flows are presented in figures and the outside vortex of the gas can be seen there. The streamlined flow of plasma-forming gas of the Arc interacts with environment, forming the external vortex flow. Thus, the free Arc is stabilized under not one, but two external flows: streamlined flow of the plasma-forming gas and the vortex flow formed by the interaction of the streamlined stream with the environment. The data with the values of the stream function also allow the determination of the amount of plasma-forming gas entrapped into the Arc.

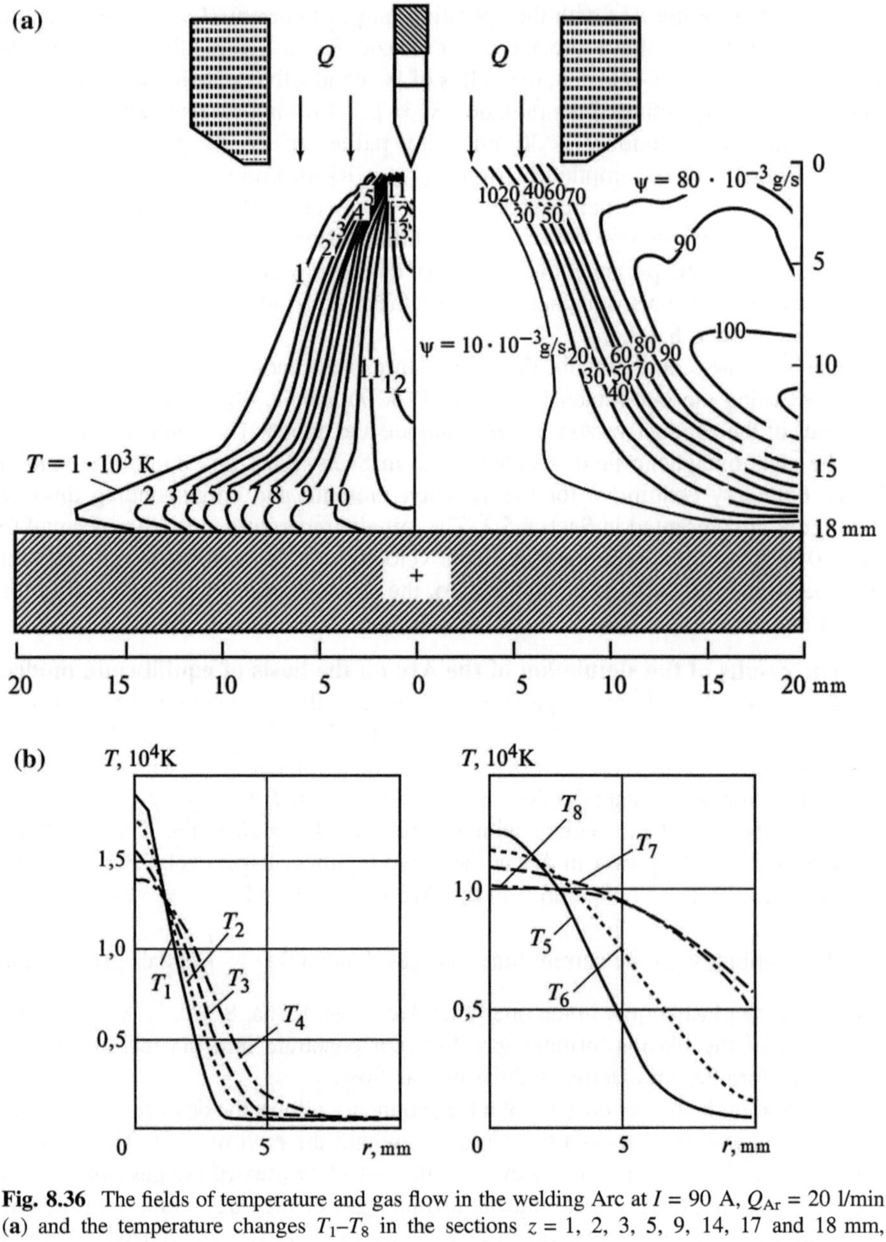

Fig. 8.36 The fields of temperature and gas flow in the welding Arc at I = 90 A, Q_{Ar} = 20 l/min (**a**) and the temperature changes T_1–T_8 in the sections z = 1, 2, 3, 5, 9, 14, 17 and 18 mm, respectively (**b**)

For the comparison of the data at the different regimes at Fig. 8.38 there are the fields of temperature and the gas flow for the same Arc with I = 160 A, Q_{Ar} = 40 l/min. The larger flow of the plasma-forming gas creates the additional

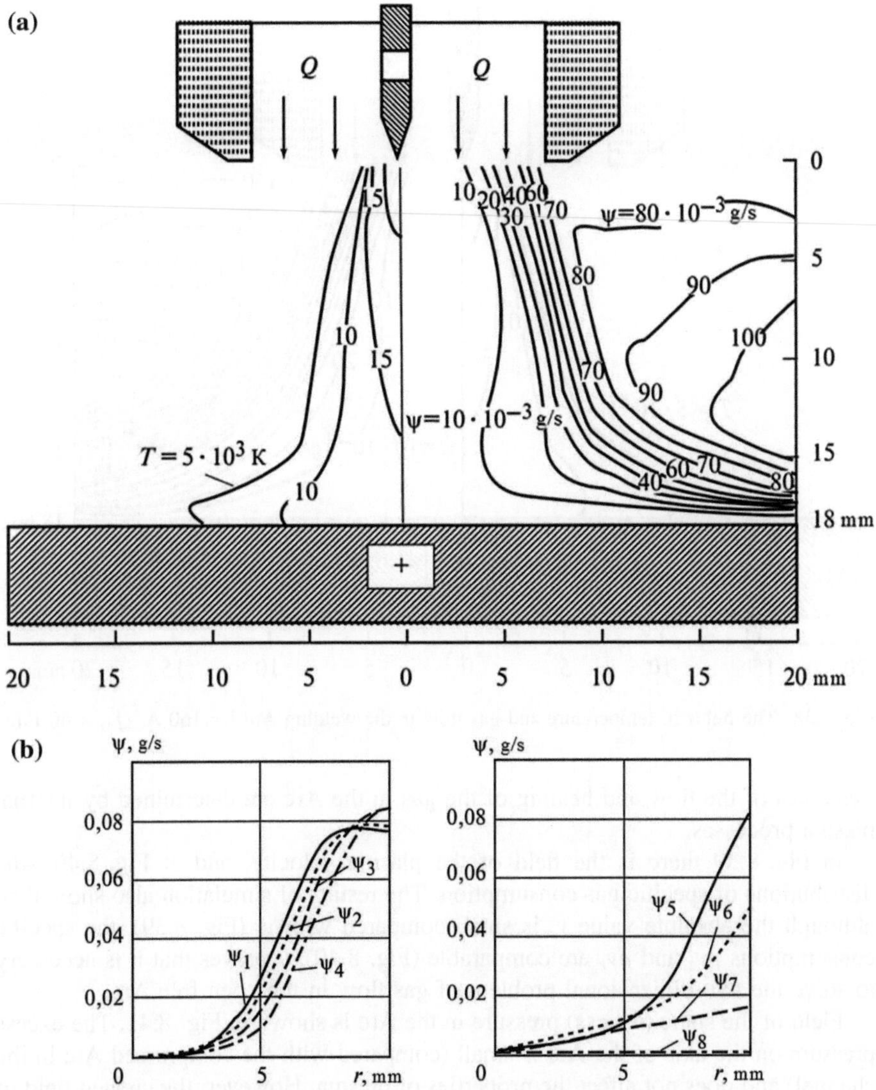

Fig. 8.37 The fields of temperature and gas flow in the welding Arc at I = 160 A, Q_{Ar} = 20 l/min (**a**) and the current function changes ψ_1–ψ_8 in the sections z = 1, 2, 3, 5, 9, 14, 17 and 18 mm, respectively (**b**)

pressure in the cathode zone and changes the shape of the temperature field in the Arc. However, the maximum values of the temperature and velocity of plasma on the axis of symmetry are virtually unchanged. The total released power in the Arc and the nature of the gas flow does not change. Consequently, the external plasma-forming flow of plasma plays only the role of stabilizing flow and the

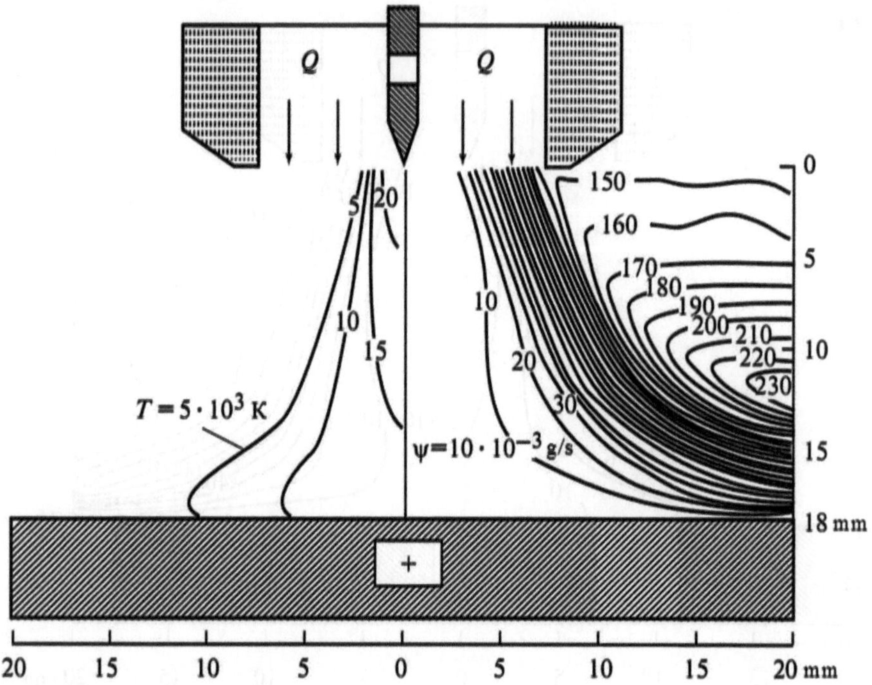

Fig. 8.38 The fields of temperature and gas flow in the welding Arc I = 160 A, Q_{Ar} = 40 l/min

dynamics of the flow and heating of the gas in the Arc are determined by internal plasma processes.

In Fig. 8.39 there is the field of the plasma velocity, and at Fig. 8.40—the distributions of specific gas consumption. The results of simulation also show that, although the absolute value v_r is small compared with v_z (Fig. 8.39), the specific consumptions ρv_z and ρv_r are comparable (Fig. 8.40), it proves that it is necessary to solve the two-dimensional problem of gas flow in the open free Arc.

Field of the static (excess) pressure in the Arc is shown in Fig. 8.41. The excess pressure on the axis of the Arc is small (compared with the compressed Arc in the channel) and does not affect the properties of plasma. However, the created field of the pressure gradient is comparable with electromagnetic force in the Arc (Figs. 8.30, 8.31, 8.42) and the gradient of static pressure can not excluded from the equations of the gas motion. The created pressure in the Arc prevents to the excessive penetration of gas into plasma, and provides the high level of temperature and conductivity of the Arc, contributing to steady and sustainable heating of the plasma flow in the Arc. Under the influence of two opposing forces—the radial component of the electromagnetic force F_r and the gradient of the static pressure $-\partial p/\partial r$, a small amount of hot gas with a temperature T = 7000–8000 K penetrates into the Arc for heating (5–10 % Q) (Figs. 8.36, 8.37 and 8.38).

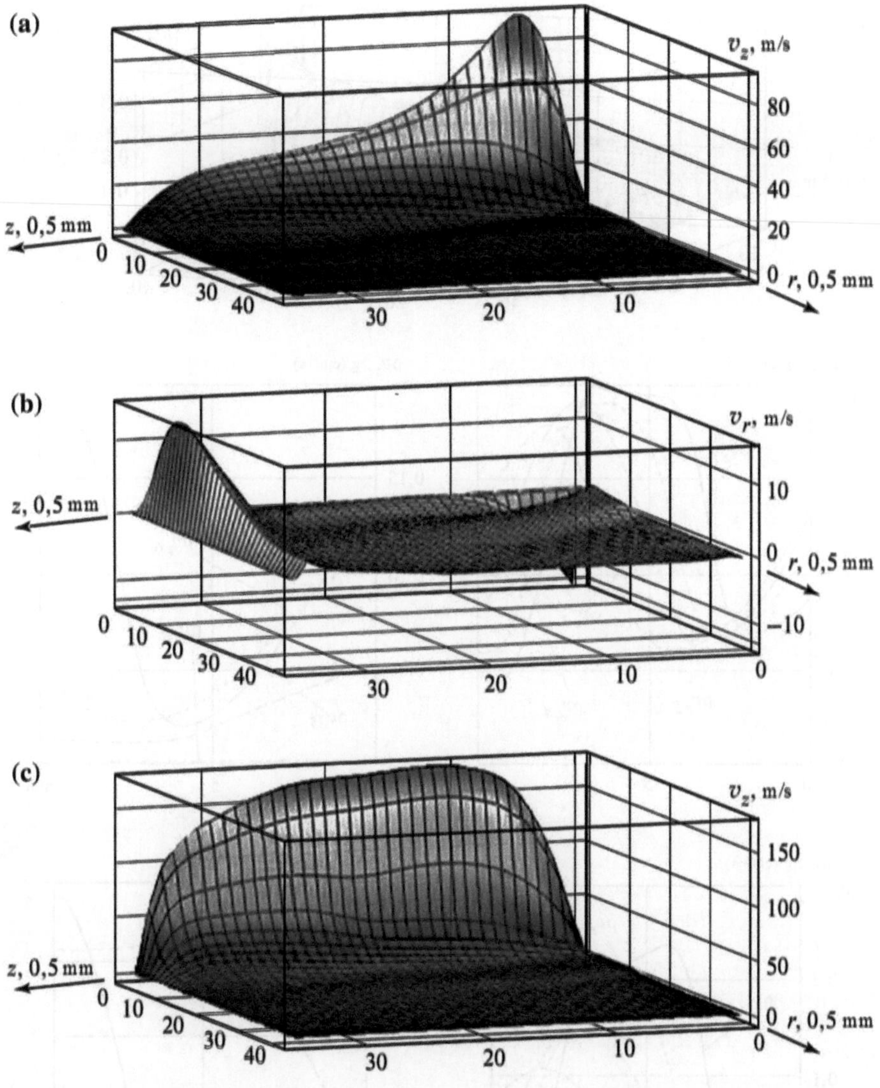

Fig. 8.39 The fields of plasma velocity components in the welding Arc at $I = 90$ A, (**a**, **b**) and 160 A (**c**) ($Q_{Ar} = 20$ l/min)

The approximate method for determination of the pressure in the Arc using the electromagnetic force $\left(\frac{\partial p}{\partial r} = F_r\right)$ is popular in literature, for example [8, 9, 21, 26, 32, 45, 46]. This estimate may not lead to the large error of the calculation of the absolute value of the pressure. However, $-\frac{\partial p}{\partial r} = -F_r$, the result force that leads to the plasma acceleration in the radial direction is equal to zero, that greatly reduces the radial velocity v_r and the amount of gas, entrained from the outside into the Arc.

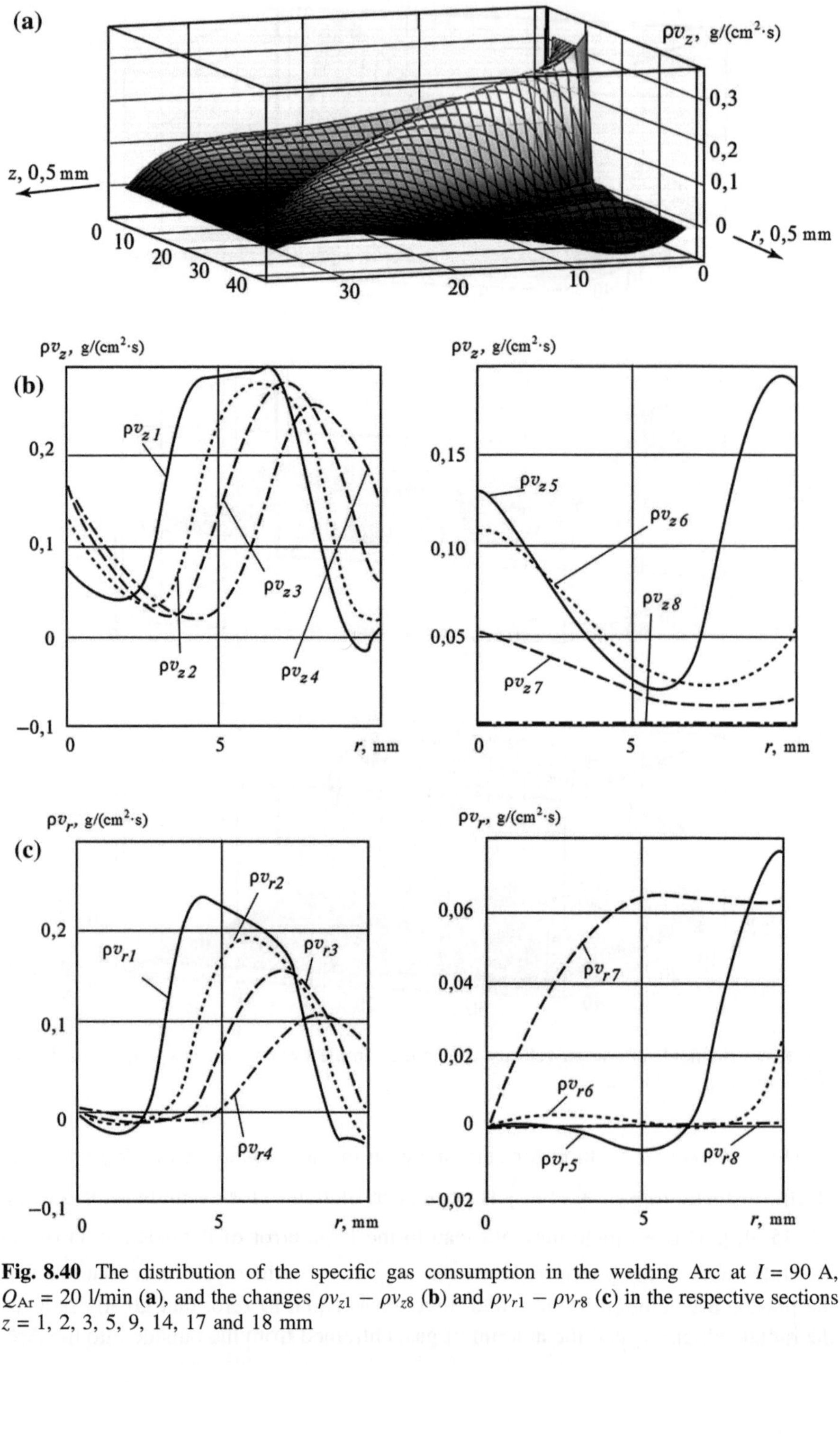

Fig. 8.40 The distribution of the specific gas consumption in the welding Arc at $I = 90$ A, $Q_{Ar} = 20$ l/min (**a**), and the changes $\rho v_{z1} - \rho v_{z8}$ (**b**) and $\rho v_{r1} - \rho v_{r8}$ (**c**) in the respective sections $z = 1, 2, 3, 5, 9, 14, 17$ and 18 mm

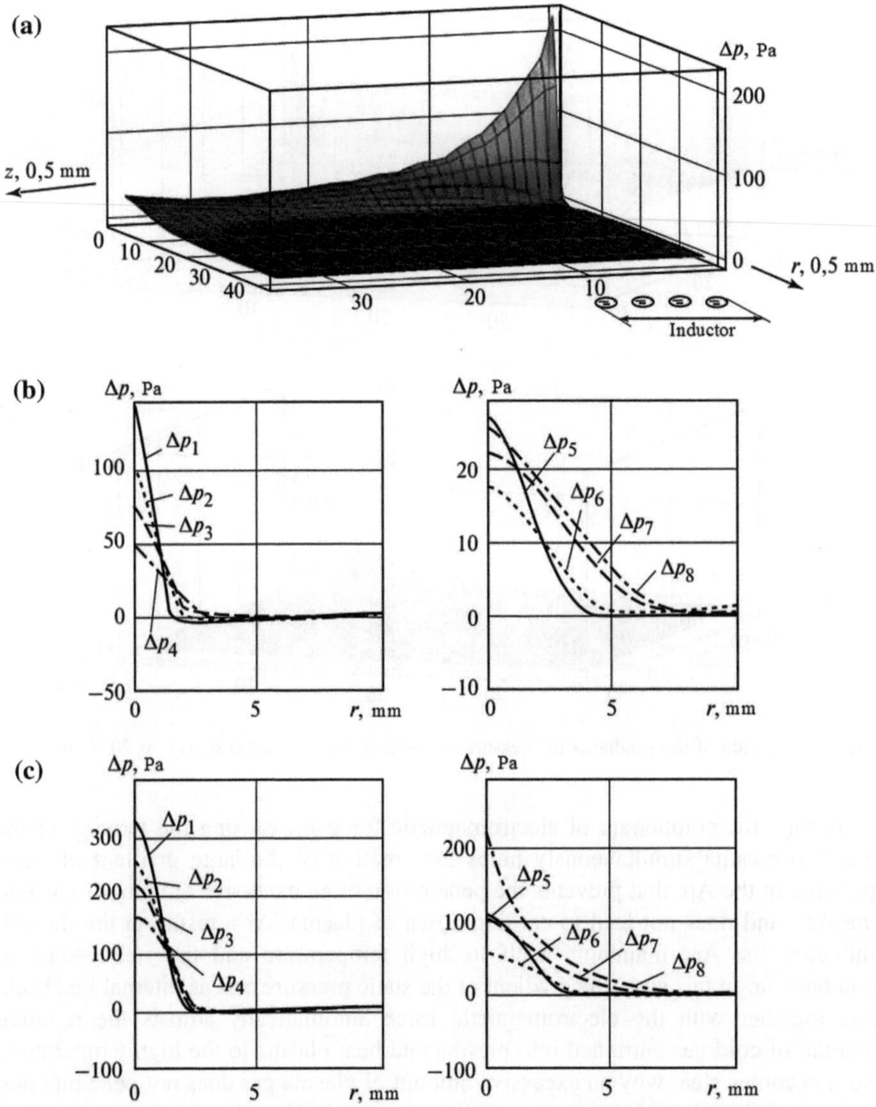

Fig. 8.41 Field of the static (excess) pressure in the welding Arc at $I = 90$ A, $Q_{Ar} = 20$ l/min (**a**) and the changes Δp_1–Δp_8 at $I = 90$ A (**b**) and 160 A (**c**) in the respective sections z = 1, 2, 3, 5, 9, 14, 17 and 18 mm

It should lead to a significant error in the calculation of the field of the flow and velocity of plasma. Furthermore, some authors ignore the gradient of static pressure that contrary, leads to overestimated value of the forces of gas acceleration flow, distortion of the fields of flow and gas velocity.

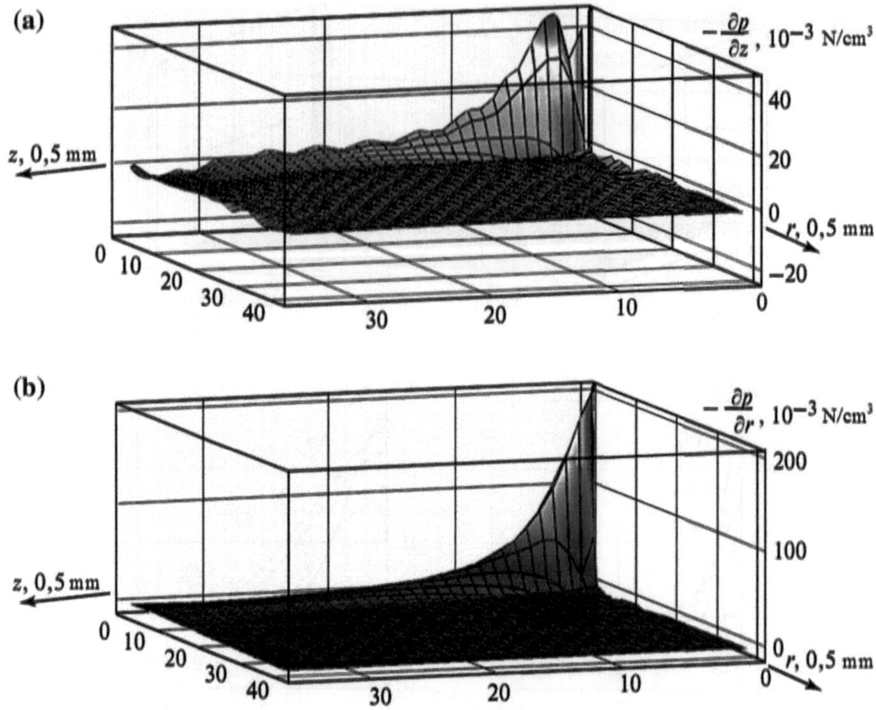

Fig. 8.42 Fields of the gradients of pressure in welding Arc at $I = 90$ A, $Q_{Ar} = 20$ l/min

In fact, the component of electromagnetic force F_r, causing the moving of the gas into plasma simultaneously helps the creation of the large gradient of static pressure in the Arc that prevents the penetration of an excessive amount of gas into the Arc, and does not lead to cooling down of plasma. As a result, at the thermal influence the Arc maintains itself its high temperature and the steady-state of combustion. In this case, the gradient of the static pressure acts as internal feedback, and together with the electromagnetic force automatically adjusts the required amount of cold gas entrained into plasma and heat plasma to the high temperature. So it becomes clear why an excessive amount of plasma gas does not penetrate into the plasma bunch and wraps around it and have the function of the plasma flow stabilization (Fig. 8.40). The simulation results show that the sum (difference) of these forces (Fig. 8.43) is equal to 3–5 % of the value of the force F_r (Fig. 8.30a) that is responsible for the acceleration of the gas in the radial direction and the amount of gas entrained into plasma.

From the standpoint of gas-dynamic the Arc operates like the magnetic pump, especially in the near-cathode zone, where there is the maximum current density, specific power and of electromagnetic force (Figs. 8.27, 8.29, 8.30, 8.31). According to the law of mass conservation the amount of gas entrained into the near-cathode area is equal to the amount of heated gas, accelerated in the Arc.

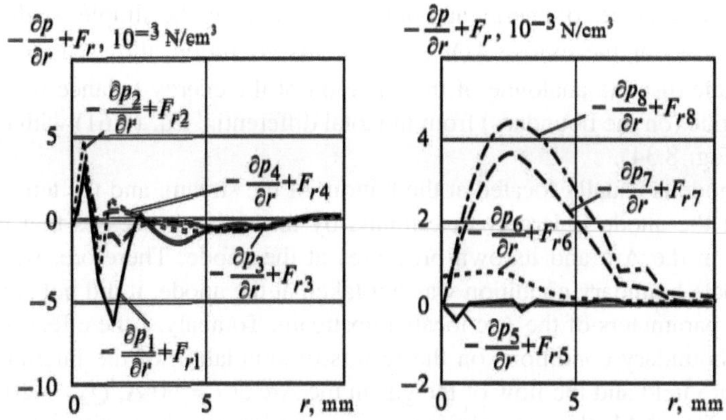

Fig. 8.43 The sum of the forces $(-\frac{\partial p_1}{\partial r}+F_{r1})$–$(-\frac{\partial p_8}{\partial r}+F_{r8})$ in open free Arc at $I = 90$ A, $Q_{Ar} = 20$ l/min in the respective sections $z = 1, 2, 3, 5, 9, 14, 17$ and 18 mm

Therefore, the accuracy of the modeling of the dynamic regime in the Arc depends on the accuracy of the determination of the total force $-\frac{\partial p}{\partial r}+F_r$ in the cathode zone.

Note also that near the axis of symmetry the gradient of static pressure $-\frac{\partial p}{\partial r}$ can be even greater than electromagnetic force $-F_r$ that leads to lockup of the gas flow in the radial direction to zero velocity at the axis. In this zone, the force F_r is sufficiently small, and the pressure generated mainly due to reciprocal movement of gas in the radial direction (Fig. 8.43).

It should be noted that often in practice, for analysis of the gas flow in the Arc the one-dimensional formulation of the problem of the electromagnetic field is used, i.e., axial component of the electromagnetic force ($F_z = 0$) is ignored and it is assumed that the acceleration of plasma in the axial direction is carried out due to the gradient of static pressure $-\frac{\partial p}{\partial z}$. The obtained results of the simulation of the Arc taking into account the two-dimensional electromagnetic field (Sect. 8.4.3) show that the absolute values of the gradient of the axial component of force of static pressure $-\frac{\partial p}{\partial z}$ (Fig. 8.42a) and electromagnetic force F_z (Fig. 8.31) are comparable. Neglecting of F_z leads to a significant reduction of the plasma velocity and to inaccurate result of calculation of the plasma and the field of its course; thus once again we prove the need to define a second component of the electromagnetic force F_z in the simulation of plasma and two-dimensional setting of electromagnetic problem of the electric Arc (Sect. 8.4).

There is the analogy in motion and heating of the gas in the electric Arc and in RF-plasma torch (Sect. 7.3.2), note that the mechanism of the penetration of gas into the plasma flow and its heating are the same in both cases.

About the Role of the Boundary Condition at the Anode The boundary condition for the temperature at the anode can be determined by the one-dimensional model of the balances of energy and field (8.37) or using the simple condition of the steady process at the exit (8.41): $\frac{\partial T}{\partial z} = 0$. This condition allows us to obtain an approximate discrete analogue of the equation of the energy balance for the points on the anode (on the boundary) from the total differential Eq. (6.61) with coefficient $a_E = 0$ (Fig. 8.34).

The anode is usually located at the bottom of the stream, and the temperature of plasma at the anode is largely determined by the active processes that take place upstream in the Arc and its own processes at the anode. Therefore, whatever the approximate boundary condition was not taken at the anode, it did not significantly affect the parameters of the Arc located upstream. To analyze the effect of different kinds of boundary conditions on the results of simulation of the internal field the temperature field and the flow of the gas in the Arc at I = 90 A, Q_{Ar} = 20 l/min are shown in Fig. 8.44, they are obtained using the boundary condition (8.41) at the anode. With this boundary condition the obtained temperature field and the gas flow near the anode are more evenly than in the case of the boundary condition (8.37), because the near-anode processes and the zone of the binding are excluded from the consideration. At a small distance (2–3 mm) from the anode the field of temperature and the gas flow are almost the same in both cases (Figs. 8.36a, 8.44), i.e., and all other parameters of the Arc are the same.

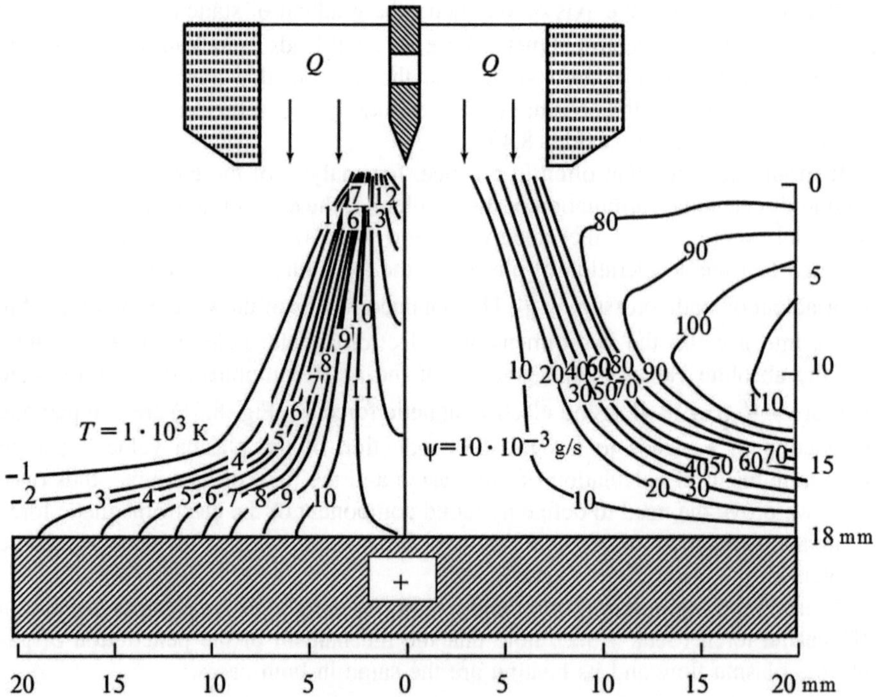

Fig. 8.44 The fields of the temperature and the gas flow in the welding Arc at I = 90 A, Q_{Ar} = 20 l/min, obtained using the boundary conditions $\frac{\partial T}{\partial z} = 0$ at the anode

The numerical verification also showed that at using of the boundary condition $\frac{\partial T}{\partial z} = 0$ at the anode, for the calculation of this zone the law of conservation of mass $(\mathrm{div}(\rho\vec{v}) = 0)$ better takes place.

2. **The simulation results of the welding Arc on the basis of the two-temperature plasma model.** The simulation results of welding Arc, derived from the two-temperature plasma model will be compared with the data obtained by the equilibrium plasma model. Terms simulation of the Argon welding Arc identical in both cases, and the results are given in the same cross-sections: z = 1, 2, 3, 5, 9, 14, 17 and 18 mm.

The temperature field of the electron gas and the function of electric current are shown in Fig. 8.45, and the fields of temperature of heavy components and the gas flow are shown in Fig. 8.46. Note that these results are almost identical to the data that are obtained on the basics of the model of equilibrium plasma (Figs. 8.23, 8.36a). The temperature distribution of the electron gas and heavy components in different sections of the Arc are shown in Fig. 8.47. For determination of the properties of plasma there are the distributions of concentration of electrons and atoms in different sections of the Arc in Fig. 8.48.

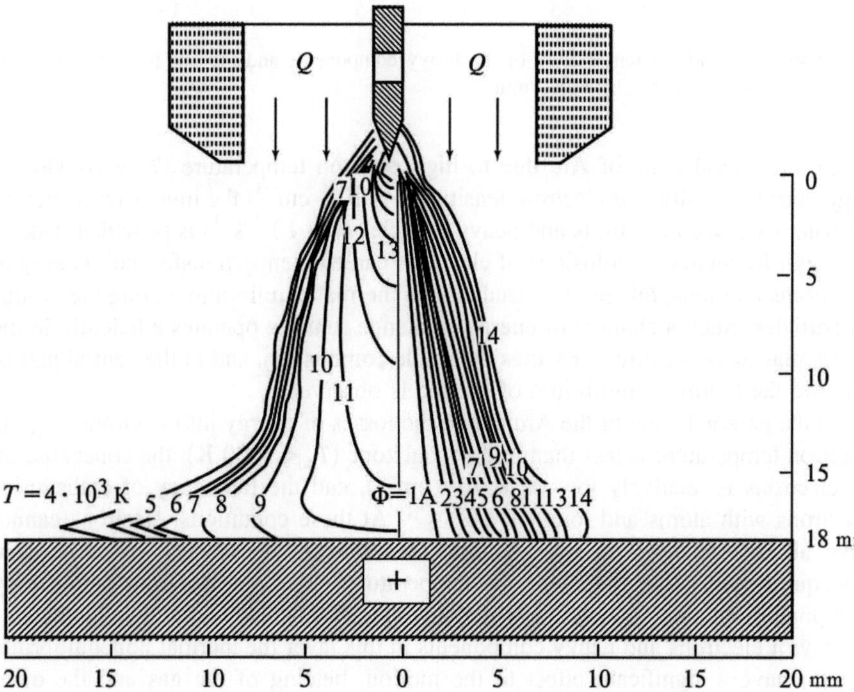

Fig. 8.45 The fields of the electron gas temperature and the function of the electric current in the welding Argon Arc at I = 90 A, Q_{Ar} = 20 l/min

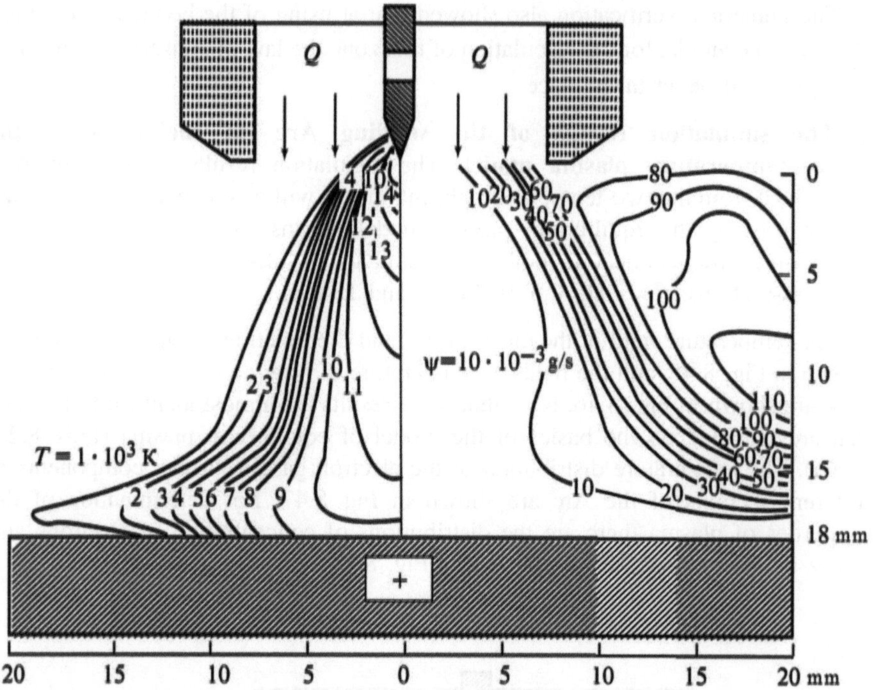

Fig. 8.46 The fields of temperature of the heavy components and the gas flow in the Argon welding Arc at $I = 90$ A, $Q_{Ar} = 20$ l/min

In the central zone of Arc due to high electron temperature ($T_e > 10{,}000$ K), high current density and electron density ($n_e > 10^{16}$ cm^{-3}) the higher frequency of collisions between electrons and heavy particles ($\nu > 10^{11}$ s^{-1}) is provided. Due to this high frequency of collisions of electrons can efficiently transfer their energy to the atoms and ions; this process leads to the thermal equilibrium among the groups of particles. Such a channel of energy exchange running operates efficiently in the dense plasma of the Arc, it ensures the stable combustion, and in the central part of the Arc the thermal equilibrium of plasma is observed.

In the external zone of the Arc due to the losses of energy into environment, the electron temperature is less than in the axial zone ($T_e < 9000$ K), the concentration of electrons is relatively low ($n_e < 10^{15}$ cm^{-3}), and the frequency of collision of electrons with atoms and ions $\nu < 10^{10}$ s^{-1}. At these conditions, electrons cannot give all the obtained energy to atoms and ions the gas stays in the thermal nonequilibrium. The difference of temperatures between electrons and heavy components in this zone is 3000–4000 K (Fig. 8.47). However, due to the low energy of electrons and heavy components in this layer the thermal nonequilibrium cannot have a significant effect to the motion, heating of the gas and the other parameters of the Arc, that is confirmed by the coincidence of the basic data of two models. Therefore, for modeling of the majority of technological processes using

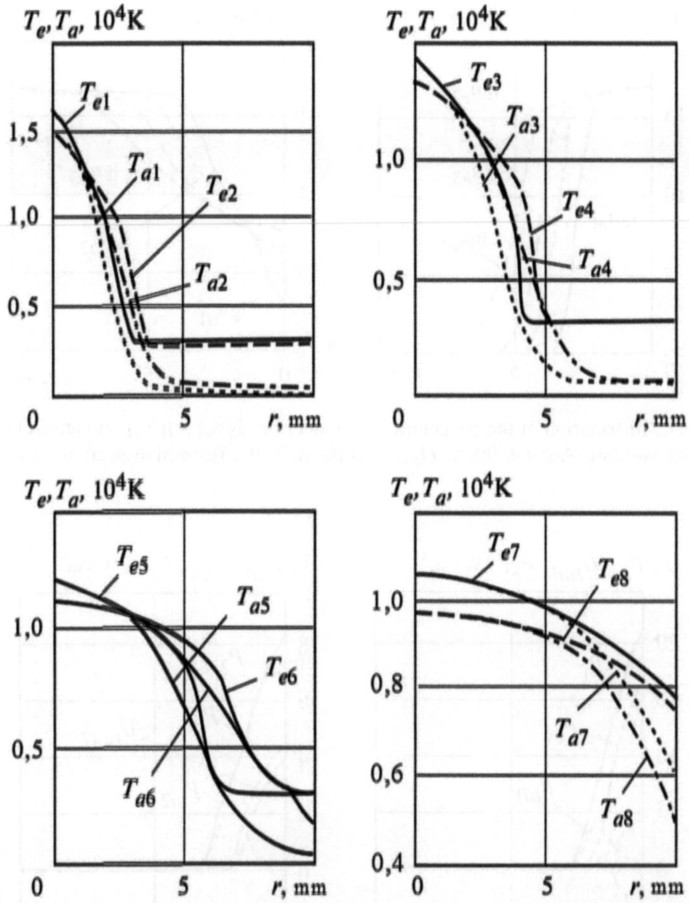

Fig. 8.47 The distribution of temperatures of electron gas T_{e1}–T_{e8} and heavy components T_{a1}–T_{a8} in the Argon welding Arc $I = 90$ A, $Q_{Ar} = 20$ l/min in the respective sections $z = 1, 2, 3, 5, 9, 14,$ 17 and 18 mm

the Arc at atmospheric pressure, including welding and melting of metals, it is enough to use the model of plasma equilibrium. Apparently, nonequilibrium state significantly affects the properties of plasma flow in the case of the small Arc current density ($I/d \ll 100$ A/cm) and at the reduction of the pressure ($p \ll 1$atm).

An important result is the balance of power in the Arc. The distribution of specific power in the different sections of the welding Arc is shown in Fig. 8.49: the power $P = \sigma E^2$ obtained by electrons from the electric field; power $P_{ai} = b(T_e - T_a)$ transmitted by electrons to heavy particles at the elastic collisions; the power losses due to radiation U_{rad}; power $P_e = \sigma E^2 - P_{ai} - U_{rad}$ of the electron gas heating, most of this power spent to heat conduction, convection and uncompensated ionization $P_I = \mathrm{div}\left(E_i n_e - \frac{5}{2} k T_e n_a\right)\vec{v}$. Thus, by these data we can to define the

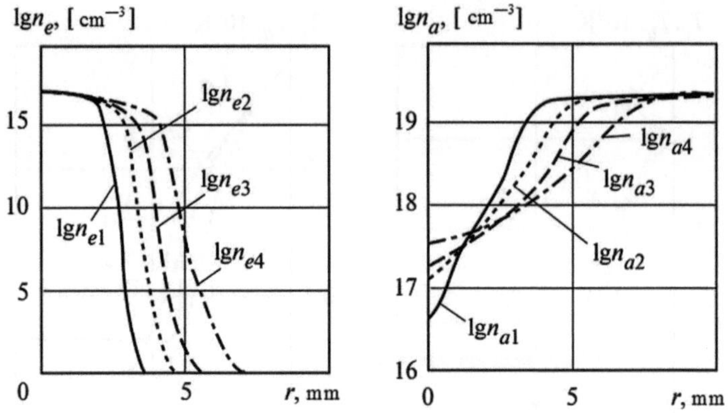

Fig. 8.48 The distribution of the concentration of electrons $\lg n_{e1} - \lg n_{e4}$ and atoms $\lg n_{a1} - \lg n_{a4}$ in the Argon welding Arc $I = 90$ A, $Q_{\mathrm{Ar}} = 20$ l/min in the respective sections $z = 1$, 2, 3 and 5 mm

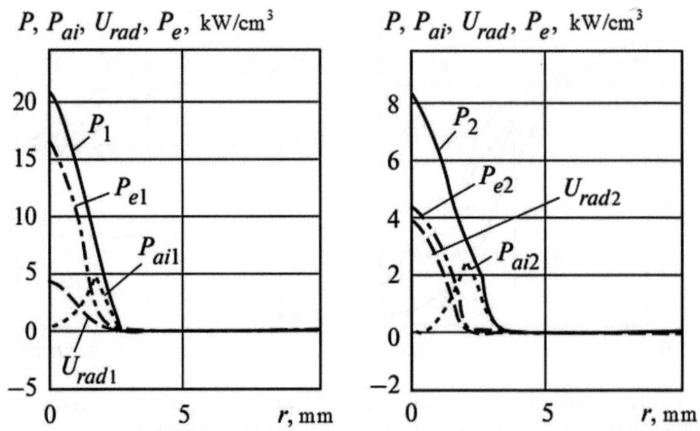

Fig. 8.49 The balance of power in the Argon welding Arc $I = 90$ A, $Q_{\mathrm{Ar}} = 20$ l/min in the sections $z = 1$ and 2 mm, respectively

role of each component in the power balance of the electric Arc; for example, the role of radiation in the balance of energy increases with increasing temperature, etc.

3. **Results of the simulation of the plasma Arc for melting of the steels**. The results of the modeling of free Arc in the chamber for melting of the steels with following data: the current $I = 1000$ A, the consumption of argon gas $Q_{\mathrm{Ar}} = 90$ l/min, the diameter of the tungsten (1.5 % ThO_2) cathode $d_c = 10$ mm, the size of the channel of plasma-forming gas $\Delta r = 1.5$ mm, the distance between the cathode and the liquid metal bath $l = 150$ mm.

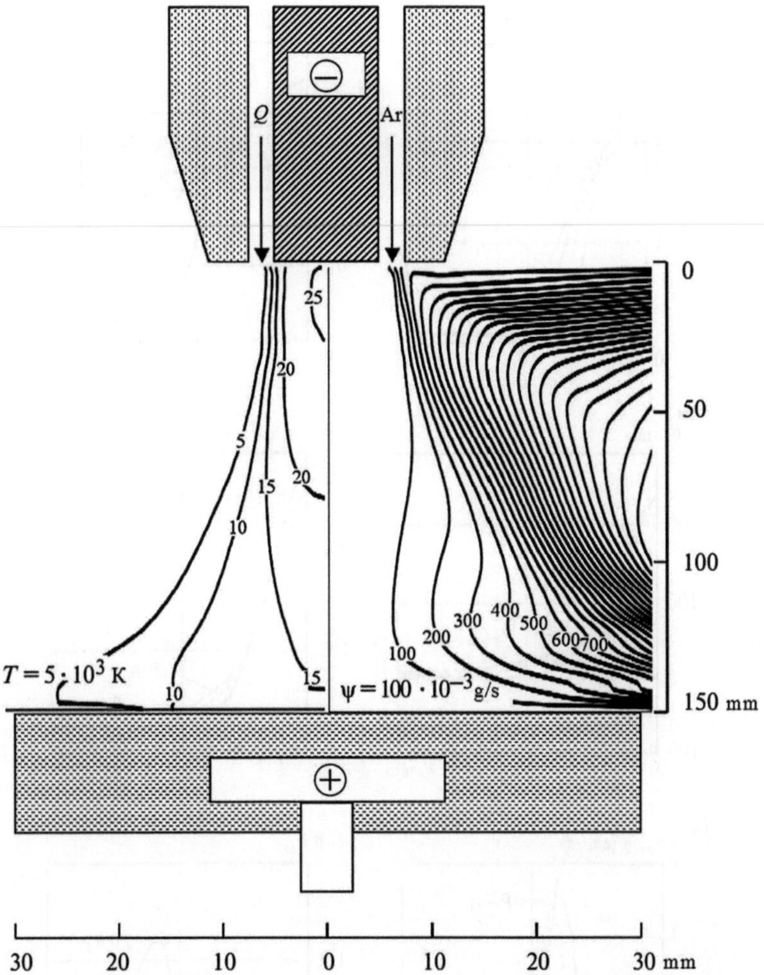

Fig. 8.50 The fields of the temperature and the flow of the gas in the chamber of the Argon plasma Arc for the melting of the steels

These obtained data of the fields of the temperature, the gas flow, the velocities and the specific consumptions of the Argon gas (Figs. 8.50, 8.51); the distribution of the intensity of electromagnetic field and specific power (Fig. 8.52), the components of the electromagnetic force F_r and the excess static pressure Δp (Fig. 8.53) and other parameters of the electric Arc allows the determination of the heating degree of the gas, to analyze the efficiency of the plasma torch, to determine the thermal efficiency of the gas heating, to estimate the heat flow into the anode, etc. The main application of the obtained results and the developed method is to use them for the theoretical study of plasma processes in the electric Arc, for design of the Arc plasma torches, for determination of the optimal and stable regime of the operation with the

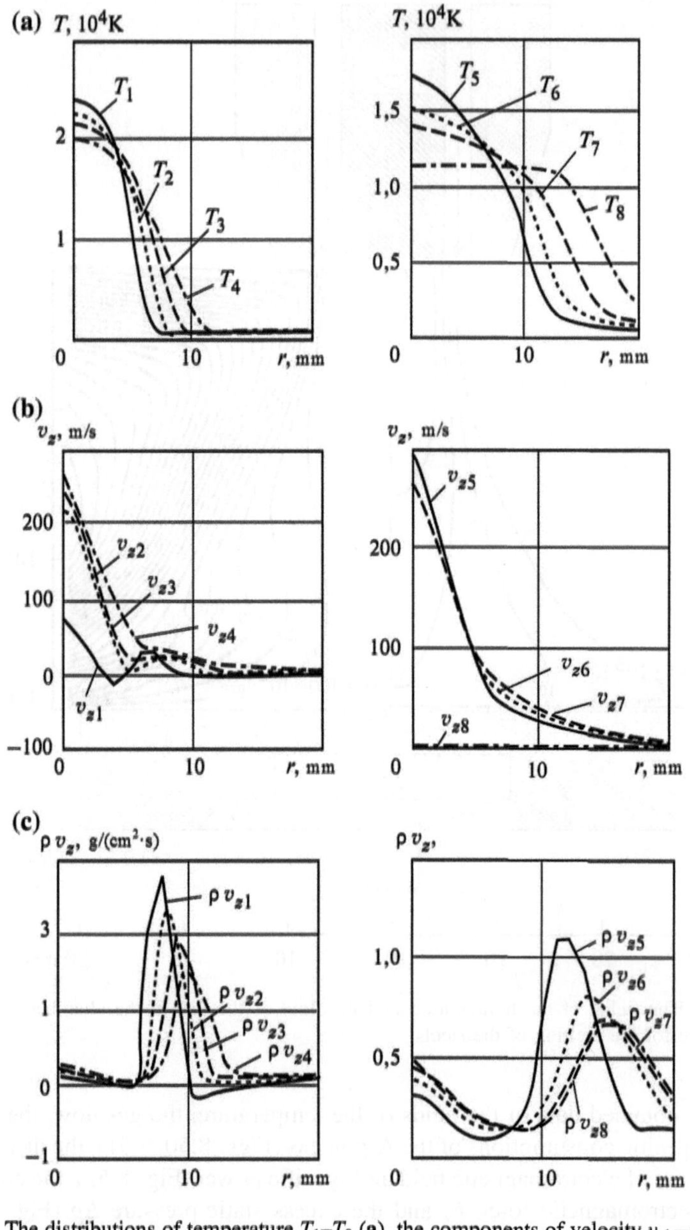

Fig. 8.51 The distributions of temperature T_1–T_8 (**a**), the components of velocity $v_{z1}-v_{z8}$ (**b**) and the specific consumptions $\rho v_{z1}-\rho v_{z8}$ (**c**) in the respective sections z = 2.5, 10, 20, 50, 80, 110, 140 and 150 mm of the chamber of the Argon plasma Arc for melting of the steels

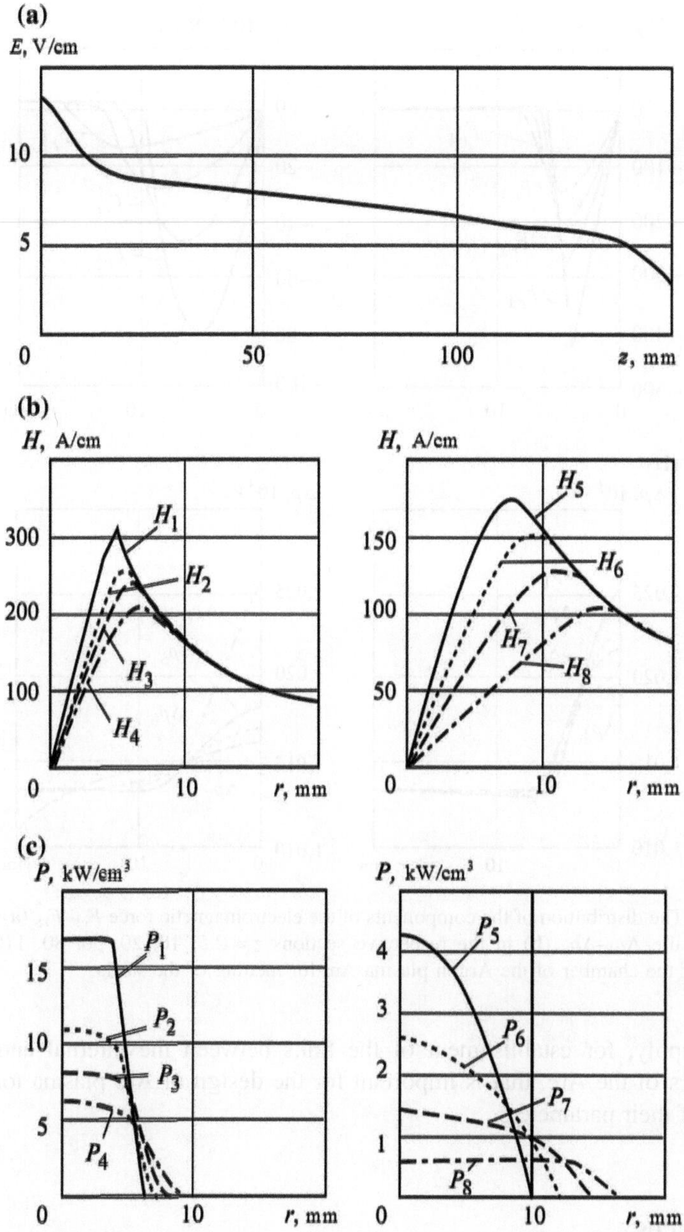

Fig. 8.52 The distributions of the intensity of electric field E on the axis (**a**), the intensities of magnetic field H_1–H_8 (**b**) and the specific power P_1–P_8 (**c**) in the respective sections z = 2.5, 10, 20, 50, 80, 110, 140 and 150 mm of the chamber of the Argon plasma Arc for the melting of the steels

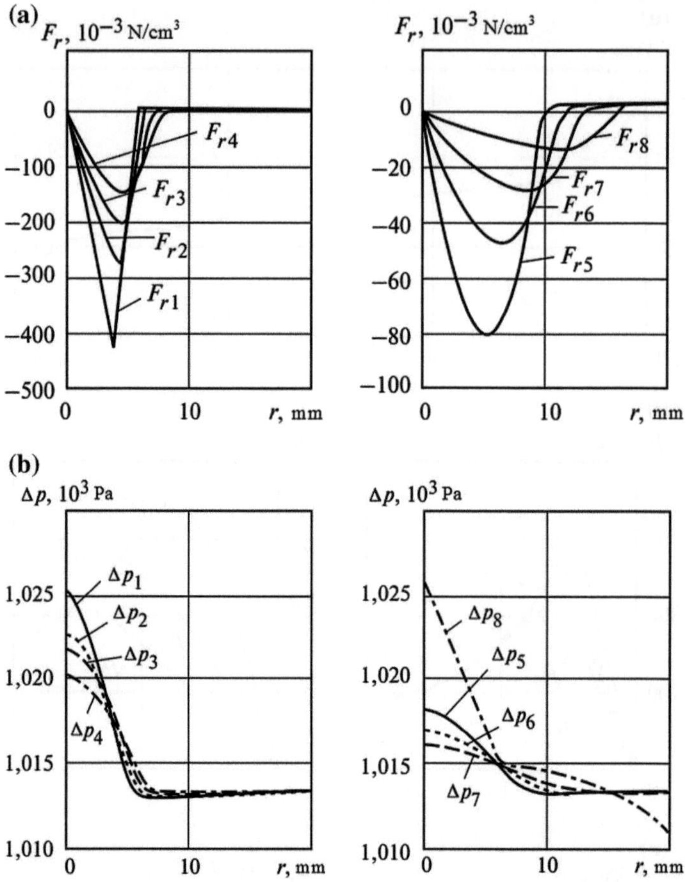

Fig. 8.53 The distribution of the components of the electromagnetic force F_{r1}–F_{r8} (**a**) and excess static pressure Δp_1–Δp_8 (**b**) in the respective sections z = 2.5, 10, 20, 50, 80, 110, 140 and 150 mm of the chamber of the Argon plasma Arc for melting of the steels

power supply, for establishment of the links between the internal and external parameters of the Arc, that is important for the design of Arc plasma torches and control of their parameters.

References

1. Dresvin S.V. Low-temperature plasma generators // Encyclopedia of Low-Temperature Plasma / Ed. by V.E. Fortov. M.: Nauka, 2000. V. II. PP 280–328.
2. Finkelberg V., Mekker G. Electric Arcs and thermal plasma. M.: Publishing House of Foreign Literature, 1961.
3. Granovsky V.L. Electric current in the gas. M.: Gostechizdat 1952.

4. Granovsky V.L. Electric current in the gas. Continuous current. M.: Nauka, 1971.
5. Modelling and calculation methods of physical and chemical processes in Low temperature plasma / Ed. L.S. Polak. M.: Nauka, 1974.
6. Anderson J.A. Transport phenomena in thermal plasma. M.: Energy 1972.
7. Novikov O.J. The stability of the electric Arc. L.: Energy 1978.
8. Zhukov M.F., Koroteev A.S., Yuryukov B.A. Applied Dynamics of thermal plasma. Novosibirsk: Nauka. 1975.
9. Zhukov M.F., Smolyakov V.J., Yuryukov B.A. Electric Arc gas heaters (plasma torches). Novosibirsk: Nauka. 1973.
10. The theory of an electrical Arc in a forced heat transfer / Ed. Zhukov M.F. Novosibirsk: Nauka. 1977.
11. The theory of thermal plasma Arc. Part 1: Mathematical Methods Plasma / M.F. Zhukov, B.A. Yuryukov, V.S. Engelsht et al. Novosibirsk. Nauka. 1987.
12. The theory of thermal plasma Arc. Part 2: Non-stationary processes and radiation heat transfer in thermal plasmas / M.F. Zhukov, B.P. Devyatov, O.J. Novikov et al. Novosibirsk. Nauka. 1987.
13. Zeldovich J.B., Raiser Y.P. Physics of Shock Waves and High-Temperature Hydrodynamic Phenomena. M.: Nauka. 1966.
14. The generation of electric Arc plasma streams / Ed. P.E. Nakoryakov. Novosibirsk: Nauka. 1987.
15. Alexandrov A.F., Ruhadze A.A. The physics of high-current electric-discharge light sources. M.: Atomizdat. 1976.
16. Dautov G.Y., Dzyuba V.L., Carp I.N. Plasma torches with stabilized electric Arcs. Kiev: Naukova Dumka. 1984.
17. Heat transfer in the Arc gas heater / A.G. Shashkoff, L. Krejci, V.I. Krylovich et al. M.: Energy. 1974.
18. Y.P. Raiser. Physics of the gas discharge. M.: Nauka, 1987.
19. Kolesnikov V.N. The Arc discharge in inert gases // Physical Optics (Proc. of Phys. Inst of Acad. Sc. of USSR.). M.: Publishing PhIAN, 1964. V. 30. pp. 66–157.
20. Asinovsky E.I., Pakhomov E.P. An analysis of the temperature field in a cylindrically symmetric pole of electric Arc // J. Thermal physics of High Temperatures. 1968. V. 6. No 2. PP. 333–336.
21. Physics and technology of Low-temperature plasma / Ed. by S.V. Dresvin. M.: Atomizdat. 1972.
22. Incropera F.P. Procedures for modeling laminar cascade Arc behavior // IEEE Transaction on Plasma Science. 1973. Vol. Ps-I, No 3. P. 3–9.
23. Panasenko L.N. The two-temperature plasma model Arc // Study of plasma devices. Minsk. 1978. PP. 23–29.
24. Averin I.B., Sinyarev G.B. The method of calculating the electric Arc blown laminar gas flow in a cylindrical channel with chemical nonequilibrium // Proc. of the VII Conf. by generators of low-temperature plasma. Alma-Ata. 1977. No 2. PP. 23–26.
25. Dresvin S.V. The two-temperature plasma model under steady purge gas through the plasma torch // J. Technical Physics Letters. 1973. No 4. PP. 3–12.
26. Voropaev A.A., Dresvin S.V. The two-temperature model of laminar Arc in a forced purge gas through the plasma torch // J. Thermal physics of High Temperatures. 1973. T. 11. No 2. PP. 333–341.
27. Donskoy A.V., Klubnikin V.S., Salangin A.A. Two-temperature simulation of Argon plasma in the channel // J. Technical Physics. 1985. V. 55. No 11. PP. 2124–2128.
28. Clarke C.J., Inkropera F.P. Thermochemical nonequilibrium in a stabilized Argon Arc plasma // J. Rocket and technical complex. 1972. V. 10. No 1. PP. 19–21.
29. Haidar J. Nonequilibrium modeling of transferred Arcs // J. Appl. Phys. 1999. Vol. 32. P. 263–272.
30. Hsu K.C., Etemadi K., Pfender E. Study of the free-burning high-intensity arcs // J. App. Phys. 1983. Vol. 54. No 3. P. 1293–1301.

31. Calculations of a free-burning electric ARC by nonequilibrium model / J. Amouroux, S. Dresvin, S. Nguyen-Kuok, A.V. Krylov // Proc. of the VI European Conference on Thermal plasma Processes. Strasbourg, 2000. P. 289–294.
32. The theory of an electrical Arc column / V.S. Engelsht, V.S. Gurovich, G.A. Desyatkov et al. Novosibirsk: Science. 1990. V.1: The Low-temperature plasma.
33. Kudinov V.V. Bobrov G.V. Coating by spraying: Theory, technology and equipment. M.: Metallurgy. 1992.
34. Research and optimization of plasma metal cutting / Y.Y. Kiselev, I.A. Protsuk et al. Kishinev: Shtiintsa. 1981.
35. Kiselev Y.Y. Energy Processes of Air plasma cutting of metals. Kishinev: Shtiintsa, 1980.
36. Donskoy A.V., Klubnikin V.S. Electrical plasma processes and installation engineering. L.: Engineering, 1979.
37. Schulz-Gulde E. // Paper of Institute of atom and molecular physics.- Hannover. Germany, 1997.
38. AIAA Paper. 1966. No 66(189). P. 1.
39. Donskoy A.V., Klubnikin V.S. // J. of Engineering Physics. 1972. V. 22. No 6.
40. Asinovsky E.I., Kirillin A.V. Encyclopedia of Low-Temperature Plasma: / Ed. Fortov V.E. V. II. M.: Nauka. 2000.
41. A study of the plasma torch with the rising current-voltage characteristics / A.S. Anshakov, M.F. Zhukov et al. // News of Sib. Dep. USSR Academy of Sciences. Ser. Tech. Sciences. 1970. No 8. Vol. 2, pp. 3–11.
42. Theory of welding processes // Ed. V.V. Frolov. M.: Higher School, 1988.
43. Lowke J.J. Calculated properties of vertical arcs stabilized by natural convection // J. App. Phys. 1979. Vol. 50. No 1. P. 147–157.
44. Kovitya P., Lowke J.J. Two-dimensional calculation of welding arc in argon // Urania, I. I. W. Doc. 1982. Vol. 53. No 11. P. 20–24.
45. Mathematical modeling of electric Arc / Ed. V.S. Engelsht. Frunze: Ilim. 1983.
46. Zhaynakov A., Engelsht V.S. Mathematical modeling of open high-current arc // Generate arc plasma streams / Ed. M.F. Zhukov. Novosibirsk: Institute of thermal physics of Sib. Dep. USSR Academy of Sciences. 1987. PP. 323–342.
47. Mechev V.S., Eroshenko L.E. Axial distribution of electric Arc temperature in Argon // Automatic Welding. 1975. No 6. PP. 14–17.

Chapter 9
Electrode Processes in the Arc Plasma Torches

In the previous chapter the investigation of the Arc plasma torches (Arc column in the channel and in the free Arc) was represented based on the model of equilibrium and nonequilibrium plasmas without considering the near electrode processes. In these models, the boundary conditions for different parameters in the original section of the Arc column were determined approximately using the balance of energy and electromagnetic field (8.37, 8.38) without taking into account the near electrode processes. However, the near electrode processes and processes in the Arc column are closely linked: they determine the number of electrons, emitted from the cathode surface, the dynamics of gas heating in the electric Arc, current, and power in the cathode region and the Arc column, and others. Therefore, the correct study of the Arc plasma torches requires an integrated approach joint with consideration of the physical processes in the electrode regions and in the Arc column.

On the other hand, the effective use of the electric Arc in technology requires the creation of Arc plasma torches with a long service life that is mainly determined by the service life of the electrode system. Results of the study of electrode processes, in addition to identifying the physical picture of the near electrode phenomena helps to solve the problem of increasing the resource of the installation. In practice, the experimental study of near electrode processes, there are a number of difficulties related primarily to the fast physical processes, the presence of the evaporating electrode material, and zone of the charge separation and others. All of this determines the necessity of using the theoretical methods for the study of electrode processes of the electric Arc, especially for Arc plasma torches, operating at atmospheric pressure in inert and molecular gases.

Theoretical study of electrode processes is often carried out by means of one-dimensional two-layer model [1–3], that is permissible at length of the electrode region, much less than its radial dimensions. Based on this model, the most important parameters of the Arc can be found, such as current density and heat fluxes to the electrodes, the power allocated at them, the voltage drop in the electrode regions, the electric field strength, and the temperature of electron and the heavy components of plasma at the cathode and anode. These values can then be

S. Nguyen-Kuok, *Theory of Low-Temperature Plasma Physics*,
Springer Series on Atomic, Optical, and Plasma Physics 95,
DOI 10.1007/978-3-319-43721-7_9

used to determine the extent of the electrodes erosion, their share of the work, as well as boundary conditions for the solution of the equation of the energy balance of the Arc column.

It should be noted that despite the numerous studies devoted to the development of models and calculation of electrode processes, the results are far from perfect, many important factors are not taken into account or are difficult to analyze, and only recently important data and suggestions on the construction of a closed model of electrode processes appeared. In the description of electrode processes, we will use mainly on the two-layer model with these latest proposals and try as much as possible reflect the physical meaning and the role of each process in the model. Much attention will also be paid to the development of methods of modeling of electrode processes, procedures of the solutions and analysis of the electrodes operating in extreme cases.

In practice, at the operating of the Arc plasma torches in inert gases, Nitrogen, Hydrogen, and mixtures the thermo-emission Arc cathodes are commonly used the tungsten-based ones [1]. In air, oxygen, nitrogen, and carbon-containing gases cathodes based on metals (Zirconium, Hafnium, Molybdenum, etc.) are used and they form the film of the compounds (oxides, nitrides, carbides) in interaction with plasma forming gases, they have high emission properties and thermal stability. The data obtained as a result of experimental and theoretical studies show that near electrode processes represent a group of closely related with each other physical processes occurring simultaneously in the near electrode region, in the body of the electrode and on the surface. Changes that occur in any of these three areas will automatically lead to changes in the other. Therefore, at the study of Arc cathodes it is necessary to consider the complex collaborative processes in the area of the near electrode, on the surface and in the material of the electrode.

9.1 Model of the Cathode Processes

Description of the Near Cathode Processes The main function of the near cathode region is to provide an enabling environment transport of charged particles between the cathode and the Arc column. A simplified picture of the phenomena in the near cathode region of the Arc plasma torches can be presented on the basis of two-layer model (Fig. 9.1).

The first layer, if we count from the cathode surface, is called the space charge layer. It has a length of the order of the Debye radius and less than the mean free path of electrons and ions l_{ei}. For example, in Argon plasma at atmospheric pressure over a temperature range $T_e = (10 - 40) \times 10^3$ K, Debye length is $r_D = 0.05$–0.02 µm, $l_{ei} = 4 - 1$ µm.

In this layer, due to the relatively large proportion of the ion current and the low mobility of ions excess positive space charge forms that causes the potential jump at

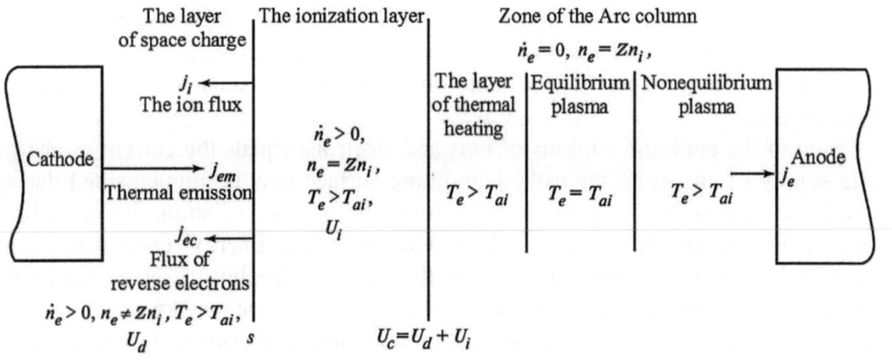

Fig. 9.1 The scheme of the electric Arc

the cathode surface (cathode potential drop). The role of the potential jump near the surface of the cathode is reduced: to accelerate the ions moving to the cathode; to ensure a sufficiently high electron emission due to the increased temperature of the surface bombarded by ions; to create at the surface of the cathode of a strong electric field, reducing the electron work function and thus facilitates the issuance; to the acceleration of the electrons emitted by the cathode to the energy levels needed to provide intensive generation of charged particles in the cathode region. This is caused by the need to create demand (in terms of energy and current balances) density of the ion current flowing to the cathode. Through this process, the proportion of the ion current in the cathode region is much larger than in the discharge column [1, 4].

The second layer, that separates the space charge layer and the Arc column, is called ionization layer. In this layer, the condition of quasi-neutrality and intensive generation of charged particles is performed and intensive generation of charged particles occurs due to the energy acquired by the electrons in the first layer. From the second layer to the first one not only ions move, but also the reverse electrons. Because of the inhibitory effect of the field only a small fraction of the electrons, that have sufficient energy to overcome the barrier, reaches the cathode surface.

Ions that are accelerated by the electric field reach the surface of the cathode and that ions are neutralized by electrons of the cathode metal. The surface emitted electrons and atoms move from the cathode. Atoms, having a first layer without collision, fall into the second one, reducing the degree of ionization [1, 3, 4]. At the same time, the electrons accelerated in the first layer, and reached to the second one, increase gradually the degree of ionization and the concentration of charged particles. A value of the concentration is close to the equilibrium at the border with plasma Arc column.

In the second layer there is a redistribution of an electron current. The results of our calculation shows that the percentage of the electron current $S_e = j_e/j$ increases from $S_e \approx 0.6$–0.9 at the cathode surface to $S_e = \mu_e/(\mu_e + \mu_i)$ in the discharge column (μ_e, μ_i—the mobility of electrons and ions respectively) [5]. As has been said, the large value of the ion current, supported by the processes occurring in the second layer, provides formation of the positive charges excess and the potential jump in the first layer. Thus, the processes in both two layers are closely related, and the calculation of the near cathode processes should be carried out jointly for both layers.

Sum of the emission currents of ions and electrons equals the current discharge and supply of energy to the cathode emitting surface may be implemented due to ion bombardment, the reverse electrons flow, the plasma radiation, the neutralization of the ions on the cathode surface, Joule heat, etc. Energy removal from the cathode is carried out basically due to the emission cooling, heat conductivity, atomic evaporation, and thermal radiation from the cathode surface.

Consider the case of the cathode with a completely smooth surface with transverse dimension much larger than the mean free paths of the plasma particles. That allows the creation of near cathode region model in the one-dimensional approximation.

In general, a rigorous theoretical study of the near cathode region can be based on the solution of the Boltzmann kinetic equations for each component of the plasma (9.1) [6].

$$\left(\vec{v}_s \cdot \nabla_r + \frac{Ze\vec{E}}{m_s} \cdot \nabla_v\right) f_s(r, v_s) = \mathrm{St}_s \tag{9.1}$$

where f_s—the distribution function of the sth plasma component, $s = a, e, i$; $\vec{v}_s$, m_s—the velocity and mass of the sth plasma component; Z—the charge number of the sth plasma component; e—the electron charge; $\vec{E}$—the intensity of electric field; St_s—collision term which takes into account both elastic and inelastic collisions between the particles of sort s with each other and with particles of other varieties.

Kinetic Boltzmann equation is written in the stationary form (9.1), since in all cases the real-time characteristic of the Arc parameters (current, voltage, gas pressure) are much longer than the plasma relaxation time. The exact solution of Eq. (9.1) causes serious difficulties due to the difficulty in determining the collision term St_s. In practice, the Eq. (9.1) is usually solved by approximate methods (Chap. 5). Currently, a two-layer model is most relevant and proven for study of near cathode processes. Here is a description of the model using the method of integral balances, which allows description of processes in the layer of space charge, in the ionization layer and on the cathode surface by means of integral equations on the basis of the current and energy balance.

The Layer of Space Charge In the space charge region with the size of the order of the Debye radius virtually there are not collisions of particles, and the flow of charged particles through this area can be considered constant. The potential distribution in the layer is described by the Poisson equation:

$$\varepsilon_0 \frac{d^2 V}{dz^2} = e(n_e - Zn_i), \quad (9.2)$$

where V—the electrostatic potential associated with the electric field intensity as $E = -\frac{dV}{dz}$, the z-axis is perpendicular to the surface of the cathode (Fig. 9.1); Z—the charge number of the ions; ε_0—dielectric permittivity in vacuum.

The solution of this Eq. (9.2) can be obtained by direct integration, if the velocity distribution of the electrons and ions on the boundary is known. At the moment there is only a rigorous solution for the case when ions moving dominate the collisions with the atoms with change their charge [7]. It is usually assumed that the distribution of electrons in the space charge region is described by the Boltzmann equation. Zero electrostatic potential is selected at the boundary of space charge layer, and the ion density is determined by solving the Boltzmann equation for ions (9.1). As a result, the electric field intensity near the cathode surface can be found from the well-known McCone equation:

$$E_c^2 = \frac{4}{\varepsilon_0}\sqrt{\frac{m_e}{2e}}\left(j_i\sqrt{\frac{m_i}{Zm_e}} - j_e\right)\sqrt{U_d}, \quad (9.3)$$

where j_i, j_e—current density of ions and electrons; m_i, m_e—mass of the ions and electrons, respectively; U_d—the voltage drop in the space charge layer.

Equation (9.3) does not take into account the fluctuations of the electric field associated with the discrete of charges of the ions. In [8] it is shown that in conditions of atmospheric pressure Arc ($j \leq 10^5$ A/cm^2, $E_k < 10^7$ V/cm), the number of additionally emitted electrons by the cathode due to the fluctuations of the electric field per ion is much less than unit.

The presence of the potential barrier at the metal-vacuum boundary prevents the output electrons from the metal. So it is necessary to give energy (work function) for electron to overcome this barrier. There are various types of emission (it depends on the nature of energy supply): thermionic, secondary electronic due to ion, photons, and excited atoms bombardment. Due to quantum mechanical effects arising from the action of the field, there is a possibility of electron tunneling through a potential barrier (field emission). Consider the current and energy balance on the surface of refractory Arc cathodes at different mechanisms of electron emission.

The Current Balance on the Cathode Surface When thermionic emission first of all it is necessary to carry heat to the cathode, due to which there occupier of the energy states of the atom lying above the Fermi level occurs. Some of these states have energy greater than the height of the potential barrier, and electrons from these levels can leave the cathode surface. Thermionic emission current density is described by the Richardson–Dushman equation [1, 3, 7]:

$$j_{\text{em}} = AT_c^2 \exp\left(-\frac{e\varphi - \Delta(e\varphi)}{kT_c}\right), \tag{9.4}$$

where A—Richardson constant, which is determined by experiment; T_c—temperature of the emitting surface of the cathode; $e\varphi$—the work function of the cathode material; k—Boltzmann constant; $\Delta(e\varphi)$—lowering the work function due to the strong electric field on the surface of the cathode—Schottky effect,

$$\Delta(e\varphi) = e\sqrt{\frac{eE_c}{4\pi\varepsilon_0}}. \tag{9.5}$$

Schottky effect becomes significant when electric field intensity is $E_c \geq 10^4$ V/cm. Work function $e\varphi$ depends on many factors: the chemical composition and structure of the material, surface conditions, etc. Existing theoretical methods do not allow reliably determination its relevance for the particular materials and working conditions. The main source of information is an experiment. Data on values of work function of different materials and characteristics of the cathodes can be found, for example, in the [1–3, 9].

As an example, Fig. 9.2 shows the dependence of the thermionic current density on temperature and work function. The lowest work function $e\varphi$ = 1.0–1.6 eV oxide cathodes have widely used in other electronic devices. They are made of alkaline metal oxide layer (e.g., an oxide sintered body containing BaO and SrO), deposited on the surface of metal wire or ribbon. Along with barium–strontium oxide cathodes other types of cathodes, including pure tungsten ones are used in vacuum devices.

On cold cathodes at very high intensity of electric fields the potential barrier is narrow and permeable to electrons and due to the tunnel effect field electron emission mechanism (cold emission) is realized. Relation between the field emission current density (A/cm^2) and the electric field intensity (V/cm) is determined by the Fowler–Nordheim equation [10]:

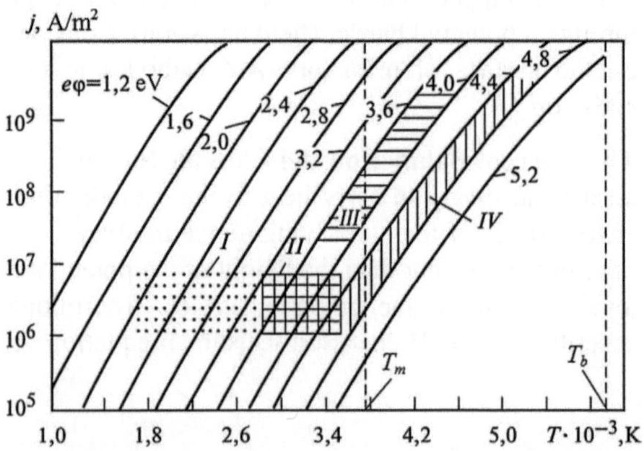

Fig. 9.2 The thermionic currents dependence of the temperature and the work function $e\varphi$ of the Arc cathodes [3]: *I* Tungsten (impregnated), *II* Tungsten (activated), *III* thermochemical, *IV* Tungsten (pure), T_m, T_b the melting and boiling temperature points of Tungsten

$$j_{\text{em}} = 1.54 \times 10^{-6} \frac{E_{\text{c}}}{\varphi} \exp\left(-6.8 \times 10^{-7} \frac{\varphi^{3/2}}{E_{\text{c}}}\right). \tag{9.6}$$

Therefore, the field electron emission (9.6) becomes significant only when $E_c > 10^7$ V/cm. The analysis of experimental data on refractory Arc cathodes and the results of further calculation show that when implemented current densities at the cathode Arc plasma torches $j = 10^2$–10^4 A/cm^2, the electric field intensity $E_c < 10^7$ V/cm. The analysis in [3, 7], also shows that, in the Arc of the secondary electron emission due to ion bombardment, and photoemission excited atoms can be neglected. Therefore, in practice for thermionic cathodes are usually limited to only emission is described by the Richardson–Dushman (9.4).

Velocity of the ions at the boundary of the space charge layer and the ionization layer v_{is} is determined by the Bohm criterion [1, 7]:

$$v_{\text{is}} = \sqrt{k(T_{\text{i}} + ZT_{\text{e}})/m_{\text{i}}}. \tag{9.7}$$

The ion current density at the cathode is

$$j_{\text{i}} = Zen_{\text{is}}v_{\text{is}}. \tag{9.8}$$

Ion concentration at the boundary of the space charge region n_{is} can be identified by the correlation equation [7], obtained from the analysis of the results of numerical calculation of the multicomponent equations for the ionization layer with a constant temperature of electrons, atoms, and ions.

$$n_{\text{is}} = n_{\text{i}\infty} \frac{0.8}{2+\alpha}, \quad \alpha = \sqrt{\frac{kT_{\text{i}}}{m_{\text{i}} D_{\text{i}\infty} k_r n_{\text{i}\infty}^2}}, \tag{9.9}$$

where $n_{\text{i}\infty}$—equilibrium ion density at the boundary of the ionization layer and the Arc column; $D_{\text{i}\infty}$, k_{r}—the diffusion coefficients of the ions and the three-body recombination at the border of the ionization layer and the Arc column (Sect. 4.4).

To determine the density of the reverse electron current j_{ec}, diffusing from the quasi-neutral plasma to the cathode, it can be assumed that the velocity of the electrons at the boundary of the space charge region is close to the average velocity at the electron temperature T_e [1, 3, 7] and their Maxwell distribution, then

$$j_{\text{ec}} = \frac{en_{\text{es}}\bar{v}_{\text{e}}}{4} \exp\left(-\frac{eU_{\text{d}}}{kT_{\text{e}}}\right), \tag{9.10}$$

where $\bar{v}_{\text{e}} = \sqrt{8kT_{\text{e}}/\pi m_{\text{e}}}$—the average thermal velocity of the electrons; $n_{\text{es}} = Zn_{\text{is}}$—the electron density at the boundary of the space charge region.

Full current density at the cathode is

$$j = j_i + j_{em} - j_{ec} = j_i + j_e. \tag{9.11}$$

The Energy Balance at the Cathode Surface Analysis of the energy balance at the cathode surface plays a key role in the study of the cathode and cathode processes, because it allows to evaluate the thermal state of the cathode and to determine its service life. Heating of the cathode is mainly carried out by the flow of ions, accelerated in the zone of the cathode potential drop. At the same time, the thermionic emission current determines the cooling of the cathode, and electrons are emitted ions are involved in the formation of the space charge determining ion flux to the cathode, and the electric field. Thus, closely interrelated processes take place in the cathode region and on the cathode surface. For drawing of the energy balance knowledge of the conditions in the area bind to the surface discharge and the space charge zone requires, it depends on many factors including the material, geometry, the cooling conditions, the cathode current level, gas type, gas pressure, and others.

The energy flux density caused by ion bombardment of the cathode surface can be obtained by assuming a Maxwell ion velocity distribution in the space charge layer [7]:

$$q_i^c = \frac{j_i}{Ze}\left[k(2T_i + ZT_e/2) + ZeU_d\right]. \tag{9.12}$$

On the boundary of the space charge layer flow density of kinetic energy of fast electrons with kinetic energy $\frac{m_e v_e^2}{2} \geq eU_d$ can be obtained by integrating with Maxwell function and equal to $\frac{j_{ec}}{e}(2kT_e + eU_d)$. By the time when electron reach the surface of the cathode its kinetic energy decreases by eU_d. Consequently, the kinetic energy flux density brought by reverse electrons to cathode is:

$$q_{ec} = \frac{j_{ec}}{e}(2kT_e + e\varphi_{eff}), \tag{9.13}$$

where $e\varphi_{eff} = e\varphi - \Delta(e\varphi)$—effective emission work function.

The kinetic energy flux density of the electrons leaving the cathode surface can be estimated by semi-Maxwell velocity distribution with a surface temperature T_c [7] and it will be equal to $2j_{em}kT_c/e$. Therefore:

$$q_{em} = \frac{j_{em}}{e}(2kT_c + e\varphi_{eff}). \tag{9.14}$$

Similarly, the energy flux density of atoms leaving the cathode as a result of the neutralization of the ions coming from the plasma will be equal to:

$$q_{\mathrm{a}} = j_{\mathrm{i}} \frac{2kT_{\mathrm{c}}}{Ze}. \tag{9.15}$$

Energy released at the cathode as a result of the neutralization of the ions is $E_{\mathrm{i}} - Ze\varphi_{\mathrm{eff}}$, E_{i}—ionization energy. Accordingly, the flow of energy, evolved at the cathode as a result of neutralization of ions, equal to:

$$q_{\mathrm{i}}^{\mathrm{i}} = \frac{j_{\mathrm{i}}}{Ze}(E_{\mathrm{i}} - Ze\varphi_{\mathrm{eff}}). \tag{9.16}$$

Energy supply due to the emission of plasma is:

$$q_R^{\mathrm{p}} = \frac{U_{\mathrm{rad}} Rl}{2(R+l)}, \tag{9.17}$$

U_{rad}—average volume coefficient of isotropic plasma radiation, it depends on the type, pressure and temperature of the plasma; l, R—the length and radius of the Arc column. For long Arcs $l > R$ we have $q_R^{\mathrm{p}} \approx \frac{U_{\mathrm{rad}}}{2} R$.

The energy flux density of thermal radiation from the surface of the cathode:

$$q_R^{\mathrm{c}} = \varepsilon \sigma T_{\mathrm{c}}^4, \tag{9.18}$$

where σ—Stefan–Boltzmann constant; ε—the degree of blackness of the cathode surface, depending on the type of material, surface conditions, temperature level, and it is determined experimentally.

The energy flux density with drawn during evaporation surface of the cathode:

$$q_m = h\dot{m}, \tag{9.19}$$

where h—specific heat of evaporation of the cathode material; $\dot{m}$—specific mass evaporation rate of the cathode discharge zone bindings.

Experimental data on the speed evaporation of the material in a binding discharge are absent. For estimate we use the results obtained under normal conditions. Our calculations show that for a tungsten cathode at a temperature in the spot $T = 3900$ K, $\dot{m} = 10^{-3}$ g/(cm^2 s), $h = 9600$ J/g, $q_m = 9.6$ W/cm^2, which is approximately two orders less than the energy flux due to ion bombardment. Furthermore, evaporated tungsten atoms can be ionized in the cathode zone, and return in the form of ions to the cathode.

The energy flux density withdrawn into the body of the cathode by heat conductivity is:

$$q_{\lambda} = \lambda \frac{\mathrm{d}T_k}{\mathrm{d}z}, \tag{9.20}$$

where λ—thermal conductivity of the cathode material.

Thus, the steady-state operation of the cathode when the temperature of the cathode and plasma does not change during this period of time, the energy balance on the surface of the cathode can be written as:

$$\begin{aligned} q = q_{\lambda} = & \frac{j_{\mathrm{i}}}{Ze}\left(k\left(2T_{\mathrm{i}} + \frac{ZT_{\mathrm{e}}}{2} - 2T_{\mathrm{c}}\right) + ZeU_{\mathrm{d}} + E_{\mathrm{i}} - Ze\varphi_{\mathrm{eff}}\right) \\ & + \frac{j_{\mathrm{ec}}}{e}(2kT_{\mathrm{e}} + e\varphi_{\mathrm{eff}}) - \frac{j_{\mathrm{em}}}{e}(2kT_{\mathrm{c}} + e\varphi_{\mathrm{eff}}) - h\dot{m} + q_R^{\mathrm{p}} - q_R^{\mathrm{c}}. \end{aligned} \tag{9.21}$$

Evaluations of author and other authors data [1, 3, 7, 8] show that the contribution to the output power flow on the surface of the cathode Arc plasma torches such streams $q_R^{\mathrm{p}}, q_R^{\mathrm{c}}$ as well as streams of emission and the secondary electron emission due to ion bombardment (because of the low strength of the field at the cathode) can be neglected.

It should be noted that when determining the internal temperature of the cathode body, Eq. (9.21) can be used as boundary conditions for solving the heat conduction equation of the cathode.

Ionization Layer The ionization layer is quasi-neutral region, adjacent to the space charge layer and the Arc column (Fig. 9.1). In this layer, there is intense ionization and redistribution of the proportion of electronic and ionic currents. The electrons emitted by the cathode and accelerated in the first layer supply necessary energy for this process.

The thickness of the ionization layer is of the order of recombination length, which is defined as the scale at which the rate of change of the concentration of charged particles due to ambipolar diffusion is comparable to the rate of ionization and recombination. For a dense plasma in the Arc discharge of high-pressure main mechanism of decay of charged particles is a triple (three-particle impact-radiation) recombination with the participation of a third body—the electron [11]. In this case, the length of the recombination can be estimated as $d = \sqrt{\frac{D_{\mathrm{amb}}}{k_{\mathrm{r}} n_{\mathrm{e}}^2}}$, where D_{amb}—the ambipolar diffusion coefficient; k_{r}—the coefficient of triple recombination [5, 12–18]. For the characteristic temperature T_{e} = 10,000–40,000 K, d = 0.1–1 mm.

In some parts inside the ionization layer the fraction of the ion current layer is the same as in the space charge region. At some distance from the ionization region (Fig. 9.1) ion current is avalanche reduced to a very small value. Since the flow of ions leaving, the ionization layer in the space charge region is much greater than the flow coming from the layer of thermal heating, ionization in the ionization layer must prevail over recombination, so a deviation from the ionization equilibrium takes place ($\dot{n}_e > 0$).

For a complete description of the processes in the ionization layer it is necessary to aware of the following phenomena: the movement and interaction of the particles, the relaxation of the electron distribution function, ionization, and recombination. Direct solution of the kinetic equations in this case is a difficult task. Therefore, in the theoretical studies used simplified models to give a mathematical description of a set of these processes. One of the following two models of the ionization layer is currently used as a rule.

The first model is based on the integral balance equations. In this model the ionization layer is assumed infinitely thin [1–3, 5]. This approximation is justified in cases where the length of the ionization of atoms by electron impact is significantly less than the mean free path of ions in the plasma. This condition is valid only when electron temperature in the ionization layer is high.

The second model can be used in those cases where the length of the ionization layer is much larger than the mean free path. This model is a system of differential equations describing the processes in the ionization layer. To get this information, method of integral–differential kinetic equation to a system of equations for the moments of the distribution function [1] is used, the plasma is considered to be in a state of partial local thermodynamic equilibrium, hold the Maxwell–Boltzmann–Saha equations (1.2, 1.4, 1.9). As is known, the first moments of the distribution functions are normal gas-dynamic variables: the concentration, velocity, temperature of the particles. Instead of the Poisson equation in the second layer the quasi-neutrality condition is used ($n_e = Zn_i$).

Gas-dynamic equations based on nonequilibrium plasma model shown in Sect. 1.3. General procedure for the solution of these equations is discussed in Sect. 6.1, and their joint solution and simulation results for the free Arc column listed in Sects. 8.5.2 and 8.5.3. The use of gas-dynamic model allows us to model simultaneously all the processes in the ionization layer and the Arc column. In this case, there is no need to separate the ionization layer (including a layer of thermal heating) from the main Arc column. The main advantage of this model is the ability to set all the parameters of the distribution of the length and radius of the ionization layer, which is not less important. However, the lack of the second model is the need for the joint solution of equations and review processes in the Arc column.

In this case, to calculate the ionization layer first model including the method of integral balance is widely used. Its main provisions are represented lower.

The method of integral balances allows description of processes in the ionization layer through a series of simple integral relations based on the balance of energy flows in this layer:

1. The energy flux density contributed to the emissive layer by electrons accelerated in the space charge layer in accordance with the foregoing is $\frac{j_{em}}{e}(2kT_c + eU_d)$.

2. Work of the electric field on the electrons moving inside the layer is equal to the voltage drop in the layer U_i, multiplied by the average density of the electron current in the layer. According to the work [7], the electron current density is equal to half the sum of the electron current density at the boundary of the space charge layer, which is equal to $j_e = j_{em} - j_{ec}$, and the electron current density at the outer edge of the ionization layer, which in the first approximation is $j = j_i + j_e$. Thus, the work of the electric field on the electrons moving inside the layer is $\frac{j_{em} - j_{ec} + j}{2} U_i$, where $U_i = \frac{kT_e}{e} \ln \frac{n_{e\infty}}{n_{es}}$—the voltage drop in the ionization layer [7]; $n_{e\infty}$ and n_{es}—the concentration of electrons in the outer border of the Arc column and the inner boundary of the space charge layer, respectively.
3. The energy flux density of the electrons introduced into the space charge layer by electrons leaving the ionization layer $\frac{j_{ec}}{e}(2kT_e + eU_d)$.
4. The energy flux density of the electrons introduced into the column of the Arc discharge in plasma by electrons leaving the ionization layer, it is $3.2 jkT_e/e$. Factor 3.2 is made of coefficient 5/2, taking into account the transfer of enthalpy due to electric current adjusted thermal diffusion coefficient which is designed for highly ionized plasma in [7].
5. Electron energy loss in the ionization of atoms equal to $\frac{j_i E_i}{Ze}$; losses in elastic collisions with heavy particles inside the layer is neglected.

Thus, the Energy balance equation for the ionization layer is as follows:

$$j_{em}(2kT_c + eU_D) + \frac{j_{em} - j_{ec} + j}{2} eU_i = j_{ec}(2kT_e + eU_d) + 3.2jkT_e + \frac{j_i E_i}{Z}.$$

Excluding the total current density j and substituting in the expression for the voltage drop in the layer U_i, we obtain the final form of the energy balance equation in the ionization layer:

$$j_{em}\left[eU_d + 2kT_c + kT_e\left(\ln\frac{n_{e\infty}}{n_{es}} - 3.2\right)\right] = j_{ec}\left[eU_d + kT_e\left(\ln\frac{n_{e\infty}}{n_{es}} - 1.2\right)\right] + j_i\left[\frac{E_i}{Z} + kT_e\left(3.2 - 0.5\ln\frac{n_{e\infty}}{n_{es}}\right)\right], \tag{9.22}$$

Taking into account the equality $U_c = U_d + U_i$, we have:

$$j_{em}(eU_c + 2kT_c - 3.2kT_e) = j_{ec}(eU_c - 1.2kT_e) + j_i\left(\frac{E_i}{Z} + 3.2kT_e - 0.5eU_i\right). \tag{9.23}$$

The remainder of the near cathode region that is external to the ionization layer is a layer of thermal heating (Fig. 9.1). The thermal heating layer, thermal nonequilibrium state is maintained as long as the temperature of the electrons and heavy particles are different.

9.2 Calculation of the Cathode Processes

The Procedure for Calculating the Cathode Region The above Eqs. (9.3–9.23) allow us to calculate a one-dimensional current transport and energy through the cathode region of the Arc. The transfer mode is specified by two parameters: the first—the temperature of the cathode surface T_c, and the second (it should be asked), for example, the potential difference cathode region U_c or electron plasma temperature T_e. All other parameters describing the transfer current and energy can be determined by solving the respective equations of the model.
The calculation procedure is as follows, for example: parameters T_c and T_e as input factors and we can solve the system of equations of ionization equilibrium for a given T_e, the temperature of the heavy particles in the plasma and pressure. From Eq. (9.9), we obtain the concentration of electrons and ions at the boundary of the space charge layer and the ionization ($n_{es} = Zn_{is}$), and from (9.7) and (9.8), the ion current density j_i in the space charge layer. The Eq. (9.23) can be solved together with the Eq. (9.10) and the Richardson–Dushman equation (9.4) for define the voltage drop in the space charge layer U_d, the reverse current density j_{ec} and the current density of electrons thermionic j_{em}. It should be emphasized that in this procedure, the decision should be a joint decision procedure McCone equation (9.3), the formulas of work effectively (9.5) and Richardson–Dushman equation (9.4). Finally, we give the current densities and all kinds of heat fluxes to the cathode, and their total values are given by (9.11), (9.21), respectively.

This solution corresponds to a one point of the current–voltage characteristics (CVC) of the cathode region. For other values of T_c and T_e should obtain the following points CVC, etc. The obtained data allow us to identify the integral characteristics of the cathode region, such as the power dissipated in the cathode, the rate of evaporation of the cathode material and the radius of the cathode spot as a function of the Arc current.

An analysis of the calculation results with the experimental data of other authors [1, 2, 19] (Fig. 9.3) allows us to conclude the approximate constancy of the current density in the cathode spot when changing the Arc current. Persistence several variables including total current density along the radius of the cathode spot allows to determine the cross section S_s and the radius R_s of the cathode spot in certain of the Arc current and its density:

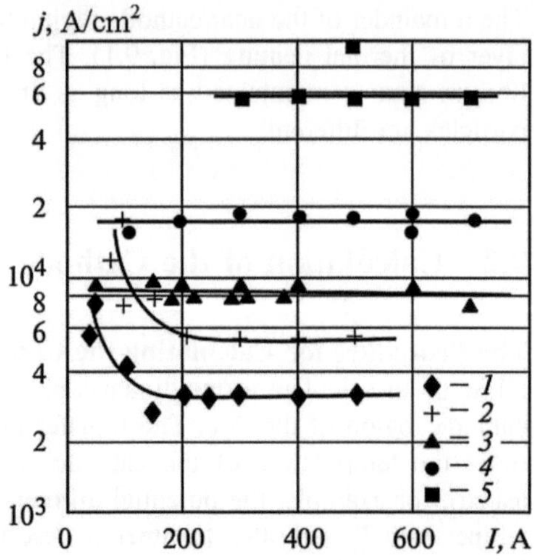

Fig. 9.3 Dependence of current density tungsten cathode from the current [19]: d_c = 5 mm—diameter of the cathode, l_c = 60 mm—length of the cathode output portion: *1* argon, *2* nitrogen; d_c = 5–8 mm, l_c = 0: *3* argon, *4* nitrogen, *5* hydrogen

$$S_s = \frac{I}{j}; \quad R_s = \sqrt{\frac{I}{\pi j}}. \tag{9.24}$$

To calculate the Arc current is given in the region of 50–1500 A, and the current density is taken for the characteristic mode of the thermionic cathode of the electric Arc at atmospheric pressure. For example, for the tungsten cathode W + Ar: $T_c = 4200$ K, $T_e = 40{,}000$ K; for thoriated tungsten cathode W − ThO_2 + Ar (1.5 % ThO_2): $T_c = 3200$ K, $T_e = 40{,}000$ K.

Power to the cathode, respectively, equal to

$$P = qS_s = q\frac{I}{j}. \tag{9.25}$$

Volt equivalent of the heat flux to the cathode is defined as:

$$V = q/j. \tag{9.26}$$

Results of Calculation of the Cathode Region of an Electric Arc We present the results of calculations cathode region of pure tungsten cathode W and thoriated tungsten cathode W(1.5 % ThO_2), typical for the majority of Arc plasma torches operating at atmospheric pressure in Argon. In the first case, the Richardson coefficient A = 6.02×10^5 A/(m K)^2, the work function $e\varphi = 4.55$ eV, the melting temperature $T_m = 3700$ K. For the second case, A = 3.00×10^4 A/(m K)^2,

$e\varphi = 2.63$ eV, the melting temperature $T_m = 2920$ K [3]. Table 9.1 gives the values of the work function of a tungsten cathode with the addition of various oxide additives.

On the outer boundary of the cathode region at atmospheric pressure and a relatively high degree of ionization and electron temperature (T_e = 20,000–40,000 K) the Argon plasma temperature of the ions, atoms, and electrons are the same ($T_e = T_i = T_a$). For a pure tungsten cathode W operating temperature range of the cathode surface is selected in accordance with the experimental data of [1, 3, 19] T_c = 3000–4400 K. For thoriated tungsten cathode W(1.5 % ThO_2): T_c = 2000–3200 K.

The dependence of the electric field intensity at the cathode surface temperature of the cathode and the plasma temperature is shown in Figs. 9.4 and 9.5. The electric field intensity decreases with increasing temperature of the cathode; this is due to the fact that as the temperature increases thermionic cathode of the cathode. As a result, for a given current of maintaining the Arc does not need a large amplification of the electric field. Similarly, an increase in the plasma temperature at the outer boundary of the cathode region of the electric field decreases, which is consistent with the physical process of the thermionic cathode.

Table 9.1 The work function of tungsten oxide with a mass content of 1–1.5 % [3]

Oxide	$e\varphi$ (eV)	Oxide	$e\varphi$ (eV)
ThO_2	3.00	Gd_2O_3	3.96
La_2O_3	2.96	Yb_2O_3	4.05
Y_2O_3	3.30	ZrO_2	3.97
Ce_2O_3	3.40	HfO_2	3.84
Nd_2O_3	3.88	–	–

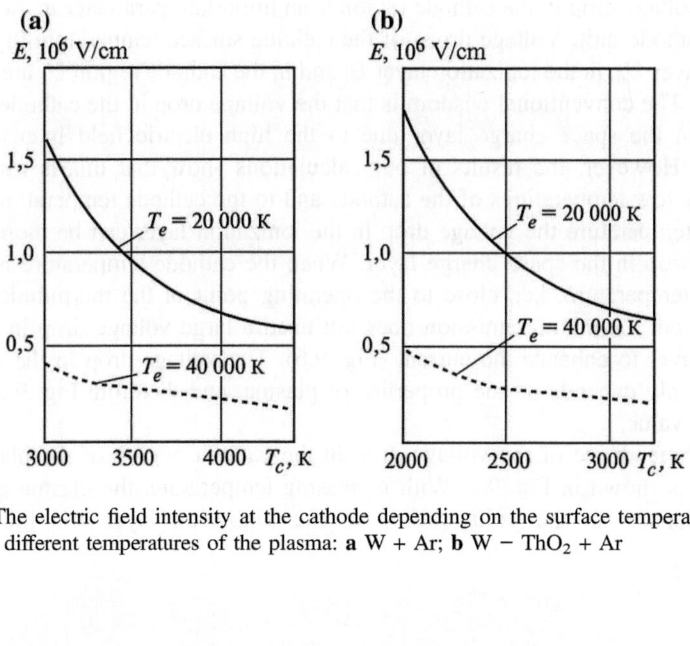

Fig. 9.4 The electric field intensity at the cathode depending on the surface temperature of the cathode at different temperatures of the plasma: **a** W + Ar; **b** W − ThO_2 + Ar

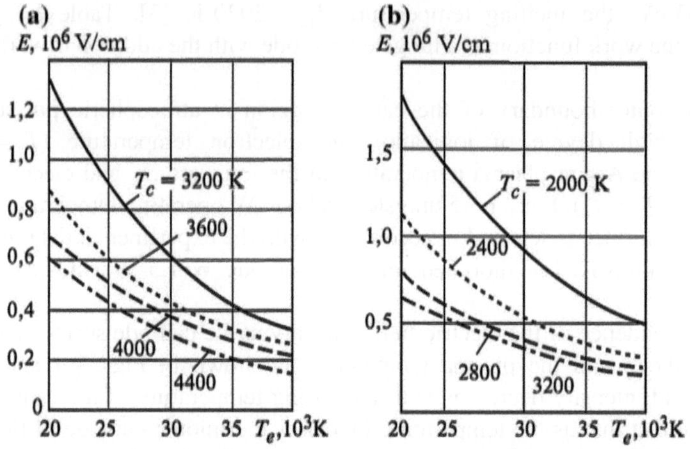

Fig. 9.5 The electric field intensity at the cathode surface, depending on the temperature of the plasma at different values of the surface temperature of the cathode: **a** W + Ar; **b** W − ThO_2 + Ar

Because of the improved thermionic thoriated tungsten cathode has a value of the electric field as a pure tungsten cathode at a much lower temperature of the cathode (T_c = 2000–3200 K compared to T_c = 3000–4400 K).

In the working temperature range of the cathode (T_c = 3000–4400 K and T_c = 2000–3200 K, respectively, for W + Ar and W − ThO_2 + Ar) and in a wide range of plasma temperatures (T_e = 20,000–40,000 K) value of the electric field intensity $E = \left(10^5 - 2 \times 10^6\right) < 10^7$ V/cm. This confirms the assumption of low current field emission (9.6) and secondary electron emission at the electric Arc (photoemission can be neglected).

The voltage drop in the cathode region is an important parameter in the selection of the cathode unit. Voltage drops of the cathode surface temperature in the space charge layer U_d, in the ionization layer U_i and in the cathode region U_c are shown in Fig. 9.6. The conventional wisdom is that the voltage drop in the cathode region is mainly in the space charge layer due to the high electric field intensity at the cathode. However, the results of our calculations show that this is true only at relatively low temperatures of the cathode and to the cathode temperature near the melting temperature the voltage drop in the ionization layer can be more than the voltage drop in the space charge layer. When the cathode temperature is near the melting temperature, i.e., close to the operating point of the thermionic cathode, high current thermionic emission does not need a large voltage drop in the space charge layer to enhance the current (Fig. 9.6). The voltage drop in the ionization layer mainly depends on the properties of plasma, and therefore Fig. 9.6, it has a constant value.

The dependence of the voltage drop in the cathode region of the plasma temperature is shown in Fig. 9.7. With increasing temperature, the plasma obtained a slight increase in the voltage drop in the ionization layer and the cathode,

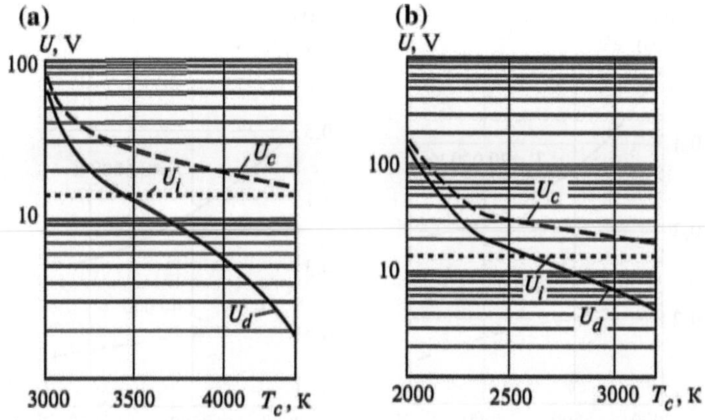

Fig. 9.6 The voltage drop in the space charge layer U_d, in the ionization layer U_i and cathode region U_c depending on the temperature of the cathode surface: **a** W + Ar; **b** W − ThO_2 + Ar

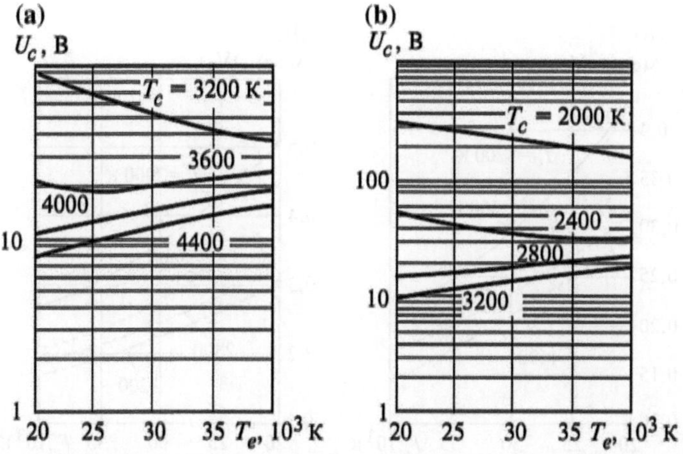

Fig. 9.7 The voltage drop in the cathode region, depending on the electron temperature for different values of the cathode surface temperature: **a** W + Ar; **b** W − ThO_2 + Ar

respectively, in the field. We see that in both cases the thermionic cathode voltage drop in the cathode region of the order of ~10–30 V, which coincides with the experimental data [1, 3].

Experimental data on the reduction of the work function of the cathode is practically absent, this value is mainly determined only by calculation. Figure 9.8 shows the dependence of the work reduction cathode on its surface temperature at different values of the electron temperature, and Fig. 9.9 shows the dependence of the work reduction cathode on the electron temperature. The average value $\Delta(e\varphi)$ ~ 0.2–0.3 eV, it is about 10 % of the work function. However, both values

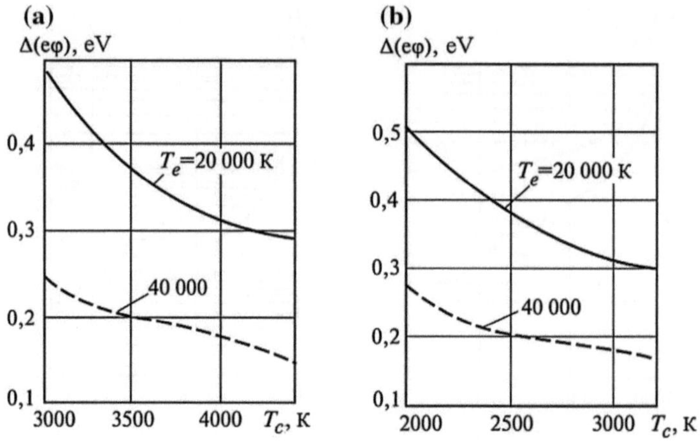

Fig. 9.8 Reducing the work function of the cathode, depending on the surface temperature of the cathode at different temperatures of the electrons: **a** W + Ar; **b** W − ThO_2 + Ar

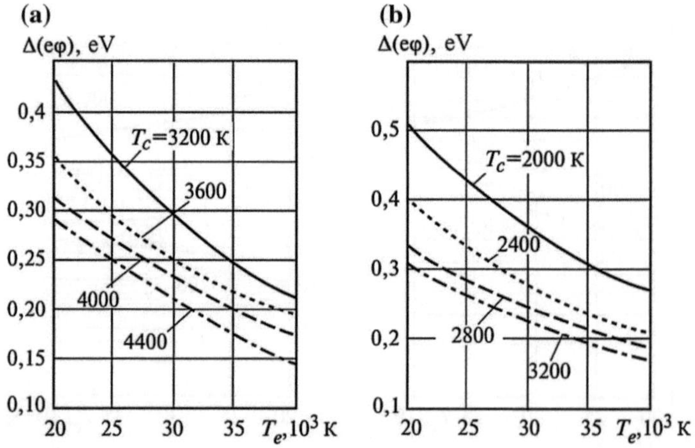

Fig. 9.9 Reducing the work function of the cathode, depending on the temperature of the electrons for different cathode surface temperatures: **a** W + Ar; **b** W − ThO_2 + Ar

included in the calculation of the exponent of thermionic emission current density, so that the accuracy of calculation of the current density and thermionic heat flux depends on the accuracy determination of value $\Delta(e\varphi)$. Since $\Delta(e\varphi)$ directly proportional to the electric field at the cathode (9.5), the variation depending $\Delta(e\varphi)$ of the temperatures of the cathode and the electrons (Fig. 9.9) is the same as that of the electric field intensity (Figs. 9.4, 9.5).

The most important parameter in the calculation of the cathode region is the current density on the cathode surface. The balance of currents in dependence on the temperature of the cathode surface shown in Fig. 9.10, data from our calculation

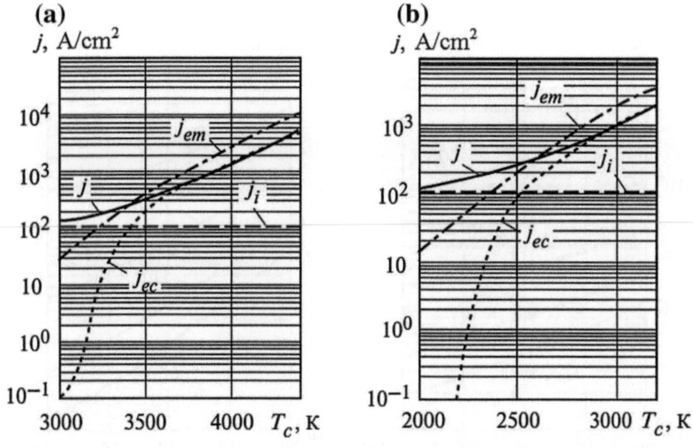

Fig. 9.10 Current density balance on the surface of the cathode, depending on the surface temperature of the cathode at a plasma temperature T_e = 40,000 K: **a** W + Ar; **b** W − ThO_2 + Ar

are practically identical with those of other authors in Fig. 9.2. In the working temperature range of both cathodes the current density is $\sim 10^2$–10^4 A/cm^2. We also see that in the working range of the temperature of the cathode surface and the electron, thermionic emission current is larger proportion in comparison with the current of reverse electron and ion current. In the area of the cold cathode (for W + Ar, $T_c < 3000$ K and for W − ThO_2 + Ar, $T_c < 2000$ K) thermionic emission current and the current of reverse electrons are small, and the ion bombardment of the surface of the cathode current prevails, due to the high electric field at the cathode (Fig. 9.4) and the voltage drop (Fig. 9.6), and plays a major role in the formation of the Arc current; for $j < 100$ A/cm^2 cathode is gradually sinking into the cold emission region, where it is necessary to take into account of the field emission and secondary emission.

The total current density at the cathode depending on the electron temperature (for different temperatures of the cathode surface) is shown in Fig. 9.11. For the same surface temperature of the cathode current density decreases with increasing the plasma temperature due to increased current density reverse electrons. From the data in Figs. 9.10 and 9.11 we can see that the current density at the cathode depends largely on the temperature of the cathode and is less dependent on the temperature of the plasma.

The proportion of electronic (thermionic emission and reverse electrons) and the ion current as a function of the temperature of the cathode surface for different values of the electron temperature is shown in Fig. 9.12. We see that a large proportion of the ion current at the cathode ($S_i = j_i/j > 10\,\%$) is more than in the Arc column, it is the result of the work of most of Arc plasma torches atmospheric pressure. How to show the data on the current balance (Fig. 9.10) and the proportion of electron and ion currents (Fig. 9.12) on the cathode surface, for the region of low temperature (cold cathode and cold emission), the proportion of ion

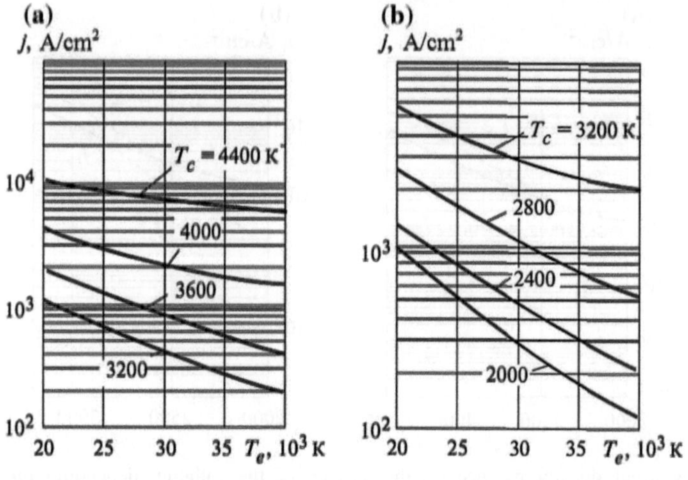

Fig. 9.11 The total current density on the cathode as a function of electron temperature, at different temperatures of the cathode surface: **a** W + Ar; b W − ThO_2 + Ar

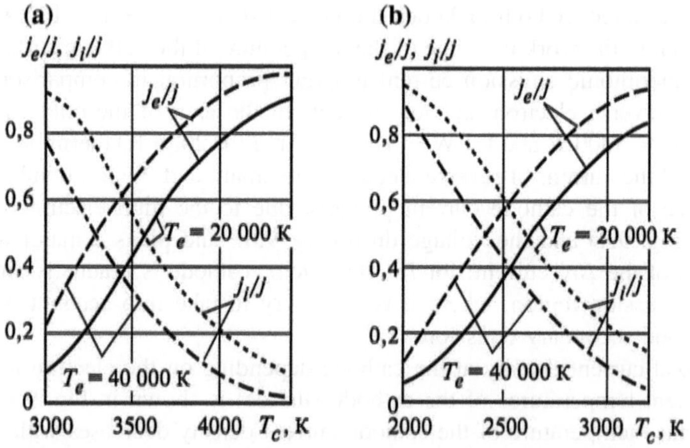

Fig. 9.12 Proportion of the electronic and ionic current as a function of the cathode surface temperature for different values of plasma temperature: **a** W + Ar; **b** W − ThO_2 + Ar

current even prevails over the electron current. At the operating temperature region of thermionic cathodes, ion current can be up to 30–50 % of the total current at the cathode.

Changing the proportion of the currents depending on the electron temperature (T_e = 20,000–40,000 K) does not exceed 20 %, and the higher the electron temperature, the greater the proportion of electron current, and good thermionic emission of electrons from the cathode surface may be provided by a high temperature of the cathode (close to the temperature melting) and the electrons

(T_e = 20,000–40,000 K). This is the condition of most thermo-emission cathode Arc plasma torches operating at atmospheric pressure for various technological applications (plasma cutting and melting of metals, plasma spraying and hardening of machine parts, plasma treatment of surfaces and materials, plasma waste disposal, etc.).

The energy balance at the cathode surface, depending on the temperature of the cathode is shown in Fig. 9.13. The energy fluxes determined from the current density and the nature of the change of these two values, depending on the temperature of the cathode and the plasma is the same. We note an important role of the heat flux of reverse electron in the energy balance at the cathode. The calculation result shows that the operating temperature range of the cathode at the electric Arc, when the temperature of the cathode is more than the melting point of the material, the energy of the reverse electron can be more, than the energy of the ion bombardment energy and thermionic emission (Fig. 9.13). This is so due to the fact that the introduction of "hot" electrons from the plasma to the lattice of the cathode material recovered their work function ($e\varphi_{eff}$) and kinetic energy ($2kT_e$). As a result, the total flow of heating the surface of the cathode may be larger than the flux of ion bombardment (Fig. 9.13).

Additional information can be obtained from analysis of the energy balance of the cathode depending on the electron temperature of the plasma (Fig. 9.14). These calculations show that the energy flux of atoms leaving the surface of the cathode, an order less than the other fluxes and in the further it can be ignored. The total

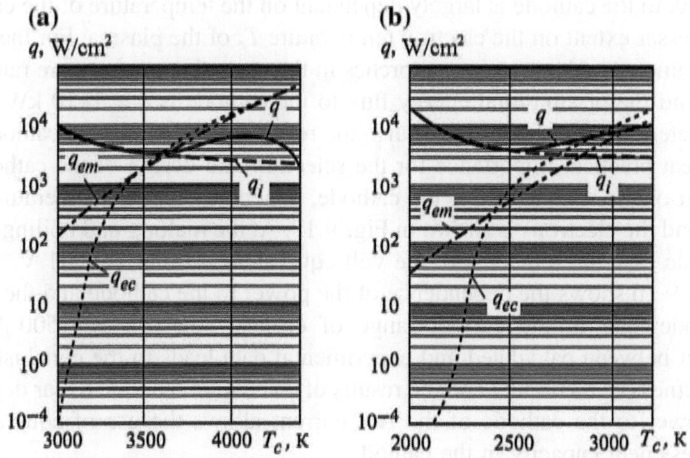

Fig. 9.13 The energy balance on the cathode surface, depending on the temperature of the cathode at the electron temperature T_e = 40,000 K: q_{em}—thermionic flux from cathode; q_{ec}—the energy flux density of reverse electrons to the cathode; q_i—the energy flux density of ions to the cathode, including the energy of recombination energy carried atoms by neutralization of the ions on the cathode surface: **a** W + Ar; **b** W − ThO_2 + Ar

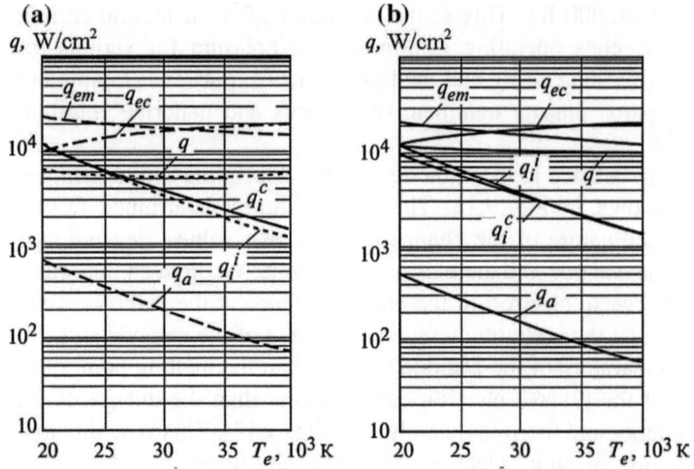

Fig. 9.14 Energy balance at the cathode, depending on the electron temperature: q_i^c—the energy flux density of ion bombardment to the cathode; q_i^i—the energy flux density resulting from the neutralization of the ions on the cathode surface; q_a—the energy flux density of atoms leaving the surface of the cathode; **a** W + Ar, T_c = 4000 K; **b** W − ThO_2 + Ar, T_c = 3200 K

energy flux to the cathode of electric Arc weakly depends on the electron temperature of the plasma and is about 5–10 kW/cm^2, which coincides with the experimental data of [1, 3]. The results of our calculations show that the total energy flux to the cathode is largely dependent on the temperature of the cathode T_c and to a lesser extent on the electron temperature T_e of the plasma. For the majority of thermionic cathode Arc plasma torches in the operating temperature range of the cathode and the plasma total energy flux to the cathode is q = 1–10 kW/cm^2.

Parameter of the cathode spot is also the result of calculating the cathode region and is great practical importance for the selection and design of the cathode. Volt equivalent of the energy flux to the cathode, depending on the temperatures of the cathode and the electrons is shown in Fig. 9.15. At the melting and boiling points of the cathode material temperature the volt equivalent is order ~10–1 V.

Figure 9.16 shows the dependence of the power to the cathode and the radius of the cathode spot in the current range of electric Arc (I = 50–1500 A). Good agreement between calculated and experimental data leads to the conclusion about the correctness of the model and the results of calculation and the linear dependence of the power to the cathode of the Arc current allows the use of equivalent volt rapid assessment capacity in the cathode.

To determine the extent of erosion of the cathode in Fig. 9.17 presents the data of the specific mass evaporation rate and heat flux of different cathode materials depending on its temperature.

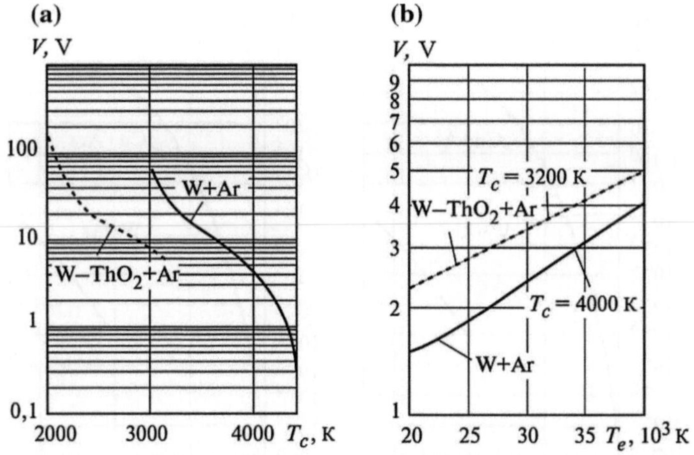

Fig. 9.15 Volt equivalent of the energy flux to the cathode, depending on the temperatures of the cathode (**a**) (with T_e = 40,000 K) and the electrons (**b**) (with T_c = 3200 and 4000 K)

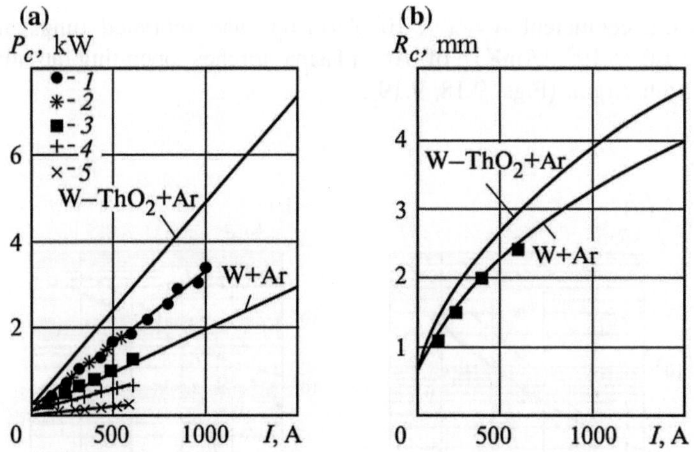

Fig. 9.16 Power transmitted to the cathode (**a**) and the radius of the cathode spot (**b**), depending on the Arc current: *lines* the calculation data; *point* the experiments; *1* l_c = 0 [20]; *2* l_c = 0, *3* l_c = 0.6 cm, *4* l_c = 1.2 cm, *5* l_c = 6 cm [19]

Tungsten cathode is usually composed of tungsten with a small proportion of different impurities. For this reason, the Richardson's coefficient *A* of the Eq. (9.4) has a different meaning, but in practice this factor is determined experimentally by measuring the current density of the Arc. We present the results of calculations of the current and energy balances in the tungsten cathode region with the

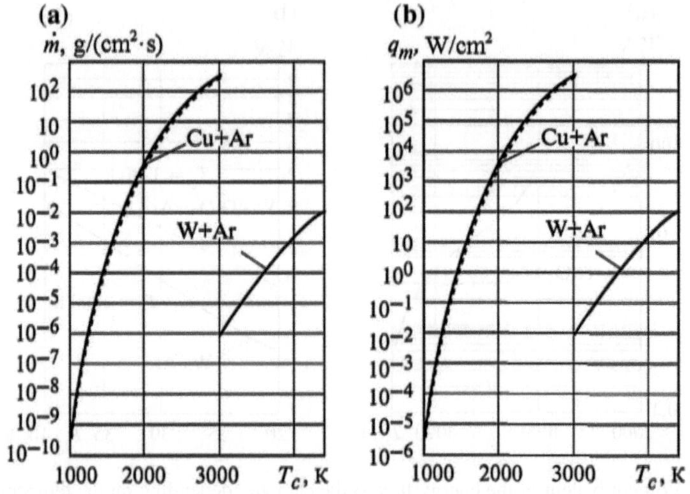

Fig. 9.17 Specific mass evaporation rate and heat flux at the cathode material evaporates depending on its temperature: 1—W + Ar; 2—Cu + Ar

Richardson's coefficient $A = 2 \times 10^6$ A/(m K)^2 and thoriated tungsten cathode with $A = 1.6 \times 10^5$ A/(mK)^2 of Arc plasma torches operating at atmospheric pressure with Argon (Figs. 9.18, 9.19).

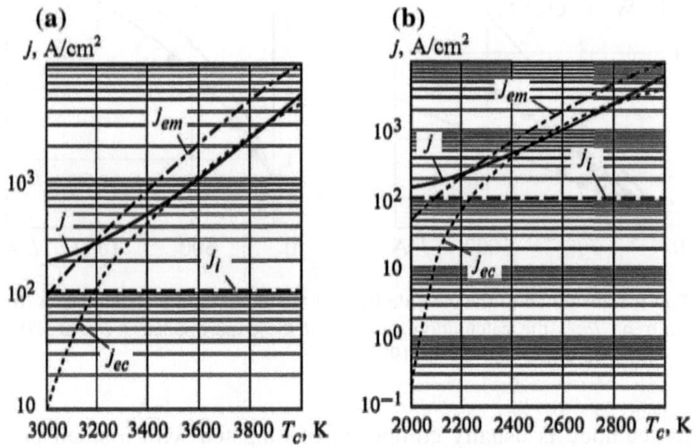

Fig. 9.18 The current balance on the surface of the cathode depending on the surface temperature of the cathode at a plasma temperature T_e = 40,000 K: **a** W + Ar; **b** W − ThO_2 + Ar

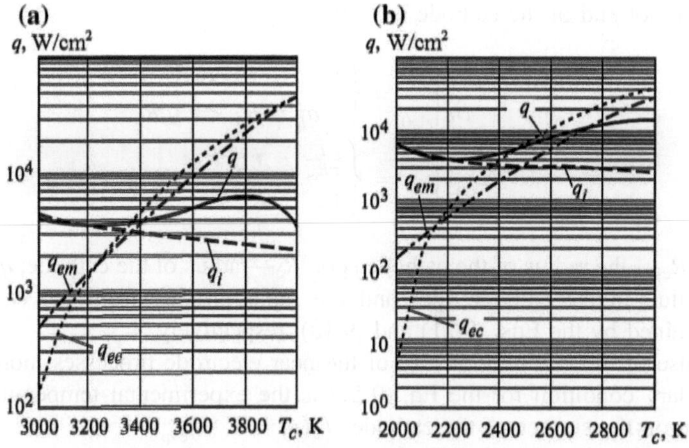

Fig. 9.19 The energy balance on the cathode surface, depending on the temperature of the cathode at the plasma temperature T_e = 40,000 K: the notation is the same as in Fig. 9.13

9.3 Thermal Problem of Heating the Cathode

Phenomena inside the cathode play a significant role in the formation of its thermal state. To determine the thermal state of the cathode the energy equation, should be solved, that in the stationary case and a cylindrical axis-symmetry coordinate system has the following form:

$$-\frac{\partial}{\partial z}\left(\lambda\frac{\partial T}{\partial z}\right)-\frac{1}{r}\frac{\partial}{\partial r}\left(r\lambda\frac{\partial T}{\partial r}\right)=\sigma E^2, \qquad (9.27)$$

where λ, σ—thermal conductivity and electrical conductivity coefficients of the cathode material, depending on temperature of the cathode; E—the electric field intensity.

Distribution of the specific power released in the form of Joule heating and current density inside the cathode can be found using equations for the electric potential V (8.19).

$$\frac{\partial}{\partial z}\left(\sigma\frac{\partial V}{\partial z}\right)+\frac{1}{r}\frac{\partial}{\partial r}\left(r\sigma\frac{\partial V}{\partial r}\right)=0;$$

and

$$S=\sigma E^2=\sigma\left[\left(\frac{\partial V}{\partial z}\right)^2+\left(\frac{\partial V}{\partial r}\right)^2\right];\; j=\sqrt{\left(\sigma\frac{\partial V}{\partial z}\right)^2+\left(\sigma\frac{\partial V}{\partial r}\right)^2}. \qquad (9.28)$$

The boundary conditions for the energy Eq. (9.27) and the electric potential (9.28) have the following form:

(a) At the hot end of the cathode ($z = 0$):

$$-\lambda \frac{\partial T}{\partial z}\bigg|_{z=0} = \begin{cases} q, & r \le R_{cs}; \\ -q_R^c, & R_{cs} < r \le R_c. \end{cases}$$

$$-\sigma \frac{\partial V}{\partial z}\bigg|_{z=0} = \begin{cases} \frac{I}{\pi R_{cs}^2}, & r \le R_{cs}; \\ 0, & R_{cs} < r \le R_c. \end{cases}$$

Here R_{cs}—the radius of the cathode spot; R_c—radius of the cathode; q, q_R^c—the total flux in the cathode spot and the radiation flux from the cathode are determined by the Eqs. (9.21) and (9.18), respectively.
Because of the complex picture of the near electrode processes more reliable boundary condition for the Eq. (9.27) is the experimental temperature distribution on the surface of the cathode: $T_c(r) = T_c(r)_{exp}$;

(b) On the cold end ($z = L$): $T = 300$ K, $V = 0$;

(c) On the axis of symmetry of the cathode ($r = 0$): $\frac{\partial T}{\partial r} = \frac{\partial V}{\partial r} = 0$;

(d) The side surface of the ($r = R_c$): $\frac{\partial T}{\partial r} = \frac{\partial V}{\partial r} = 0$. For a water-cooled cathode temperature and electric potential on the cathode surface can be defined or determined experimentally.

A discrete analogue of the Eq. (9.28) given in Sect. 8.4.1. For the solution of the energy Eq. (9.27) it is possible to use the well-known method of control volume, described in Sect. 6.1. With the above boundary condition the Eqs. (9.27, 9.28) can be solved on the grid line [$z1$, $r2$], shifted by a half step along the radius relative to the grid [$z1$, $r1$] (Fig. 9.20). The counting of grid coordinates [$z1$, $r1$] is from the center on the surface of the cathode ($z1_0 = r1_0 = 0$).

Power balance of the control volume and its discrete analog can be obtained by integrating an Eq. (9.27) around all of the each control volume (Fig. 9.20). Where

Fig. 9.20 Grid lines and control volume for the solution of the energy equation of the cathode

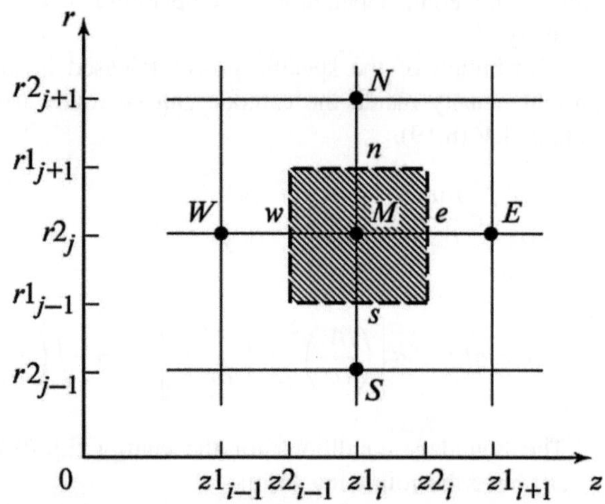

the source function is constant inside control volume $\Delta V = \frac{(z_e - z_w)(r_n - r_s)(r_n + r_s)}{2}$ and it is a linear function of temperature $S = S_c + S_p T$, and the coefficients of S_c and S_p are defined like (6.56). We have:

$$-\int_{r_s}^{r_n}\int_{z_w}^{z_e} \frac{\partial}{\partial z}\left(\lambda \frac{\partial T}{\partial z}\right) dz r dr - \int_{z_w}^{z_e}\int_{r_s}^{r_n} \frac{1}{r}\frac{\partial}{\partial r}\left(r\lambda \frac{\partial T}{\partial r}\right) r dr dz = \int_{z_w}^{z_e}\int_{r_s}^{r_n} S r dr dz$$

The integral of each term on the left side of this equation contains the first derivative of the same order, so you can accurately produce its integration at least once, we finally obtain:

$$-\int_{r_s}^{r_n}\left[\left(\lambda\frac{\partial T}{\partial z}\right)_e - \left(\lambda\frac{\partial T}{\partial z}\right)_w\right] r dr - \int_{z_w}^{z_e}\left[\left(r\lambda\frac{\partial T}{\partial r}\right)_n - \left(r\lambda\frac{\partial T}{\partial r}\right)_s\right] dz$$
$$= (S_{cM} + S_{pM}T_M)\Delta V.$$

In order to implement the integration of the last equation admits that her integrands take the average values in the points of *e*, *w*, *n*, *s*, respectively, and the first derivatives are approximated as:

$$\begin{cases} \left(\lambda\frac{\partial T}{\partial z}\right)_e = \lambda_e \frac{T_E - T_M}{z_E - z_M}; \ \left(\lambda\frac{\partial T}{\partial z}\right)_w = \lambda_w \frac{T_M - T_W}{z_M - z_W}; \\ \left(\lambda\frac{\partial T}{\partial r}\right)_n = \lambda_n \frac{T_N - T_M}{r_N - r_M}; \ \left(\lambda\frac{\partial T}{\partial r}\right)_s = \lambda_s \frac{T_M - T_S}{r_M - r_S}. \end{cases}$$

Substituting these expressions in the last integral equation, we obtain a general discrete analog for cathode energy Eq. (9.27) in the form:

$$a_M T_M = a_E T_E + a_W T_W + a_N T_N + a_S T_S + b_M, \tag{9.29}$$

where

$$\begin{cases} a_E = \frac{\lambda_e (r_n - r_s)(r_n + r_s)}{2(z_E - z_M)} > 0; \quad a_W = \frac{\lambda_w (r_n - r_s)(r_n + r_s)}{2(z_M - z_W)} > 0; \\ a_N = \frac{\lambda_n r_n (z_e - z_w)}{r_N - r_M} > 0; \quad a_S = \frac{\lambda_s r_s (z_e - z_w)}{r_M - r_S} > 0; \\ a_M = a_E + a_W + a_N + a_S - S_{pM}\Delta V > 0; \quad b_M = S_{cM}\Delta V. \end{cases}$$

Note that all the discrete analog coefficients of (9.29) are positive and satisfy the convergence criterion (6.49). So for example, the temperature distribution inside the body of tungsten is shown in Fig. 9.21.

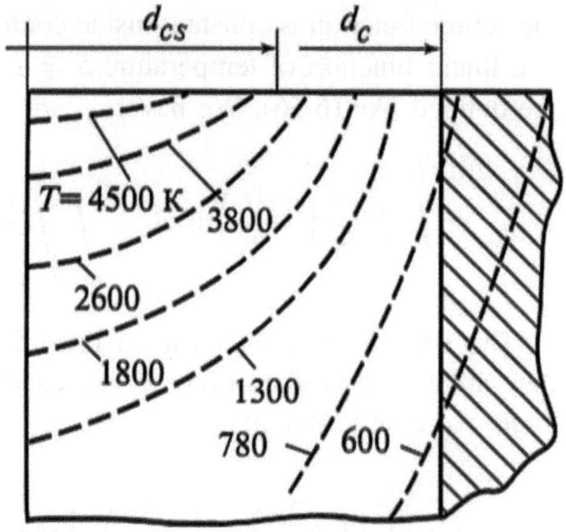

Fig. 9.21 The temperature distribution inside the tungsten cathode ($d_{cs} = 6$ mm, $d_c = 10$ mm)

Diffusion of Additive Material to the Surface of the Cathode Introduction of activating additives (Table 9.1), reducing the work function, leads to a decrease in the required operating temperature of the cathode. However, the rapid diffusion of the additive material to the surface of the cathode and thus of its concentration within a short period of time causes an increase in the work function and correspondingly increase the temperature of the cathode surface.

In this regard, the main objective of increasing the cathode performance and service life of the electric Arc is the reduction in the rate of release of the activating additive outside and to ensure effective recharge its surface layers from the deep layers of the cathode material [1].

In theoretical terms, this problem can be reduced to the study processes of diffusion additives within the body of the cathode, output it to the surface and evaporation from it. The corresponding diffusion equation of additives from the depths of the cathode is:

$$\frac{\partial n}{\partial t} = \frac{\partial}{\partial z}\left(D\frac{\partial n}{\partial z}\right) + \frac{1}{r}\frac{\partial}{\partial r}\left(rD\frac{\partial n}{\partial r}\right), \tag{9.30}$$

where n—the concentration of the additive in the cathode; D—diffusion coefficient of the additive.

Equation (9.30) has the general form of nonstationary heat conduction equation and can be solved on the grid [$z1$, $r2$] (Fig. 9.20). A discrete analog of this equation for a fully implicit scheme [21] is as follows:

$$a_M n_M = a_E n_E + a_W n_W + a_N n_N + a_S n_S + b_M,$$
$$\begin{cases} a_E = \frac{D_e(r_n - r_s)(r_n + r_s)}{2(z_E - z_M)} > 0; \quad a_W = \frac{D_w(r_n - r_s)(r_n + r_s)}{2(z_M - z_W)} > 0; \\ a_N = \frac{D_n r_n (z_e - z_w)}{r_N - r_M} > 0; \quad a_S = \frac{D_s r_s (z_e - z_w)}{r_M - r_S} > 0; \\ a_M = a_E + a_W + a_N + a_S + \frac{\Delta V}{\Delta t} > 0; \quad b_M = \frac{\Delta V}{\Delta t} n_M^0. \end{cases} \tag{9.31}$$

where $\Delta V = \frac{(z_e - z_w)(r_n - r_s)(r_n + r_s)}{2}$—the control volume; Δt—time step; n_M^0—the value of the concentration of the additive in the node M at the previous moment; n_M, n_E, n_W, n_N, n_S—additive concentration in the respective nodes of the control volume (Fig. 9.20) in the given time.

9.4 Model of the Anode Processes

In the cathode region more than 99 % of the electrons that are incident on the surface of the anode, comes to the potential drop of the arc column and serve as charge carriers in this area. This fact is illustrated by the value of specific erosion (average 5×10^{-7} g/C) of copper water-cooled cylindrical anode used in the Arc plasma torches. Even if we assume that all atoms that are evaporated from the anode surface are fully once ionized, the ion current will be a few one of thousandths of the total current [20]. Thus, the Arc area immediately adjacent to the anode and the charges are transferred only by electrons, resulting in the formation in this region of the negative space charge. It is this uncompensated charge that determines the anode potential drop. It usually decreases with increasing current and at the transition to the currents in the hundreds and thousands of amperes is 5–1 V [22]. The value of the anode potential drop is generally determined by experiment, the data about it can be found in [20].

Near the surface of the anode separation of the electron temperature and the temperature of ions and neutral particles is observed and, consequently, the breach of thermodynamic equilibrium occurs. If it is assumed that plasma is in a state of PLTE, the equation of energy balance for the anode can be written as follows:

$$q_a = \frac{j_e}{e}\left(eU_a + e\varphi_{\text{eff}} + \frac{5kT_e}{2}\right) + q_R^p + q_c + q_T - q_R^a - q_m. \tag{9.32}$$

In this case, as for the energy balance at the cathode, the below relations will be shown.

1. The energy flux of electrons that was brought to the surface of the anode,

$$q_a^e = \frac{j_e}{e}\left(eU_a + e\varphi_{\text{eff}} + \frac{5kT_e}{2}\right),$$

where the first term $j_e U_a$ corresponds to the kinetic energy obtained by electrons in the near anode voltage drop, the second $j_e\varphi_{\text{eff}}$—the potential energy of electrons arriving at the anode surface and neutralizing the positive ions of metal in the entrance to the lattice $e\varphi_{\text{eff}} = e\varphi - \Delta(e\varphi)$—effective work function, $\Delta(e\varphi) = e\sqrt{\frac{eE_a}{4\pi\varepsilon_0}}$—reduction of the work function due to the electric field (E_a)

on anode surface—the Schottky effect; third term determines the enthalpy of the electron, that corresponds to the temperature of the plasma on the boundary of the near anode potential drop and the energy acquired by electrons in motion inside the area. Last energy, reflecting the contribution of the thermal energy of the electrons, is somewhat different from $\frac{3}{2}kT_e$, and electrons cannot save the Maxwell distribution in the area of near anode potential drop [20].

2. The energy flux of radiation from the plasma to the surface of the anode $q_R^p = \frac{U_{rad}Rl}{2(R+l)}$, where R, l—the average radius and Arc length.
3. The energy flux of radiation from the surface of the anode $q_R^a = \varepsilon\sigma T_a^4$, where ε—the degree of blackness of the anode; σ—the Stefan–Boltzmann constant; T_a—the surface temperature of the anode.
4. The energy flux brought to the surface of the anode by plasma convection $q_c = \rho v_z H_p St$, where ρv_z—specific mass flow of plasma; St—the Stanton's number; v_z—the plasma velocity at the distance of the mean free path of the anode surface; H_p—the plasma enthalpy of its temperature at the anode surface.
5. The energy flux brought to the surface of the anode due to heat conduction can be expressed in the case of large temperature gradient in the boundary layer through the difference of enthalpies [20]:

$$q_T = \frac{\lambda_a}{c_{pa}} \frac{H_p - H_a}{\Delta z},$$

where H_a, λ_a, c_{pa}—the enthalpy, thermal conductivity and specific heat of plasma, respectively at a temperature of the surface anode; Δz—the average length of near anode layer.

6. The energy flux used to evaporate the material of the anode $q_m = h\dot{m}$, where h—the specific heat of vaporization of the anode material; $\dot{m}$—the specific mass evaporation rate of the anode material.

One of the most important technological parameters using Arc plasma torches is the power transmitted to anode. To find this parameter, it is necessary to integrate the energy flux to the anode (9.32) for the cross section of anode. An approximate equation for this power is:

$$P_a = I\left(U_a + \varphi + \frac{5kT_e}{2e}\right) + P_R^p + P_c + P_T - P_R^a - P_m. \tag{9.33}$$

If the contribution of the last terms in (9.33) is small, then we have a linear relationship between the power transmitted to the anode and the current Arc.

Representation about the power transmitted to the anode at atmospheric pressure can be obtained from the experimental data given in Fig. 9.22. Most details about the power are represented in [19, 20, 23]. The results of these experiments [19, 23, 24] have showed that the power slightly depends on the electrode material and the anode is a merely passive collector of current. Volt equivalent of the heat flux to the anode

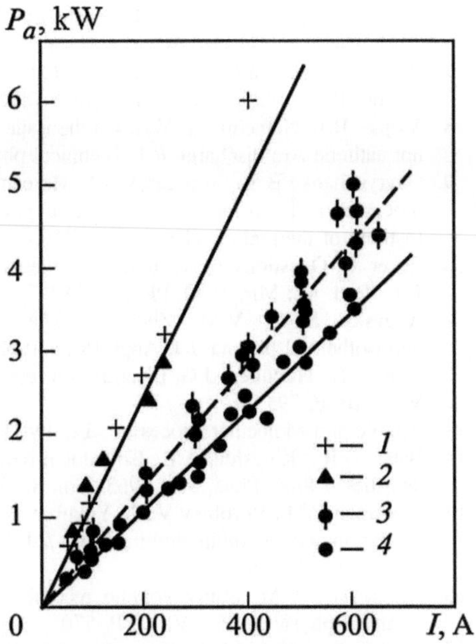

Fig. 9.22 The power transmitted to the anode [19]: *1* in the air; *2* in the nitrogen; *3* in the argon (contracted binding); *4* the same (diffuse binding)

is not dependent on the current and in diffuse attachment of the Arc in argon is 5.85 V (Fig. 9.22, point 4). In the [25] some the less value of volt equivalent for the energy flux through the anode spot to the copper electrode (5 V) is indicated at the range of the current I = 10–200 A.

In the contracted binding the volt equivalent of power transferred to anode increases to 7.35 V (Fig. 9.22, point 3). The current density at the anode in a diffuse attachment of the Arc to the anode $j \leq 10^3$ A/cm^2, and at the transition to the contracted mode it increases to $j \approx 10^4$–10^5 A/cm^2 [19]. Experiments [19] show that in nitrogen and in the air the anode binding is always contracted and volt equivalent of heat flux approximately 12.5 V (Fig. 9.22, point 1, 2). The energy flux density in this case reaches 10^5–10^6 W/cm^2. Thus, for the model of anode region the role of the last terms in Eq. (9.33) is small and it can be neglected in the calculation of energy fluxes toward the anode.

References

1. Near-electrode processes in an Arc discharge / M.F. Zhukov, N.P. Kozlov, A.V. Pustogarov et al. Novosibirsk: Science. 1982.
2. Thermochemical cathodes / M.F. Zhukov, A.V. Pustogarov et al. Novosibirsk: Science. 1985.
3. Gordeev V.F., Pustogarov A.V. Thermionic cathodes Arc. M.: Energoatomizdat. 1988.
4. The Encyclopedia of Low-Temperature Plasma / Ed. V.E. Fortov. V. I. M.: Nauka, 2000.
5. M. Mitchner, C. Kruger. Partially ionized gases. M.: Mir, 1976.

6. Kozlov N.P., Khvesyuk V.I. On the theory of cathode processes Arcs // J. Technical physics. 1971. Part I: pp. 2135–2141; Part II: pp. 2142–2150.
7. Belinov M.S., Marotta A. A model of the cathode region of atmospheric pressure Arcs // J. App. Phys. 1995. Vol. 28. PP. 1869–1882.
8. Mojzes B.J., Nemchinskii V.A. On the issue of abnormally high emission of non-evaporable hot cathode Arc discharge // J. Technical physics. 1974. V. 14. PP. 2539–2547.
9. Gavryushenko B.S., Pustogarov A.V. Research electrodes of plasma torches // Near-electrode processes and erosion of electrodes of plasma torches / Ed. M.F. Zhukov. Novosibirsk: Institute of thermal physics. 1977. PP. 95–122.
10. Ecker G. Questions of vacuum Arc theory // Vacuum Arc: Trans. from English. / Ed. J. Laffetri. M.: Mir. 1982. PP. 267–384.
11. Alievsky M.Y., V.M. Zhdanov. The transport equations for a many-component non-isothermal plasma // J. Appl. Phys. 1963. № 5. pp. 11–17.
12. Hinnov E., Hirchberg J.G. Electron-Ion recombination in dense plasmas // Phys. Rev. 1962. Vol. 126. P. 795.
13. Atomic and Molecular Processes / Ed. by D. Bates. M.: Mir, 1964.
14. Bates D.R., Kingston A.E. Collisional-radiative recombination at low temperatures and densities // Proc. Phys. Soc. 1963. Vol. 83. P. 43.
15. Biberman L.M., Vorobiev V.M., Yakubov N.T. Nonequilibrium Low-temperature plasma IV ionization and recombination functions // J. Thermal physics of High Temperatures. 1969. V. 7. № 4. PP. 593–603.
16. Aleskovsky Y.M. Study volume recombination in a Cesium plasma // J. Experimental technical physics. 1963. V. 17. P. 570.
17. Cool T.A., Zukoski E.E. Recombination, ionization and nonequilibrium electrical conductivity in seeded plasmas // Phys. Fluids. 1966. Vol. 9. P. 780.
18. Desai S.V. Corcoran W.H. Recombination of electrons and ions in an atmospheric argon plasma // J. Quant. Spectrosc. Radiat. Transfer. 1969. Vol. 9. PP. 1371–1386.
19. Near-electrode processes and erosion of electrodes of plasma torches // Collected Articles / Ed. M.F. Zhukov. Novosibirsk: Institute of thermal physics. 1977.
20. Zhukov M.F., Koroteev A.S., Yuryukov B.A. Applied Dynamics of thermal plasma. Novosibirsk: Nauka. 1975.
21. Patankar S. Numerical methods for solving problems of heat transfer and fluid dynamics: M .: Energoatomizdat 1984.
22. Finkelberg V., Mekker G. Electric Arcs and thermal plasma. M.: Publishing House of Foreign Literature, 1961.
23. A study of thermal, electric characteristics of plasma and erosion of the anode / Jeenbaev J. et al. // News of Sib. Dep. USSR Academy of Sciences. Ser. Tech. Dandaron B., Urbak E.K. Investigation of the thermal regime of the rod tungsten cathode // Abs. V Conf. by generators of Low-temperature plasma. Novosibirsk. 1972. Part II. PP. 40–43.
24. Sciences. 1973. № 3. Vol. 1. PP. 3–6.
25. Bykhovsky D.G. Plasma cutting. L.: Engineering. 1972.

Chapter 10
Heat Exchange and Movement of the Solid Particles in the Plasma

The high-energy plasma flow is widely used for the treatment of particles and for spraying. Elucidation of the physical pattern of heating and movement of particles in plasma, as well as determination of the technological parameters of the process, such as the thermal efficiency, the degree of heating, temperature and velocity of particles in plasma, the optimal heating regime can be realized based on a study of heat transfer and movement of solid particles, including their phase conversion in the plasma flow. This problem is particularly important in connection with the development of a new branch of plasma chemistry that uses a plasma stream for recycling and disposal of chemical, medical and industrial waste, chemicals warfare and toxic substances. Recently a unique device are designed to solve such problems on the basis of RF and arc plasma torches with hundreds kW of power, working with such mediums as air, methane and propane. Study of heat transfer and movement of solid (dust) particles in plasma is also important to study the aerodynamics of supersonic gas-dispersion streams of plasma, and determination of the structure and development of atmospheres of many galaxies, and to simulate the conditions of entry of spacecrafts and satellites systems in the dense atmosphere of the Earth.

Description of the processes of heat exchange and movement of solid particles in plasma using the existing concepts is possible in principle in the preparation of the complete system of equations for conservation of mass, energy and momentum, using the chemical reactions and phase transitions. However, at the solving of such a system of equations for even more simple than plasma objects there are serious difficulties. Therefore, the choice of the mathematical formulation of the task should be determined as the results of preliminary analysis of the role of individual factors and the complexity of calculations from that ate ultimately very important at the use of mathematical models. It should be noted that many important results can be obtained with a model estimate of the role of individual factors in greatly simplified

S. Nguyen-Kuok, *Theory of Low-Temperature Plasma Physics*,
Springer Series on Atomic, Optical, and Plasma Physics 95,
DOI 10.1007/978-3-319-43721-7_10

conditions, particularly if the processes are considers that are limited only by heating and melting of materials, and there is no the aim to obtain a product of a given particle size and composition during a complex physic-chemical conversion of feedstock.

Currently, most mathematical models of plasma processes with dispersed phase is based on the one-dimensional approximation, describing changes of the internal temperature field of a spherical particle in an isothermal non-gradient plasma flow, or non-gradient heating of particles in the flow with variable parameters. In many cases, such a limited statement of the task is justified and allows obtaining of information sufficient for proper selection of the values of the basic technological parameters. However, using the high-power plasma flow and a working area with a large size, as well as in the processing of poly-dispersion materials for industrial production both internal and external parameters for the raw material particles should be noted. The consideration of the kinetics of the product formation processes requires also our greater attention. All of these things are connected with the inevitable complexity of the mathematical description.

In such complex processes as receiving of the powders and their compounds with the specified quality in restoring of the dispersed materials in plasma flow, it is difficult to predict the result. The estimation of the role of individual factors can be found only in the course of researches. The most important of these parameters are: the composition and properties of the initial gas and the dispersed phases of the products; initial temperatures and velocities, as well as consumptions of the gas and material; the structure of plasma flux; method of the entrance and size of the particle of material; possible chemical reactions in the system; conditions of the heat exchange and assumption of the flux in the surrounding space. For the solving of the technological task of the particles treatment in plasma it is necessary to determine: the efficiency of the technological process; the quality of the processed product; optimal regimes for processing of material and operation of the installation; the maximum mass factor for the processing of particles at the given parameters of the plasma installation and others.

To construct a model that takes into account the interaction between plasma and the groups of particles, and the creation of the technique of their modeling, it is necessary to consider the following tasks: the movement and heating of the particles in given fields of temperature and velocity of plasma in view of phase transformations during melting and evaporation; movement and heating of plasma loaded by the flow of particles and their interaction. In the theoretical study the fundamental equations of the model for the calculation of heating and motion of particles in the plasma flow are non-gradient and (or) the gradient equation of the particles heating; the equations of motion of particles; equation of heating, taking into account the phase transformation at melting and evaporation of the particles. Example of calculation is performed for the plasma treatment of MgO particles in the air RF plasma torch (1.76 MHz–60 kW).

10.1 The Basic Approaches to the Description of the Heat Exchange and Movement of the Particles in the Plasma

10.1.1 The Criteria of Heat Transfer and Gas Dynamics of Particles

The heating of particles in the plasma flow is mainly determined by the active heat exchange and is estimated by the Nusselt number (*Nu*), that is the ratio of convective to conductive heat transfer across (normal to) the boundary. Because of the lack of direct experimental data of the heating of small free-moving particles in plasma flow in the range of values of *Nu* there are some difficulties. The equations of Rants–Marshall are mostly used for spherical particles in [1, 2] where the determining is usually the temperature of plasma flow.

$$Nu = 2 + 0.6\,Re^{0.5}\,Pr^{0.33}, \tag{10.1}$$

where *Re*, *Pr*—the numbers of Reynolds and Prandtl.

The temperature difference between plasma and surface of particles in plasma flow is often essential, and the conditions of heat transfer depend on the parameters of the boundary layer on the surface of particles, in expression (10.1), it was proposed to introduce a parameter $\left(\frac{\rho_p \mu_p}{\rho_{ps}\mu_{ps}}\right)^{0.2}$, here ρ_p, μ_p—the density and viscosity of plasma; ρ_{ps}, μ_{ps}—the density and viscosity of plasma at the surface temperature of particles. Here and after index «*s*» corresponds to the parameters of particles, index «*p*» corresponds to the parameters of plasma, index «*ps*»—to parameters of plasma at the surface temperature of particles. Experimental data of the heating of assigned spherical particles with diameter of 0.15–1 mm in plasma jets [3] confirm the opportunity to use then Eq. (10.1), using $\frac{\lambda_{ps}}{\lambda_p}$ in the form:

$$Nu = 2\frac{\lambda_{ps}}{\lambda_p} + 0.6 Re_{ps}^{0.5} Pr_{ps}^{0.33} \left(\frac{\rho_p \mu_p}{\rho_{ps}\mu_{ps}}\right)^{0.2}. \tag{10.2}$$

Using of $\frac{\lambda_{ps}}{\lambda_p}$ clarifies the possibility of reducing of *Nu* to the level less than two, not only for irregular shaped bodies but also for spherical particles at high temperature gradients.

The Reynolds number is the ratio of the intensities of convective and viscous flow is defined as:

$$Re = \frac{\rho_p \left| v_p - v_s \right| d_s}{\mu_p}, \tag{10.3}$$

where v_p, v_s—the velocity of plasma and particles, respectively; d_s—diameter of particles.

The Prandtl number as the ratio of the intensity of the conductive and viscous flows is calculated by equation

$$Pr = \frac{\mu_p c_{pp}}{\lambda_p}. \quad (10.4)$$

There are more than two dozen of equations for determination of the aerodynamic resistance C_d in depending on Re, obtained by calculation and experimentally (mainly for enshrined particles larger than 1 mm). Most equations for the calculation of $C_d(Re)$ is:

$$C_d(Re) = \frac{24}{Re} + ARe^{-B},$$

where the coefficients A and B depend on the specific conditions and are determined experimentally.

Many factors influence to the ratio C_d, a number of papers [1–7] is devoted to analyze of these factors. The following equation for calculation of $C_d(Re)$ for the sphere (the so-called “standard” curve of resistance), obtained in the cases of a steady isothermal flow. It has been well studied and confirmed by numerous experiments for a wide range of Re numbers.

$$C_d(Re) = \frac{16.6}{Re^{0.75}} + 0.2. \quad (10.5)$$

The number of Bio characterize the intensity of heat exchange of particles with a plasma flow

$$Bi = \frac{\alpha d_s}{2\lambda_s}, \quad (10.6)$$

where α—coefficient of heat transfer of particles; λ_s, d_s—the thermal conductivity and diameter of the particle.

The number of Fourier for heat homochronicity is defined as

$$Fo = \frac{4\lambda_s t}{\rho_s c_{ps} d_s^2}, \quad (10.7)$$

where t—time.

The other dependences can be found in the references about the heat transfer [5–7]. In the phase changes in the processes of melting and evaporation the dependence of heat transfer of particles in the plasma flow should change. In [1, 2, 8], there are the equations for calculation of Nu and C_d for evaporation of particle, but their use is hampered due to the lack of reliable experimental data.

10.1.2 The Non-gradient Heating and Movement of the Particles in the Plasma

Model of non-gradient heating (Fig. 10.1) can be used to study of the heating of small particles ($d_s \ll 50$ μm) with high thermal conductivity λ_s, i.e., when the number of Bio (10.6), that characterize the intensity of heat exchange with plasma flow particles, is much less than one [1, 2]. Similar conditions may take place at the low heat transfer coefficient α. In this case the temperature gradient of the particle is small and it is believed that the particle in the whole volume has the same temperature. For the model calculation of heating the spherical shape of particles is usually assumed (Fig. 10.1).

The equation of non-gradient heating of particle is:

$$\rho_s c_{ps} \frac{dT_s}{dt} = \frac{6}{d_s}\left[\alpha(T_p - T_s) - \varepsilon\sigma T_s^4\right] + U_{rad}, \tag{10.8}$$

where ρ_s, c_{ps}—the density and specific heat of particles, respectively; T_p—the temperature of plasma at the external edge of the particle; T_s—the temperature of the particles; ε—the degree of blackness of particles; σ—the constant of Stefan–Boltzmann; U_{rad}—the specific power of the plasma radiation.

The total heat flux that is going to the heating of particles,

$$q_{\sum} = \alpha(T_p - T_s) + U_{rad}\frac{d_s}{6} - \varepsilon\sigma T_s^4,$$

where $\alpha(T_p - T_s) = q$ is the heat flux coming from plasma to the heating of particles by convection and thermal conductivity; $U_{rad}\frac{d_s}{6} = q_{rad}$—the heat flux from the plasma emission to the surface of particle; $\varepsilon\sigma T_s^4 = q_{\sigma s}$—the heat flux of the emission from the surface of particle. The total flux of energy to the surface of

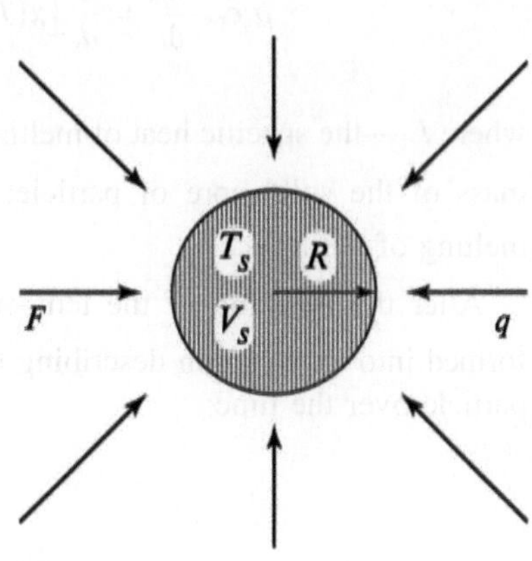

Fig. 10.1 The scheme of non-gradient heating of particle in plasma

particle determines the velocity of changes of its temperature that is expressed by the left side of Eq. (10.8).

The heat transfer coefficient depends on many factors and is determined by the dynamics of heat transfer between plasma and particle. In most of the works it is calculated by the Nusselt number (10.2) as:

$$\alpha = Nu\frac{\lambda_{ps}}{d_s}. \tag{10.9}$$

Analytical Solution of the Equation the Non-gradient Heating Analytical solution of Eq. (10.8) exists only at all its constant coefficients and can be justified for a sufficiently small period of time when all the parameters of particles and plasma can be considered permanent. For simplicity we assume that the contribution of the radiation flux into the total flux that is going into the heating of particles is insignificant and it can be neglected. Then the analytic solution of Eq. (10.8) with the initial condition $T_s|_{t=0} = T_{s0}$ has the form

$$T_s(t) = T_p\left(1 - e^{-kt}\right) + T_{s0}e^{-kt}, \tag{10.10}$$

where the time constant $k = \frac{6\alpha}{\rho_s c_{ps} d_s}$.

The total time of heating of particle to the melting temperature T_{s1} is calculated using the equation

$$t_0 = \frac{1}{k}\ln\left(\frac{T_p - T_{s0}}{T_p - T_{s1}}\right) = \frac{\rho_s c_{ps} d_s^2}{6Nu\lambda_{ps}}\ln\left(\frac{T_p - T_{s0}}{T_p - T_{s1}}\right). \tag{10.11}$$

After the reaching of the melting temperature of heating the energy due to a classical representation is consumed only the destruction of the molecular structures and the temperature of particle remain constant. Thus, the general equation of non-gradient heating of particle (10.8), taking into account the melting phase is as:

$$\rho_s c_{ps}\frac{dT_s}{dt} = \frac{6}{d_s}\left[\alpha\left(T_p - T_s\right) - \varepsilon\sigma T_s^4\right] + \frac{6L_1}{\pi d_s^3}\frac{dm_s}{dt}, \tag{10.12}$$

where L_1—the specific heat of melting of the material of particles; $m_s = \rho_s\frac{\pi d_s^3}{6}$—the mass of the solid core of particle; $\frac{6L_1}{\pi d_s^3}\frac{dm_s}{d_t}$—the specific power expended to the melting of particles.

After the reaching of the temperature of melting $\left(\frac{dT_s}{d_t} = 0\right)$ (10.12) is transformed into an equation describing the change in the diameter of the solid core of particle over the time:

$$\frac{\mathrm{d}d_s}{\mathrm{d}t} = -2\frac{\alpha(T_p - T_{s1}) - \varepsilon\sigma T_{s1}^4}{\rho_s L_1} = C_1.$$

At the constant transfer coefficients of plasma and particles with diameter of solid core of particle in dependence on time is defined as

$$d_s = -C_1 t + d_{s0}, \tag{10.13}$$

where d_{s0}—the diameter of particle at the beginning of melting.

Time of the full melting, respectively, is calculated using the equation

$$t_1 = -\frac{d_{s0}}{C_1} = \frac{\rho_s d_{s0}^2 L_1}{2\left[Nu\lambda_{ps}(T_p - T_{s1}) - \varepsilon\sigma T_{s1}^4\right]}. \tag{10.14}$$

A total time required for a full heating and melting of particles is:

$$t = t_0 + t_1 = \frac{\rho_s d_{s0}^2}{2}\left[\frac{c_{ps}}{3Nu\lambda_{ps}}\ln\left(\frac{T_p - T_{s0}}{T_p - T_{s1}}\right) + \frac{L_1}{Nu\lambda_{ps}(T_p - T_{s1}) - \varepsilon\sigma T_{s1}^4}\right]. \tag{10.15}$$

Calculation of non-gradient heating of particles on the basis of this simplified heat balance equation, despite its approximation, allows in some cases without cumbersome calculations to make estimation of the velocity of heating of particles and to carry out the selection of the optimal regime of the technological process.

In real cases, when solid particles move inside the plasma flow with sharply varying parameters, the assumption of the constancy of the parameters of plasma and particles in the analytical solution of (10.10, 10.11) and (10.13–10.15) is not justified. At the same time, a numerical method for the solution of heating of particles and for obtaining of heat flow going into the heating of particle should be used, it is necessary to determine the coordinates of location of the particle and the speed of its movement in flight, i.e., it is necessary to solve the equation of motion of particles.

Movement of the Solid Particles in Plasma Particles moving at the high speed plasma flow, are not only actively heated, but are quickly accelerated. Different forces acts to the particle in the plasma flow: the force of aerodynamic resistance F_a, the Magnus force F_M, the force of thermophoresis F_t, gravity F_g, Archimedes force F_A, the force of the apparent mass F_{am}, the force of Basse F_B, electrostatic forces, the force of light pressure and others [1]. The results of calculation of these forces for typical plasma jets of RF and Arc plasma torches are shown in Table 10.1.

The data in Table 10.1 show, the basic forces that determine the motion of particles in plasma flow are the force of aerodynamic resistance and the force of gravity, and the force of thermophoresis should be considered only for very small

Table 10.1 The results of the calculation of forces acting on particles of Al_2O_3 in jets of RF and Arc plasma torches

Parameters	Argon						Air					
	$T_p = 10{,}000$ K, $v_p = 20$ m/s			$T_p = 13{,}000$ K, $v_p = 400$ m/s			$T_p = 7000$ K, $v_p = 20$ m/s			$T_p = 9000$ K, $v_p = 400$ m/s		
d_s (μm)	1	10	100	1	10	100	1	10	100	1	10	100
F_a (N)	6×10^{-8}	6×10^{-7}	6×10^{-6}	8.3×10^{-7}	7.5×10^{-6}	5.6×10^{-5}	4×10^{-8}	4×10^{-7}	4×10^{-6}	1.1×10^{-5}	10^{-5}	8×10^{-5}
F_A (N)	6×10^{-18}	6×10^{-15}	6×10^{-12}	8×10^{-7}	8×10^{-14}	8×10^{-11}	4×10^{-18}	4×10^{-15}	4×10^{-12}	5×10^{-17}	5×10^{-14}	5×10^{-11}
F_g (N)	2×10^{-14}	2×10^{-11}	2×10^{-8}	2×10^{-14}	2×10^{-11}	2×10^{-8}	2×10^{-14}	2×10^{-11}	2×10^{-8}	2×10^{-14}	2×10^{-11}	2×10^{-8}
F_t (N)	8×10^{-10}	2×10^{-9}	2×10^{-8}	5.7×10^{-8}	5.7×10^{-6}	5.7×10^{-5}	9×10^{-10}	10^{-9}	10^{-8}	1.3×10^{-8}	6×10^{-8}	6×10^{-7}
F_M (N)	4×10^{-12}	4×10^{-10}	4×10^{-8}	7×10^{-10}	7×10^{-8}	7×10^{-6}	2×10^{-12}	2×10^{-10}	2×10^{-8}	7×10^{-10}	7×10^{-8}	7×10^{-6}
F_{am} (N)	2×10^{-17}	2×10^{-14}	2×10^{-11}	2×10^{-15}	2×10^{-12}	2×10^{-9}	10^{-17}	10^{-14}	10^{-11}	10^{-15}	10^{-14}	10^{-9}
F_B (N)	9×10^{-12}	9×10^{-10}	9×10^{-8}	2×10^{-10}	2×10^{-8}	2×10^{-6}	6×10^{-12}	6×10^{-10}	6×10^{-8}	2×10^{-10}	2×10^{-8}	2×10^{-6}

particles ($d_s < 1$ μm). Other forces are much smaller in absolute values, and can be neglected in simulation of motion of particles in the plasma stream. Two-dimensional equations of motion of particles in the plasma flow recorded in a cylindrical coordinate system are:

$$\begin{cases} \rho_s \frac{dv_{sz}}{dt} = 0.75 C_{dz} \frac{\rho_p (v_{pz} - v_{sz}) |v_{pz} - v_{sz}|}{d_s} + \rho_s g_z; \\ \rho_s \frac{dv_{sr}}{dt} = 0.75 C_{dr} \frac{\rho_p (v_{pr} - v_{sr}) |v_{pr} - v_{sr}|}{d_s}. \end{cases} \tag{10.16}$$

Here v_{sz}, v_{sr}, v_{pz}, v_{pr}—the axial and radial components of the velocity of particle and plasma, respectively; C_{dz}, C_{dr}—the coefficients of aerodynamic resistance in two directions.

10.1.3 The Gradient Heating Particles in the Plasma

The assumption of the absence of a temperature gradient inside the particle is true only in the region of small number of Bio ($Bi < 0.1$) and for sufficiently large values of the Fourier number ($Fo \gg 1$). In this case, the kinetics of heating of particles depends weakly on heat-inertial properties and is almost completely determined by the conditions of the external heat transfer.

In those cases when the intensity of heating and speed of passage (numbers of *Bi*, *Fo*) define a substantial temperature gradient inside the particle, for calculation of its temperature in dependence of time the following equation should be solved, it is written in a spherical coordinate system

$$\rho_s c_{ps} \frac{\partial T_s(r,t)}{\partial t} = \frac{1}{r^2} \frac{\partial}{\partial r} \left(r^2 \lambda_s \frac{\partial T_s(r,t)}{\partial r} \right). \tag{10.17}$$

The initial condition $T_s(r,0) = T_{s0}$ and the boundary conditions of the third kind are accepted.

$$\frac{\partial T_s(0,t)}{\partial r} = 0; \quad \lambda_s \frac{\partial T_s(R,t)}{\partial r} = \alpha [T_p - T_s(R,t)] + U_{rad} R/3 - \varepsilon \sigma T_s^4(R,t), \tag{10.18}$$

where $R = \frac{d_s}{2}$—the radius of particle; $\alpha [T_p - T_s(R,t)]$—the heat flow (convective and conductive) from plasma to the surface of particles; $U_{rad} R/3$—the heat flux from plasma radiation coming to the particle; $\varepsilon \sigma T_s^4(R,t)$—the heat flux from the radiation emitted by the surface of particles.

The Numerical Solution of the Equation of Gradient Heating of Particles
Equation of the gradient heating (10.17) is the equation of transient heat conduction, the solution of it can be obtained using the method of control volume.

The energy balance inside each of the control volume (Fig. 10.2) can be obtained by integrating of the Eq. (10.17) for the control volume and time:

$$\int_{r_w}^{r_e}\int_{t}^{t+\Delta t} \rho_s c_{ps}\frac{\partial T_s}{\partial t}\mathrm{d}t r^2 \mathrm{d}r = \int_{t}^{t+\Delta t}\int_{r_w}^{r_e} \frac{1}{r^2}\frac{\partial}{\partial r}\left(r^2\lambda_s\frac{\partial T_s}{\partial r}\right) r^2 \mathrm{d}r\mathrm{d}t. \tag{10.19}$$

In the integration of the Eq. (10.19), the following assumptions are used: inside the control volume (r_w, r_e) and at the time interval (t, $t + \Delta t$) the value $\rho_s c_{ps}$ takes a constant value corresponding to the central point M and at time t; the temperature of particle in the transient member $\frac{\partial T_s(r,t)}{\partial t}$ is constant, as the central point M corresponding for the interval (r_w, r_e).

If the left and right side of the Eq. (10.19) is once integrated (for convenience we denote $T_s \equiv T$), we have:

$$\frac{(\rho_s c_{ps})_M (r_e^3 - r_w^3)}{3}\int_{t}^{t+\Delta t}\frac{\partial T_M}{\partial t}\mathrm{d}t = \int_{t}^{t+\Delta t} r_e^2\lambda_{se}\frac{T_E - T_M}{r_E - r_M}\mathrm{d}t - \int_{t}^{t+\Delta t} r_w^2\lambda_{sw}\frac{T_M - T_W}{r_M - r_W}\mathrm{d}t. \tag{10.20}$$

We use the following approximations for the change in time from t to $t + \Delta t$ of the temperatures T_M, T_E, T_W : $\int_t^{t+\Delta t} T_M \mathrm{d}t = \left[\varepsilon T_M + (1-\varepsilon)T_M^0\right]\Delta t$, where ε—a certain weight coefficient, that varies from 0 to 1. Here and below we use the notations T_M, T_E, T_W as the temperature at the time $t + \Delta t$, and T_M^0, T_E^0, T_W^0—at the

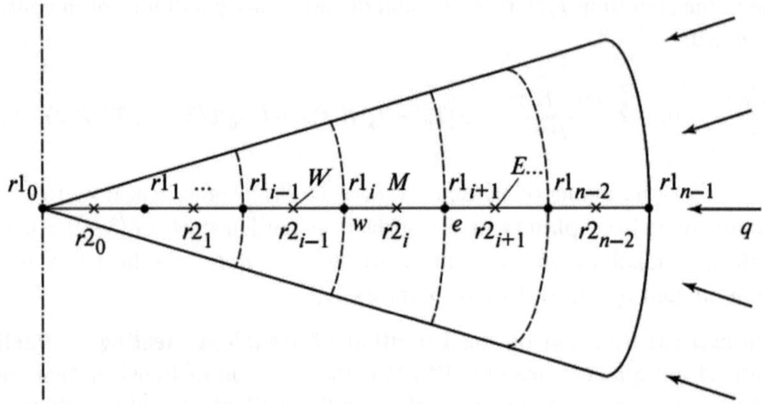

Fig. 10.2 Grid and the control volumes for the solving of the equation of gradient heating of particles

time t. Using the similar relations for the integral $T_{\rm E}$, $T_{\rm W}$ from t to $t + \Delta t$, from the Eq. (10.20) we find:

$$\frac{(\rho_s c_{ps})_M (r_e^3 - r_w^3)}{3\Delta t}(T_M - T_M^0) = \varepsilon\left[r_e^2\lambda_{se}\frac{T_E - T_M}{r_E - r_M} - r_w^2\lambda_{sw}\frac{T_M - T_W}{r_M - r_W}\right] + (1-\varepsilon)\left[r_e^2\lambda_{se}\frac{T_E^0 - T_M^0}{r_E - r_M} - r_w^2\lambda_{sw}\frac{T_M^0 - T_W^0}{r_M - r_W}\right].$$

The result is a discrete analogue of the gradient heating of particles (10.17) in the form:

$$a_M T_M = a_E\left(\varepsilon T_E + (1-\varepsilon)T_E^0\right) + a_W\left(\varepsilon T_W + (1-\varepsilon)T_W^0\right) + \left[a_M^0 - (1-\varepsilon)(a_E + a_W)\right]T_M^0, \quad (10.21)$$

where

$$\begin{cases} a_E = \frac{r_e^2\lambda_{se}}{r_E - r_M} > 0; \quad a_W = \frac{r_w^2\lambda_{sw}}{r_M - r_W} > 0; \\ a_M^0 = \frac{(\rho_s c_{ps})_M (r_e^3 - r_w^3)}{3\Delta t} > 0; \quad a_M = \varepsilon(a_E + a_W) + a_M^0 > 0. \end{cases}$$

For certain values of the weigh coefficients ε have known schemes for solving of the transient equation of heat conduction. In particular, ε = *0* we get the explicit scheme for ε = 0.5—implicit scheme of Crank–Nicolson for ε = 1—a fully implicit scheme. The basic principle of a fully implicit scheme is that in the step of time the temperature is assumed to the new value of T_M. Thus, the thermal conductivity should be recalculated using T_M in the iterative process in the same way as in the solution of the stationary case. The analysis of these schemes in [9] shows that the sustainable solution of discrete analogues is guaranteed using only a fully implicit scheme, i.e., if ε = 1, and this scheme is taken for solving of our task. The discrete analogue for a fully implicit scheme have the form:

$$a_M T_M = a_E T_E + a_W T_W + a_M^0 T_M^0, \quad (10.22)$$

where

$$\begin{cases} a_E = \frac{r_e^2\lambda_{se}}{r_E - r_M} > 0; \quad a_W = \frac{r_w^2\lambda_{sw}}{r_M - r_W} > 0; \\ a_M^0 = \frac{(\rho_s c_{ps})_M (r_e^3 - r_w^3)}{3\Delta t} > 0; \quad a_M = a_E + a_W + a_M^0 > 0. \end{cases}$$

Note that if $\Delta t \to \infty$ from (10.22) the discrete analogue of the stationary equation of heat conductivity can be obtained.

Using the boundary conditions (10.18), the discrete analogue can be obtained for the boundary points near the center and the surface of particles.

For a point near the center of the particle $r = r2_0$ (Fig. 10.2):

$$a_{\mathrm{M}}T_{\mathrm{M}} = a_{\mathrm{E}}T_{\mathrm{E}} + a_{\mathrm{M}}^{0}T_{\mathrm{M}}^{0}, \tag{10.23}$$

Here, from the boundary condition in the center of the particle the discrete analogue $a_W = 0$.

For a point near the surface of the particle $r = r2_{n-2}$ (Fig. 10.2):

$$a_{\mathrm{M}}T_{\mathrm{M}} = a_{\mathrm{W}}T_{\mathrm{W}} + a_{\mathrm{M}}^{0}T_{\mathrm{M}}^{0} + q_{\mathrm{R}}R^{2}, \tag{10.24}$$

where $q_{\mathrm{R}} = \alpha\left[T_{\mathrm{p}} - T_{\mathrm{s}}(R,t)\right] - \varepsilon\sigma T_{\mathrm{s}}^{4}(R,t) + U_{\mathrm{rad}}R/3$—the total heat flux at the surface of the particle.

10.1.4 Evaporation of Particles in the Plasma

Evaporation of particles in the plasma flow is an important technological process that should be studied in many problems of plasma chemistry. In particular, evaporation of particles is an indispensable condition for obtaining of the materials with small disperse, such as by direct evaporation, and then using of chemical reactions. The mechanism of evaporation processes of crystalline solids in sufficient considered in details [10–13]. It can be presented as step process consisting in transition of atoms and molecules through the different states at the surface of the border crystal–vapor and through the adsorption layer into the vapor phase. In the case of liquid, taking into account its intermediate phase state between solid and gaseous two mechanisms of evaporation are possible—the direct and step through the adsorption layer. The mechanism of evaporation is mainly determined by the activation energy and at high heat fluxes particle vaporization starts before reaching of the boiling point. The resulting vapors create the steam jacket around the particle, consisting of a mixture of heavy vapor molecules and gas molecules. As a result, the process has to be stabilized by the heat balance. Thermal energy supplied from plasma to the particle, will be consumed to its heating, evaporation and heating of the vapor to the temperature of plasma and in the steady regime—to the evaporation and heating at a constant temperature of particles.

According to J. Stefan, the evaporation process can be represented the diffusion of vapor from the particle surface, above this surface there is a saturated vapor, into the environment gas. The processes of heat and mass transfer in the plasma flow are similar to each other; to the heat transfer due to the molecular thermal conductivity corresponds to molecular diffusion, convection heat transfer—convective diffusion. Thus, the theoretical and experimental results obtained in the study of heat transfer processes and methods of their investigation can be directly applied to the processes of diffusion, and vice versa.

Basic Equations and Criteria In case of simultaneous heating, melting and evaporation of the particle the equation of non-gradient heating can be written as:

$$\rho_s c_{ps}\frac{dT_s}{dt} = \frac{6}{d_s}\left[\alpha(T_p - T_s) - \varepsilon\sigma T_s^4 - L_2 j\right] + \frac{6L_1}{\pi d_s^3}\frac{dm_s}{dt}, \tag{10.25}$$

where L_1, L_2—the specific heat of melting and evaporation of the material of particles, respectively; $m_s = \rho_s\frac{\pi d_s^3}{6}$—the mass of the solid core of particles; j—the density of the flux from the surface of the vaporized material.

The density of vapor stream from the surface of vaporized material, i.e., the quantity of substances transported through the unit of surface area per unit of time, is often expressed by Fick's law [14]:

- in a stationary medium:

$$\vec{j} = -D\text{grad}C; \tag{10.26}$$

- in a mobile environment:

$$\vec{j} = -D\text{grad}C + \vec{v}C, \tag{10.26a}$$

where C—the concentration of the evaporated substance; D—the diffusion coefficient; $\vec{v}$—the difference between velocity of medium and of particle.

The differential equation of diffusion in a stationary medium has the form

$$\frac{\partial C}{\partial t} = \text{div}\,(D\text{grad}C) + q, \tag{10.27}$$

where q—density of sources of the substance, i.e., the quantity of the substance formed by chemical reactions in the unit of volume per the unit of time.

The diffusion equation in the mobile environment, respectively, has the form:

$$\frac{\partial C}{\partial t} + \text{div}\,(C\vec{v}) = \text{div}\,(D\text{grad}C) + q. \tag{10.27a}$$

In the cases when the resulting amount of substance can be taken as $q = 0$ and for the stationary processes $\frac{\partial C}{\partial t} = 0$, the diffusion Eq. (10.27a) can be written as:

$$\text{div}(C\vec{v}) = \text{div}(D\text{grad}C). \tag{10.28}$$

The velocity of evaporation of particles I, i.e., the quantity of substance m, carried through the surface with area S per unit of time is defined as:

$$I = -\frac{dm}{dt} = jS. \tag{10.29}$$

Solution of the problem of particles evaporation is reduced to the determination of the evaporation velocity I or the flux density j. Depending on the size of particles, pressure, and composition of the medium the evaporation can be at diffusion (the Knudsen number $Kn < 1$), free-molecular ($Kn > 1$) or transient regime [1, 14].

Quasi-stationary Evaporation of Particles Fixed in Relation to the Environment In the case where the particle is immobile in the environment, and its radius r_0 is much higher than the mean free path of the vapor molecules l, the vapor concentration at the surface of the particle is equal to the equilibrium concentration C_0. This equation of Maxwell [14] is valid:

$$I = I_0 = 4\pi r_0 D(C_0 - C_\infty), \tag{10.30}$$

Here $C = C_0$ at $r = r_0$, $C = C_\infty$ at $r = \infty$.

In this case, the velocity of particles evaporation is completely determined by the velocity of steam diffusion into plasma, and we have a purely diffusive regime of evaporation.

Vapor concentration at a distance r from the center of particle is calculated using the equation

$$C = C_\infty + \frac{r_0}{r}(C_0 - C_\infty). \tag{10.31}$$

Particle radius in dependence of time t can be defined as:

$$r = \sqrt{r_0^2 - \frac{2D}{\rho_s}(C_0 - C_\infty)t}. \tag{10.32}$$

If we assume that the vapor of substance conforms to the laws of ideal gases, and express the concentration of vapor C using its partial pressure $p : C = \frac{pM}{RT}$, where M—the molecular mass of the evaporating substance in a gaseous state; R—the gas constant; T—the temperature of vapor, the Maxwell equation takes the form of

$$I_0 = 4\pi r_0 D\frac{(p_0 - p_\infty)M}{RT}. \tag{10.33}$$

Here the vapor pressure $p = p_0$ at $r = r_0$, $p = p_\infty$ at $r = \infty$.

As first pointed out by J. Stefan, near the surface of the evaporating body the hydrodynamic flow (the Stefan's flow) of the gaseous medium from the surface should exist. In this case, the velocity of the particles evaporation can be determined approximately [1] using the equation

$$I = I_0\left(1 + \frac{p_0 + p_\infty}{2p_p - p_0 - p_\infty}\right), \tag{10.34}$$

Here p_p—the pressure of the plasma.

Influence of the boundary layer on the velocity of evaporation can be accounted if we assume that the equation of Fick (10.26) and the ensuing Eq. (10.26a) are applicable only at a distance of more than $\Delta \sim l$ from the surface of the particles, and in the layer near the wall with the thickness Δ the exchange of molecules of vapor smoothly occurs in vacuum. The thickness of boundary layer Δ is the distance from the particle surface, where evaporated molecules have its first collision with other molecules, and the vapor density decreases from C_0 near the surface of particle to C_∞ in the plasma flow. The velocity of evaporation of the particle in this case can be defined as

$$I = \frac{I_0}{r_0/(r_0 + \Delta) + D/(r_0 \varepsilon \alpha)}, \tag{10.35}$$

Here $\varepsilon = \frac{1}{4}\sqrt{\frac{8kT}{\pi M}}$—one fourth of the average velocity of the vapor molecules; α—coefficient of vapor condensation.

With increasing ratio l/r_0 the diffusion regime of evaporation is gradually transformed into kinetic and at large l/r_0 the diffusion finish to play a role in the process of evaporation [11].

Quasi-stationary Evaporation of Particles Moving in Relation to the Environment The greatest practical interest is the process of evaporation of particles moving relative to plasma by the force of aerodynamic resistance and other forces.

The accounting of all factors influencing to the evaporation of particles complicated its calculations, very often various criterial dependences of the numbers of Reydnols *Re*, Nusselt *Nu*, Sherwood *Sh*, Schmidt *Sc* and others are taken. They take into account the phase transformations of particles and different from the usual equations by some correction factors. Thus, the velocity of the particles evaporation taking into account the boundary layer in the moving flow of plasma is determined as [1, 7]:

$$I = I_0\left(1 + \beta Re^{1/2} Sc^{1/3}\right), \tag{10.36}$$

where $\beta = \frac{\Delta}{l}$; I_0—the velocity of evaporation obtained by Maxwell's Eq. (10.30) or (10.33).

Similarly, comparing the models (when the particles are fixed and when they are addicted by the flow of plasma) for large values of *Re* and the thickness of the boundary layer $\Delta \ll r_0$, we have the following equation for determination of the velocity of evaporation of particles moving in relation to the flow of plasma:

$$I = \frac{I_0\left(1 + \beta Re^{1/2} Sc^{1/3}\right)}{1 + \frac{D}{\varepsilon\alpha(2r_0/Sh - \beta l)}}. \tag{10.37}$$

At very high velocities of the flow, when the thickness of the diffusion boundary layer is smaller than the mean free path of gas molecules ($\Delta < l$), almost all evaporated molecules fly away by the gas flow, i.e., evaporation will move with the same speed as vacuum:

$$I = 4\pi r_0^2 D C_0 \varepsilon\alpha. \tag{10.38}$$

For estimate of the velocity of evaporation in the transition regime, the equation of Sherman [1, 8] can be used.

$$I = \frac{I_d I_k}{I_d + I_k}. \tag{10.39}$$

Here I_d, I_k—the velocity of evaporation in the diffusion and kinetic regimes.

Experimental Methods for the Study of Evaporation In practice, for determination of the velocity of evaporation of the material of particles the experimental data are often used. In the experimental study of evaporation and dissociation of materials the following methods are most frequently used: the high-temperature mass spectrometry (combination of the Knudsen method with mass spectrometric analysis of the products of evaporation); dynamic, such as method of flow; static; method of boiling points, etc.

The flux density j in view of the evaporated material is determined by measuring pressure of the saturated vapor at this surface at a known value of temperature. It is believed that vapor molecules on the surface of the particles are in equilibrium and have the Maxwell velocity distribution. Then we have the following equation for calculation of the flux density from evaporation:

$$j = \frac{p}{\sqrt{2\pi RT/M}}. \tag{10.40}$$

There is a large number of experimental data for determination of the saturated vapor pressure for different materials in literature, and they can be used for the calculation of the flux density from evaporation. However, it should be noted that the data of the evaporation of particles at high temperatures in many cases are not available.

Experiments [15] show that the process of evaporation of material of particles at high temperature plasma is often accompanied by complex chemical transformations, dissociation and condensation of molecules. For example, at the evaporation of a magnesium oxide MgO the dissociation mainly occurs in two ways:

$$MgO_{(s)} = Mg_{(g)} + 0.5O_{2(g)};$$
$$MgO_{(s)} = Mg_{(g)} + O_{(g)};$$

and in a lesser degree extent by sublimation according to the scheme:

$$MgO_{(s)} = MgO_{(g)},$$

Here the solid phase is signified by the index "s", and index "g"—the gaseous one.

As a result of experimental measurements [15] at the temperature of particles $T_s = 1900$ K content of molecules of vapor is obtained on the surface of $MgO_{(s)}$: Mg—52.27 %, O_2—32.91 %, O—13.82 %, MgO—1 %. The velocities of the chemical reactions in plasma are different, and their mathematical descriptions are rather cumbersome and time-consuming, so here they are not given here.

10.2 Calculation of the Heating and Movement of the Particles with the Given Parameters of the Plasma Flow

10.2.1 Installation for Plasma Treatment of Powder Materials

The plasma method for processing of refractory material has a number of advantages over mechanical crushing. Firstly, the mechanical crushing method can lead to further pollution, thereby require the additional chemical cleaning. Secondly, mechanical crushing to small and ultra-small particles with diameter less than 30–40 μm is a difficult and sometimes impossible task. Thirdly, the problem of chemical cleaning of materials and making them spherical cannot be solved by using this method.

Compared with the traditional method for processing of materials the plasma method does not have these disadvantages and, furthermore, allows obtaining of high technical and economical indicators. At present, for the treatment of powder particles the plasma jets of RF and Arc plasma torches are often used. The jets of Arc plasma torches have a relatively small diameter, high velocity of flow with a large temperature gradient. However, the presence of electrodes and its erosion in some cases does not allow obtaining of the pure product. RF plasma torches with equal power to the Arc plasma torches allow obtaining of the plasma jet with greater diameter with and small speed of flow and a large spatial region of uniform temperature. Small speeds of RF plasma also significantly alleviate the problem of particles entering into the plasma jet, and provide them to stay longer in the jet in compare with Arc plasma. Furthermore, the RF plasma, uncontaminated by the materials of electrode, allows processing and obtaining of the pure powder material.

As an example, the description of RF-installation and plasma technology for processing of MgO particles, that are widely used in industry. A significant disadvantage of currently produced industrial electrical MgO is a large specific surface of the powder that leads to significant reduction of electrical characteristics of tubular electric heaters produced using MgO and have very strong requirements on the chemical composition of the powder, especially the composition of CaO and Fe_2O_3.

The proposed process for treatment of the particles is a preparation of melted spherical grains of MgO in RF plasma jet. Spheroidization of the MgO periclase particles is produced in air RF plasma with the power of 60 kW (Fig. 10.3). Installation allows receiving of MgO powder consisting of spherical particles with a minimum specific surface, high electrical parameters even with high content of calcium and iron impurities, minimum abrasion and high turnover.

Fig. 10.3 Air RF plasma installation (1.76 MHz–60 kW) and a melting chamber for the processing of MgO periclase particles in a stream with the distributed entrance of particles at the perimeter of plasma torch

The jet of air plasma coming out of RF plasma torch with metal cut water-cooled chamber, consisting of some of sections, assembled with clamps on the housing of the discharge chamber (Fig. 10.3). The discharge chamber is placed in a quartz cover and induction plasma is generated in a channel by a RF inductor. There is a straight-jet air supply in the plasma torch. The plasma forming air is supplied through the central tube and the thermal protection gas is formed by a coaxial quartz tube and gas-former. Thus, the air forms a thermally protective layer of the moving cold gas between plasma and the walls of plasma torch. This air supply significantly increases the efficiency of the plasma torch and makes its work stable. Basic parameters of RF installation for processing of particles are shown below:

Main features of the feed medium:	
Voltage (V)	3 × 380
Frequency (Hz)	50
Input power (kW)	90
Power of the oscillation circuit (kW)	60
Operating frequency (MHz)	1.76 ± 0.044
The plasma-forming and transport gas	Argon, air
Air consumption (m^3/h)	1.8–3.5
Cooling water consumption (m^3/h), no less than	4.5
Productivity (kg/h), for the particles with size of 100 μm	~8

It was found that the plasma installation, working using a once-side scheme of entrance of materials into plasma, it is impractical to use in industrial environments due to low technical and economic factors. The consumption of the power in its use is 33–59 kW h/kg of the finished product at a performance of 1.1–2.6 kg/h. In this case, it was decided to change the scheme for entrance of MgO particles from the once-side to the reverse with the spouted layer (Fig. 10.3). Powder particles of MgO with diameter 100–200 μm are fed downwards to the plasma jet directed upwards. The difference of this method of entrance of particles compared with input by once-side or horizontal scheme is that the particles flying toward the plasma jet, is inhibited there and stops at a certain point and then starts to move in the opposite direction and is come into the melting chamber (Fig. 10.3). The scheme of RF installation for the thermal treatment of the particles is shown in Fig. 10.4.

Such organization of "plasma fountain" significantly increases the residence time of the particles in plasma compared with once-side or horizontal scheme, that allows melting of even the larger particles. Due to greater residence time of the particles in a plasma jet they more efficiently extract energy of the jet for its heating. Thus, at the fixed power of the jet per unit of time more particles can be melted and the best performance of the process can be achieved.

The electrical MgO produced by Bogdanovich's refractory plant was taken for the researches. For all the investigated MgO a number of analyzes was made: characteristics of chemical and grain compositions, liquid absorption, fluidity,

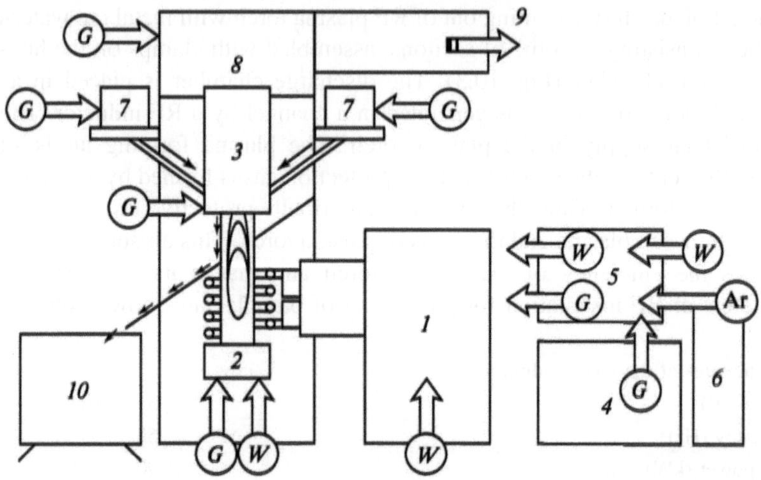

Fig. 10.4 Scheme of the RF installation for the thermal treatment of the particles: *1* RF generator; *2* RF plasma torch; *3* melting chamber; *4* compressor making the necessary consumption of plasma-forming and transport gas *G* (the air); *5* rotameters unit for control the consumption of water *W* and air *G*; *6* cylinder with Argon for ignition of the plasma torch; *7* the unit for entrance of initial material feed (feeder); *8* pre-filter; *9* exit to the main filter and ventilation system; *10* collecting hopper of processed particles

density after tapping; the specific, volume resistivity of insulation and leakage current at voltage $U = 500$ V are measured for experienced electric heaters.

The results showed that treatment with MgO periclase counter submitting it to the plasma allows to obtain a finished product with high electrical insulating properties at the level of the best periclase of foreign companies and technical and economic parameters are in several times higher than in one-side feed of powder into the plasma jet. Basic parameters of periclase particles before treatment and after it are shown in Tables 10.2, 10.3 and 10.4.

The sufficient degree of spheroidization of the periclase grains was reached at their treatment in this installation (80 ± 10) %. The physical and chemical parameters of obtained periclase MgO are better than material PPE-1 M and the achieved performance 70 % spheroidization of 8–3 kg/h for MgO fractions with diameter 100–200 μm.

10.2.2 Solution for Equations of the Heating and Movement Particles

The equations of motion and non-gradient heating of particles is a system of nonlinear ordinary differential equations and can be written in the Cauchy form as follows [16, 17]:

Table 10.2 Influence of plasma treatment on the chemical composition of MgO periclase particles

Class of MgO periclase particles	Material	Chemical composition (%)						
		CaO	SiO_2	Al_2O_3	Fe_2O_3	C	MgO	ppp
Third	Original	1.68	1.55	0.31	0.11	0.079	96.11	0.16
First and second	After plasma treatment	1.62	1.70	0.05	0.07	0.031	96.47	0.06

Table 10.3 Changes of the electrical-isolating characteristics of the MgO particles after plasma treatment

Class of MgO periclase particles	Material	Thermal load of heating element (W/cm²)					
		6		8		10	
		R (MOm)	*I* (mA)	*R* (MOm)	*I* (mA)	*R* (MOm)	I (mA)
Third	Original	4.70	0.35	0.47	1.20	0.10	3.32
First and second	After plasma treatment	26.7	0.20	2.70	0.53	0.53	1.31

Table 10.4 Change of specification of MgO periclase particles after plasma treatment

Class of MgO periclase particles	Material	Liquid absorption (%)	Dielectric strength (kV/mm)	Fluidity (g/s)	Density (g/cm³)
Third	Original	2.5	0.8	0.6	230
First and second	After plasma treatment	1.5	1.5	0.8	251

$$\begin{cases} \frac{dT_s}{d_t} = \frac{1}{\rho_s c_{ps}}\left[\frac{6}{d_s}\left(\alpha(T_p - T_s) - \varepsilon\sigma T_s^4\right) + U_{rad}\right]; \\ \frac{dv_{sz}}{d_t} = 0.75C_{dz}\frac{\rho_p\left(v_{pz}-v_{sz}\right)\left|v_{pz}-v_{sz}\right|}{\rho_s d_s} + g_z; \\ \frac{dv_{sr}}{d_t} = 0.75C_{dr}\frac{\rho_p\left(v_{pr}-v_{sr}\right)\left|v_{pr}-v_{sr}\right|}{\rho_s d_s}. \end{cases} \tag{10.41}$$

To solve the system of Eq. (10.41), the following initial boundary conditions are taken: $t = 0$, $T_s(0) = T_{s0}$, $v_{sz}(0) = v_{sz0}$, $v_{sr}(0) = v_{sr0}$, $z(0) = z_0$, $r(0) = r_0$.

In order to obtain the coefficients of (10.41) depending on time it is necessary to determine the coordinates of the particle by the equations

$$z(t) = \int_0^t v_{sz}(t)dt; \quad r(t) = \int_0^t v_{sr}(t)dt. \tag{10.42}$$

The task of Cauchy reduces to integrating the system of differential equations in the general form:

$$\frac{dY_j}{dx} = F_j(x, Y_j), \tag{10.43}$$

where Y_j—vector of unknown variables; F_j—vector of the right part of the system of differential equations; x—independent variable; $j = 1{-}N$—serial number of the equation; N—the number of equations.

The solution of system (10.43) with the initial conditions $x = x_0$, $Y_j(x_0) = Y_{j0}$ leads to finding the dependencies, passing through the points, defined by these conditions. The order of the numerical integration defines the order of solution method. The method of Runge–Kutta of fourth order is the most widely-used method for solving of the system (10.43) with a constant integration step h = const. Its advantages are the high accuracy (error of the method $R \sim h^5$) and reduced risk of occurrence of instability of the solution [18]. Algorithm realization of Runge–Kutta method is the cyclic calculations of $Y_{j,i+1}$ to every $i + 1$—step using the following equations:

$$\begin{cases} K_{1j} = hF_j(x_i,\ Y_{ji});\ K_{2j} = hF_j(x_i + 0.5h,\ Y_{ji} + 0.5K_{1j}); \\ K_{3j} = hF_j(x_i + 0.5h,\ Y_{ji} + 0.5K_{2j});\ K_{4j} = hF_j(x_i + 0.5h,\ Y_{ji} + K_{3j}); \\ Y_{j,i+1} = Y_{ji} + \frac{1}{6}(K_{1j} + 2K_{2j} + 2K_{3j} + K_{4j}). \end{cases} \tag{10.44}$$

In the case of gradient heating two tasks should be solved at the same time (step time): first—the solution of equations of motion of a particle of the Cauchy task with the definition of its position and the heat flux going into the heating of particles; second—solution of the equation of gradient heating (10.17) with the boundary condition (10.18) for finding the temperature distribution inside the particle at any time. Discrete analogue for equation of the gradient heating are obtained in the form (10.21–10.24). The following are the results of calculation of motion and heating of the particles in a plasma flow in cases of gradient and non-gradient heating of particles.

The Result of the Calculation of the Movement and the Heating of the Particles in a Given Plasma Flow To calculate the movement and the heating of the particles in a given flow of plasma it is necessary to define firstly all its parameters. The developed above procedure (Chaps. 6 and 7) the simulation of air plasma in RF plasma torch (1.76 MHz–60 kW) is carried out, the main parameters of it are given in Tables 7.3 and 7.4, and the temperature field and the plasma flow are shown in Fig. 10.5.

The obtained field of temperature and flow (speed) at the simulation of Aair plasma in RF plasma torch (1.76 MHz–60 kW) are used to study the motion and heating of MgO periclase particles. Initial data for the particles of MgO are: diameter d_s = 100, 150, 200 μm; density ρ_s = 3580 kg/m^3; the melting temperature T_{s1} = 2900–3100 K [15, 19] in dependence of MgO periclase chemical

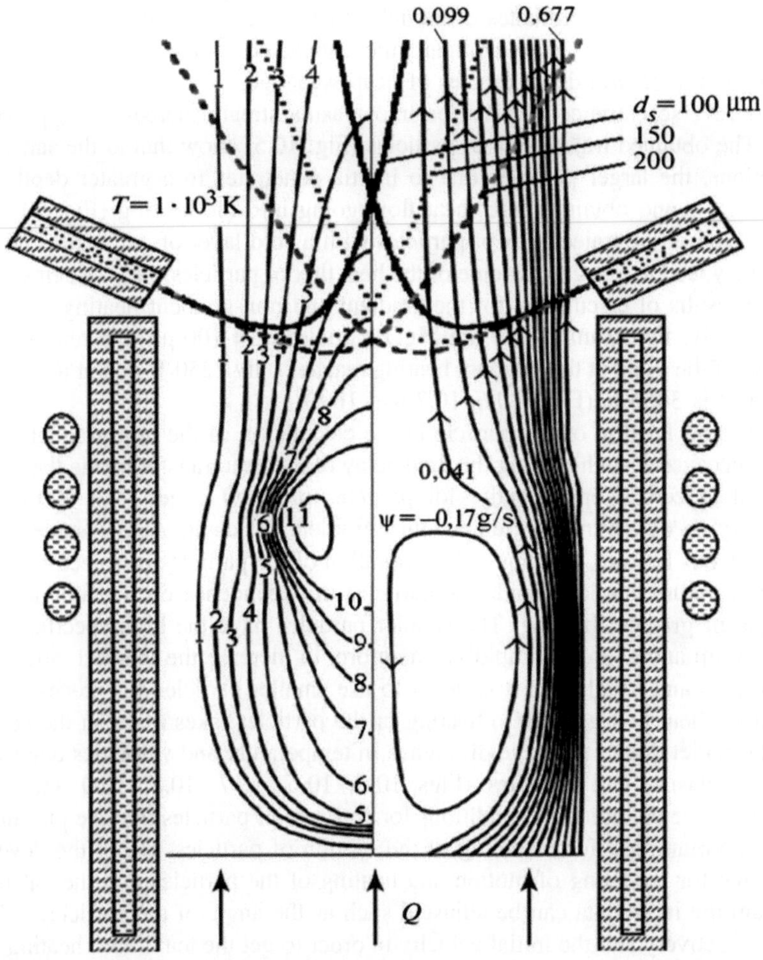

Fig. 10.5 The trajectories of the particles MgO periclase (d_s = 100, 150, 200 μm; $v_{sz0} = -4$ m/s, $v_{sr0} = -1$ m/s) in the air RF plasma torch (1.76 MHz–60 kW)

composition; initial components of velocity $v_{sz0} = -4$ m/s, $v_{sr0} = -1$ m/s; the initial temperature T_{s0} = 300 K. In this case the coordinate of input point z_0 = 540 mm (40 mm from the edge of the plasma torch), r_0 = 30 mm.

In the calculation all the local parameters of the particles were determined in a given plasma flow: temperature and velocity of particles, the heat flux going into heating of the particles, the temperature distribution inside the particles, the flight time, the trajectory of motion; melting and evaporation points et al. The obtained data allow find the optimal parameters depending on the particular material. In order to optimize the field of temperature and the velocity of plasma processing are analyzed, the size of particles, that should be melted at the end of plasma jet. In the decision such criteria of heat exchange are also received: *Re*, *Nu*, C_d, and others, in dependence on the coordinates of particles that makes it possible with the help of the developed technique to compare the calculated and experimental data and determine the criteria dependencies of heat exchange.

The reverse entrance of particles into plasma stream increases the processing time. The obtained trajectories of particles (Fig. 10.5) show that at the same initial conditions, the larger particles due to inertia penetrates to a greater depth of the plasma flow and obtain greater heat flow going into the heating (Fig. 10.6). The intersection of the trajectories of particles with a cold layer of gas at the end of the flight may lead to a negative value of the heat flux of particles (i.e., to their cooling).

The results of calculation for the gradient and non-gradient heating differ quite significantly. For example, for the MgO particles d_s = 100 μm the temperature at the end of the flight at the gradient heating reaches only 2250 K and at non-gradient heating it is 3050 K (Figs. 10.6, 10.7 and 10.8).

The rapid heating of the particle in the calculation of the non-gradient heating occurs because at such heating the time delay of the heat transfer inside the particles is equal to zero. Consequently, for particles of MgO larger than 100 μm it is necessary to determine the temperature gradient inside it. At the same time, at simultaneous heating, melting and evaporation of the particles it is necessary to get the temperature gradient inside the particles in dependence on the time and solve the task of gradient heating. The smaller particles have the high specific surface, and heat up and evaporate rapidly. Therefore, to increase the thermal efficiency of the process and installation it is better to use smaller particles for processing. The maximum heat flux going into heating of the particles takes place in the center of the plasma jet due to the large difference in temperature and velocities between the flow of plasma and particles (Figs. 10.5, 10.6, 10.7, 10.8, 10.9 and 10.10). Therefore, the definition of conditions for entrance of particles into the plasma zone is an important task for modeling of the motion of particles. Using the developed technique for modeling of motion and heating of the particles and the calculation program the input data can be adjusted such as the angle of the particles delivery, place of delivery and the initial velocity in order to get the maximum heating of the particles. Losses of energy by radiation of particle is much less than the heat flow coming to its heating by plasma, so in the further calculations the energy losses of particle by radiation are negligible.

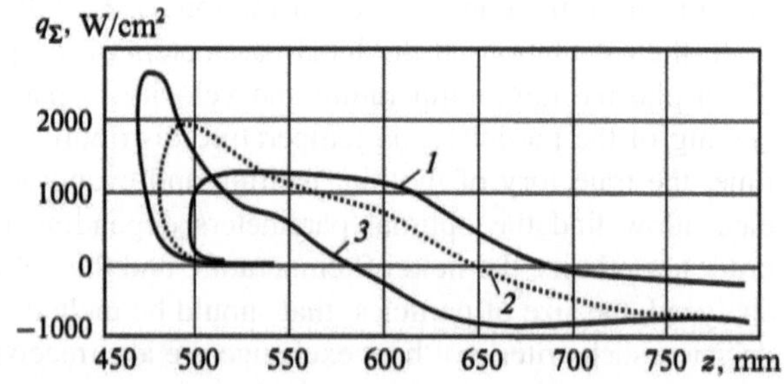

Fig. 10.6 Change of the total heat flux going to the heating of particles MgO in the air RF plasma torch (1.76 MHz–60 kW): *1* d_s = 100 μm, *2* d_s = 150 μm, *3* d_s = 200 μm

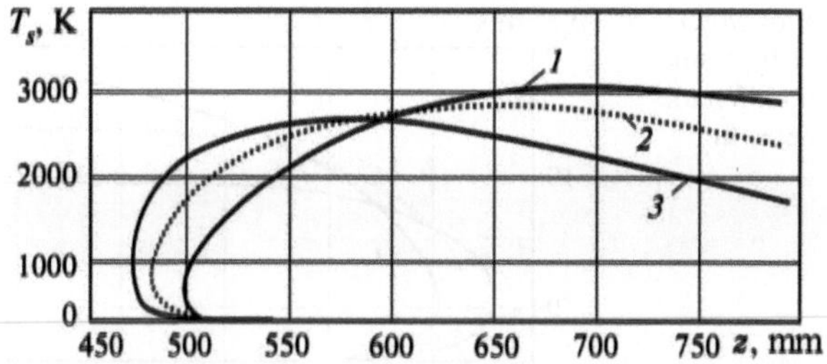

Fig. 10.7 Change of heating temperature (non-gradient heating) of particles MgO in the air RF plasma torch (1.76 MHz–60 kW): notations are the same as in Fig. 10.6

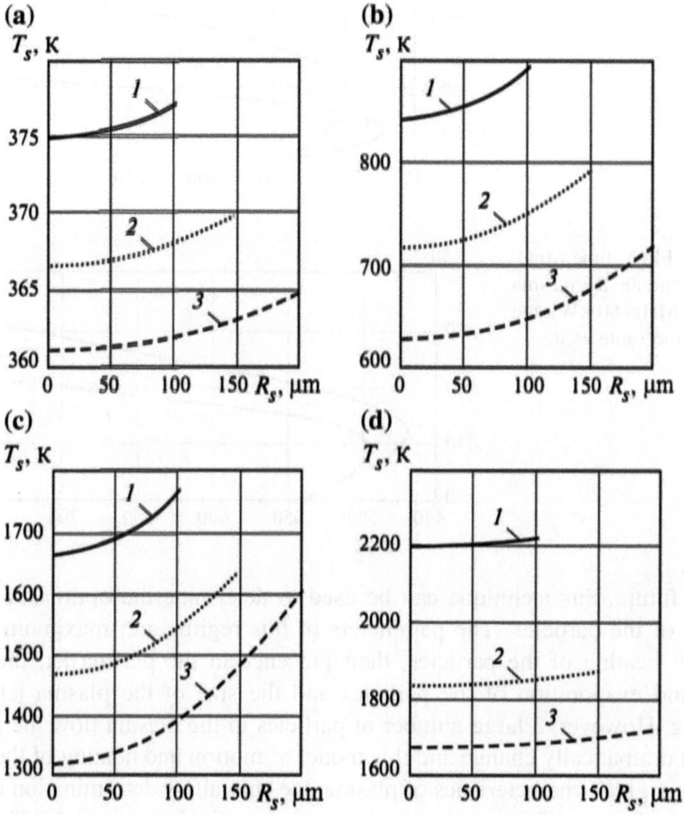

Fig. 10.8 Temperature distribution inside the particles MgO (gradient heating) of different diameter at different times during the motion in the air RF plasma torch (1.76 MHz–60 kW): **a** t = 15 ms, **b** t = 25 ms, **c** t = 35 ms, **d** t = 45 ms; the other notations are the same as in Fig. 10.6

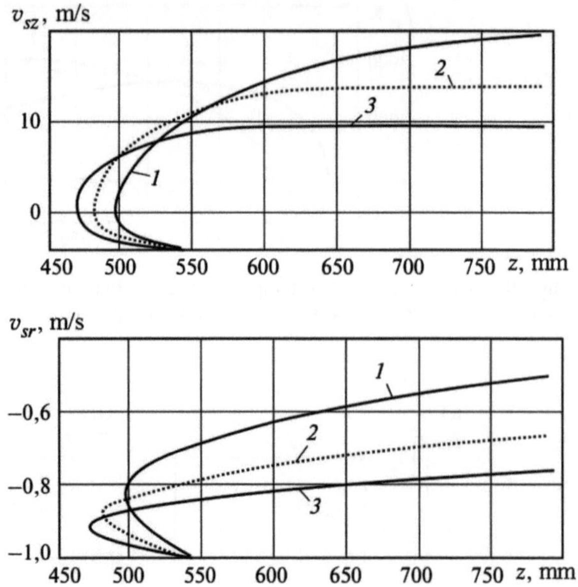

Fig. 10.9 Velocity of MgO particles in the air RF plasma torch (1.76 MHz–60 kW): the notation is the same as in Fig. 10.6

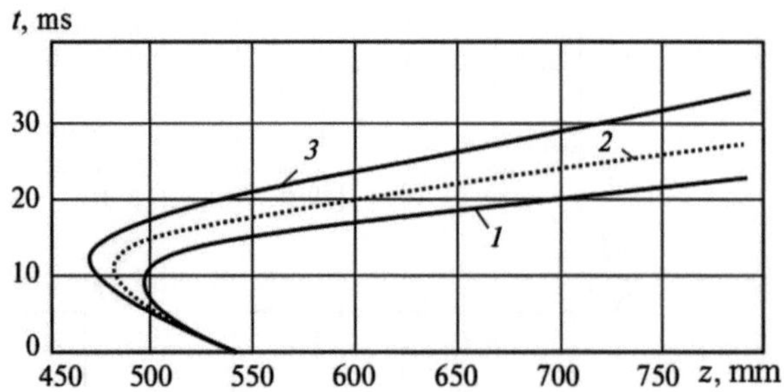

Fig. 10.10 Flight time MgO particles in the air RF plasma torch (1.76 MHz–60 kW): the notation is the same as in Fig. 10.6

In the future, this technique can be used to determine the optimal regimes for treatment of the particles. The parameters of this regime are: maximum heat flux going into heating of the particles, their presence in the plasma jet, the point of melting and evaporation of the particles and the size of the plasma jet for their processing. However, a large number of particles in the plasma flow the properties of plasma dramatically change and this model of motion and heating of the particles in plasma at given characteristics of plasma does not allow determination of the real parameters of a two-phase medium: plasma and particulate materials. Therefore, it is necessary to put and solve the task of taking into account the interaction of the particles with the plasma flow.

10.3 The Model of Plasma, Loaded the Flow of Solid Particles

In order to account the interaction of particles with plasma flow and to determine the parameters for the processing of the dispersed raw materials it is necessary to build a model of plasma, loaded the flow of solid particles. Imagine the model [16, 20], that allows not only calculation of the velocity of acceleration and heating of dispersed particles in plasma, but also to consider the reverse effect of groups of particles to the parameters of plasma—cooling down and braking of the latter. The method that will be used for solving of the proposed model can be called the method of quasi-single particles (like a single particle), because the behavior of a large number of solid particles in plasma will be considered without taking into account of their mutual influence (collisions, screening, etc.). Each particle is placed in the reporting layer interacts with plasma, "unaware" the presence of other particles. This model allows us to describe the interaction of the flow of solid particles with plasma and from a technological point of view makes it possible to determine the important parameters of particle processing—limiting mass consumption of material and efficiency of the process and the installation as a whole, to choose the best parameters for obtaining of product with the required quality.

The model of plasma, loaded by the flow of particles includes the following equations: equations of momentum and heating of the particles; equation of momentum of plasma in view of the loss of momentum at the particles acceleration; equation of the energy balance in plasma taking into account the losses of power to the heating of the particles; continuity equations for all the gas and particulate flow; the system of equations for the description of the electromagnetic field of plasma; the system of equations for determination of the composition and thermal properties of plasma and particles.

The system of equations for the description of electromagnetic field of RF and Arc plasma torches and its solutions are discussed above (Sects. 7.2, 7.3 and 8.3). Methods for determination of the composition and properties of plasma is given in Sects. 4.1, 4.2 and 4.3. In many cases, the thermal properties of plasma particles are known in advance and are shown in the tables in dependence on the temperature, so the problem of obtaining of the properties of plasma particles at a certain temperature is reduced to the approximation task. Below we present the basic equations of the model of two flows of plasma particles in a two-dimensional cylindrical coordinate system.

We assume that all the particles at this point and at this time have the same average temperatures and velocities and do not interact to each other. Thus, it is possible to apply the equations of heating and of motion of a single particle to determine the parameters (temperature, velocity) of all the flow of particles. Also note that the proposed model is valid only for the plasma flow that has a much higher spatial resolution compared with the processed particles. Further in equations for simplicity we omit all the indices «*p*», referring to plasma.

1. *Equation of heating of the particles*

$$\rho_s c_{ps}\frac{dT_s}{dt}=\frac{6}{d_s}\left[\alpha(T_p-T_s)-\varepsilon\sigma T_s^4\right]+U_{rad}. \qquad (10.45)$$

2. *The momentum equations of particles*

$$\rho_s\frac{dv_{sz}}{dt}=0.75C_{dz}\frac{\rho(v_z-v_{sz})|v_z-v_{sz}|}{d_s}+\rho_s g_z, \qquad (10.46)$$

$$\rho_s\frac{dv_{sr}}{dt}=0.75C_{dr}\frac{\rho(v_r-v_{sr})|v_r-v_{sr}|}{d_s}. \qquad (10.47)$$

3. *The equation of balance energy in plasma* taking into account the loss of power to the heating of the particles:

$$\rho c_p\left[v_z\frac{\partial T}{\partial z}+v_r\frac{\partial T}{\partial r}\right]=\frac{\partial}{\partial z}\left(\lambda\frac{\partial T}{\partial z}\right)+\frac{1}{r}\frac{\partial}{\partial r}\left(r\lambda\frac{\partial T}{\partial r}\right)+\sigma E^2-U_{rad}$$
$$-\frac{6}{d_s}\left[\alpha(T-T_s)-\varepsilon\sigma T_s^4\right]G_{sn}. \qquad (10.48)$$

4. *The momentum equations of the plasma flow*, taking into account the loss of momentum due to the acceleration of particles.

$$\frac{\partial}{\partial z}(\rho v_z v_z)+\frac{1}{r}\frac{\partial}{\partial r}(r\rho v_r v_z)=-\frac{\partial p}{\partial z}+F_{Bz}+F_{\mu z}+\rho g_z$$
$$-0.75C_{dz}\frac{\rho(v_z-v_{sz})|v_z-v_{sz}|}{d_s}G_{sn}; \qquad (10.49)$$

$$\frac{\partial}{\partial z}(\rho v_z v_r)+\frac{1}{r}\frac{\partial}{\partial r}(r\rho v_r v_r)=-\frac{\partial p}{\partial r}+F_{Br}+F_{\mu r}$$
$$-0.75C_{dr}\frac{\rho(v_r-v_{sr})|v_r-v_{sr}|}{d_s}G_{sn}, \qquad (10.50)$$

where F_{Bz}, F_{Br}—the components of the electromagnetic force; $F_{\mu z}$, $F_{\mu r}$—components of viscous forces; G_{sn}—the loading coefficient of the material.

The coefficient G_{sn} is defined as the volume ratio between the processed material and plasma (the volume occupied by the particles in the unit of plasma volume). In practice, G_{sn} is not often given as input data, but the mass consumption G_s of material, that should be treated in the plasma flow (the performance of processing of the material). Let the processed particle with density ρ_s are fed and uniformly distributed in the plasma flow with an average diameter D and the average velocity in the initial section v_{s0}. The relationship between G_{sn} and G_s is defined as:

$$G_{sn} = \frac{G_s}{900\rho_s \pi D^2 v_{s0}}. \tag{10.51}$$

Note that G_{sn} is a dimensionless coefficient. Relevant units of the following values: G_s—kg/h, ρ_s—kg/m^3, D—m, v_{s0}—m/s.

In the equation of balance energy of plasma (10.48) a new member appears, taking into account the losses of power for heating of the particles. Similarly, in the equations of motion of the plasma flow (10.49, 10.50) the members $0.75C_{dz}\frac{\rho(v_z - v_{sz})|v_z - v_{sz}|}{d_s}G_{sn}$ and $0.75C_{dr}\frac{\rho(v_r - v_{sr})|v_r - v_{sr}|}{d_s}G_{sn}$ correspond to the losses of momentum in due to the particle acceleration in two directions, z and r.

The solution of Eqs. (10.45–10.47) using the method of Runge–Kutta is presented above (Sect. 10.2.2), method for solving of the momentum equations and the equation of energy balance in plasma (10.48–10.50) is given in Sect. 6.1. The sequential procedure for solving of the equations, analysis of stability and the organization of computing procedure are given in Sects. 6.2, 6.3.

Thus, the joint solution of the Eqs. (10.45–10.50) allows the determination of not only the parameters of plasma and particles but also the effects of cooling and braking of plasma at a given load coefficient of material. By adjusting the coefficient G_{sn}, we can find the limiting material consumption G_s, the efficiency and effectiveness of the process.

Normally, the limit mass consumption of material G_s for this plasma unit must ensure the achievement of the melting temperature of the particles T_{s1} at the end of the plasma flow and the preservation of high temperature of the plasma flow in the discharge ($T > 7000$–8000 K).

The mass consumption of the material is the most important parameter in the design of plasma installations for the processing of particles, and the determination of its efficiency, and the product cost. The dependence of the heating temperature of

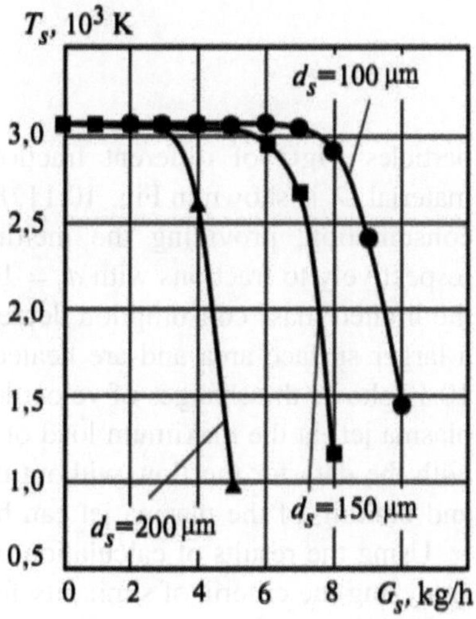

Fig. 10.11 The temperature of the heating of particles in dependence on the load coefficient of MgO periclase at their processing in the air RF plasma torch (1.76 MHz–60 kW)

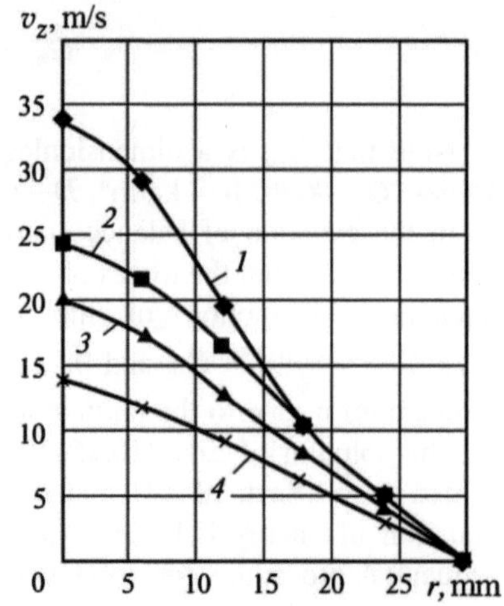

Fig. 10.12 Changes of the gas velocity (z = 800 mm) in the air RF plasma torch (1.76 MHz–60 kW) at the maximum loading of MgO periclase (G_s = 8, 6, 3 kg/h, for the fractions d_s = 100, 150, 200 μm, respectively): *1* without loading of particles, *2* d_s = 200 μm, *3* d_s = 150 μm, *4* d_s = 100 μm

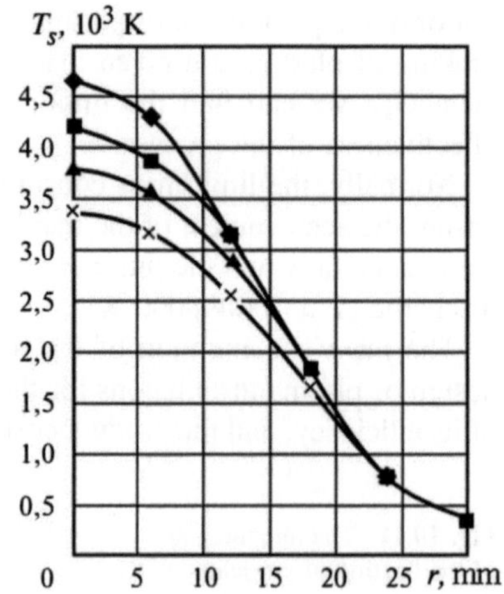

Fig. 10.13 Change of the gas temperature (z = 800 mm) in the air RF plasma torch (1.76 MHz–60 kW) at the maximum loading of MgO periclase (G_s = 8, 6, 3 kg/h, respectively, for the fractions d_s = 100, 150, 200 μm): the notation is the same as in Fig. 10.12

particles MgO of different fractions depending on the load coefficient of the material G_s is shown in Fig. 10.11. From this figure it is possible to find the limiting consumption, providing the melting point of the particles (G_s = 8, 6, 3 kg/h respectively to fractions with d_s = 100, 150, 200 μm). With increasing particle size the limited mass consumption decreases. This is because the smaller particles have a larger surface area and are heated faster than larger particles. Figure 10.12 and 10.13 shows the changes of velocities and temperatures of the gas (at the end of the plasma jet) at the maximum load of MgO periclase particles. Comparing these data with the data for the flow without downloading of particles, the degree of cooling and braking of the plasma jet can be found.

Using the results of calculation of motion of particles and its heating, as well as analyzing the criteria of similarity for a heat transfer and gas dynamics the flying of

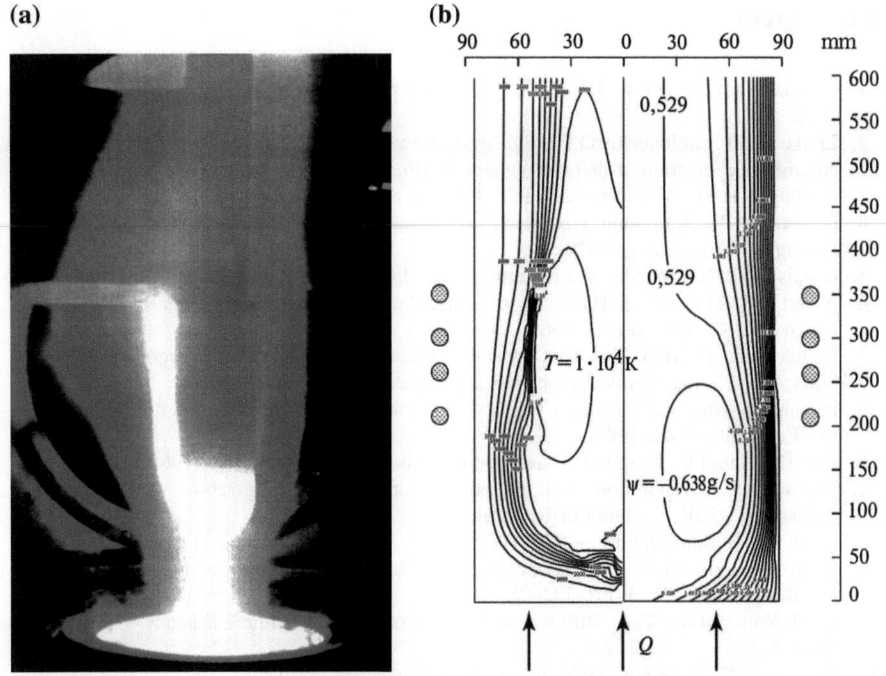

Fig. 10.14 The air plasma jet of RF plasma torch (0.44 MHz–1000 kW) at the testing of the frontal part of the metal body (**a**) and the field of temperature and the flow of air plasma in its channel (Q = 6000 l/min, U_1 = 9.45 kV, I_1 = 1200 A, P_{str} = 400 kW) (**b**)

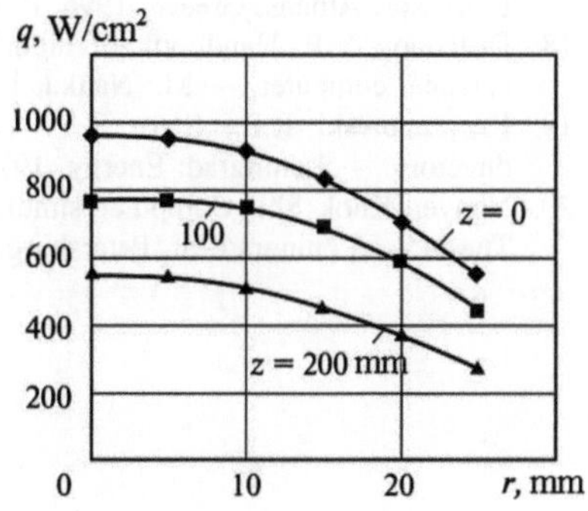

Fig. 10.15 Change of the density of the heat flux going into the heating of the front part of the metal body with diameter d = 50 mm in the flow of the air RF plasma torch (0.44 MHz–1000 kW): z—the distance from the edge

bodies in the plasma flow can also be simulated. For example, the heat flux was determined, that heat the front part of the metal body with a diameter d = 50 mm in the jet of the air RF plasma torch (0.44 MHz–1000 kW). The main results of the simulation are shown in Figs. 10.14 and 10.15.

References

1. Tsvetkov Y.V., Panfilov S.A. Low-temperature plasma in the recovery process. M.: Nauka. 1980.
2. Zhukov M.F., Solonenko O.P. High dusty powder material in the jet treatment processes. Novosibirsk: Institute of thermal physics of Sib. Dep. USSR Academy of Sciences, 1990.
3. Coating Plasma / V.V. Kudinov, P.Y. Pekshev et al. M.: Nauka. 1990.
4. Donskoy A.V., Klubnikin V.S. Electrical plasma processes and installation engineering. — Leningrad: Engineering, 1979.
5. Gorbis E.R. Heat transfer and hydrodynamics dispersed through streams. M.: Energy. 1970.
6. Eckert E.R., Drake R.M. The theory of heat and mass transfer / Trans. from English. Ed. A.V. Lykov. M.: Gosenergoizdat. 1961.
7. Deitch M.E., Philipov G.A. Gas dynamics of two-phase media. M.: Energoizdat. 1981.
8. Suris A.L. Plasma-chemical processes and devices. M.: Chemistry, 1989.
9. Patankar S. Numerical methods for solving problems of heat transfer and fluid dynamics. — M.: Energoatomizdat 1984.
10. Hirs D., Pound G. Evaporation and condensation. — M.: Metallurgy, 1966.
11. Fuchs N.A. Evaporation and growth of droplets in a gaseous medium. — M.: Publishing USSR Academy of Sciences, 1958.
12. Nesis E.I. Boiling liquids. — M.: Nauka. 1973.
13. Knakke O., Stransky I.N. Evaporation mechanism // The success of Metal Physics. — M.: Metallurgy, 1966. V. 3. PP. 222–282.
14. Frank-Kamenetskiy D.A. Diffusion and Heat Transfer in Chemical Kinetics. — M.: Nauka. 1987.
15. Kazenas E.K., Tsvetkov Yu.V. Evaporation oxides. — M.: Nauka, 1997.
16. Dresvin S.V., Panfilov S.A., Nguyen-Kuok Shi. Calculation of heating solids and the granules in the plasma Arcs // Proc. Ros. Scientific and Technical. Conf. "Advanced technology materials handling processes." SPb. 1995. PP. 5–11.
17. Study of the metallurgical silicon powder treatment by thermal RF plasma / S. Magnaval, D. Morvan, J. Amouroux et al. // Proc. of the IV European Conference on Thermal plasma Processes. Athens. Greece, 1996. P. 453–459.
18. Deaconov V.P. Handbook of algorithms and programs in the language BASIC for the personal computer. — M.: Nauka, 1989.
19. Krzyzanowski R.E., Stern Z.Y. Thermal properties of nonmetallic materials (oxides) directory. — Leningrad: Energy. 1973.
20. Nguyen-Kuok Shi. Computer simulation of the flow of solid particles with plasma: PhD Thesis. — Leningrad: St. Petersburg State Technical University, 1992.

Chapter 11
The Features of the Experimental Methods and Automated Diagnostic Systems of RF and Arc Plasma Torches

11.1 The Experimental Setup and Automated Diagnostic System for the Study of the RF and Arc Plasma Torches

The experimental setup and diagnostic system for the study of RF plasma torches for spectral analysis are presented in Fig. 11.1. The RF generator has the operating frequency of 27.12 MHz and power ~7 kW. The plasma-forming gas is Argon. The RF generator allows the adjustment of the voltage applied to the inductor, and the launch of the plasma torch (plasma ignition) using a high-voltage discharge. The operating parameters of generator are: the grid current—200 mA, the anode current—0.7 A, voltage at the anode—2.4–3 kW.

Channel of RF plasma torch is a cylindrical quartz tube with internal diameter of 26 mm. The plasma forming gas is supplied through the central channel 19 (Fig. 11.1) with the internal diameter of 6 mm, the protective gas (Argon) is introduced tangentially through the holes in the bottom side of the quartz tube of plasma torch 24. The water-cooled copper induction coil *1* with two–three windings provides the passage of the high-frequency current from RF generator (27.12 MHz) on its internal side. The scheme of RF plasma torch, type of RF discharge, the geometric dimensions of inductor with a quartz tube are shown in Figs. 11.2, 11.3 and 11.4.

The used in operating RF plasma torch allows obtaining plasma with maximum electron temperature (10–11) 10^3 K, the electron concentration of the order (1–3) 10^{16} cm^{-3}, and degree of ionization about 12–13 %. One of the features of RF plasma is nonuniform change in the radial distribution of power density caused by the skin effect in RF discharge (Sect. 7.2.3). The most significant specific energy is carried in the external annular area of RF discharge. Therefore, the temperature and the electron density in this region may exceed their values on the discharge axis. The RF plasma emission spectrum consists of continuous emission (continuum) and radiation in atomic and ionic lines.

S. Nguyen-Kuok, *Theory of Low-Temperature Plasma Physics*,
Springer Series on Atomic, Optical, and Plasma Physics 95,
DOI 10.1007/978-3-319-43721-7_11

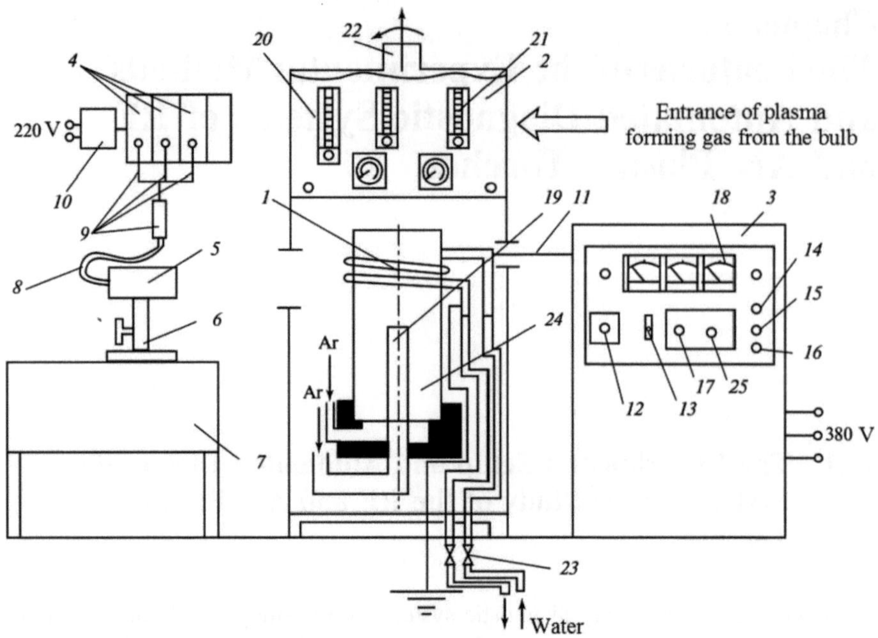

Fig. 11.1 Scheme of the experimental setup for diagnostic of RF plasma torches: *1*, *2* inductor and the body of RF plasma torch; *3* RF generator, *4* three-channel fiber optic spectrometer, *5* collimation lens, *6* reuter, *7* optical table, *8* fiber optic cable, *9* fiber optic splitter, *10* computer, *11* waveguide; *12* switch to turn on the air cooling of RF generator, *13* switch for the cathode filament in a lamp of RF generator, *14* button for the power supply to anode, *15* the same to remove the voltage, *16* button for applying of a high voltage to "ignite" RF discharge, *17* switch to turn on the electronic unit for control of RF discharge, *18* the block devices for measuring of the net current, current and voltage at the anode (lamp), *19* quartz axial channel, *20* rotameter of annular flow of protective gas, *21* rotameter of axial flow of the plasma gas, *22* nipple for gas exhaust from plasma torch to atmosphere, *23* valve for cooling water supply to inductor, *24* the quartz channel of plasma torch, *25* the handle of fine adjustment of the anode voltage

For the determination of point for the spectral measurement the end of the optical cable 8, directed to plasma, was installed in the desired position by means of a laser. To do this, the laser is applied to the second end of the optical cable, and a laser beam determines the desired point for diagnostic in RF or Arc plasma torches.

The automated system of the spectral measurements for obtaining complex spectral characteristics of plasma radiation in different zones of RF and Arc plasma torches determine the plasma parameters (concentration of excited particles, electron density, temperature, etc.). The system of spectral measurements include spectral devices of high class of accuracy, the receivers of optical radiation with a wide spectral range of sensitivity, means of transmission, registration, and processing of information. In a first variant the system consists of the diffraction spectrograph DFS-452 and multichannel optical spectrum analyzers MOAS-2, that include two photo-head with a photodiode CCD rulers Toshiba TDS1250A

Fig. 11.2 Scheme of RF plasma torch

Fig. 11.3 Argon RF plasma torch 27.12 MHz

Fig. 11.4 The geometrical dimensions of inductor with a quartz tube

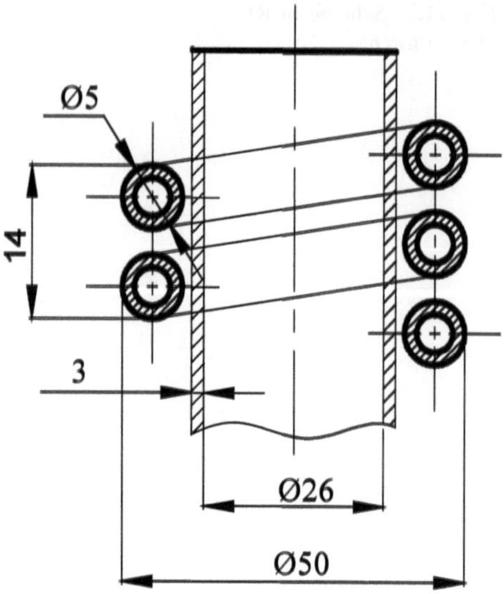

(CCD charge-coupled device), and printed circuit board with the ISA interface for connection with computer. Photo-heads are fixed on the plate corresponding to the shape and size of the cassette of spectrograph that is set to the place of cassette. Photo-heads are connected with the interface board by a cable. Each photodiode ruler contains 3724 light-sensitive elements with length of 8 μm and height of 200 μm (the total length of the line is 30 mm). The size of the spectral sensitivity range is 180–1000 nm. The time of signal accumulation is given by period of the survey lines and can be varied between 80 and 20,000 ms. The 12-bit analog-to-digital converter is included in each photo-head, so only digital signals, that provide a sufficiently high noise interference protection of the system, are transmitted to computer by an interface cable. For operating with the system MOAS-2 a special program is used.

The spectrograph DFS-452 allows working in all over the range of sensitivity of rulers. In the work with a diffraction grating with 600 lines per 1 mm the inverse linear dispersion is 1.6 nm/mm or 0.013 nm per cell of ruler and in the work with a diffraction grating with 1200 lines per 1 mm—respectively, 0.8 nm/mm and 0.0065 nm cell of ruler.

One of the rulers is disposed along the direction of dispersion of the spectrograph and serves for detection of the spectrum, and the other—perpendicularly—is used for the determination of intensity depending on the height of the gap at the predetermined wavelength.

All received data are transmitted to the personal computer. Program for operating with the system MOAS-2 allows us to survey any of the photodiode rulers or both simultaneously with arbitrarily set accumulation time once or cyclically. It also allows display signal from any ruler or two signals from both of the rulers simultaneously. Thereafter, data are written to the disk for further processing.

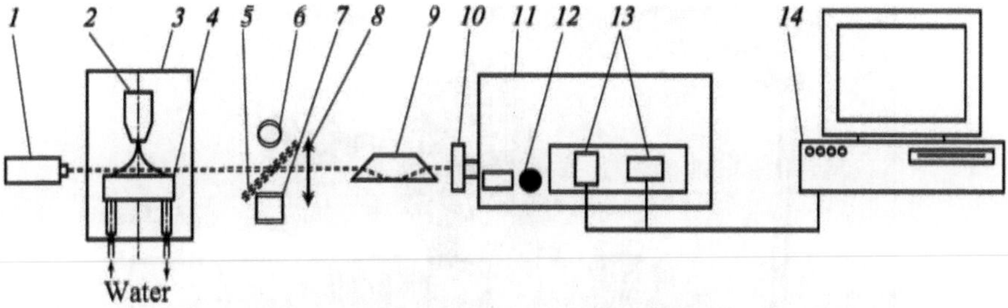

Fig. 11.5 Scheme of the automated system of spectral measurements for Arc plasma torches: *1* alignment laser, *2* cathode, *3* camera with two quartz optical windows, *4* anode, *5* rotating mirror, *6* lamp with hollow cathode LT-2, *7* band tungsten lamp, *8* condenser, *9* Dove's prism, *10* split of spectrograph, *11* spectrograph, *12* drum of wavelengths, *13* CCD ruler, *14* computer

Scheme of automated system for the spectral measurements is shown in Fig. 11.5. The general view of the stands and automated system for the study of free arc and RF plasma torches is represented in Fig. 11.6.

11.1.1 Adjusting Elements in Optical System

The purpose of adjustment is the location of selected area of plasma on the optical axis ($z = z_0$, $r = 0$) and quartz condenser with a focal length $f = 150$ mm, providing the projection of increased sharp image on the middle of spectrograph slit with height H. Alignment is implemented with a laser in the visible range using the selected diffraction grating (600 or 1200 lines per mm) with open slit $\Delta_{\text{int}} = 200$–400 μm in the following sequence:

1. In the absence of condenser the hit of laser beam to the middle of diffraction grating in spectrograph is provided. This is achieved by visual observation from the cassette part of DFS-452 of the movement of light spot on the surface of diffraction grating (as the moving of laser);
2. Using the adjustment screws of the Arc chamber the z coordinate of the Arc is approximately hold up on the optical axis, defined by the laser beam;
3. According to the equation of thin lens $\frac{1}{f} = \frac{1}{a} + \frac{1}{b}$ the position of condenser on the optical axis is selected, it provides the selected increase $k = b/a$.

The next adjustable element is the gas-discharge lamp with a hollow cathode LT-2, emitting the spectrum of atoms CuI in the wavelength range 200–600 nm and the spectrum of atoms NeI in the range 600–900 nm. This lamp is used to perform two important tasks of spectral studies:

1. For the calibration of the wavelength scales of spectrograph, i.e., for establishing communication between the wavelengths on the limb of drum λ_l in DFS-452 (approximately corresponding to the wavelength in the middle of the focal plane

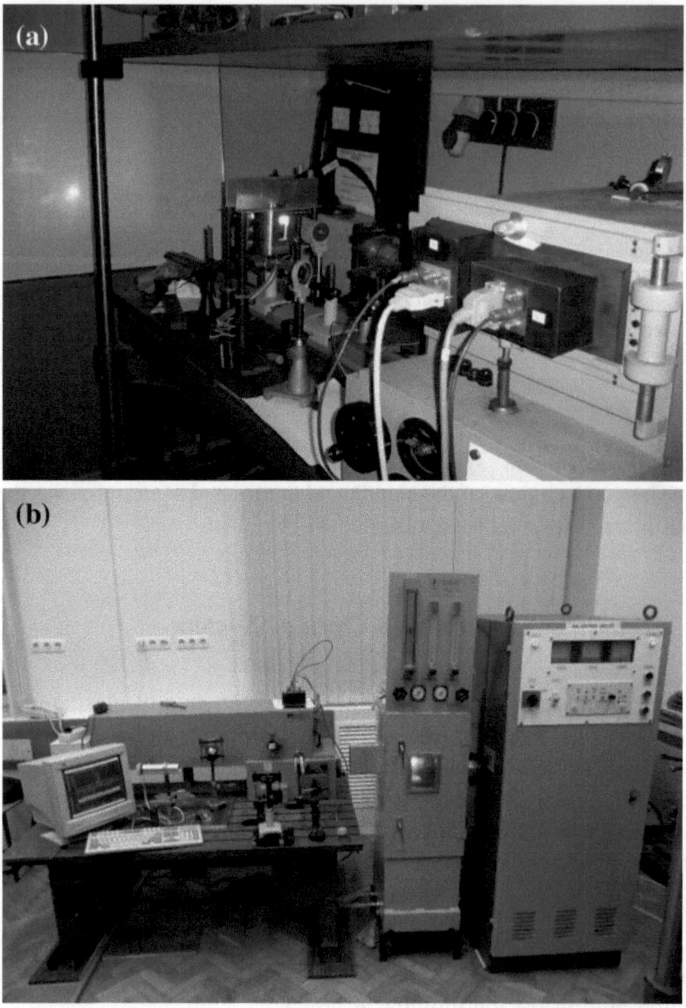

Fig. 11.6 Experimental setup and automated diagnostic system of free Arc plasma torch (**a**) and RF plasma torch (**b**)

of device) and a wavelength at the CCD ruler. This calibration is carried out by the registration of spectrum of the lamp LT-2 emission at a fixed wavelength λ_l and its decryption using a database on the position of the radiating lines of copper atoms CuI (the example of such a calibration is shown in Fig. 11.7).

2. For measuring the hardware function of the used optical system. The hardware functions is formed by the basic elements of spectrograph—an entrance slit and the diffraction grating—and is included in the monitoring contour of the spectral lines in the form of a convolution of physical contour of the observed lines and

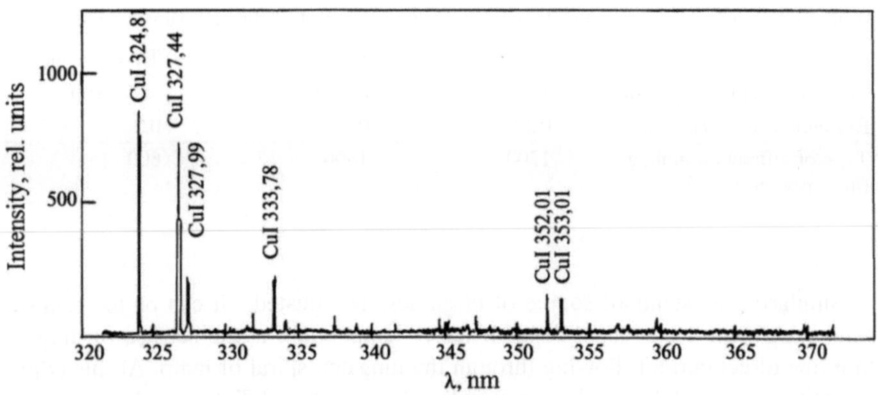

Fig. 11.7 The identification of the emission spectrum for copper CuI

hardware functions. A feature of radiation for the lamps with hollow cathode is a small physical width of the spectral lines of CuI and NeI, due to the small Stark component ($n_e \leq 10^{12}$ cm^{-3}, $\delta\lambda_{St} \ll 0.01$ Å), and the Doppler width of the lines ($T_a \geq 0.05$ eV $\ll T_e, \delta\lambda_D \leq 0.04$ Å). Therefore, registration of the spectrum of this lamp with a narrow entrance slit of spectrograph $\Delta_{ent}D$ (Δ_{ent}—opening of the split, mm; D—the reverse linear dispersion, Å/mm), less than the expected physical line width of ArI and ArII in Argon plasma $\delta_{1/2} : \Delta_{ent}D < \delta_{1/2} \approx 0.2$–2 Å allows obtaining of the hardware function of the measuring system in the entire range of wavelength 250–950 nm. Chosen spectrograph slit width should be coordinated with the spectral resolution of the system. In this case, the resolution is equal to the interval, falling at one photosensitive element of CCD ruler (one pixel with width $\Delta_{pix} = 8$ μm) $\Delta\lambda_{1200} = 8 \times 10^{-3} \times 8 = 0.064$ Å and $\Delta\lambda_{600} = 8 \times 10^{-3} \times 16 = 0.13$ Å for diffraction gratings with 1200 and 600 lines per mm, respectively.

For full use of the resolution of system for registration of narrow spectral lines the disclosure of spectrograph slit should be chosen smaller than the pixel size $\Delta_{ent} < \Delta_{pix} = 8$ μm, and for the reliable determination of the hardware function of the measuring system the contours of the spectral lines of CuI with high ratio of signal to noise should be obtained.

The value of the hardware function is installed as the mean arithmetic value of the half-widths of several measured line contours:

$$\delta_{hf} = \frac{\sum_1^N \delta_{1/2}}{N} \quad (11.1)$$

By analyzing the spectrum for CuI in the wavelength range 320–570 nm, it was found that hardware function $\delta_{hf} = 0.2$ Å.

Table 11.1 Features of the channels in spectrometer Avaspect of the company Avantes

Characteristics	First channel	Second channel	Third channel
The wavelength range (nm)	195–469	455–593	487–1100
Resolution (nm), not less	0.18	0.12	0.3
Type of diffraction grating (lines per mm)	1200	1800	600

Similarly, the standard source of brightness is adjusted—it can be the tungsten band lamp with a graduated dependence of temperature at the surface of tungsten from the direct current, flowing through the tungsten spiral of lamp. At this adjustment the mirror and the condenser, as in the case of a lamp LT-2, provide a projection of the sharp image of tungsten spiral to the entrance split of spectrograph DFS-452.

Adjustment of the spectrum analyzer MOAS-2 is to provide a contact of the laser line on the horizontal CCD ruler. This is carried out (when the system MOAS-2 is switched on) by moving the cassette part of spectrograph by the switch toggle "up–down" for moving the laser signal that is getting on the CCD ruler (at the minimum split of DFS–452 $\Delta_{\mathrm{pix}} \leq 10$ μm). To implement this procedure, it is necessary to use the calibration characteristic of the system connecting the values of wavelength at the limb of spectrograph with a wavelength at the center of the horizontal CCD ruler. This curve depends on the chosen diffraction grating (1200 or 600 lines per mm) and on the orientation of the system MOAS-2 that allows two positions in the focal plane of DFS-452: using the horizontal ruler in the long-wave (right) or short-wave (left) part of the spectrum by turning the system to angle 180°.

Instead of DFS-452 the spectrometer Avaspect of the company Avantes, equipped with three channels can be also used. The characteristics of the channels are shown in Table 11.1.

For all the channels of spectrometer the necessary spectrometer calibration was made that provided the absolute values of the radiation intensity. The spectrometer was connected to computer using the USB cable and working with it was carried out using the software «Avasoft». It allows checking of changes in spectrum or in its parts at the real time, obtaining the values of intensity or wavelength for the individual points and record the resulting data in formats suitable for further decoding the spectrum and the determination of plasma parameters.

11.2 The Basic Methods of the Plasma Diagnostics

The main task for the diagnostics of plasma is to determine the local values of the following parameters:

- concentrations of atoms n_{a}, electrons n_{e}, and ions n_{i};
- concentration of excited atoms $n_{\mathrm{a}}^{*}\left(E_{\mathrm{a}}^{*}\right)$ and ions $n_{\mathrm{i}}^{*}\left(E_{\mathrm{i}}^{*}\right)$;

- temperature of electrons T_e and heavy particles T_a;
- distribution function of the electron energy $f_e(\varepsilon)$;
- intensity of electric E and magnetic fields H;
- Volt–ampere characteristics of the Arc and others.

Establishing relationships between the plasma parameters with its spectral characteristics in a mathematical sense is an inverse problem, the accuracy of its solutions is often linked to the right of the model for the description of study plasma. The main source of information for us will be the absolute and relative intensities of the spectral lines and continuum.

For RF and Arc plasma torches a model of local thermodynamic equilibrium (LTE) can be used, and the Maxwell, Saha, and Boltzmann equations are valid. In the spectral diagnostics, the following basic methods can be used:

1. Determination of the electron temperature from analysis of the relative distributions of excited atoms and ions as a function of the excitation energy. The range of changes in the excitation energy of atoms and ions should be approximately equal to kT_e and more.
2. Calculation of temperature of the electrons from the decay of continuum intensity at short wavelengths ($\lambda \leq 300$ nm). In this case, obtained dependences of the absolute intensity of continuum from the radiation frequency are approximated using the model of local thermodynamic equilibrium: $\varepsilon(\nu) \sim \exp(-h\nu/kT_e)$.
3. Determination of the concentration of electrons using the Stark width of the lines for atoms and ions. This method does not require any models describing plasma—the necessary condition for its application is the predominant role of electrons (and ions) in the broadening of the studied spectral lines, as well as commensurate (or exceed) of the Stark width of investigated line with its hardware broadening. The movement of particles in dense plasma and associated with it change in the distance between them lead to rapid changes in the electric field around each of emitting particles. As a result, the energy levels of atoms (ions) are displaced by different distances and it leads to broadening of the spectral lines in the emission spectrums of plasma. The Stark broadening allows to estimate the concentration of charged particles in plasma.

The studied plasma in RF and Arc plasma torches even in the axial region maybe not in a state of full thermodynamic equilibrium primarily due to the fact that induction and Arc discharges are the "open" systems. The exhaust of radiation and heat (due to gradients of parameters), and the presence of an external electric field leads plasma to the state different from the state of full thermodynamic equilibrium. However, in areas with significant concentrations of electrons (these particles have "lion's share" of the energy of the field E, try to maintain a state of thermodynamic equilibrium) the conditions of local thermodynamic equilibrium can be ensured. In such conditions, the following basic equations are valid: normalized per unit distribution function of the electron energy is described by Maxwell equation; behavior of excited atoms and ions is described using the Boltzmann equation at the electron temperature; the ionization equilibrium in plasma is described by Saha equation at the ionization temperature, equal to the electron temperature.

Numerous calculations and experiments [1–3] show that in the case of spatially inhomogeneous plasma in the areas of highest values of electron density the Maxwell distribution of electrons $f_e(\varepsilon) \approx f_{Me}(\varepsilon)$ inside the core of function (ε = 0–7 eV) can be expected. When the degree of plasma ionization $\alpha_i = \frac{n_e}{n_a + n_i} > 10^{-2}$ the population of the upper excited states of atoms and ions are often close to the Boltzmann values and ionization equilibrium is described using the Saha equation.

Method of Relative Intensities for the Determination of Electron Temperature The electron temperature is one of the main plasma parameters. Its value is necessary to calculate the composition of plasma and its transfer coefficients, for the determination of electrical, magnetic, energy, and thermal characteristics of the discharge. Furthermore, the temperature of electrons allows us also to find the interaction cross sections of particles in plasma, its thermodynamic properties, and many fundamental physical parameters, such as ionization and recombination coefficients, thermal diffusion coefficients, and others.

According to the literature, for example, linear and continuous emission of Argon plasma at atmospheric pressure in the studied range of temperature and wavelength is optically thin. Therefore, the intensity of the spectral line, W/(cm^3 sr), at the transition of atom (ion) from the *k*th level to the *i*th level is expressed by

$$I_{ki} = \frac{1}{4\pi} h\nu_{ki} A_{ki} n_k = \int\limits_0^\infty i_{ki}(\lambda)\mathrm{d}\lambda, \qquad (11.2)$$

where ν_{ki}—the frequency of the line center ($\nu_{ki} = \nu_o = c/\lambda_0$), A_{ki}—the probability of spontaneous transition from the *k*th level into the *i*th level (the first Einstein coefficient); n_k—population of emitting (upper) level of transition. Equation (11.2) is accurate and then

$$\tau_0 = \int\limits_0^L k_0 \mathrm{d}l \ll 1 \qquad (11.3)$$

Here τ_0 is the optical thickness of plasma layer with depth *L*, it should be small. In (11.3) k_0—absorption coefficient at the center of the studied line.

Thus, the ratio (11.2) allows us to analyze the populations of excited states of atoms and ions with different excitation energies E_k, E_i. In the context of local thermodynamic equilibrium the Boltzmann law is valid, it binds the concentration of particles of the excited levels $\frac{n_k}{g_k} = \frac{n_i}{g_i}\exp\left(-\frac{E_k - E_i}{kT_e}\right)$, where g_k, g_i—the statistical weight of the particles of kth and ith levels. Therefore, the analysis of experimental curves $\frac{n_k}{g_k} = f(E_k)$ allows the determination of electron temperature using:

- relative intensity of two lines in this class (ArI or ArII):

$$kT_e = f\left(\frac{I_{ki}}{I_{mn}}\right) = \frac{E_k - E_m}{\ln\left(\frac{I_{mn}\nu_{ki}A_{ki}g_k}{I_{ki}\nu_{mn}A_{mn}g_m}\right)}. \tag{11.4}$$

- relative intensities of lines with different multiplicity of ionization (Saha and Boltzmann equations are valid):

$$\frac{I_{ki}^{i}}{I_{mn}^{a}} = kT_e^{3/2}\frac{1}{n_e}\exp\left(-\frac{E_m^a - E_k^i - I_i^a}{kT_e}\right). \tag{11.5}$$

In this case, for the determination of electron temperature knowledge of their concentration n_e is required.

The analysis of dependences $\frac{n_k}{g_k}(E_k)$ for atomic and ionic lines is the most informative. It is possibly covering the greatest variation range of the excitation energy for the particles, because the determining error for T_e type from the expression (11.4) $\delta T_e \sim \frac{T_e}{\Delta E_k}$ is reduced with the increase of this range. In the same direction, it operates a statistical set of values n_k/g_k as much as possible the number of excited states. Evaluation is both a method of checking the correctness of the Boltzmann law to communicate the populations of the levels of particles.

The Abel Transformation for the Determination of Radial Distribution of Plasma Parameters Abel transformation is the standard procedure of transition from the chordates intensities to the radial ones if there is axial symmetry of the emitting object. The relationship between the intensities is described using the Abel integral equation.

In the case of the light source with cylindrical or spherical symmetry the observed intensities can be transformed into the radial distribution of plasma intensity. Let the z-axis to be the axis of symmetry for the cylindrical discharge. The x-axis of the circular plasma disk is aimed to the observer I (Fig. 11.8). The observer I is looking along the line, parallel to the x-axis and passing at the distance y from the center. The local emissivity of plasma at a distance r from the axis is noted as $\varepsilon(r)$. The observer I sees the projection [integral $I(y)$] as axial-symmetric function $\varepsilon(r)$ along the direction of observation. The function $I(y)$ corresponds to the radiation intensity integrated over x at the distance y from the plane xz. Then the function $I(y)$ is linked to $\varepsilon(r)$ by the integral Abel transformation:

$$I(y) = 2\int_0^{\sqrt{R_0^2 - y^2}} \varepsilon(r)\mathrm{d}x = 2\int_y^{R_0} \frac{\varepsilon(r)r\mathrm{d}r}{\sqrt{r^2 - y^2}}, \tag{11.6}$$

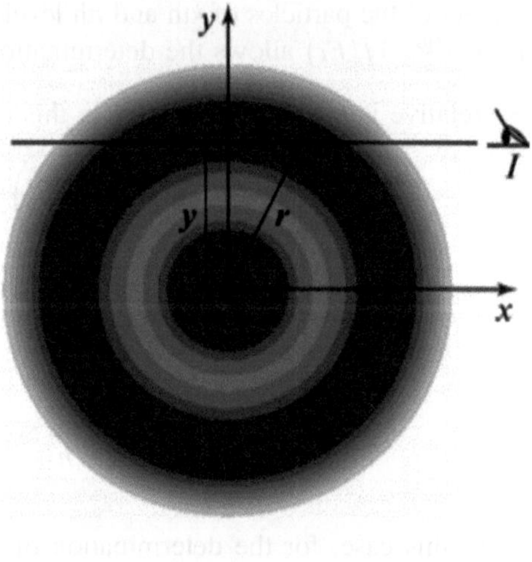

Fig. 11.8 The partition of the plasma column into the cylinders

where R_0—radius, $\varepsilon(r) = 0$. Note that the variable r is associated with x and y as $r^2 = x^2 + y^2$, if integrating it does not change the sign, then the integrand is an even function.

Transformation of Eq. (11.6) gives equation for the required value $\varepsilon(r)$:

$$\varepsilon(r) = -\frac{1}{\pi}\int_{r}^{R_0} \frac{dI(y)}{dy}\frac{dy}{\sqrt{r^2 - y^2}}. \tag{11.7}$$

Equation (11.7) is back of Eq. (11.6), it is used to restore the function $\varepsilon(r)$ using its projection $I(y)$.

Included in the limits of integration radius R_0 is the desired optical radius—it corresponds to the vanishing of the experimental intensity $I(y)$. As it is adopted, we provide the calculation using (11.7) in relative coordinates r/R_0, where r—the current radius.

Carrying out the Abel transformation allows obtaining the relative radial distributions from the chordates distributions that are registered during the experiment. However, radial distribution of plasma parameters is of the most interest. Thus, there is the problem of transition from the radial distributions of the relative intensities of radiation to the radial distribution of the main parameters of plasma. The basis for the solution of this problem is the following relationship for the absolute intensity of the continuous radiation, containing single and double ions:

$$\varepsilon^{ff} = \text{const}\left(n_e(T_e)n_i^+(T_e) + 4n_e(T_e)n_i^{++}(T_e)\right)\frac{\exp(-h\nu/T_e)}{\sqrt{T_e}}. \tag{11.8}$$

Equilibrium plasma composition $\left[n_e(T_e),\ n_i^+(T_e),\ n_i^{++}(T_e)\right]$ can be obtained from the system of Saha equations. If we know the temperature of electrons in the axial zone T_{e0} (from the spectrometric measurements) and the equilibrium composition of plasma, it is possible to build a dependence $\frac{\varepsilon^{ff}(T_e)}{\varepsilon^{ff}(T_{e0})}$, compare it with the existing radial relative intensity distribution of radiation, received at wavelength corresponding to the continuum, we can obtain the dependence $T_e(r)$. Further, the known distribution $T_e(r)$ and the composition of plasma equilibrium allow the building of distributions $n_e(r)$, $n_i^+(r)$, $n_i^{++}(r_e)$, $n_a(r)$.

The Method for Determination of the Electron Density in Plasma The electron density can be found from the absolute intensity of the continuum at any (nonequilibrium) plasma state. The contour of spectral lines at the emission of low-temperature plasma contains information about the density of electrons (broadening of line due to the quadratic Stark effect), and temperature of the heavy particles (Doppler effect due to thermal motion of the studied atoms or ions). As it follows from the data in [4], in the area with the density of electrons in 10^{16}–10^{17} cm^{-3} the Stark width of atomic lines is 0.1–3 Å, i.e., it lies in the area of available spectral resolution, and the width of ion lines, typically, in 2–3 times more narrow. As for the Doppler width of the lines ArI and ArII, it is in the temperature range (10–20) 10^3 K even if $\lambda_0 = 800$ nm and $\delta\lambda_D = 2\frac{\lambda_0}{c}\sqrt{\frac{2kT}{M}\ln 2} \leq 0.15$ Å or less than the hardware functions of the measuring system.

Therefore, the practical work with the contours of the spectral lines—to analyze the mixed (Voigt) contours of lines, the Lorentz component of them corresponds to the Stark broadening of lines, and the Gaussian component is the sum of the Doppler effect and hardware broadening: $\delta\lambda_G = \sqrt{\delta\lambda_D^2 + \delta\lambda_{hf}^2}$.

Thus, right selection of specific spectral lines of atoms and ions can successfully help to solve two problems of the spectral diagnostics:

- Measuring T_e using the analysis of dependence $\frac{n_k}{g_k} = f(E_k)$;
- Measuring n_e using the Lorentz component of the Voigt profiles of the spectral lines.

One of the main spectral characteristics that it is necessary to be obtained from the recorded spectrums is width of the spectral lines at the half of its height. It is very important to identify this characteristic correctly. Hardware circuit of the recording system has a Gaussian form:

$$I_G(\lambda) = \frac{I_0}{\sqrt{\pi}\delta\lambda'_G}\exp\left[-\left(\frac{\lambda - \lambda_0}{\delta\lambda'_G}\right)^2\right], \tag{11.9}$$

where λ—wavelength; I_0—intensity of radiation in the center of line; λ_0—wavelength at the center of the line; $\delta\lambda_G = 2\sqrt{\ln 2}\,\delta\lambda'_G$—the line width at the half of height.

Lorentz contour describes the summary influence of several mechanisms of broadening and several widths: the natural width, the width of the Van der Waals $\delta\lambda_V$ (turbulence of the radiating level by the neighboring excited states, not optically connected with the ground state of atom or ion), and the width of the resonance broadening $\delta\lambda_r$, due to the interaction with the ground state, or bound states (for example, the first excited state). In the conditions of low-temperature plasma at atmospheric pressure only Stark broadening mechanism plays an important role. The broadening of the contour due to the quadratic Stark effect has the Lorentz form:

$$I_{\mathrm{L}}(\lambda) = \frac{I_0}{2\pi} \frac{\delta\lambda_{\mathrm{L}}}{(\lambda - \lambda_0)^2 + (\delta\lambda_{\mathrm{L}}/2)^2}, \tag{11.10}$$

where $\delta\lambda_{\mathrm{L}}$—the line width at the half of height for the Lorentz contour.

Thus, the studied experiment lines are described by two contours—the Gaussian and the Lorentz. The resulting line is the superposition of these contours and looks as the Voigt contour:

$$I_{\mathrm{F}}(\lambda) = I_0 \frac{a}{\pi} \int\limits_{-\infty}^{+\infty} \frac{\exp(-y^2)}{(u - y)^2 + a^2} \mathrm{d}y, \tag{11.11}$$

where $I_{\mathrm{F}}(\lambda)$—the intensity of Voigt contour in dependence of the wavelength; $u = \frac{\lambda - \lambda_0}{\delta\lambda'_{\mathrm{G}}}$; $a = \frac{\delta\lambda_{\mathrm{L}}}{2\delta\lambda'_{\mathrm{G}}}$.

This expression using the Voigt contour in the future is a function of wavelength that is defined using four parameters: the intensity of radiation I_0 and the wavelength λ_0 in the center of spectral line, the Gaussian $\delta\lambda_{\mathrm{G}}$, and the Lorentz $\delta\lambda_{\mathrm{L}}$ half-widths $I_{\mathrm{F}}(\lambda) = f(\lambda, I_0, \lambda_0, \delta\lambda_{\mathrm{G}}, \delta\lambda_{\mathrm{L}})$.

The unknown quantity is the half-width of the Stark broadening that should be determined from the available experimental Voigt contour. For this the half-width $\delta\lambda_{\mathrm{F}}$ of the Voigt contour is determined by the direct measurement. After the determination of this value and if the value $\delta\lambda_{\mathrm{G}}$ is known, the value of the half-width of the Stark broadening using the expression shown in Fig. 11.9 [5] can be found.

The method for the determination of electrons concentration is based on the comparison of known tabular data on the Stark broadening constants [6] with the line width at the half of height for the Lorentz contour, obtained in experiment. The concentration is calculated from the equation

$$n_{\mathrm{e}}^{\exp}(\delta\lambda_{\mathrm{L}}) = \frac{n_{\mathrm{e}}(\delta_{\mathrm{St}})}{\delta_{\mathrm{St}}} \delta\lambda_{\mathrm{L}}, \tag{11.12}$$

where δ_{St}—the Stark width at the half of height of the spectral line $n_{\mathrm{e}}(\delta_{\mathrm{St}})$—electron density that is corresponding to the Stark width at the half of height, taken from the reference data for a given temperature of plasma; $\delta\lambda_{\mathrm{L}}$—the experimental

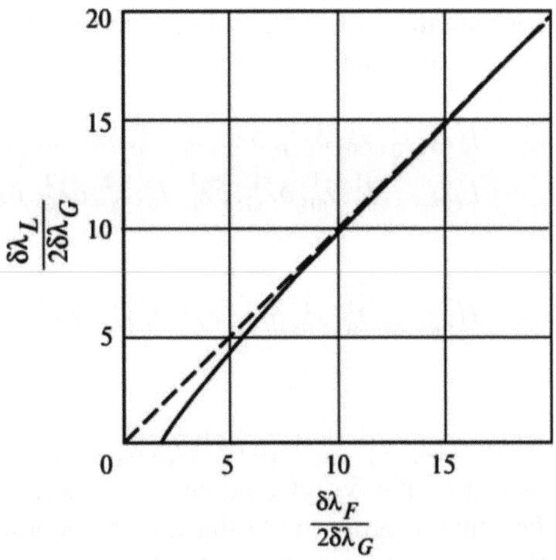

Fig. 11.9 The dependence of $\frac{\delta\lambda_L}{2\delta\lambda_G}$ from $\frac{\delta\lambda_F}{2\delta\lambda_G}$

Stark width at the half of height of the spectral line. The obtained data on the electron density can be compared with the results of solving the Saha equation.

Explanation of Superposed Lines Because each line represents the Voigt contour, then the superposition of several spectral lines of the intensity of the overall circuit I(λ) will consist of the intensities of all lines and continuum emission intensity:

$$I(\lambda) = S(\lambda) + \sum_k I_F^k(\lambda); \quad (11.13)$$

where k—the notation of the spectral line.

Using the dependence of intensity for each of the four Voigt contours from four parameters (I_0, λ_0, $\delta\lambda_L$, $\delta\lambda_G$) the expression for the function of total contour consisting of P lines, can be written as

$$\begin{aligned} I(\lambda, I_0^1, \lambda_0^1, \delta\lambda_G^1, \delta\lambda_L^1, I_0^2, \lambda_0^2, \delta\lambda_G^2, \delta\lambda_L^2, \ldots, I_0^P, \lambda_0^P, \delta\lambda_G^P, \delta\lambda_L^P, S(\lambda)) = S(\lambda) \\ + I_F(\lambda, \lambda_0^1, I_0^1, \delta\lambda_G^1, \delta\lambda_L^1) + I_F(\lambda, I_0^2, \lambda_0^2, \delta\lambda_G^2, \delta\lambda_L^2) + \cdots \\ + I_F(\lambda, I_0^P, \lambda_0^P, \delta\lambda_G^P, \delta\lambda_L^P), \end{aligned} \quad (11.14)$$

where the superscript of parameters I_0, λ_0, $\delta\lambda_L$, $\delta\lambda_G$ denotes the number of spectral line, and registered contour is the described function by the functions of the intensity of continuum and the Voigt contour s, each of them is defined by four parameters I_0, λ_0, $\delta\lambda_L$, $\delta\lambda_G$.

The spectrometer presents the data about spectrum in the following format $\langle\lambda_i^{\exp}\rangle - \langle I_i^{\exp}\rangle$, where $\lambda_i^{\exp}$—the wavelength corresponding to the ith pixel in the CCD detector, $I_i^{\exp}$—the intensity of radiation registered by the ith pixel. For the decryption the recorded spectrum is divided into sections, each of them contains a certain number of data points, and for each region containing P lines, the following system of equations is solved:

$$
\begin{aligned}
& I\left(\lambda_k, I_0^1, \lambda_0^1, \delta\lambda_G^1, \delta\lambda_L^1, I_0^2, \lambda_0^2, \delta\lambda_G^2, \delta\lambda_L^2, \ldots, I_0^P, \lambda_0^P, \delta\lambda_G^P, \delta\lambda_L^P, S\right) = I_k^{\exp}; \\
& I\left(\lambda_{k+1}, I_0^1, \lambda_0^1, \delta\lambda_G^1, \delta\lambda_L^1, I_0^2, \lambda_0^2, \delta\lambda_G^2, \delta\lambda_L^2, \ldots, I_0^P, \lambda_0^P, \delta\lambda_G^P, \delta\lambda_L^P, S\right) = I_{k+1}^{\exp}; \\
& \vdots \\
& I\left(\lambda_{k+M}, I_0^1, \lambda_0^1, \delta\lambda_G^1, \delta\lambda_L^1, I_0^2, \lambda_0^2, \delta\lambda_G^2, \delta\lambda_L^2, \ldots, I_0^P, \lambda_0^P, \delta\lambda_G^P, \delta\lambda_L^P, S\right) = I_{k+M}^{\exp},
\end{aligned}
\tag{11.15}
$$

where $I\left(\lambda_k, I_0^1, \lambda_0^1, \delta\lambda_G^1, \delta\lambda_L^1, I_0^2, \lambda_0^2, \delta\lambda_G^2, \delta\lambda_L^2, \ldots, I_0^P, \lambda_0^P, \delta\lambda_G^P, \delta\lambda_L^P, S\right)$—the analytical record of the Voigt contour; λ—experimental wavelength; k—the number of the first line of array for M lines, corresponding to a given area, for the substitution to the system of equations. In the range of decryption of part of the spectrum the intensity of continuum can be considered constant [function $S(\lambda)$ can be replaced by constant S].

Parameters I_0, λ_0, $\delta\lambda_L$, $\delta\lambda_G$, S are taken as unknown variables. If the number of Eq. (11.15) (i.e., the experimental points of the plot) is equal to the number of variables in the system or exceeds it, the system can be solved. However, the exact solution of the system is not possible to get, because the experimental values of intensity cannot be exactly equal to the theoretical one due to the influence of random errors, the noise in the CCD detector, and a number of other factors, the account for all of them is impossible. However, an approximate solution can be found, it provides the best agreement of analytical contour with the experimental one. The special function Minerr of the mathematical package MathCAD is used to find the approximate solution. It differs from other functions for solutions of equations that iterate until the solution stops "improving", and when it occurs, returns the best result. The decryption of recorded line using three superposed lines of spectrum is presented in Fig. 11.10.

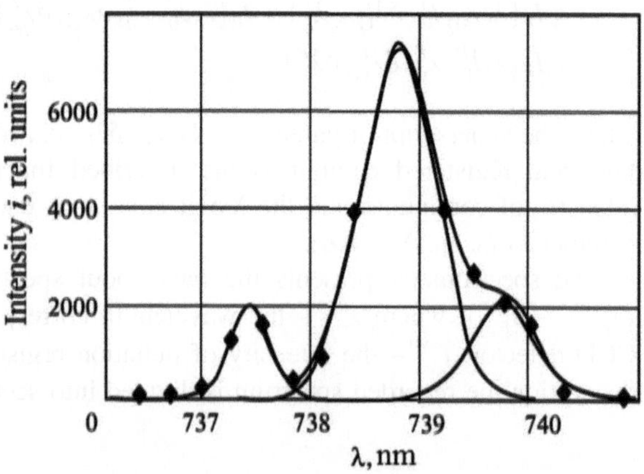

Fig. 11.10 The result of decryption for three superposed lines

This algorithm allows theoretical decryption of almost any of the spectrums. Basically the number of lines for consideration is limited only by the speed of computer and by the number of data points belonging to each of lines in this area. At least four points that do not lie on the level of continuum should be matched to one line, i.e., for superposed lines, the number of points should be equal to $4R$, where R—the number of superposed lines. This should be taken from the type of Voigt function that consists of four parameters I_0, λ_0, $\delta\lambda_L$, $\delta\lambda_G$ which correspond to the kind of line and one of them S—for her "rise above the x-axis." Processing all the spectrum gives the decoded lines with specific parameters of the contours and lines from the reference data associated with them and contains the necessary constants for further calculations. Using this information, it is sufficient to substitute the required data into the equations for the determination of electron temperature by the method of relative intensities and the concentration of electrons by the Stark broadening.

Note that the presence of a greater number of points of continuum increase the accuracy of the finding of theoretical circuit, particularly if there is a sufficiently large number of points that belong to a single line, there is no need to solve the problem of determination of its parameters with the help of a system of equations, because the spline, constructed using the experimental points, describes it quite significantly.

The Absolute Calibration of the Intensity of Radiation and Registration of the Spectrum Using the Standard of Brightness For obtaining the absolute values of intensity of the spectral lines and the continuous radiation of plasma it is necessary to compare the registered in experiment relative intensities of the radiation with intensities of the standard source. In the wavelength range of 320–1000 nm such a source can be the calibrated tungsten band lamp with the known dependence of brightness temperature in the pyrometric wavelength ($\lambda \approx 650$ nm) from the current flowing through the band. The tungsten lamp with small dimensions with $T_{br} = 2400$–2500 K at currents of 3–3.5 A was used in our work.

The calculations of absolute spectral intensity of radiation, W/(cm^3 nm sr), are performed by the relation:

$$\varepsilon_\lambda^{abs} = \frac{i_\lambda^{sp}}{i_\lambda^{st}} I_{\lambda,T}^{st} \frac{k}{R_0}, \tag{11.16}$$

where i_λ^{sp}—the recorded in relative units spectral intensity of plasma emission; i_λ^{st}—the intensity of the radiation of standard lamp, expressed in relative units; $I_{\lambda,T}^{st}$—the spectral intensity of radiation for the standard tungsten band W/(cm^2 nm sr); R_0—length of radiator in the direction of observer, cm; k—coefficient, that takes into account differences in recording of the spectrums for plasma and standard lamp.

The dependence $I_{\lambda,T}^{st}$ from the frequency is obtained using the Planck's law and the literature data about the effect of temperature and wavelength to the emissivity of tungsten:

$$I_{\lambda,T}^{\mathrm{st}} = \frac{2hc^2}{\lambda^5\left[\exp\left(\frac{hc}{\lambda T_{\mathrm{br}}}-1\right)\right]}\frac{\varepsilon_\lambda^w(T_{\mathrm{br}})}{\varepsilon_{\mathrm{pyr}}^w(T_{\mathrm{br}})}, \tag{11.17}$$

where $\varepsilon_\lambda^w(T_{\mathrm{br}})$—the degree of black for the tungsten surface at the temperature T_{br}, and the wavelength λ, $\varepsilon_{\mathrm{pyr}}^w(T_{\mathrm{br}})$—the same at pyrometric wavelength.

For the correctness of the procedure of absolute calibration of spectrum the standard source should be placed on the optical axis of our system instead of discharge, maintaining the position of condenser as the same. It will project the sharp image of a vertical tungsten band to the entrance slit of spectrograph. The registration of the spectrum of a standard source should be carried out at the same spectral ranges as for plasma spectrums. Thus, we ensure the full compliance of registration schemes for plasma and for the standard source. In the spectrum range of λ = 200–300 nm, the deuterium lamp can be used with continuous radiation as a standard source.

The Method of Measuring the Magnetic Field of Plasma Magnetic probe can be used for measuring intensity of the magnetic field of RF plasma torch. In the magnetic probe method the induction of the magnetic field is characterized by electromotive force at the magnetic probe. The electromotive force is recorded using oscilloscope.

The magnetic probe is a coil of several turns of a conductive material that is placed into the investigated magnetic field. The flow of the vector of magnetic field induction for one turn is defined as

$$\Phi = B_\perp S_{\mathrm{c}}, \tag{11.18}$$

where $B_\perp$—the projection of the vector of magnetic field induction on the normal to the surface turns; S_{c}—the square of the windings in the coil.

For coils of identical turns, if we take the magnetic field inside it as uniform, electromotive force is defined as the product of the changes in time for Φ and the number of turns N:

$$\varepsilon = -\frac{\mathrm{d}\Phi}{\mathrm{d}t}N. \tag{11.19}$$

If to place a magnetic probe into the region near inductor of the plasma torch, the induced magnetic field will create electromotive force inside this probe and its change can be fixed with oscilloscope. The peak value of the magnetic induction can be found using the equation

$$B = \frac{\varepsilon}{2\pi f S_{\mathrm{eff}}}, \tag{11.20}$$

where S_{eff}—effective square of the cross section in the coil; f—the frequency of oscillation of the field. The intensity of magnetic field is defined as

$$H = \frac{B}{\mu_0 \mu}. \tag{11.21}$$

In plasma $\mu = 1$,

$$H = \frac{\varepsilon}{8\pi^2 10^{-7} f S_{\text{eff}}}. \tag{11.22}$$

The experimental value of electromotive force used in the calculation is defined as the root mean square of the oscilloscopic waveform:

$$\varepsilon = \sqrt{\frac{1}{t_2 - t_1} \int_{t_1}^{t_2} \varepsilon_{\text{os}}^2 \, dt}, \tag{11.23}$$

where t_1, t_2—the start and the final time of readings; ε_{os}—the instantaneous electromotive force, obtained from the oscilloscopic waveform.

11.3 Measurements of Plasma Parameters in the RF and Arc Plasma Torches

Registration of the Emission Spectrums of Argon Plasma Atoms Ar, as well as other atoms of inert gases (except He), are filled with p-shell ($4p^6$), that is especially stable. Therefore, for the excitation of one of the p-electrons the high energy of levels is required (the ionization potential of the Ar atoms E_I = 15.76 eV, and the excitation energy of the lowest excited state E_1 = 13.08 eV). Due to the significant energies of excitation the atom (and ion) spectrum decays into Argon spectrum lying in the wavelength $\lambda < 180$ nm corresponding to a transition from the excited to the ground level, and on the spectrum, that lies in the infrared, visible, and near-ultraviolet ranges and appearing in the transitions from one excited state to another one [5, 7].

The emission spectrum of Argon plasma at atmospheric pressure in the range of λ = 400–450 nm (Fig. 11.11) at the temperature 16 × 10^3 K represents some of continuous emission spectrums arising in free–free and free–bound transitions of electrons to the field of ions, and a large number of spectral lines of atoms (spectrum of ArI) and ions (spectrum of ArII).

The Program of the Automated Processing of Plasma Emission Spectrums The automated program for the registration and processing of the spectrum of plasma is carried out in mathematical package MathCAD 14 of the professional firm Mathsoft. The program is a set of procedures that are common for all the processed spectrums. It consists of two modules, processing the information received from two CCD lines (spectrum itself and the radial distribution). The work

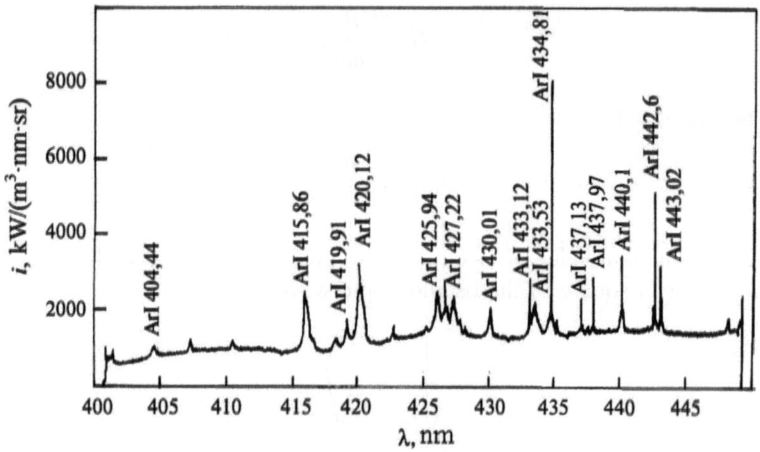

Fig. 11.11 The emission spectrum of Argon plasma in the open free Arc in the range λ = 400–450 nm at distance from the cathode tip z = 2 mm (I = 80 A, a diffraction grating—600 lines per mm)

with spectrum chord distribution is processed at first in order to obtain an optical radius of the cable R_0, and to use the conversions of Abel. The value of optical radius of image is written into the file «R0» in the ASCII symbols and then is used in a second module of the program for the determination of absolute values of the intensities of lines.

Processing the spectrum in the program starts with the input of the following data:

1. Parameters of the experimental spectrum;
2. Spectrum of the standard (tungsten bands);
3. Database for the lines of the studied spectrum;
4. Number of points in continuum;
5. Coordinates of points in continuum;
6. Coordinates of base points are sets of the start and the final wavelengths (in nanometers) of the studied range;
7. Conditions of the experiment (the slit width and the registration time of the standard spectrum and plasma).

Explanation of the Spectrum At the second stage of processing a comparison of two lines of the experimental spectrum with lines from a database of spectrums [4] is carried out. This comparison is carried out by entering the experimental and table base points. The comparison can be performed separately for two atomic or ionic lines or in their totality. As a result of decoding all of the experimental lines should coincide with the tabulated data (Fig. 11.12).

Determination of Absolute Intensities in Emission Spectrum of Plasma The third step is determination of local intensities, i.e., the transfer of the relative

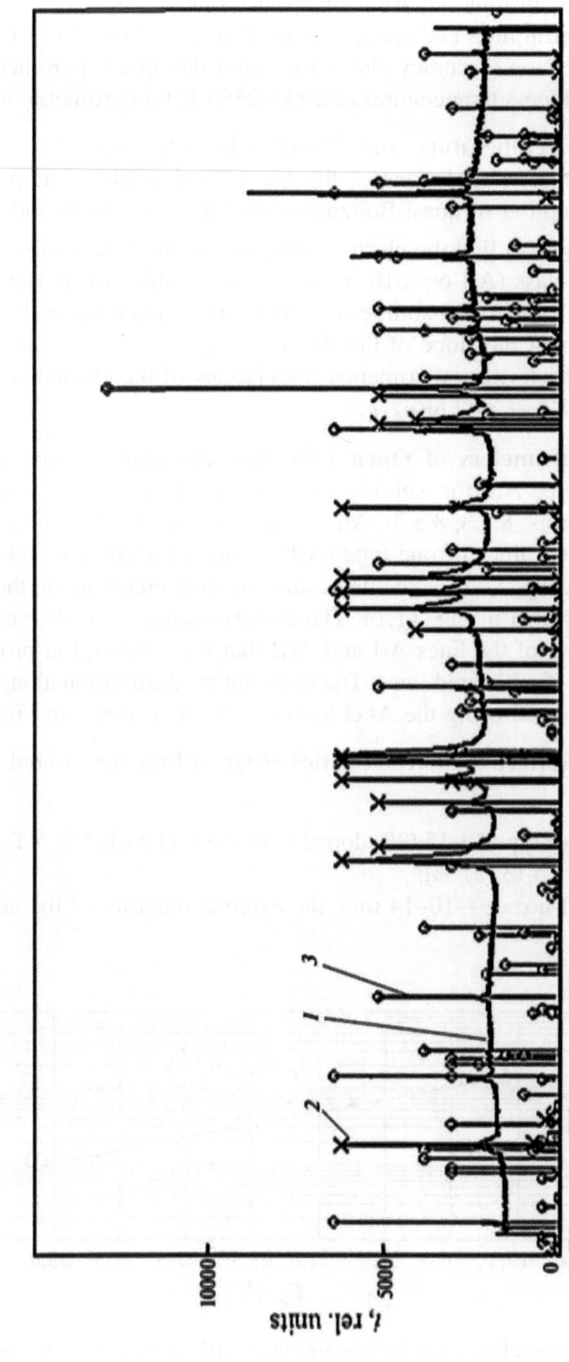

Fig. 11.12 Decoded spectrum of Argon plasma in open free Arc in the wavelength range $\lambda = 400$–450 nm for $z = 2$ mm, $I = 80$ A: *1* experimental spectrum, *2* lines ArI, *3* lines ArII

intensities into absolute units [W/(cm^3 nm sr)]. In the program the procedure for determination of the absolute intensities is carried out automatically. The absolute intensity of the spectral lines and continuum used tungsten band lamp TRSH-2850, calibrated with precision (accuracy class of 0.01) of the optical pyrometer EOP-66 in the range of brightness temperatures of 2000–2550 K for pyrometric wavelength.

Determination of Temperature and Electron Density Analysis of the experimental dependences of populations of the kth excited levels of atoms and ions $\frac{n_k}{g_k} = f(E_k)$ in assumption of quasi-Boltzman distribution of atoms and ions of the energies E_k, allows us to find the electron temperature from the relative intensities of the lines in this class (ArI or ArII). Using a large number of registered lines of ionized argon with known probabilities of transition the temperature is determined by taking into account the slope of the line $\ln(n_k/g_k) = f(E_k)$ drawn through the points corresponding to the experimental populations of the excited states of particles of the class: $kT_e = tg^{-1} \ln(n_k/g_k)$.

Measuring the Parameters of Open Free Arc The main results of the measurement of open free Arc for welding are presented here, data of simulation are presented above (Sects. 8.4.3, 8.5.3). An example of determination of the electrons temperature along the lines of one type (ArII) using the method of relative intensities is shown in Fig. 11.13 and the results of determination of the electrons concentration are shown in Fig. 11.14. The electron density was determined using the Stark broadening of the lines ArI and ArII that was obtained in program from the Voigt contours of registered lines. The temperature distribution along the radius of the free arc, obtained using the Abel transformation is shown in Fig. 11.15.

Measurements of External Characteristics of Open Free Arc Initial parameters of Argon Arc:

Tungsten cathodes of type VI-15 (30) doped with Y_2O_3 [1.5 (3) %]; VT-15 (30, 50) doped with ThO_2 [1.5 (3, 5) %];
Internal diameter of nozzle—10–14 mm, the external diameter of the nozzle—15–19 mm;

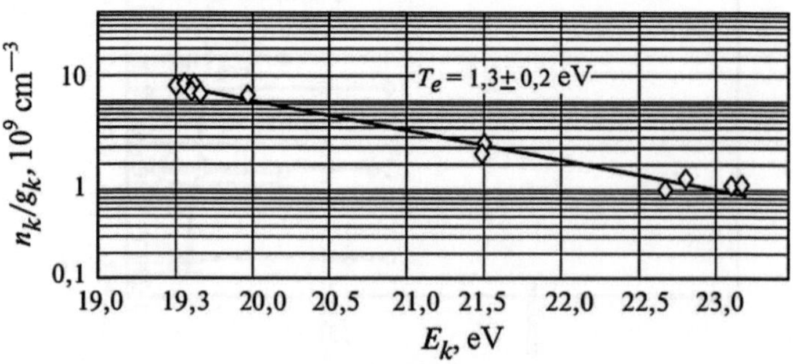

Fig. 11.13 Determination of T_e using the spectral lines ArII on the axis of the open free Arc (z = 2 mm)

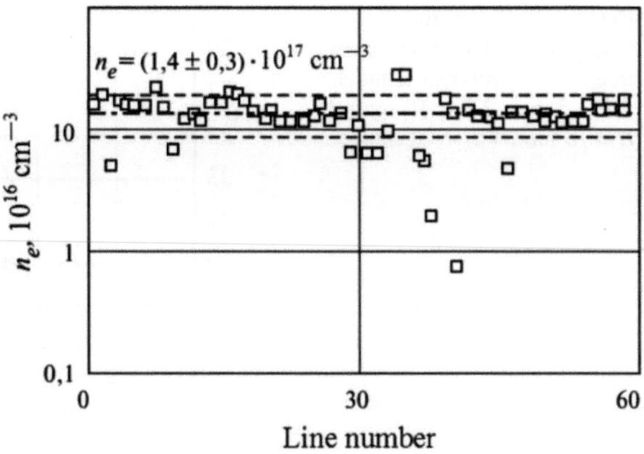

Fig. 11.14 Determination of n_e at the axis of the open free Arc after processing 60 lines ArI and ArII (z = 2 mm, diffraction grating—600 lines per mm)

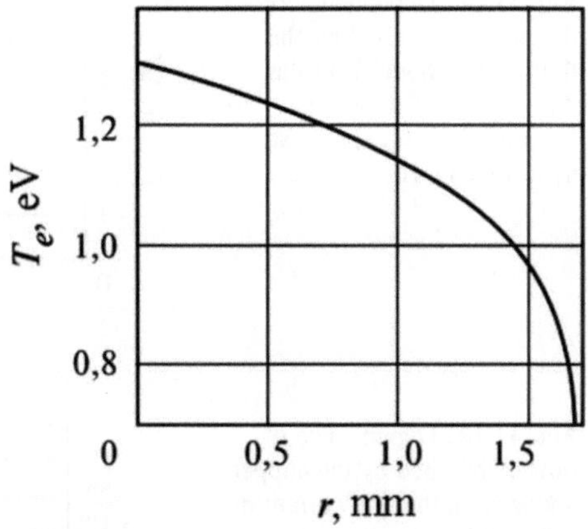

Fig. 11.15 The radial distribution of T_e, obtained by processing the chord distribution of intensity in continuum (λ = 420 nm) using the Abel transformations in sections z = 2 mm

Diameter of the tungsten cathode d_c = 2–1.5 mm;
Cathode length—83 mm, length of the water-cooled parts—50 mm;
The anode plate made of copper with internal water-cooled size 67 × 67 × 13 mm;
Arc length L = 5–20 mm;
Plasma gas consumption Q_{Ar} = 20–40 l/min;
Consumption of the cooling water of anode and cathode G_w = 40 g/s;
Initial temperature of cooling water $T_c^0 = 6.2$ °C, $T_a^0 = 5.8$ °C
Power supply with a range of current adjustment I = 40–160 A and voltage U = 12–28 V.
The current–voltage characteristics of a free arc of different lengths are shown in Fig. 11.16.

The power transmitted to electrodes (Fig. 11.17, 11.18) is found using the calorimetric method. At first, using a thermocouple (thermocouples) the input temperature of the cooling water of cathode and anode are measured at a predetermined consumption, then the power transmitted by electrode is determined using

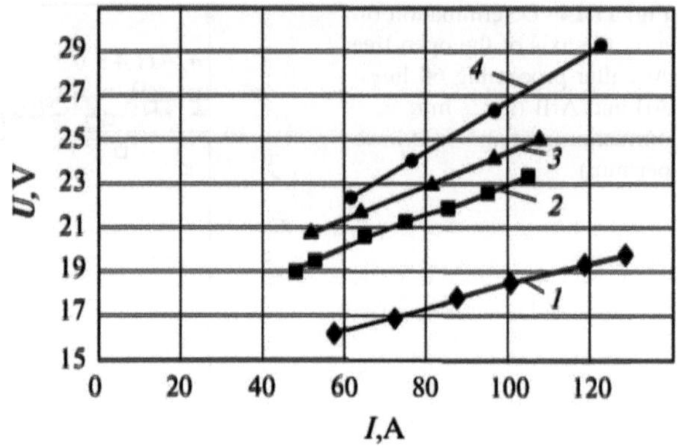

Fig. 11.16 The current–voltage characteristics of open free Arc for different lengths: *1* $L = 5$ mm, *2* $L = 10$ mm; *3* $L = 15$ mm, *4* $L = 18$ mm

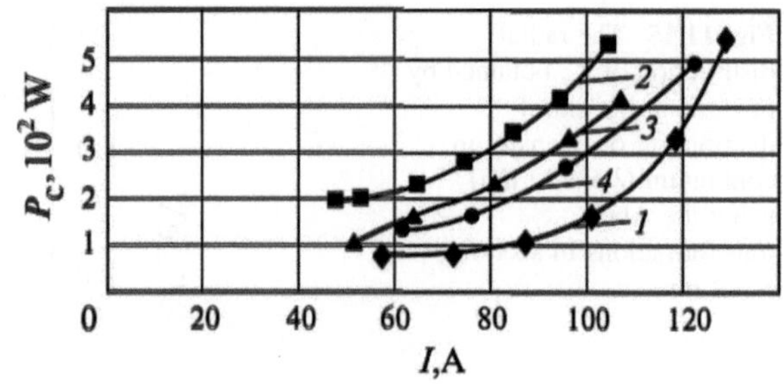

Fig. 11.17 Dependence of the power absorbed by the thoriated cathode from the Arc current of different lengths: the notes are the same as in Fig. 11.16

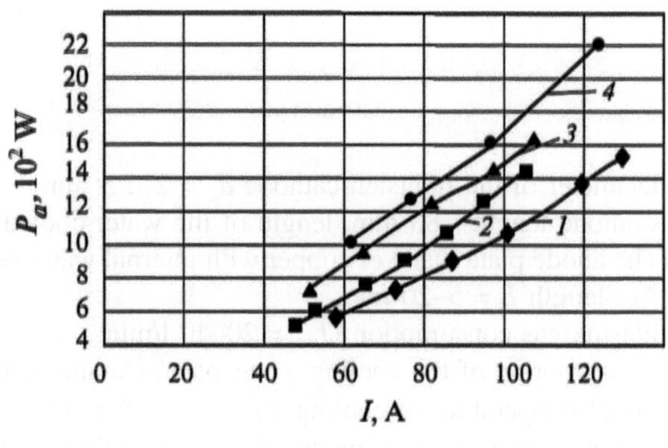

Fig. 11.18 Dependence of power absorbed by the copper anode from the Arc current of different lengths: the notes are the same as in Fig. 11.16

the equation: $P = c_p G_w (T - T^0)$, here $c_p = 4.18$ J/(g K)—the specific heat of the water; G_w—the consumption of the cooling water of anode and cathode, g/s; T^0—the initial temperature of the cooling water; T—the measured temperature of the water at the exit of plasma torch.

Determination of Temperature and Density of RF Plasma The temperature and density are determined using the lines of the degree of ionization of ArI RF plasma at atmospheric pressure of plasma torch with frequency of 27.12 MHz (Fig. 11.3),

Table 11.2 Characteristics of some transitions of Argon atoms ArI (NIST ASD [4]) for the channel No. 3 of spectrometer Avaspec

Wavelength (nm)	$g_k A_{ki}$, 10^{-6} s^{-1}	Excitation energy E_k (eV)	Experimental integral intensity of the spectral line, relative units
675.2834	9.65	14.7425	165.117
696.5431	19.20	13.3278	1566
703.0251	13.40	14.8388	302.715
706.7218	19.00	13.3022	1837
720.698	7.44	15.0220	168.239
727.2936	5.49	13.3278	1103
735.3293	6.70	14.7805	305.968
737.2118	17.00	14.7570	783.99
738.398	42.40	13.3022	5731
772.4207	35.10	13.3365	10,750

intended for spectral analysis. For the determination of electron temperature using the method of relative intensities the obtained spectrums ArI with wavelengths 550–1000 nm were used, the information about them has been taken from the database of NIST [4]. For obtaining the distribution of electrons temperature the spectrums were recorded at some points of plasma in one section of inductor. The used spectrometer Avaspect of the company Avantes had the built calibration for wavelength and absolute intensities that is not allowed to determine the dependence of its sensitivity from the wavelength and to consider it in the resulting spectrum. The probabilities of some transitions of the argon atoms, as well as statistical weight and the energies of the upper levels of these transitions are shown in Table 11.2.

The temperature distribution along the radius for the cross section in the center of inductor (z = 10 mm), of RF plasma torch (27.12 MHz) is obtained using the Abel transformation and it is given in Fig. 11.19. Calculated temperature field for comparison is shown in Fig. 11.20.

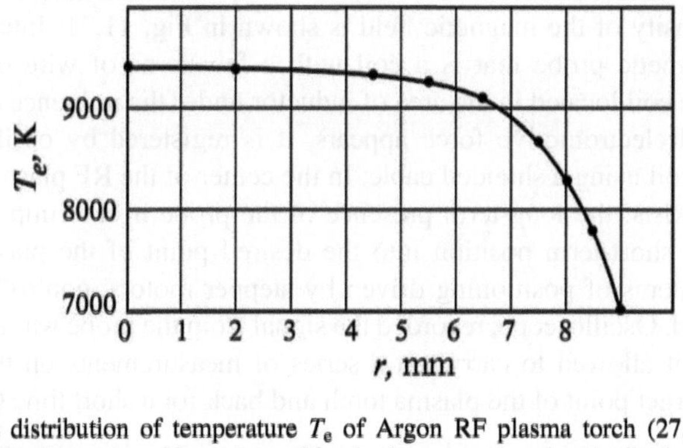

Fig. 11.19 The distribution of temperature T_e of Argon RF plasma torch (27.12 MHz) in the central section of inductor (z = 10 mm), resulting in the processing the distribution of intensity for continuum (λ = 700 nm) using Abel conversions

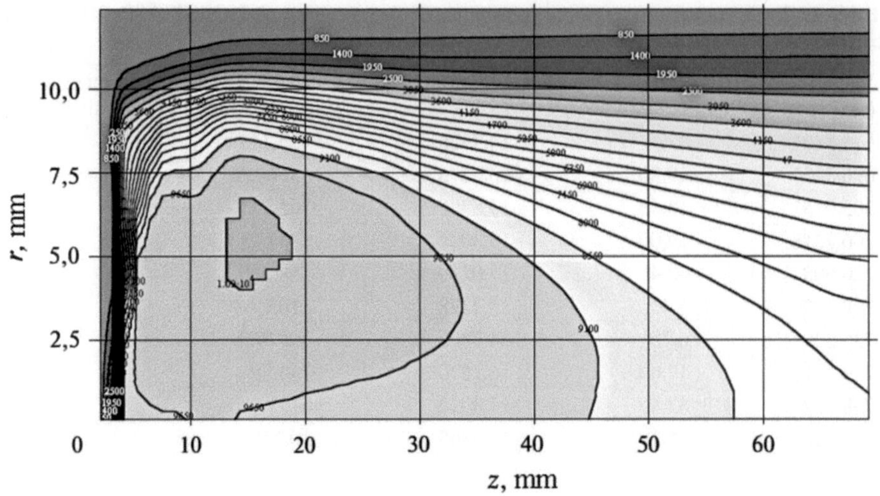

Fig. 11.20 Calculated temperature field T_e, K in Argon RF plasma torch (27.12 MHz)

For the determination of density of electrons the same spectrums were used for the electron temperature. The calculated density was compared with its theoretical value obtained by Saha equations taking into account experimentally determined temperature. To estimate the density of electrons, for example, the spectral line $\lambda = 549$, 567 nm with strongly expressed Stark broadening (about 0.3 nm) was used to make it possible to neglect other components and to consider the broadening of half-width in the experiment as a Lorentz one without additional amendments. As a result, electron density, $n_e = 2.24 \times 10^{16}$ cm^{-3} was defined by the Stark broadening of the lines but the calculated value of electron density for argon plasma in RF plasma torch (27.12 MHz) $n_e = 2.11 \times 10^{16}$ cm^{-3} at the temperature $T = 10{,}200$ K in the center of inductor ($z = 10$ mm, $r = 0$ mm).

Measuring of the Intensity of Magnetic Field in the RF Plasma Torch Using the Magnetic Probe Method The scheme of experimental setup for the determination of intensity of the magnetic field is shown in Fig. 11.21. Intensity is found using the magnetic probe that is a coil with a few turns of wire of a refractory material. If the coil located in the area of inductor under the influence of the induced magnetic field electromotive force appears, it is registered by oscilloscope, connected to the coil using a shielded cable. In the center of the RF plasma temperature 10×10^3 K exists, the long-term presence of the probe in it is impossible. In this regard, for its short-term position into the desired point of the plasma torch two coordinate systems of positioning driven by stepper motors, controlled by a computer, was used. Oscilloscope, recorded the signal from the probe was also controlled by computer, it allowed to carry out a series of measurements on the path of the probe to the target point of the plasma torch and back for a short time (less than 2 s).

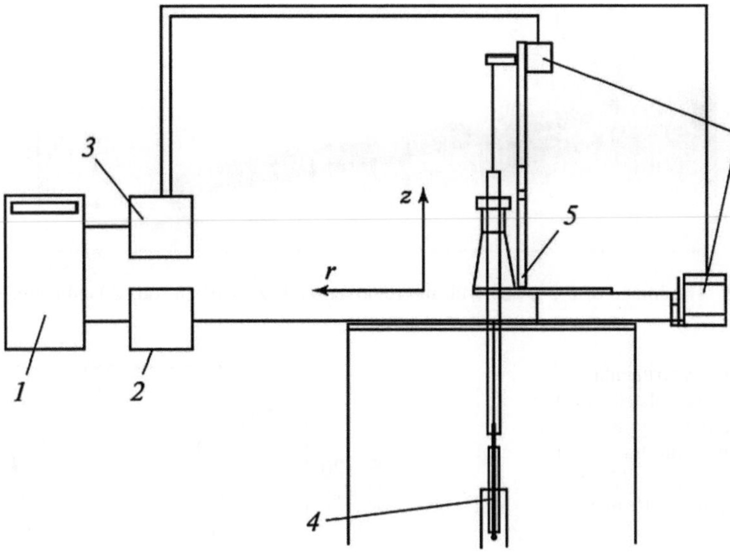

Fig. 11.21 Scheme of the experimental setup for determination of the intensity of magnetic field in RF plasma torch: *1* PC, *2* digital oscilloscope, *3 LPT* controller of stepper motors, *4* RF plasma torch with a magnetic probe, *5* positioning of probe, *6* step motors

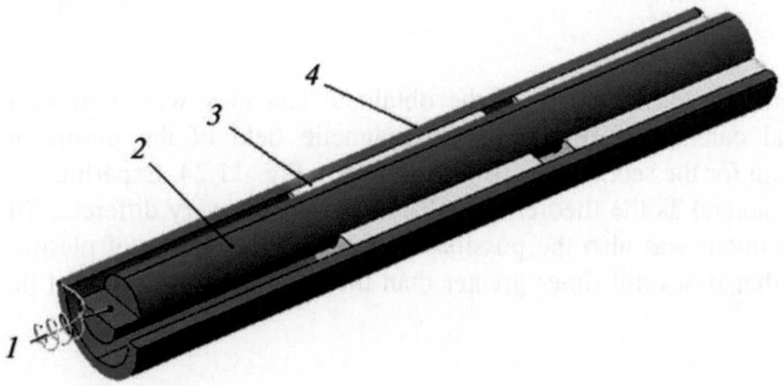

Fig. 11.22 The section of the probe: *1* tungsten spiral, *2* tungsten rod, *3* ceramic insulators, *4* lanthanum screen

Image of bottom part of magnetic probe in section and with removed coil are shown in Figs. 11.22, 11.23. Screen 4 (Fig. 11.22) is needed to protect the stem from interference and it covers the stem all over the entire length, where the influence of the magnetic field is strong. The rod *2* inside the screen is passed through ceramic cylinder *3*, isolating it from the screen. Not having time to warm up during the experiment to the temperature of destruction, the rod can also take the heat from the probe.

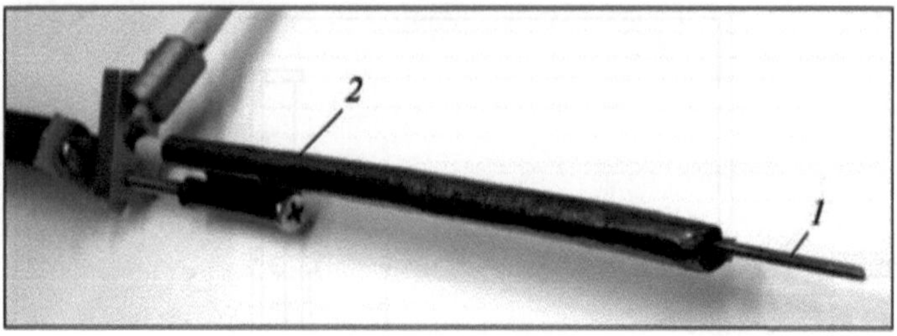

Fig. 11.23 The image of the probe with the removed coil: *1* tungsten rod, *2* lanthanum screen

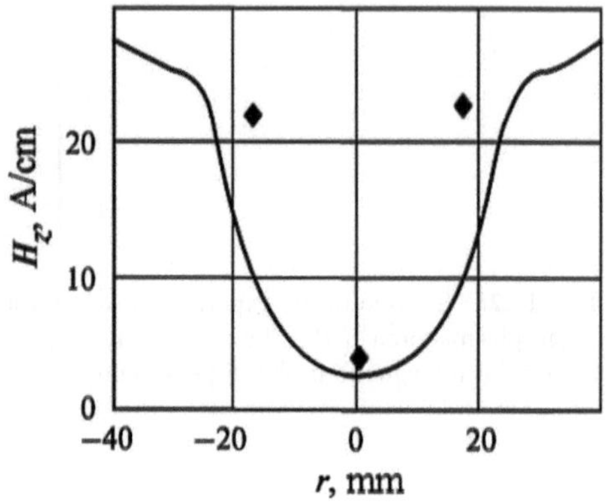

Fig. 11.24 Experimental (points) and calculated (*line*) data of the intensity of the magnetic field on the axis of RF plasma torch (27.12 MHz) at the section $z = 10$ mm

To estimate the accuracy of the obtained data they were compared with the theoretical calculation results of the magnetic field of the plasma torch. This comparison for the section $z = 10$ mm shown in Fig. 11.24. Experimental data have the same spread as the theoretical values but they are very different. The result of this experiment was also the possibility of probe diagnostics of plasma with temperature that is several times greater than the melting temperature of the probe.

References

1. Schulz-Gulde E. // Paper of Institute of atom and molecular physics. — Hannover. Germany, 1997.
2. AIAA Paper. 1966. № 66(189). P. 1.
3. Donskoy A.V., Klubnikin V.S. // J. of Engineering Physics. 1972. V. 22. No.6.
4. NIST Atomic Spectra Database. Lines Data (http://physics.nist.gov/cgi-bin/AtData/main_asd).
5. Sobelman I.I. Introduction to the theory of atomic spectra. — M.: Nauka. 1977.
6. Automated System for plasma emission spectra / E.H. Isakayev, V.F. Chinnov et al. // Instruments and Experimental Techniques. 2001. № 1. PP. 47–53.
7. Frisch S.E. Optical spectra of atoms. — M.-L.: Fizmatgiz 1963.

Appendix A

The values of the statistical weights sum, affinity energy of the electron and the ionization energy of the atoms and ions of single, double and triple ionization:

Z—sequence number in the periodic table; A—atomic weight; g_{-1}, g_0, g_1, g_2, g_3—the statistical weight sum of electrons, atoms, and ions of single, double and triple ionization; I_{-1}—the energy of the electron affinity, eV; I_0, I_1, I_2, I_3—ionization energy of the atoms and ions of single, double and triple ionization, eV.

S. Nguyen-Kuok, *Theory of Low-Temperature Plasma Physics*,
Springer Series on Atomic, Optical, and Plasma Physics 95,
DOI 10.1007/978-3-319-43721-7

Z	A	Elements	g_{-1}	I_{-1}	g_0	I_0	g_1	I_1	g_2	I_2	g_3	I_3
1	1	H	1	0.754	2	13.599						
1	2	D	1	–	2	13.602						
1	3	T	1	–	2	13.604						
2	3	He	2	0.076	1	24.586	2	54.418				
2	4	He	2	–	1	24.588	2	54.418				
3	6.94	Li	1	0.61	2	5.392	1	75.641	2	122.45		
4	9.01	Be	2	0.30	1	9.323	2	18.211	1	153.90	2	217.72
5	10.81	B	1	–	2	8.298	1	25.155	2	37.931	1	259.38
6	12.01	C	4	1.27	1	11.260	2	24.384	1	47.89	2	64.49
7	14.01	N	5	–	4	14.534	1	29.602	2	47.45	1	77.47
8	16	O	4	1.465	5	13.618	4	35.118	1	54.936	2	77.414
9	19	F	1	3.4	4	17.423	5	34.971	4	62.71	1	87.14
10	20.18	Ne	2	–	1	21.565	4	40.963	5	63.46	4	97.12
11	23	Na	1	0.548	2	5.139	1	47.287	4	71.620	5	98.92
12	24.31	Mg	2	–	1	7.646	2	15.035	1	80.144	4	109.27
13	27	Al	1	0.5	2	5.986	1	18.829	2	28.448	1	119.99
14	28.1	Si	4	1.39	1	8.152	2	16.346	1	33.493	2	45.142
15	31	P	5	0.77	4	10.487	1	19.770	2	30.203	1	51.444
16	32.06	S	4	2.077	5	10.360	4	23.338	1	34.83	2	47.305
17	35.45	Cl	1	3.62	4	12.986	5	23.814	4	39.61	1	53.47
18	39.95	Ar	2	–	1	15.760	4	27.630	5	40.911	4	59.81
19	39.1	K	1	0.501	2	4.341	1	31.63	4	45.81	5	60.91
20	40.08	Ca	4	–	1	6.113	2	11.872	1	50.913	4	67.3
21	45	Sc	5	0.19	4	6.562	3	12.800	4	24.757	1	73.49
22	47.9	Ti	4	0.08	5	6.82	4	13.58	5	27.49	4	43.27

(continued)

(continued)

Z	A	Elements	g_{-1}	L_1	g_0	I_0	g_1	I_1	g_2	I_2	g_3	I_3
23	51	V	7	0.53	4	6.74	1	14.66	4	29.31	5	46.71
24	52	Cr	6	0.67	7	6.766	6	16.50	1	31.0	4	49.2
25	55	Mn	9	–	6	7.434	7	15.640	6	33.67	1	51.2
26	56	Fe	10	0.16	9	7.902	10	16.188	9	30.65	6	54.8
27	59	Co	9	0.66	10	7.86	9	17.084	10	33.5	9	51.3
28	58.71	Ni	2	1.16	9	7.637	6	18.169	9	35.3	10	54.9
29	63.55	Cu	1	1.23	2	7.726	1	20.293	6	36.84	9	57.4
30	65.37	Zn	2	–	1	9.394	2	17.964	1	39.72	6	59.57
31	69.72	Ga	1	0.5	2	5.999	1	20.515	2	30.7	1	64.2
32	72.59	Ge	4	1.2	1	7.900	2	15.935	1	34.2	2	45.7
33	74.92	As	5	0.8	4	9.789	1	18.59	2	28.4	1	50.1
34	78.96	Se	4	2.0	5	9.752	4	21.16	1	30.82	2	42.95
35	79.9	Br	1	3.37	4	11.814	5	21.81	4	35.90	1	47.3
36	83.8	Kr	2	–	1	14.000	4	24.360	5	36.95	4	52.5
37	85.47	Rb	1	0.468	2	4.177	1	27.290	4	39.2	5	52.6
38	87.62	Sr	4	–	1	5.695	2	11.030	1	42.88	4	56.28
39	88.91	Y	5	0.31	4	6.217	1	12.24	2	20.525	1	60.61
40	91.22	Zr	2	0.43	5	6.837	4	13.13	5	23.1	4	34.419
41	92.91	Nb	7	0.89	2	6.88	1	14.32	4	25.0	5	37.7
42	95.94	Mo	6	0.75	7	7.099	6	16.16	1	27.2	4	46.4
43	[99]	Tc	11	1.0	6	7.28	7	15.26	6	29.5	1	43
44	101.1	Ru	10	1.5	11	7.366	10	16.76	9	28.5	6	46.5
45	102.9	Rh	1	1.14	10	7.46	9	18.08	10	31.1	9	45.6
46	106.4	Pd	2	0.56	1	8.336	6	19.43	9	32.9	10	48.8

(continued)

(continued)

Z	A	Elements	g_{-1}	I_{-1}	g_0	I_0	g_1	I_1	g_2	I_2	g_3	I_3
47	107.9	Ag	1	1.3	2	7.576	1	21.49	6	34.8	9	52
48	112.4	Cd	2	–	1	8.994	2	16.908	1	37.47	6	55
49	114.8	In	1	0.5	2	5.786	1	18.87	2	28.0	1	57.0
50	118.7	Sn	4	1.25	1	7.344	2	14.632	1	30.50	2	40.74
51	121.8	Sb	5	1.05	4	8.609	1	16.53	2	25.32	1	44.16
52	127.6	Te	4	1.9	5	9.010		18.8		30.6		37.9
53	126.9	I	1	3.07	4	10.451	5	19.131	4	33.0	1	41.7
54	131.3	Xe	2	–	1	12.130	4	20.98	5	31.0	4	45
55	132.9	Cs	1	0.471	2	3.894	1	23.15	4	33.4	5	46
56	137.3	Ba	4	–	1	5.212	2	10.004	1	35.8	4	47
57	138.9	La	9	0.5	4	5.577	5	11.1	4	19.18	1	49.9
58	140.1	Ce	10	–	9	5.539	8	10.8	9	20.20	6	36.76
59	140.9	Pr	9	–	10	5.47	9	10.6	10	21.62	9	38.98
60	144.2	Nd	6	–	9	5.525	8	10.7	3	22.1	10	40.4
61	[141]	Pm	1	–	6	5.58	5	10.9	6	22.3	9	41.0
62	150.4	Sm	8	–	1	5.644	2	11.1	1	23.4	6	41.4
63	152	Eu	5	–	8	5.670	9	11.24	8	24.9	1	42.7
64	157.3	Gd	16	–	5	6.150	6	12.1	5	20.6	8	44.0
65	159	Tb	17	–	16	5.864		11.5	16	21.9	13	39.4
66	162.5	Dy	16	–	17	5.939		11.7	17	22.8	16	41.4
67	164.9	Ho	13	–	16	6.022		11.8	16	22.8	17	42.5
68	167.3	Er	8	–	13	6.108		11.9	13	22.7	16	42.7
69	168.9	Tm	1	–	8	6.184		12.1	8	23.7	13	42.7
70	173	Yb	4	–	1	6.254	2	12.18	1	25.05	8	43.6

(continued)

(continued)

Z	A	Elements	g_{-1}	L_1	g_0	I_0	g_1	I_1	g_2	I_2	g_3	I_3
71	175	Lu	5	–	4	5.426	1	13.9	2	20.96	1	45.25
72	178.5	Hf	4	–	5	6.8	4	14.9	5	23.3	4	33.4
73	180.9	Ta	1	0.32	4	7.89	3	16.2		22.3		33.1
74	183.9	W	6	0.816	1	7.98	2	17.7		24.1		35.4
75	186.2	Re	9	0.15	6	7.88	7	16.6		26		37.7
76	190.2	Os	10	1.4	9	8.73	10	17		25		40
77	192.2	Ir	7	1.57	10	9.05	11	17		27		39
78	195.1	Pt	2	2.13	7	8.96	6	18.56		28.5		41.1
79	197	Au	1	2.31	2	9.226	1	20.5	6	34		43
80	200.6	Hg	2	–	1	10.438	2	18.76		34.2	6	46
81	204.4	Tl	1	0.5	2	6.108	1	20.43	2	29.85		50.7
82	207.2	Pb	4	1.2	1	7.417	2	15.033	1	31.94	2	42.23
83	209	Bi	5	1.0	4	7.286	1	16.7	2	25.56	1	45.3
84	[210]	Po	4	1.6	5	8.417		19.4		27.3		38
85	[210]	At	1	–	4	9.0		20.1		29.3		41
86	[222]	Rn	2	–	1	10.75		21.4		29.4		43.8
87	[223]	Fr	1	–	2	4.0		22.5		33.5		43
88	[226]	Ra	4	–	1	5.279	2	10.15		34.3		46.4
89	[227]	Ac	5	–	4	5.2	1	11.75	2	20		49
90	232	Th		–	5	6.1	4	11.9	9	18.3	6	28.7
91	[231]	Pa		–		6.0	9				9	
92	238	U		–		6.194	10	11.9		20	10	37
93	[237]	Np	1	–		6.266	13				9	
94	[244]	Pu	8	–	1	6.06	2				6	

(continued)

(continued)

Z	A	Elements	g_{-1}	I_{-1}	g_0	I_0	g_1	I_1	g_2	I_2	g_3	I_3
95	[243]	Am		–	8	6.0	9		8		13	
96	[247]	Cm	16	–		6.02	8				8	
97	[247]	Bk	17	–	16	6.23	17				13	
98	[252]	Cf	16	–	17	6.30					16	
99	[254]	Es	13	–	16	6.42			16		17	
100	[257]	Fm		–	13	6.5					16	
101	[257]	Md		–	8	6.6					13	
102	[255]	No		–	1	6.6			1		8	

Appendix B

The bracket integrals for the determination of the viscosity of the monocomponent plasma

$$\left[S_{5/2}^{(0)}(W_1^2)\,\mathbf{W}_1\overset{\circ}{\mathbf{W}}_1, S_{5/2}^{(0)}(W_1^2)\,\mathbf{W}_1\overset{\circ}{\mathbf{W}}_1\right]_1 = 4\Omega_1^{(2,2)} \tag{B.1}$$

$$\left[S_{5/2}^{(0)}(W_1^2)\,\mathbf{W}_1\overset{\circ}{\mathbf{W}}_1, S_{5/2}^{(1)}(W_1^2)\,\mathbf{W}_1\overset{\circ}{\mathbf{W}}_1\right]_1 = 7\Omega_1^{(2,2)} - 2\Omega_1^{(2,3)} \tag{B.2}$$

$$\left[S_{5/2}^{(0)}(W_1^2)\,\mathbf{W}_1\overset{\circ}{\mathbf{W}}_1, S_{5/2}^{(2)}(W_1^2)\,\mathbf{W}_1\overset{\circ}{\mathbf{W}}_1\right]_1 = \frac{63}{8}\Omega_1^{(2,2)} - \frac{9}{2}\Omega_1^{(2,3)} + \frac{1}{2}\Omega_1^{(2,4)} \tag{B.3}$$

$$\left[S_{5/2}^{(0)}(W_1^2)\,\mathbf{W}_1\overset{\circ}{\mathbf{W}}_1, S_{5/2}^{(3)}(W_1^2)\,\mathbf{W}_1\overset{\circ}{\mathbf{W}}_1\right]_1 = \frac{231}{32}\Omega_1^{(2,2)} - \frac{99}{16}\Omega_1^{(2,3)} + \frac{11}{8}\Omega_1^{(2,4)} - \frac{1}{12}\Omega_1^{(2,5)} \tag{B.4}$$

$$\left[S_{5/2}^{(0)}(W_1^2)\,\mathbf{W}_1\overset{\circ}{\mathbf{W}}_1, S_{5/2}^{(4)}(W_1^2)\,\mathbf{W}_1\overset{\circ}{\mathbf{W}}_1\right]_1 = \frac{3003}{512}\Omega_1^{(2,2)} - \frac{429}{64}\Omega_1^{(2,3)} + \frac{143}{64}\Omega_1^{(2,4)} - \frac{13}{48}\Omega_1^{(2,5)} + \frac{1}{96}\Omega_1^{(2,6)} \tag{B.5}$$

$$\left[S_{5/2}^{(1)}(W_1^2)\,\mathbf{W}_1\overset{\circ}{\mathbf{W}}_1, S_{5/2}^{(1)}(W_1^2)\,\mathbf{W}_1\overset{\circ}{\mathbf{W}}_1\right]_1 = \frac{301}{12}\Omega_1^{(2,2)} - 7\Omega_1^{(2,3)} + \Omega_1^{(2,4)} \tag{B.6}$$

$$\left[S_{5/2}^{(1)}(W_1^2)\,\mathbf{W}_1\overset{\circ}{\mathbf{W}}_1, S_{5/2}^{(2)}(W_1^2)\,\mathbf{W}_1\overset{\circ}{\mathbf{W}}_1\right]_1 = \frac{1365}{32}\Omega_1^{(2,2)} - \frac{321}{16}\Omega_1^{(2,3)} + \frac{25}{8}\Omega_1^{(2,4)} - \frac{1}{4}\Omega_1^{(2,5)} \tag{B.7}$$

S. Nguyen-Kuok, *Theory of Low-Temperature Plasma Physics*,
Springer Series on Atomic, Optical, and Plasma Physics 95,
DOI 10.1007/978-3-319-43721-7

$$\left[S_{5/2}^{(1)}(W_1^2)\,\mathbf{W}_1\overset{o}{\mathbf{W}}_1, S_{5/2}^{(3)}(W_1^2)\,\mathbf{W}_1\overset{o}{\mathbf{W}}_1\right]_1 = \frac{6699}{128}\Omega_1^{(2,2)} - \frac{297}{8}\Omega_1^{(2,3)} + \frac{385}{48}\Omega_1^{(2,4)} - \frac{5}{6}\Omega_1^{(2,5)} + \frac{1}{24}\Omega_1^{(2,6)} \tag{B.8}$$

$$\left[S_{5/2}^{(1)}(W_1^2)\,\mathbf{W}_1\overset{o}{\mathbf{W}}_1, S_{5/2}^{(4)}(W_1^2)\,\mathbf{W}_1\overset{o}{\mathbf{W}}_1\right]_1 = \frac{109{,}109}{2048}\Omega_1^{(2,2)} - \frac{52{,}767}{1024}\Omega_1^{(2,3)} + + \frac{11{,}869}{768}\Omega_1^{(2,4)} - \frac{2405}{1152}\Omega_1^{(2,5)} + \frac{59}{384}\Omega_1^{(2,6)} - \frac{1}{192}\Omega_1^{(2,7)} \tag{B.9}$$

$$\left[S_{5/2}^{(2)}(W_1^2)\,\mathbf{W}_1\overset{o}{\mathbf{W}}_1, S_{5/2}^{(2)}(W_1^2)\,\mathbf{W}_1\overset{o}{\mathbf{W}}_1\right]_1 = \frac{25{,}137}{256}\Omega_1^{(2,2)} - \frac{1755}{32}\Omega_1^{(2,3)} + \frac{381}{32}\Omega_1^{(2,4)} - \frac{9}{8}\Omega_1^{(2,5)} + \frac{1}{16}\Omega_1^{(2,6)} + \frac{1}{2}\Omega_1^{(4,4)} \tag{B.10}$$

$$\left[S_{5/2}^{(2)}(W_1^2)\,\mathbf{W}_1\overset{o}{\mathbf{W}}_1, S_{5/2}^{(3)}(W_1^2)\,\mathbf{W}_1\overset{o}{\mathbf{W}}_1\right]_1 = \frac{155{,}925}{1024}\Omega_1^{(2,2)} - \frac{56{,}727}{512}\Omega_1^{(2,3)} + \frac{3795}{128}\Omega_1^{(2,4)} - \frac{249}{64}\Omega_1^{(2,5)} + \frac{17}{64}\Omega_1^{(2,6)} - \frac{1}{96}\Omega_1^{(2,7)} + \frac{11}{8}\Omega_1^{(4,4)} - \frac{1}{4}\Omega_1^{(4,5)} \tag{B.11}$$

$$\left[S_{5/2}^{(2)}(W_1^2)\,\mathbf{W}_1\overset{o}{\mathbf{W}}_1, S_{5/2}^{(4)}(W_1^2)\,\mathbf{W}_1\overset{o}{\mathbf{W}}_1\right]_1 = \frac{3{,}072{,}069}{16{,}384}\Omega_1^{(2,2)} - \frac{719{,}433}{4096}\Omega_1^{(2,3)} + \frac{237{,}237}{4096}\Omega_1^{(2,4)} - \frac{4901}{512}\Omega_1^{(2,5)} + \frac{891}{1024}\Omega_1^{(2,6)} - \frac{35}{768}\Omega_1^{(2,7)} + \frac{1}{768}\Omega_1^{(2,8)} + \frac{143}{64}\Omega_1^{(4,4)} - \frac{13}{16}\Omega_1^{(4,5)} + \frac{1}{16}\Omega_1^{(4,6)} \tag{B.12}$$

$$\left[S_{5/2}^{(3)}(W_1^2)\,\mathbf{W}_1\overset{o}{\mathbf{W}}_1, S_{5/2}^{(3)}(W_1^2)\,\mathbf{W}_1\overset{o}{\mathbf{W}}_1\right]_1 = \frac{1{,}168{,}167}{4096}\Omega_1^{(2,2)} - \frac{245{,}025}{1024}\Omega_1^{(2,3)} + \frac{236{,}005}{3072}\Omega_1^{(2,4)} - \frac{1573}{128}\Omega_1^{(2,5)} + \frac{859}{768}\Omega_1^{(2,6)} - \frac{11}{192}\Omega_1^{(2,7)} + \frac{1}{576}\Omega_1^{(2,8)} + \frac{1463}{288}\Omega_1^{(4,4)} - \frac{11}{8}\Omega_1^{(4,5)} + \frac{1}{8}\Omega_1^{(4,6)} \tag{B.13}$$

$$\left[S^{(3)}_{5/2}(W_1^2)\,\mathbf{W}_1\overset{\circ}{\mathbf{W}}_1, S^{(4)}_{5/2}(W_1^2)\,\mathbf{W}_1\overset{\circ}{\mathbf{W}}_1\right]_1 = \frac{26{,}951{,}925}{65{,}536}\Omega_1^{(2,2)} - \frac{13{,}505{,}349}{32{,}768}\Omega_1^{(2,3)} + \frac{7{,}661{,}797}{49{,}152}\Omega_1^{(2,4)} - \frac{738{,}569}{24{,}576}\Omega_1^{(2,5)} + \frac{41{,}005}{12{,}288}\Omega_1^{(2,6)} - \frac{4175}{18{,}432}\Omega_1^{(2,7)} + \frac{85}{9216}\Omega_1^{(2,8)} - \frac{1}{4608}\Omega_1^{(2,9)} + \frac{23{,}881}{2304}\Omega_1^{(4,4)} - \frac{4745}{1152}\Omega_1^{(4,5)} + \frac{37}{64}\Omega_1^{(4,6)} - \frac{1}{32}\Omega_1^{(4,7)} \tag{B.14}$$

$$\left[S^{(4)}_{5/2}(W_1^2)\,\mathbf{W}_1\overset{\circ}{\mathbf{W}}_1, S^{(4)}_{5/2}(W_1^2)\,\mathbf{W}_1\overset{\circ}{\mathbf{W}}_1\right]_1 = \frac{711{,}736{,}025}{1{,}048{,}576}\Omega_1^{(2,2)} - \frac{50{,}053{,}575}{65{,}536}\Omega_1^{(2,3)} + + \frac{64{,}969{,}619}{196{,}608}\Omega_1^{(2,4)} - \frac{3{,}630{,}289}{49{,}152}\Omega_1^{(2,5)} + \frac{953{,}681}{98{,}304}\Omega_1^{(2,6)} - \frac{29{,}393}{36{,}864}\Omega_1^{(2,7)} + \frac{4699}{110{,}592}\Omega_1^{(2,8)} - \frac{13}{9216}\Omega_1^{(2,9)} + \frac{1}{36{,}864}\Omega_1^{(2,10)} + \frac{470{,}041}{18{,}432}\Omega_1^{(4,4)} - \frac{28{,}223}{2304}\Omega_1^{(4,5)} + \frac{5341}{2304}\Omega_1^{(4,6)} - \frac{13}{64}\Omega_1^{(4,7)} + \frac{1}{128}\Omega_1^{(4,8)} + \frac{1}{96}\Omega_1^{(6,6)} \tag{B.15}$$

Appendix C

The bracket integrals for the determination of the viscosity of the multicomponent plasma

$$\left[S_{5/2}^{(0)}(W_1^2)\,\mathbf{W}_1\overset{o}{\mathbf{W}}_1, S_{5/2}^{(0)}(W_2^2)\,\mathbf{W}_2\overset{o}{\mathbf{W}}_2\right]_{12}(M_1M_2)^{-1} = -\frac{80}{3}\Omega_{12}^{(1,1)} + 8\Omega_{12}^{(2,2)} \quad \text{(C.1)}$$

$$\left[S_{5/2}^{(0)}(W_1^2)\,\mathbf{W}_1\overset{o}{\mathbf{W}}_1, S_{5/2}^{(1)}(W_2^2)\,\mathbf{W}_2\overset{o}{\mathbf{W}}_2\right]_{12}(M_1^2M_2)^{-1}$$
$$= -\frac{280}{3}\Omega_{12}^{(1,1)} + \frac{112}{3}\Omega_{12}^{(1,2)} + 28\Omega_{12}^{(2,2)} - 8\Omega_{12}^{(2,3)} \quad \text{(C.2)}$$

$$\left[S_{5/2}^{(0)}(W_1^2)\,\mathbf{W}_1\overset{o}{\mathbf{W}}_1, S_{5/2}^{(2)}(W_2^2)\,\mathbf{W}_2\overset{o}{\mathbf{W}}_2\right]_{12}(M_1^3M_2)^{-1}$$
$$= -210\Omega_{12}^{(1,1)} + 168\Omega_{12}^{(1,2)} - 24\Omega_{12}^{(1,3)} + 63\Omega_{12}^{(2,2)} - 36\Omega_{12}^{(2,3)} + 4\Omega_{12}^{(2,4)} \quad \text{(C.3)}$$

$$\left[S_{5/2}^{(0)}(W_1^2)\,\mathbf{W}_1\overset{o}{\mathbf{W}}_1, S_{5/2}^{(3)}(W_2^2)\,\mathbf{W}_2\overset{o}{\mathbf{W}}_2\right]_{12}(M_1^4M_2)^{-1}$$
$$= -385\Omega_{12}^{(1,1)} + 462\Omega_{12}^{(1,2)} - 132\Omega_{12}^{(1,3)} + \frac{88}{9}\Omega_{12}^{(1,4)} \quad \text{(C.4)}$$
$$+ \frac{231}{2}\Omega_{12}^{(2,2)} - 99\Omega_{12}^{(2,3)} + 22\Omega_{12}^{(2,4)} - \frac{4}{3}\Omega_{12}^{(2,5)}$$

$$\left[S_{5/2}^{(0)}(W_1^2)\,\mathbf{W}_1\overset{o}{\mathbf{W}}_1, S_{5/2}^{(4)}(W_2^2)\,\mathbf{W}_2\overset{o}{\mathbf{W}}_2\right]_{12}(M_1^5M_2)^{-1}$$
$$= -\frac{5005}{8}\Omega_{12}^{(1,1)} + 1001\Omega_{12}^{(1,2)} - 429\Omega_{12}^{(1,3)}$$
$$+ \frac{572}{9}\Omega_{12}^{(1,4)} - \frac{26}{9}\Omega_{12}^{(1,5)} + \frac{3003}{16}\Omega_{12}^{(2,2)} \quad \text{(C.5)}$$
$$- \frac{429}{2}\Omega_{12}^{(2,3)} + \frac{143}{2}\Omega_{12}^{(2,4)} - \frac{26}{3}\Omega_{12}^{(2,5)} + \frac{1}{3}\Omega_{12}^{(2,6)}$$

S. Nguyen-Kuok, *Theory of Low-Temperature Plasma Physics*,
Springer Series on Atomic, Optical, and Plasma Physics 95,
DOI 10.1007/978-3-319-43721-7

$$\left[S^{(1)}_{5/2}(W_1^2)\,\mathbf{W}_1\overset{o}{\mathbf{W}}_1, S^{(1)}_{5/2}(W_2^2)\,\mathbf{W}_2\overset{o}{\mathbf{W}}_2\right]_{12}(M_1^2M_2^2)^{-1}$$
$$= -\frac{1540}{3}\Omega_{12}^{(1,1)} + \frac{784}{3}\Omega_{12}^{(1,2)} - \frac{128}{3}\Omega_{12}^{(1,3)} + \frac{602}{3}\Omega_{12}^{(2,2)} - 56\Omega_{12}^{(2,3)} + 8\Omega_{12}^{(2,4)} - 16\Omega_{12}^{(3,3)} \tag{C.6}$$

$$\left[S^{(1)}_{5/2}(W_1^2)\,\mathbf{W}_1\overset{o}{\mathbf{W}}_1, S^{(2)}_{5/2}(W_2^2)\,\mathbf{W}_2\overset{o}{\mathbf{W}}_2\right]_{12}(M_1^3M_2^2)^{-1} = -1575\Omega_{12}^{(1,1)} + 1218\Omega_{12}^{(1,2)}$$
$$- 276\Omega_{12}^{(1,3)} + 24\Omega_{12}^{(1,4)} + \frac{1365}{2}\Omega_{12}^{(2,2)} - 321\Omega_{12}^{(2,3)} + 50\Omega_{12}^{(2,4)} - 4\Omega_{12}^{(2,5)} - 72\Omega_{12}^{(3,3)} + 16\Omega_{12}^{(3,4)} \tag{C.7}$$

$$\left[S^{(1)}_{5/2}(W_1^2)\,\mathbf{W}_1\overset{o}{\mathbf{W}}_1, S^{(3)}_{5/2}(W_2^2)\,\mathbf{W}_2\overset{o}{\mathbf{W}}_2\right]_{12}(M_1^4M_2^2)^{-1} = -\frac{7315}{2}\Omega_{12}^{(1,1)} + 4004\Omega_{12}^{(1,2)}$$
$$- 1254\Omega_{12}^{(1,3)} + \frac{1496}{9}\Omega_{12}^{(1,4)} - \frac{80}{9}\Omega_{12}^{(1,5)} + \frac{6699}{4}\Omega_{12}^{(2,2)} - 1188\Omega_{12}^{(2,3)} + \frac{770}{3}\Omega_{12}^{(2,4)} - \frac{80}{3}\Omega_{12}^{(2,5)}$$
$$+ \frac{4}{3}\Omega_{12}^{(2,6)} - 198\Omega_{12}^{(3,3)} + 88\Omega_{12}^{(3,4)} - 8\Omega_{12}^{(3,5)} \tag{C.8}$$

$$\left[S^{(1)}_{5/2}(W_1^2)\,\mathbf{W}_1\overset{o}{\mathbf{W}}_1, S^{(4)}_{5/2}(W_2^2)\,\mathbf{W}_2\overset{o}{\mathbf{W}}_2\right]_{12}(M_1^5M_2^2)^{-1} = -\frac{115{,}115}{16}\Omega_{12}^{(1,1)} + \frac{83{,}083}{8}\Omega_{12}^{(1,2)}$$
$$- \frac{8723}{2}\Omega_{12}^{(1,3)} + \frac{7007}{9}\Omega_{12}^{(1,4)} - \frac{611}{9}\Omega_{12}^{(1,5)} + \frac{22}{9}\Omega_{12}^{(1,6)} + \frac{109{,}109}{32}\Omega_{12}^{(2,2)} - \frac{52{,}767}{16}\Omega_{12}^{(2,3)}$$
$$+ \frac{11{,}869}{12}\Omega_{12}^{(2,4)} - \frac{2405}{18}\Omega_{12}^{(2,5)} + \frac{59}{6}\Omega_{12}^{(2,6)} - \frac{1}{3}\Omega_{12}^{(2,7)} - 429\Omega_{12}^{(3,3)} + 286\Omega_{12}^{(3,4)} - 52\Omega_{12}^{(3,5)} + \frac{8}{3}\Omega_{12}^{(3,6)} \tag{C.9}$$

$$\left[S^{(2)}_{5/2}(W_1^2)\,\mathbf{W}_1\overset{o}{\mathbf{W}}_1, S^{(2)}_{5/2}(W_2^2)\,\mathbf{W}_2\overset{o}{\mathbf{W}}_2\right]_{12}(M_1^3M_2^3)^{-1} = -\frac{24{,}255}{4}\Omega_{12}^{(1,1)}$$
$$+ 5670\Omega_{12}^{(1,2)} - 1746\Omega_{12}^{(1,3)} + 216\Omega_{12}^{(1,4)} - 12\Omega_{12}^{(1,5)} + \frac{25{,}137}{8}\Omega_{12}^{(2,2)} - 1755\Omega_{12}^{(2,3)}$$
$$+ 381\Omega_{12}^{(2,4)} - 36\Omega_{12}^{(2,5)} + 2\Omega_{12}^{(2,6)} - 492\Omega_{12}^{(3,3)} + 144\Omega_{12}^{(3,4)} - 16\Omega_{12}^{(3,5)} + 16\Omega_{12}^{(4,4)} \tag{C.10}$$

$$\left[S^{(2)}_{5/2}(W_1^2)\,\mathbf{W}_1\overset{o}{\mathbf{W}}_1, S^{(3)}_{5/2}(W_2^2)\,\mathbf{W}_2\overset{o}{\mathbf{W}}_2\right]_{12}(M_1^4M_2^3)^{-1} = -\frac{135{,}135}{8}\Omega_{12}^{(1,1)}$$
$$+ \frac{79{,}695}{4}\Omega_{12}^{(1,2)} - 7722\Omega_{12}^{(1,3)} + 1320\Omega_{12}^{(1,4)} - 106\Omega_{12}^{(1,5)} + 4\Omega_{12}^{(1,6)} + \frac{155{,}925}{16}\Omega_{12}^{(2,2)}$$
$$- \frac{56{,}727}{8}\Omega_{12}^{(2,3)} + \frac{3795}{2}\Omega_{12}^{(2,4)} - 249\Omega_{12}^{(2,5)} + 17\Omega_{12}^{(2,6)} - \frac{2}{3}\Omega_{12}^{(2,7)} - 1815\Omega_{12}^{(3,3)}$$
$$+ \frac{2398}{3}\Omega_{12}^{(3,4)} - 124\Omega_{12}^{(3,5)} + 8\Omega_{12}^{(3,6)} + 88\Omega_{12}^{(4,4)} - 16\Omega_{12}^{(4,5)} \tag{C.11}$$

$$\begin{aligned}\Big[S_{5/2}^{(2)}(W_1^2)\,\mathbf{W}_1\overset{\circ}{\mathbf{W}}_1, S_{5/2}^{(4)}(W_2^2)\,\mathbf{W}_2\overset{\circ}{\mathbf{W}}_2\Big]_{12}(M_1^5M_2^3)^{-1} &= -\frac{2{,}477{,}475}{64}\Omega_{12}^{(1,1)} + \frac{909{,}909}{16}\Omega_{12}^{(1,2)} \\ &- \frac{433{,}719}{16}\Omega_{12}^{(1,3)} + \frac{11{,}869}{2}\Omega_{12}^{(1,4)} - \frac{2613}{4}\Omega_{12}^{(1,5)} + 37\Omega_{12}^{(1,6)} - \Omega_{12}^{(1,7)} + \frac{3{,}072{,}069}{128}\Omega_{12}^{(2,2)} \\ &- \frac{719{,}433}{32}\Omega_{12}^{(2,3)} + \frac{237{,}237}{32}\Omega_{12}^{(2,4)} - \frac{4901}{4}\Omega_{12}^{(2,5)} + \frac{891}{8}\Omega_{12}^{(2,6)} - \frac{35}{6}\Omega_{12}^{(2,7)} + \frac{1}{6}\Omega_{12}^{(2,8)} \\ &- \frac{9867}{2}\Omega_{12}^{(3,3)} + \frac{9152}{3}\Omega_{12}^{(3,4)} - \frac{1924}{3}\Omega_{12}^{(3,5)} + 64\Omega_{12}^{(3,6)} - \frac{8}{3}\Omega_{12}^{(3,7)} + 286\Omega_{12}^{(4,4)} - 104\Omega_{12}^{(4,5)} \\ &+ 8\Omega_{12}^{(4,6)}\end{aligned} \tag{C.12}$$

$$\begin{aligned}\Big[S_{5/2}^{(3)}(W_1^2)\,\mathbf{W}_1\overset{\circ}{\mathbf{W}}_1, S_{5/2}^{(3)}(W_2^2)\,\mathbf{W}_2\overset{\circ}{\mathbf{W}}_2\Big]_{12}(M_1^4M_2^4)^{-1} &= -\frac{875{,}875}{16}\Omega_{12}^{(1,1)} + \frac{297{,}297}{4}\Omega_{12}^{(1,2)} \\ &- \frac{69{,}267}{2}\Omega_{12}^{(1,3)} + \frac{22{,}264}{3}\Omega_{12}^{(1,4)} - \frac{2425}{3}\Omega_{12}^{(1,5)} + 44\Omega_{12}^{(1,6)} - \frac{32}{27}\Omega_{12}^{(1,7)} + \frac{1{,}168{,}167}{32}\Omega_{12}^{(2,2)} \\ &- \frac{245{,}025}{8}\Omega_{12}^{(2,3)} + \frac{236{,}005}{24}\Omega_{12}^{(2,4)} - 1573\Omega_{12}^{(2,5)} + \frac{859}{6}\Omega_{12}^{(2,6)} - \frac{22}{3}\Omega_{12}^{(2,7)} + \frac{2}{9}\Omega_{12}^{(2,8)} \\ &- \frac{33{,}737}{4}\Omega_{12}^{(3,3)} + \frac{13{,}310}{3}\Omega_{12}^{(3,4)} - \frac{8290}{9}\Omega_{12}^{(3,5)} + 88\Omega_{12}^{(3,6)} - 4\Omega_{12}^{(3,7)} + \frac{5852}{9}\Omega_{12}^{(4,4)} - 176\Omega_{12}^{(4,5)} \\ &+ 16\Omega_{12}^{(4,6)} - \frac{32}{3}\Omega_{12}^{(5,5)}\end{aligned} \tag{C.13}$$

$$\begin{aligned}\Big[S_{5/2}^{(3)}(W_1^2)\,\mathbf{W}_1\overset{\circ}{\mathbf{W}}_1, S_{5/2}^{(4)}(W_2^2)\,\mathbf{W}_2\overset{\circ}{\mathbf{W}}_2\Big]_{12}(M_1^5M_2^4)^{-1} &= -\frac{18{,}293{,}275}{128}\Omega_{12}^{(1,1)} + \frac{14{,}559{,}545}{64}\Omega_{12}^{(1,2)} \\ &- \frac{3{,}990{,}129}{32}\Omega_{12}^{(1,3)} + \frac{1{,}559{,}701}{48}\Omega_{12}^{(1,4)} - \frac{107{,}939}{24}\Omega_{12}^{(1,5)} + \frac{4057}{12}\Omega_{12}^{(1,6)} - \frac{713}{54}\Omega_{12}^{(1,7)} + \frac{7}{27}\Omega_{12}^{(1,8)} \\ &+ \frac{26{,}951{,}925}{256}\Omega_{12}^{(2,2)} - \frac{13{,}505{,}349}{128}\Omega_{12}^{(2,3)} + \frac{7{,}661{,}797}{192}\Omega_{12}^{(2,4)} - \frac{738{,}569}{96}\Omega_{12}^{(2,5)} + \frac{41{,}005}{48}\Omega_{12}^{(2,6)} \\ &- \frac{4175}{72}\Omega_{12}^{(2,7)} + \frac{85}{36}\Omega_{12}^{(2,8)} - \frac{1}{18}\Omega_{12}^{(2,9)} - \frac{221{,}507}{8}\Omega_{12}^{(3,3)} + \frac{655{,}655}{36}\Omega_{12}^{(3,4)} - \frac{41{,}873}{9}\Omega_{12}^{(3,5)} \\ &+ \frac{5386}{9}\Omega_{12}^{(3,6)} - \frac{122}{3}\Omega_{12}^{(3,7)} + \frac{4}{3}\Omega_{12}^{(3,8)} + \frac{23{,}881}{9}\Omega_{12}^{(4,4)} - \frac{9490}{9}\Omega_{12}^{(4,5)} + 148\Omega_{12}^{(4,6)} - 8\Omega_{12}^{(4,7)} \\ &- \frac{208}{3}\Omega_{12}^{(5,5)} + \frac{32}{3}\Omega_{12}^{(5,6)}\end{aligned} \tag{C.14}$$

$$\left[S_{5/2}^{(4)}\left(W_1^2\right)\mathbf{W}_1\overset{\circ}{\mathbf{W}}_1, S_{5/2}^{(4)}\left(W_2^2\right)\mathbf{W}_2\overset{\circ}{\mathbf{W}}_2\right]_{12}\left(M_1^5M_2^5\right)^{-1} = -\frac{428{,}402{,}975}{1024}\Omega_{12}^{(1,1)} + \frac{47{,}562{,}515}{64}\Omega_{12}^{(1,2)}$$
$$-\frac{29{,}713{,}255}{64}\Omega_{12}^{(1,3)} + \frac{6{,}779{,}773}{48}\Omega_{12}^{(1,4)} - \frac{758{,}173}{32}\Omega_{12}^{(1,5)} + \frac{27{,}287}{12}\Omega_{12}^{(1,6)} - \frac{13{,}213}{108}\Omega_{12}^{(1,7)} + \frac{91}{27}\Omega_{12}^{(1,8)}$$
$$-\frac{5}{108}\Omega_{12}^{(1,9)} + \frac{711{,}736{,}025}{2048}\Omega_{12}^{(2,2)} - \frac{50{,}053{,}575}{128}\Omega_{12}^{(2,3)} + \frac{64{,}969{,}619}{384}\Omega_{12}^{(2,4)} - \frac{3{,}630{,}289}{96}\Omega_{12}^{(2,5)}$$
$$+\frac{953{,}681}{192}\Omega_{12}^{(2,6)} - \frac{29{,}393}{72}\Omega_{12}^{(2,7)} + \frac{4699}{216}\Omega_{12}^{(2,8)} - \frac{13}{18}\Omega_{12}^{(2,9)} + \frac{1}{72}\Omega_{12}^{(2,10)} - \frac{1{,}705{,}275}{16}\Omega_{12}^{(3,3)}$$
$$+\frac{2{,}879{,}591}{36}\Omega_{12}^{(3,4)} - \frac{289{,}133}{12}\Omega_{12}^{(3,5)} + \frac{33{,}982}{9}\Omega_{12}^{(3,6)} - \frac{3061}{9}\Omega_{12}^{(3,7)} + \frac{52}{3}\Omega_{12}^{(3,8)} - \frac{4}{9}\Omega_{12}^{(3,9)}$$
$$+\frac{470{,}041}{36}\Omega_{12}^{(4,4)} - \frac{56{,}446}{9}\Omega_{12}^{(4,5)} + \frac{10{,}682}{9}\Omega_{12}^{(4,6)} - 104\Omega_{12}^{(4,7)} + 4\Omega_{12}^{(4,8)} - \frac{5096}{9}\Omega_{12}^{(5,5)}$$
$$+\frac{416}{3}\Omega_{12}^{(5,6)} - \frac{32}{3}\Omega_{12}^{(5,7)} + \frac{16}{3}\Omega_{12}^{(6,6)} \tag{C.15}$$

$$\left[S_{5/2}^{(0)}\left(W_1^2\right)\mathbf{W}_1\overset{\circ}{\mathbf{W}}_1, S_{5/2}^{(0)}\left(W_1^2\right)\mathbf{W}_1\overset{\circ}{\mathbf{W}}_1\right]_{12} = \frac{80}{3}\left(M_1M_2\right)\Omega_{12}^{(1,1)} + 8\left(M_2^2\right)\Omega_{12}^{(2,2)} \tag{C.16}$$

$$\left[S_{5/2}^{(0)}\left(W_1^2\right)\mathbf{W}_1\overset{\circ}{\mathbf{W}}_1, S_{5/2}^{(1)}\left(W_1^2\right)\mathbf{W}_1\overset{\circ}{\mathbf{W}}_1\right]_{12}$$
$$= \frac{280}{3}\left(M_1M_2^2\right)\Omega_{12}^{(1,1)} - \frac{112}{3}\left(M_1M_2^2\right)\Omega_{12}^{(1,2)} + 28\left(M_2^3\right)\Omega_{12}^{(2,2)} - 8\left(M_2^3\right)\Omega_{12}^{(2,3)} \tag{C.17}$$

$$\left[S_{5/2}^{(0)}\left(W_1^2\right)\mathbf{W}_1\overset{\circ}{\mathbf{W}}_1, S_{5/2}^{(2)}\left(W_1^2\right)\mathbf{W}_1\overset{\circ}{\mathbf{W}}_1\right]_{12} = 210\left(M_1M_2^3\right)\Omega_{12}^{(1,1)} - 168\left(M_1M_2^3\right)\Omega_{12}^{(1,2)}$$
$$+24\left(M_1M_2^3\right)\Omega_{12}^{(1,3)} + 63\left(M_2^4\right)\Omega_{12}^{(2,2)} - 36\left(M_2^4\right)\Omega_{12}^{(2,3)} + 4\left(M_2^4\right)\Omega_{12}^{(2,4)} \tag{C.18}$$

$$\left[S_{5/2}^{(0)}\left(W_1^2\right)\mathbf{W}_1\overset{\circ}{\mathbf{W}}_1, S_{5/2}^{(3)}\left(W_1^2\right)\mathbf{W}_1\overset{\circ}{\mathbf{W}}_1\right]_{12} = 385\left(M_1M_2^4\right)\Omega_{12}^{(1,1)} - 462\left(M_1M_2^4\right)\Omega_{12}^{(1,2)}$$
$$+132\left(M_1M_2^4\right)\Omega_{12}^{(1,3)} - \frac{88}{9}\left(M_1M_2^4\right)\Omega_{12}^{(1,4)} + \frac{231}{2}\left(M_2^5\right)\Omega_{12}^{(2,2)} - 99\left(M_2^5\right)\Omega_{12}^{(2,3)}$$
$$+22\left(M_2^5\right)\Omega_{12}^{(2,4)} - \frac{4}{3}\left(M_2^5\right)\Omega_{12}^{(2,5)} \tag{C.19}$$

$$\left[S_{5/2}^{(0)}\left(W_1^2\right)\mathbf{W}_1\overset{\circ}{\mathbf{W}}_1, S_{5/2}^{(4)}\left(W_1^2\right)\mathbf{W}_1\overset{\circ}{\mathbf{W}}_1\right]_{12} = \frac{5005}{8}\left(M_1M_2^5\right)\Omega_{12}^{(1,1)} - 1001\left(M_1M_2^5\right)\Omega_{12}^{(1,2)}$$
$$+429\left(M_1M_2^5\right)\Omega_{12}^{(1,3)} - \frac{572}{9}\left(M_1M_2^5\right)\Omega_{12}^{(1,4)} + \frac{26}{9}\left(M_1M_2^5\right)\Omega_{12}^{(1,5)} + \frac{3003}{16}\left(M_2^6\right)\Omega_{12}^{(2,2)}$$
$$-\frac{429}{2}\left(M_2^6\right)\Omega_{12}^{(2,3)} + \frac{143}{2}\left(M_2^6\right)\Omega_{12}^{(2,4)} - \frac{26}{3}\left(M_2^6\right)\Omega_{12}^{(2,5)} + \frac{1}{3}\left(M_2^6\right)\Omega_{12}^{(2,6)} \tag{C.20}$$

$$\left[S_{5/2}^{(1)}(W_1^2)\,\mathbf{W}_1\overset{\circ}{\mathbf{W}}_1, S_{5/2}^{(1)}(W_1^2)\,\mathbf{W}_1\overset{\circ}{\mathbf{W}}_1\right]_{12} = \frac{1540}{3}\left(\frac{4}{11}M_1^3M_2 + \frac{7}{11}M_1M_2^3\right)\Omega_{12}^{(1,1)}$$
$$-\frac{784}{3}(M_1M_2^3)\Omega_{12}^{(1,2)} + \frac{128}{3}(M_1M_2^3)\Omega_{12}^{(1,3)} + \frac{602}{3}\left(\frac{22}{43}M_1^2M_2^2 + \frac{21}{43}M_2^4\right)\Omega_{12}^{(2,2)} \tag{C.21}$$
$$-56(M_2^4)\Omega_{12}^{(2,3)} + 8(M_2^4)\Omega_{12}^{(2,4)} + 16(M_1M_2^3)\Omega_{12}^{(3,3)}$$

$$\left[S_{5/2}^{(1)}(W_1^2)\,\mathbf{W}_1\overset{\circ}{\mathbf{W}}_1, S_{5/2}^{(2)}(W_1^2)\,\mathbf{W}_1\overset{\circ}{\mathbf{W}}_1\right]_{12} = 1575\left(\frac{8}{15}M_1^3M_2^2 + \frac{7}{15}M_1M_2^4\right)\Omega_{12}^{(1,1)}$$
$$-1218\left(\frac{8}{29}M_1^3M_2^2 + \frac{21}{29}M_1M_2^4\right)\Omega_{12}^{(1,2)} + 276(M_1M_2^4)\Omega_{12}^{(1,3)} - 24(M_1M_2^4)\Omega_{12}^{(1,4)}$$
$$+\frac{1365}{2}\left(\frac{44}{65}M_1^2M_2^3 + \frac{21}{65}M_2^5\right)\Omega_{12}^{(2,2)} - 321\left(\frac{44}{107}M_1^2M_2^3 + \frac{63}{107}M_2^5\right)\Omega_{12}^{(2,3)}$$
$$+50(M_2^5)\Omega_{12}^{(2,4)} - 4(M_2^5)\Omega_{12}^{(2,5)} + 72(M_1M_2^4)\Omega_{12}^{(3,3)} - 16(M_1M_2^4)\Omega_{12}^{(3,4)} \tag{C.22}$$

$$\left[S_{5/2}^{(1)}(W_1^2)\,\mathbf{W}_1\overset{\circ}{\mathbf{W}}_1, S_{5/2}^{(3)}(W_1^2)\,\mathbf{W}_1\overset{\circ}{\mathbf{W}}_1\right]_{12} = \frac{7315}{2}\left(\frac{12}{19}M_1^3M_2^3 + \frac{7}{19}M_1M_2^5\right)\Omega_{12}^{(1,1)}$$
$$-4004\left(\frac{6}{13}M_1^3M_2^3 + \frac{7}{13}M_1M_2^5\right)\Omega_{12}^{(1,2)} + 1254\left(\frac{4}{19}M_1^3M_2^3 + \frac{15}{19}M_1M_2^5\right)\Omega_{12}^{(1,3)}$$
$$-\frac{1496}{9}(M_1M_2^5)\Omega_{12}^{(1,4)} + \frac{80}{9}(M_1M_2^5)\Omega_{12}^{(1,5)} + \frac{6699}{4}\left(\frac{22}{29}M_1^2M_2^4 + \frac{7}{29}M_2^6\right)\Omega_{12}^{(2,2)}$$
$$-1188\left(\frac{11}{18}M_1^2M_2^4 + \frac{7}{18}M_2^6\right)\Omega_{12}^{(2,3)} + \frac{770}{3}\left(\frac{11}{35}M_1^2M_2^4 + \frac{24}{35}M_2^6\right)\Omega_{12}^{(2,4)} - \frac{80}{3}(M_2^6)\Omega_{12}^{(2,5)}$$
$$+\frac{4}{3}(M_2^6)\Omega_{12}^{(2,6)} + 198(M_1M_2^5)\Omega_{12}^{(3,3)} - 88(M_1M_2^5)\Omega_{12}^{(3,4)} + 8(M_1M_2^5)\Omega_{12}^{(3,5)} \tag{C.23}$$

$$\left[S_{5/2}^{(1)}(W_1^2)\,\mathbf{W}_1\overset{\circ}{\mathbf{W}}_1, S_{5/2}^{(4)}(W_1^2)\,\mathbf{W}_1\overset{\circ}{\mathbf{W}}_1\right]_{12} = \frac{115{,}115}{16}\left(\frac{16}{23}M_1^3M_2^4 + \frac{7}{23}M_1M_2^6\right)\Omega_{12}^{(1,1)}$$
$$-\frac{83{,}083}{8}\left(\frac{48}{83}M_1^3M_2^4 + \frac{35}{83}M_1M_2^6\right)\Omega_{12}^{(1,2)} + \frac{8723}{2}\left(\frac{24}{61}M_1^3M_2^4 + \frac{37}{61}M_1M_2^6\right)\Omega_{12}^{(1,3)}$$
$$-\frac{7007}{9}\left(\frac{8}{49}M_1^3M_2^4 + \frac{41}{49}M_1M_2^6\right)\Omega_{12}^{(1,4)} + \frac{611}{9}(M_1M_2^6)\Omega_{12}^{(1,5)} - \frac{22}{9}(M_1M_2^6)\Omega_{12}^{(1,6)}$$
$$+\frac{109{,}109}{32}\left(\frac{88}{109}M_1^2M_2^5 + \frac{21}{109}M_2^7\right)\Omega_{12}^{(2,2)} - \frac{52{,}767}{16}\left(\frac{88}{123}M_1^2M_2^5 + \frac{35}{123}M_2^7\right)\Omega_{12}^{(2,3)}$$
$$+\frac{11{,}869}{12}\left(\frac{44}{83}M_1^2M_2^5 + \frac{39}{83}M_2^7\right)\Omega_{12}^{(2,4)} - \frac{2405}{18}\left(\frac{44}{185}M_1^2M_2^5 + \frac{141}{185}M_2^7\right)\Omega_{12}^{(2,5)}$$
$$+\frac{59}{6}(M_2^7)\Omega_{12}^{(2,6)} - \frac{1}{3}(M_2^7)\Omega_{12}^{(2,7)} + 429(M_1M_2^6)\Omega_{12}^{(3,3)} - 286(M_1M_2^6)\Omega_{12}^{(3,4)}$$
$$+52(M_1M_2^6)\Omega_{12}^{(3,5)} - \frac{8}{3}(M_1M_2^6)\Omega_{12}^{(3,6)} \tag{C.24}$$

$$\left[S_{5/2}^{(2)}(W_1^2)\,\mathbf{W}_1\overset{\circ}{}\mathbf{W}_1, S_{5/2}^{(2)}(W_1^2)\,\mathbf{W}_1\overset{\circ}{}\mathbf{W}_1\right]_{12} = \frac{24{,}255}{4}\left(\frac{8}{77}M_1^5M_2 + \frac{48}{77}M_1^3M_2^3 + \frac{21}{77}M_1M_2^5\right)\Omega_{12}^{(1,1)}$$
$$- 5670\left(\frac{8}{15}M_1^3M_2^3 + \frac{7}{15}M_1M_2^5\right)\Omega_{12}^{(1,2)} + 1746\left(\frac{28}{97}M_1^3M_2^3 + \frac{69}{97}M_1M_2^5\right)\Omega_{12}^{(1,3)}$$
$$- 216(M_1M_2^5)\Omega_{12}^{(1,4)} + 12(M_1M_2^5)\Omega_{12}^{(1,5)} + \frac{25{,}137}{8}\left(\frac{24}{133}M_1^4M_2^2 + \frac{88}{133}M_1^2M_2^4 + \frac{21}{133}M_2^6\right)\Omega_{12}^{(2,2)}$$
$$- 1755\left(\frac{44}{65}M_1^2M_2^4 + \frac{21}{65}M_2^6\right)\Omega_{12}^{(2,3)} + 381\left(\frac{52}{127}M_1^2M_2^4 + \frac{75}{127}M_2^6\right)\Omega_{12}^{(2,4)} - 36(M_2^6)\Omega_{12}^{(2,5)}$$
$$+ 2(M_2^6)\Omega_{12}^{(2,6)} + 492\left(\frac{14}{41}M_1^3M_2^3 + \frac{27}{41}M_1M_2^5\right)\Omega_{12}^{(3,3)} - 144(M_1M_2^5)\Omega_{12}^{(3,4)} +$$
$$+ 16(M_1M_2^5)\Omega_{12}^{(3,5)} + 16(M_1^2M_2^4)\Omega_{12}^{(4,4)} \tag{C.25}$$

$$\left[S_{5/2}^{(2)}(W_1^2)\,\mathbf{W}_1\overset{\circ}{}\mathbf{W}_1, S_{5/2}^{(3)}(W_1^2)\,\mathbf{W}_1\overset{\circ}{}\mathbf{W}_1\right]_{12} = \frac{135{,}135}{8}\left(\frac{8}{39}M_1^5M_2^2 + \frac{24}{39}M_1^3M_2^4\right.$$
$$\left. + \frac{7}{39}M_1M_2^6\right)\Omega_{12}^{(1,1)} - \frac{79{,}695}{4}\left(\frac{8}{115}M_1^5M_2^2 + \frac{72}{115}M_1^3M_2^4 + \frac{35}{115}M_1M_2^6\right)\Omega_{12}^{(1,2)}$$
$$+ 7722\left(\frac{20}{39}M_1^3M_2^4 + \frac{19}{39}M_1M_2^6\right)\Omega_{12}^{(1,3)} - 1320\left(\frac{4}{15}M_1^3M_2^4 + \frac{11}{15}M_1M_2^6\right)\Omega_{12}^{(1,4)}$$
$$+ 106(M_1M_2^6)\Omega_{12}^{(1,5)} - 4(M_1M_2^6)\Omega_{12}^{(1,6)} + \frac{155{,}925}{16}\left(\frac{24}{75}M_1^4M_2^3 + \frac{44}{75}M_1^2M_2^5 + \frac{7}{75}M_2^7\right)\Omega_{12}^{(2,2)}$$
$$- \frac{56{,}727}{8}\left(\frac{24}{191}M_1^4M_2^3 + \frac{132}{191}M_1^2M_2^5 + \frac{35}{191}M_2^7\right)\Omega_{12}^{(2,3)} + \frac{3795}{2}\left(\frac{74}{115}M_1^2M_2^5\right.$$
$$\left. + \frac{41}{115}M_2^7\right)\Omega_{12}^{(2,4)} - 249\left(\frac{30}{83}M_1^2M_2^5 + \frac{53}{83}M_2^7\right)\Omega_{12}^{(2,5)} + 17(M_2^7)\Omega_{12}^{(2,6)} - \frac{2}{3}(M_2^7)\Omega_{12}^{(2,7)}$$
$$+ 1815\left(\frac{924}{1815}M_1^3M_2^4 + \frac{891}{1815}M_1M_2^6\right)\Omega_{12}^{(3,3)} - \frac{2398}{3}\left(\frac{28}{109}M_1^3M_2^4 + \frac{81}{109}M_1M_2^6\right)\Omega_{12}^{(3,4)}$$
$$+ 124(M_1M_2^6)\Omega_{12}^{(3,5)} - 8(M_1M_2^6)\Omega_{12}^{(3,6)} + 88(M_1^2M_2^5)\Omega_{12}^{(4,4)} - 16(M_1^2M_2^5)\Omega_{12}^{(4,5)} \tag{C.26}$$

$$\left[S_{5/2}^{(2)}(W_1^2)\,\mathbf{W}_1\overset{\circ}{}\mathbf{W}_1, S_{5/2}^{(4)}(W_1^2)\,\mathbf{W}_1\overset{\circ}{}\mathbf{W}_1\right]_{12}$$
$$= \frac{2{,}477{,}475}{64}\left(\frac{16}{55}M_1^5M_2^3 + \frac{32}{55}M_1^3M_2^5 + \frac{7}{55}M_1M_2^7\right)\Omega_{12}^{(1,1)}$$
$$- \frac{909{,}909}{16}\left(\frac{16}{101}M_1^5M_2^3 + \frac{64}{101}M_1^3M_2^5 + \frac{21}{101}M_1M_2^7\right)\Omega_{12}^{(1,2)}$$
$$+ \frac{433{,}719}{16}\left(\frac{16}{337}M_1^5M_2^3 + \frac{208}{337}M_1^3M_2^5 + \frac{113}{337}M_1M_2^7\right)\Omega_{12}^{(1,3)}$$
$$- \frac{11{,}869}{2}\left(\frac{40}{83}M_1^3M_2^5 + \frac{43}{83}M_1M_2^7\right)\Omega_{12}^{(1,4)} + \frac{2613}{4}\left(\frac{16}{67}M_1^3M_2^5 + \frac{51}{67}M_1M_2^7\right)\Omega_{12}^{(1,5)}$$
$$- 37(M_1M_2^7)\Omega_{12}^{(1,6)} + (M_1M_2^7)\Omega_{12}^{(1,7)} + \frac{3{,}072{,}069}{128}\left(\frac{144}{341}M_1^4M_2^4 + \frac{176}{341}M_1^2M_2^6 + \frac{21}{341}M_2^8\right)\Omega_{12}^{(2,2)}$$

$$
\begin{aligned}
&-\frac{719{,}433}{32}\left(\frac{144}{559}M_1^4M_2^4+\frac{352}{559}M_1^2M_2^6+\frac{63}{559}M_2^8\right)\Omega_{12}^{(2,3)}\\
&+\frac{237{,}237}{32}\left(\frac{48}{553}M_1^4M_2^4+\frac{384}{553}M_1^2M_2^6+\frac{121}{553}M_2^8\right)\Omega_{12}^{(2,4)}\\
&-\frac{4901}{4}\left(\frac{224}{377}M_1^2M_2^6+\frac{153}{377}M_2^8\right)\Omega_{12}^{(2,5)}+\frac{891}{8}\left(\frac{272}{891}M_1^2M_2^6+\frac{619}{891}M_2^8\right)\Omega_{12}^{(2,6)}\\
&-\frac{35}{6}\left(M_2^8\right)\Omega_{12}^{(2,7)}+\frac{1}{6}\left(M_2^8\right)\Omega_{12}^{(2,8)}+\frac{9867}{2}\left(\frac{14}{23}M_1^3M_2^5+\frac{9}{23}M_1M_2^7\right)\Omega_{12}^{(3,3)}\\
&-\frac{9152}{3}\left(\frac{7}{16}M_1^3M_2^5+\frac{9}{16}M_1M_2^7\right)\Omega_{12}^{(3,4)}+\frac{1924}{3}\left(\frac{7}{37}M_1^3M_2^5+\frac{30}{37}M_1M_2^7\right)\Omega_{12}^{(3,5)}\\
&-64\left(M_1M_2^7\right)\Omega_{12}^{(3,6)}+\frac{8}{3}\left(M_1M_2^7\right)\Omega_{12}^{(3,7)}\\
&+286\left(M_1^2M_2^6\right)\Omega_{12}^{(4,4)}-104\left(M_1^2M_2^6\right)\Omega_{12}^{(4,5)}+8\left(M_1^2M_2^6\right)\Omega_{12}^{(4,6)}
\end{aligned}
\tag{C.27}
$$

$$
\begin{aligned}
&\left[S_{5/2}^{(3)}(W_1^2)\,\mathbf{W}_1\overset{\circ}{\mathbf{W}}_1,\,S_{5/2}^{(3)}(W_1^2)\,\mathbf{W}_1\overset{\circ}{\mathbf{W}}_1\right]_{12}\\
&=\frac{875{,}875}{16}\left(\frac{64}{2275}M_1^7M_2+\frac{792}{2275}M_1^5M_2^3+\frac{1188}{2275}M_1^3M_2^5+\frac{231}{2275}M_1M_2^7\right)\Omega_{12}^{(1,1)}\\
&-\frac{297{,}297}{4}\left(\frac{8}{39}M_1^5M_2^3+\frac{24}{39}M_1^3M_2^5+\frac{7}{39}M_1M_2^7\right)\Omega_{12}^{(1,2)}\\
&+\frac{69{,}267}{2}\left(\frac{152}{2099}M_1^5M_2^3+\frac{1320}{2099}M_1^3M_2^5+\frac{627}{2099}M_1M_2^7\right)\Omega_{12}^{(1,3)}\\
&-\frac{22{,}264}{3}\left(\frac{12}{23}M_1^3M_2^5+\frac{11}{23}M_1M_2^7\right)\Omega_{12}^{(1,4)}+\frac{2425}{3}\left(\frac{676}{2425}M_1^3M_2^5+\frac{1749}{2425}M_1M_2^7\right)\Omega_{12}^{(1,5)}\\
&-44\left(M_1M_2^7\right)\Omega_{12}^{(1,6)}+\frac{32}{27}\left(M_1M_2^7\right)\Omega_{12}^{(1,7)}\\
&+\frac{1{,}168{,}167}{32}\left(\frac{272}{5057}M_1^6M_2^2+\frac{2376}{5057}M_1^4M_2^4+\frac{2178}{5057}M_1^2M_2^6+\frac{231}{5057}M_2^8\right)\Omega_{12}^{(2,2)}\\
&-\frac{245{,}025}{8}\left(\frac{24}{75}M_1^4M_2^4+\frac{44}{75}M_1^2M_2^6+\frac{7}{75}M_2^8\right)\Omega_{12}^{(2,3)}\\
&+\frac{236{,}005}{24}\left(\frac{2744}{21{,}455}M_1^4M_2^4+\frac{14{,}652}{21{,}455}M_1^2M_2^6+\frac{4059}{21{,}455}M_2^8\right)\Omega_{12}^{(2,4)}\\
&-1573\left(\frac{90}{143}M_1^2M_2^6+\frac{53}{143}M_2^8\right)\Omega_{12}^{(2,5)}+\frac{859}{6}\left(\frac{298}{859}M_1^2M_2^6+\frac{561}{859}M_2^8\right)\Omega_{12}^{(2,6)}-\frac{22}{3}M_2^8\Omega_{12}^{(2,7)}\\
&+\frac{2}{9}M_2^8\Omega_{12}^{(2,8)}+\frac{33{,}737}{4}\left(\frac{328}{3067}M_1^5M_2^3+\frac{1848}{3067}M_1^3M_2^5+\frac{891}{3067}M_1M_2^7\right)\Omega_{12}^{(3,3)}\\
&-\frac{13{,}310}{3}\left(\frac{28}{55}M_1^3M_2^5+\frac{27}{55}M_1M_2^7\right)\Omega_{12}^{(3,4)}+\frac{8290}{9}\left(\frac{1076}{4145}M_1^3M_2^5+\frac{3069}{4145}M_1M_2^7\right)\Omega_{12}^{(3,5)}\\
&-88\left(M_1M_2^7\right)\Omega_{12}^{(3,6)}+4\left(M_1M_2^7\right)\Omega_{12}^{(3,7)}+\frac{5852}{9}\left(\frac{34}{133}M_1^4M_2^4+\frac{99}{133}M_1^2M_2^6\right)\Omega_{12}^{(4,4)}\\
&-176\left(M_1^2M_2^6\right)\Omega_{12}^{(4,5)}+16\left(M_1^2M_2^6\right)\Omega_{12}^{(4,6)}+\frac{32}{3}\left(M_1^3M_2^5\right)\Omega_{12}^{(5,5)}
\end{aligned}
\tag{C.28}
$$

$$
\begin{aligned}
&\left[S_{5/2}^{(3)}(W_1^2)\,\mathbf{W}_1\overset{\circ}{\mathbf{W}}_1, S_{5/2}^{(4)}(W_1^2)\,\mathbf{W}_1\overset{\circ}{\mathbf{W}}_1\right]_{12} \\
&= \frac{18{,}293{,}275}{128}\left(\frac{256}{3655}M_1^7M_2^2 + \frac{1584}{3655}M_1^5M_2^4 + \frac{1584}{3655}M_1^3M_2^6 + \frac{231}{3655}M_1M_2^8\right)\Omega_{12}^{(1,1)} \\
&- \frac{14{,}559{,}545}{64}\left(\frac{256}{14{,}545}M_1^7M_2^2 + \frac{4752}{14{,}545}M_1^5M_2^4 + \frac{7920}{14{,}545}M_1^3M_2^6 + \frac{1617}{14{,}545}M_1M_2^8\right)\Omega_{12}^{(1,2)} \\
&+ \frac{3{,}990{,}129}{32}\left(\frac{1744}{9301}M_1^5M_2^4 + \frac{5808}{9301}M_1^3M_2^6 + \frac{1749}{9301}M_1M_2^8\right)\Omega_{12}^{(1,3)} \\
&- \frac{1{,}559{,}701}{48}\left(\frac{688}{10{,}907}M_1^5M_2^4 + \frac{6864}{10{,}907}M_1^3M_2^6 + \frac{3355}{10{,}907}M_1M_2^8\right)\Omega_{12}^{(1,4)} \\
&+ \frac{107{,}939}{24}\left(\frac{4288}{8303}M_1^3M_2^6 + \frac{4015}{8303}M_1M_2^8\right)\Omega_{12}^{(1,5)} - \frac{4057}{12}\left(\frac{1120}{4057}M_1^3M_2^6 + \frac{2937}{4057}M_1M_2^8\right)\Omega_{12}^{(1,6)} \\
&+ \frac{713}{54}(M_1M_2^8)\Omega_{12}^{(1,7)} - \frac{7}{27}(M_1M_2^8)\Omega_{12}^{(1,8)} \\
&+ \frac{26{,}951{,}925}{256}\left(\frac{1088}{8975}M_1^6M_2^3 + \frac{4752}{8975}M_1^4M_2^5 + \frac{2904}{8975}M_1^2M_2^7 + \frac{231}{8975}M_2^9\right)\Omega_{12}^{(2,2)} \\
&- \frac{13{,}505{,}349}{128}\left(\frac{1088}{31{,}481}M_1^6M_2^3 + \frac{14{,}256}{31{,}481}M_1^4M_2^5 + \frac{14{,}520}{31{,}481}M_1^2M_2^7 + \frac{1617}{31{,}481}M_2^9\right)\Omega_{12}^{(2,3)} \\
&+ \frac{7{,}661{,}797}{192}\left(\frac{15{,}728}{53{,}579}M_1^4M_2^5 + \frac{32{,}208}{53{,}579}M_1^2M_2^7 + \frac{5643}{53{,}579}M_2^9\right)\Omega_{12}^{(2,4)} \\
&- \frac{738{,}569}{96}\left(\frac{6224}{56{,}813}M_1^4M_2^5 + \frac{38{,}544}{56{,}813}M_1^2M_2^7 + \frac{12{,}045}{56{,}813}M_2^9\right)\Omega_{12}^{(2,5)} \\
&+ \frac{41{,}005}{48}\left(\frac{24{,}472}{41{,}005}M_1^2M_2^7 + \frac{16{,}533}{41{,}005}M_2^9\right)\Omega_{12}^{(2,6)} \\
&- \frac{4175}{72}\left(\frac{1304}{4175}M_1^2M_2^7 + \frac{2871}{4175}M_2^9\right)\Omega_{12}^{(2,7)} + \frac{85}{36}(M_2^9)\Omega_{12}^{(2,8)} - \frac{1}{18}(M_2^9)\Omega_{12}^{(2,9)} \\
&+ \frac{221{,}507}{8}\left(\frac{328}{1549}M_1^5M_2^4 + \frac{924}{1549}M_1^3M_2^6 + \frac{297}{1549}M_1M_2^8\right)\Omega_{12}^{(3,3)} \\
&- \frac{655{,}655}{36}\left(\frac{328}{4585}M_1^5M_2^4 + \frac{2772}{4585}M_1^3M_2^6 + \frac{1485}{4585}M_1M_2^8\right)\Omega_{12}^{(3,4)} \\
&+ \frac{41{,}873}{9}\left(\frac{1538}{3221}M_1^3M_2^6 + \frac{1683}{3221}M_1M_2^8\right)\Omega_{12}^{(3,5)} - \frac{5386}{9}\left(\frac{614}{2693}M_1^3M_2^6 + \frac{2079}{2693}M_1M_2^8\right)\Omega_{12}^{(3,6)} \\
&+ \frac{122}{3}(M_1M_2^8)\Omega_{12}^{(3,7)} - \frac{4}{3}(M_1M_2^8)\Omega_{12}^{(3,8)} + \frac{23{,}881}{9}\left(\frac{68}{167}M_1^4M_2^5 + \frac{99}{167}M_1^2M_2^7\right)\Omega_{12}^{(4,4)} \\
&- \frac{9490}{9}\left(\frac{68}{365}M_1^4M_2^5 + \frac{297}{365}M_1^2M_2^7\right)\Omega_{12}^{(4,5)} + 148(M_1^2M_2^7)\Omega_{12}^{(4,6)} \\
&- 8(M_1^2M_2^7)\Omega_{12}^{(4,7)} + \frac{208}{3}(M_1^3M_2^6)\Omega_{12}^{(5,5)} - \frac{32}{3}(M_1^3M_2^6)\Omega_{12}^{(5,6)}
\end{aligned}
\tag{C.29}
$$

$$
\begin{aligned}
&\left[S^{(4)}_{5/2}(W_1^2)\,\mathbf{W}_1\overset{\circ}{\mathbf{W}}_1, S^{(4)}_{5/2}(W_1^2)\,\mathbf{W}_1\overset{\circ}{\mathbf{W}}_1\right]_{12} = \frac{428{,}402{,}975}{1024}\\
&\quad\times\left(\begin{array}{l}\frac{640}{85{,}595}M_1^9M_2+\frac{13{,}312}{85{,}595}M_1^7M_2^3\\ +\frac{41{,}184}{85{,}595}M_1^5M_2^5+\frac{27{,}456}{85{,}595}M_1^3M_2^7+\frac{3003}{85{,}595}M_1M_2^9\end{array}\right)\Omega_{12}^{(1,1)}\\
&\quad-\frac{47{,}562{,}515}{64}\left(\frac{256}{3655}M_1^7M_2^3+\frac{1584}{3655}M_1^5M_2^5+\frac{1584}{3655}M_1^3M_2^7+\frac{231}{3655}M_1M_2^9\right)\Omega_{12}^{(1,2)}\\
&\quad+\frac{29{,}713{,}255}{64}\left(\frac{3776}{207{,}785}M_1^7M_2^3+\frac{68{,}016}{207{,}785}M_1^5M_2^5+\frac{113{,}256}{207{,}785}M_1^3M_2^7+\frac{22{,}737}{207{,}785}M_1M_2^9\right)\Omega_{12}^{(1,3)}\\
&\quad-\frac{6{,}779{,}773}{48}\left(\frac{688}{3647}M_1^5M_2^5+\frac{2288}{3647}M_1^3M_2^7+\frac{671}{3647}M_1M_2^9\right)\Omega_{12}^{(1,4)}\\
&\quad+\frac{758{,}173}{32}\left(\frac{11{,}280}{174{,}963}M_1^5M_2^5+\frac{111{,}488}{174{,}963}M_1^3M_2^7+\frac{52{,}195}{174{,}963}M_1M_2^9\right)\Omega_{12}^{(1,5)}\\
&\quad-\frac{27{,}287}{12}\left(\frac{1120}{2099}M_1^3M_2^7+\frac{979}{2099}M_1M_2^9\right)\Omega_{12}^{(1,6)}\\
&\quad+\frac{13{,}213}{108}\left(\begin{array}{l}\frac{3944}{13{,}213}M_1^3M_2^7+\\ \frac{9269}{13{,}213}M_1M_2^9\end{array}\right)\Omega_{12}^{(1,7)}-\frac{91}{27}(M_1M_2^9)\Omega_{12}^{(1,8)}+\frac{5}{108}(M_1M_2^9)\Omega_{12}^{(1,9)}\\
&\quad+\frac{711{,}736{,}025}{2048}\left(\begin{array}{l}\frac{10{,}624}{711{,}025}M_1^8M_2^2+\frac{169{,}728}{711{,}025}M_1^6M_2^4+\frac{370{,}656}{711{,}025}M_1^4M_2^6+\\ \qquad+\frac{151{,}008}{711{,}025}M_1^2M_2^8+\frac{9009}{711{,}025}M_2^{10}\end{array}\right)\Omega_{12}^{(2,2)}\\
&\quad-\frac{50{,}053{,}575}{128}\left(\frac{1088}{8975}M_1^6M_2^4+\frac{4752}{8975}M_1^4M_2^6+\frac{2904}{8975}M_1^2M_2^8+\frac{231}{8975}M_2^{10}\right)\Omega_{12}^{(2,3)}\\
&\quad+\frac{64{,}969{,}619}{384}\left(\frac{16{,}064}{454{,}333}M_1^6M_2^4+\frac{204{,}464}{454{,}333}M_1^4M_2^6+\frac{209{,}352}{454{,}333}M_1^2M_2^8+\frac{24{,}453}{454{,}333}M_2^{10}\right)\Omega_{12}^{(2,4)}\\
&\quad-\frac{3{,}630{,}289}{96}\left(\frac{6224}{21{,}481}M_1^4M_2^6+\frac{12{,}848}{21{,}481}M_1^2M_2^8+\frac{2409}{21{,}481}M_2^{10}\right)\Omega_{12}^{(2,5)}\\
&\quad+\frac{953{,}681}{192}\left(\frac{102{,}480}{953{,}681}M_1^4M_2^6+\frac{636{,}272}{953{,}681}M_1^2M_2^8+\frac{214{,}929}{953{,}681}M_2^{10}\right)\Omega_{12}^{(2,6)}\\
&\quad-\frac{29{,}393}{72}\left(\frac{1304}{2261}M_1^2M_2^8+\frac{957}{2261}M_2^{10}\right)\Omega_{12}^{(2,7)}+\frac{4699}{216}\left(\frac{1384}{4699}M_1^2M_2^8+\frac{3315}{4699}M_2^{10}\right)\Omega_{12}^{(2,8)}\\
&\quad-\frac{13}{18}(M_2^{10})\Omega_{12}^{(2,9)}+\frac{1}{72}(M_2^{10})\Omega_{12}^{(2,10)}+\frac{1{,}705{,}275}{16}\left(\begin{array}{l}\frac{368}{11{,}925}M_1^7M_2^3+\frac{4264}{11{,}925}M_1^5M_2^5+\\ \frac{6006}{11{,}925}M_1^3M_2^7+\frac{1287}{11{,}925}M_1M_2^9\end{array}\right)\Omega_{12}^{(3,3)}\\
&\quad-\frac{2{,}879{,}591}{36}\left(\frac{328}{1549}M_1^5M_2^5+\frac{924}{1549}M_1^3M_2^7+\frac{297}{1549}M_1M_2^9\right)\Omega_{12}^{(3,4)}\\
&\quad+\frac{289{,}133}{12}\left(\frac{4856}{66{,}723}M_1^5M_2^5+\frac{39{,}988}{66{,}723}M_1^3M_2^7+\frac{21{,}879}{66{,}723}M_1M_2^9\right)\Omega_{12}^{(3,5)}\\
&\quad-\frac{33{,}982}{9}\left(\frac{614}{1307}M_1^3M_2^7+\frac{693}{1307}M_1M_2^9\right)\Omega_{12}^{(3,6)}+\frac{3061}{9}\left(\frac{682}{3061}M_1^3M_2^7+\frac{2379}{3061}M_1M_2^9\right)\Omega_{12}^{(3,7)}\\
&\quad-\frac{52}{3}(M_1M_2^9)\Omega_{12}^{(3,8)}+\frac{4}{9}(M_1M_2^9)\Omega_{12}^{(3,9)}+\\
&\quad+\frac{470{,}041}{36}\left(\frac{232}{3287}M_1^6M_2^4+\frac{1768}{3287}M_1^4M_2^6+\frac{1287}{3287}M_1^2M_2^8\right)\Omega_{12}^{(4,4)}\\
&\quad-\frac{56{,}446}{9}\left(\frac{68}{167}M_1^4M_2^6+\frac{99}{167}M_1^2M_2^8\right)\Omega_{12}^{(4,5)}+\frac{10{,}682}{9}\left(\frac{1012}{5341}M_1^4M_2^6+\frac{4329}{5341}M_1^2M_2^8\right)\Omega_{12}^{(4,6)}\\
&\quad-104(M_1^2M_2^8)\Omega_{12}^{(4,7)}+4(M_1^2M_2^8)\Omega_{12}^{(4,8)}+\frac{5096}{9}\left(\frac{10}{49}M_1^5M_2^5+\frac{39}{49}M_1^3M_2^7\right)\Omega_{12}^{(5,5)}\\
&\quad-\frac{416}{3}(M_1^3M_2^7)\Omega_{12}^{(5,6)}+\frac{32}{3}(M_1^3M_2^7)\Omega_{12}^{(5,7)}+\frac{16}{3}(M_1^4M_2^6)\Omega_{12}^{(6,6)}
\end{aligned}
\tag{C.30}
$$

Appendix D

The bracket integrals to determine the diffusion coefficient, thermal diffusion, electrical and thermal conductivity of the monocomponent plasma

$$\left[S^{(p)}_{3/2}(W_1^2)\mathbf{W}_1, S^{(q)}_{3/2}(W_1^2)\mathbf{W}_1\right]_1 = 0; \quad p,q=0 \tag{D.1}$$

$$\left[S^{(1)}_{3/2}(W_1^2)\mathbf{W}_1, S^{(1)}_{3/2}(W_1^2)\mathbf{W}_1\right]_1 = 4\Omega_1^{(2,2)} \tag{D.2}$$

$$\left[S^{(1)}_{3/2}(W_1^2)\mathbf{W}_1, S^{(2)}_{3/2}(W_1^2)\mathbf{W}_1\right]_1 = 7\Omega_1^{(2,2)} - 2\Omega_1^{(2,3)} \tag{D.3}$$

$$\left[S^{(1)}_{3/2}(W_1^2)\mathbf{W}_1, S^{(3)}_{3/2}(W_1^2)\mathbf{W}_1\right]_1 = \frac{63}{8}\Omega_1^{(2,2)} - \frac{9}{2}\Omega_1^{(2,3)} + \frac{1}{2}\Omega_1^{(2,4)} \tag{D.4}$$

$$\left[S^{(1)}_{3/2}(W_1^2)\mathbf{W}_1, S^{(4)}_{3/2}(W_1^2)\mathbf{W}_1\right]_1 = \frac{231}{32}\Omega_1^{(2,2)} - \frac{99}{16}\Omega_1^{(2,3)} + \frac{11}{8}\Omega_1^{(2,4)} - \frac{1}{12}\Omega_1^{(2,5)} \tag{D.5}$$

$$\left[S^{(1)}_{3/2}(W_1^2)\mathbf{W}_1, S^{(5)}_{3/2}(W_1^2)\mathbf{W}_1\right]_1 = \frac{3003}{512}\Omega_1^{(2,2)} - \frac{429}{64}\Omega_1^{(2,3)} + \frac{143}{64}\Omega_1^{(2,4)} - \frac{13}{48}\Omega_1^{(2,5)} + \frac{1}{96}\Omega_1^{(2,6)} \tag{D.6}$$

$$\left[S^{(2)}_{3/2}(W_1^2)\mathbf{W}_1, S^{(2)}_{3/2}(W_1^2)\mathbf{W}_1\right]_1 = \frac{77}{4}\Omega_1^{(2,2)} - 7\Omega_1^{(2,3)} + \Omega_1^{(2,4)} \tag{D.7}$$

$$\left[S^{(2)}_{3/2}(W_1^2)\mathbf{W}_1, S^{(3)}_{3/2}(W_1^2)\mathbf{W}_1\right]_1 = \frac{945}{32}\Omega_1^{(2,2)} - \frac{261}{16}\Omega_1^{(2,3)} + \frac{25}{8}\Omega_1^{(2,4)} - \frac{1}{4}\Omega_1^{(2,5)} \tag{D.8}$$

$$\left[S^{(2)}_{3/2}(W_1^2)\mathbf{W}_1, S^{(4)}_{3/2}(W_1^2)\mathbf{W}_1\right]_1 = \frac{4389}{128}\Omega_1^{(2,2)} - \frac{429}{16}\Omega_1^{(2,3)} + \frac{55}{8}\Omega_1^{(2,4)} - \frac{5}{6}\Omega_1^{(2,5)} + \frac{1}{24}\Omega_1^{(2,6)} \tag{D.9}$$

S. Nguyen-Kuok, *Theory of Low-Temperature Plasma Physics*,
Springer Series on Atomic, Optical, and Plasma Physics 95,
DOI 10.1007/978-3-319-43721-7

$$\left[S_{3/2}^{(2)}(W_1^2)\mathbf{W}_1, S_{3/2}^{(5)}(W_1^2)\mathbf{W}_1\right]_1 = \frac{69{,}069}{2048}\Omega_1^{(2,2)} - \frac{35{,}607}{1024}\Omega_1^{(2,3)} + \frac{3003}{256}\Omega_1^{(2,4)} - \frac{715}{384}\Omega_1^{(2,5)} + \frac{59}{384}\Omega_1^{(2,6)} - \frac{1}{192}\Omega_1^{(2,7)} \tag{D.10}$$

$$\left[S_{3/2}^{(3)}(W_1^2)\mathbf{W}_1, S_{3/2}^{(3)}(W_1^2)\mathbf{W}_1\right]_1 = \frac{14{,}553}{256}\Omega_1^{(2,2)} - \frac{1215}{32}\Omega_1^{(2,3)} + \frac{313}{32}\Omega_1^{(2,4)} - \frac{9}{8}\Omega_1^{(2,5)} + \frac{1}{16}\Omega_1^{(2,6)} + \frac{1}{6}\Omega_1^{(4,4)} \tag{D.11}$$

$$\left[S_{3/2}^{(3)}(W_1^2)\mathbf{W}_1, S_{3/2}^{(4)}(W_1^2)\mathbf{W}_1\right]_1 = \frac{81{,}081}{1024}\Omega_1^{(2,2)} - \frac{34{,}155}{512}\Omega_1^{(2,3)} + \frac{2717}{128}\Omega_1^{(2,4)} - \frac{211}{64}\Omega_1^{(2,5)} + \frac{17}{64}\Omega_1^{(2,6)} - \frac{1}{96}\Omega_1^{(2,7)} + \frac{11}{24}\Omega_1^{(4,4)} - \frac{1}{12}\Omega_1^{(4,5)} \tag{D.12}$$

$$\left[S_{3/2}^{(3)}(W_1^2)\mathbf{W}_1, S_{3/2}^{(5)}(W_1^2)\mathbf{W}_1\right]_1 = \frac{1{,}486{,}485}{16{,}384}\Omega_1^{(2,2)} - \frac{389{,}961}{4096}\Omega_1^{(2,3)} + \frac{150{,}293}{4096}\Omega_1^{(2,4)} - \frac{3653}{512}\Omega_1^{(2,5)} + \frac{779}{1024}\Omega_1^{(2,6)} - \frac{35}{768}\Omega_1^{(2,7)} + \frac{1}{768}\Omega_1^{(2,8)} + \frac{143}{192}\Omega_1^{(4,4)} - \frac{13}{48}\Omega_1^{(4,5)} + \frac{1}{48}\Omega_1^{(4,6)} \tag{D.13}$$

$$\left[S_{3/2}^{(4)}(W_1^2)\mathbf{W}_1, S_{3/2}^{(4)}(W_1^2)\mathbf{W}_1\right]_1 = \frac{525{,}525}{4096}\Omega_1^{(2,2)} - \frac{127{,}413}{1024}\Omega_1^{(2,3)} + \frac{48{,}323}{1024}\Omega_1^{(2,4)} - \frac{1155}{128}\Omega_1^{(2,5)} + \frac{247}{256}\Omega_1^{(2,6)} - \frac{11}{192}\Omega_1^{(2,7)} + \frac{1}{576}\Omega_1^{(2,8)} + \frac{77}{48}\Omega_1^{(4,4)} - \frac{11}{24}\Omega_1^{(4,5)} + \frac{1}{24}\Omega_1^{(4,6)} \tag{D.14}$$

$$\left[S_{3/2}^{(4)}(W_1^2)\mathbf{W}_1, S_{3/2}^{(5)}(W_1^2)\mathbf{W}_1\right]_1 = \frac{10{,}975{,}965}{65{,}536}\Omega_1^{(2,2)} - \frac{6{,}239{,}805}{32{,}768}\Omega_1^{(2,3)} + \frac{1{,}378{,}663}{16{,}384}\Omega_1^{(2,4)} - \frac{157{,}651}{8192}\Omega_1^{(2,5)} + \frac{10{,}391}{4096}\Omega_1^{(2,6)} - \frac{1229}{6144}\Omega_1^{(2,7)} + \frac{85}{9216}\Omega_1^{(2,8)} - \frac{1}{4608}\Omega_1^{(2,9)} + \frac{2431}{768}\Omega_1^{(4,4)} - \frac{169}{128}\Omega_1^{(4,5)} + \frac{37}{192}\Omega_1^{(4,6)} - \frac{1}{96}\Omega_1^{(4,7)} \tag{D.15}$$

$$\left[S_{3/2}^{(5)}(W_1^2)\mathbf{W}_1, S_{3/2}^{(5)}(W_1^2)\mathbf{W}_1\right]_1 = \frac{257{,}041{,}785}{1{,}048{,}576}\Omega_1^{(2,2)} - \frac{20{,}383{,}935}{65{,}536}\Omega_1^{(2,3)} + \frac{10{,}212{,}345}{65{,}536}\Omega_1^{(2,4)} - \frac{673{,}803}{16{,}384}\Omega_1^{(2,5)} + \frac{210{,}763}{32{,}768}\Omega_1^{(2,6)} - \frac{2561}{4096}\Omega_1^{(2,7)} + \frac{1409}{36{,}864}\Omega_1^{(2,8)} - \frac{13}{9216}\Omega_1^{(2,9)} + \frac{1}{36{,}864}\Omega_1^{(2,10)} + \frac{15{,}015}{2048}\Omega_1^{(4,4)} - \frac{2873}{768}\Omega_1^{(4,5)} + \frac{571}{768}\Omega_1^{(4,6)} - \frac{13}{192}\Omega_1^{(4,7)} + \frac{1}{384}\Omega_1^{(4,8)} + \frac{1}{480}\Omega_1^{(6,6)} \tag{D.16}$$

Appendix E

The bracket integrals to determine the diffusion coefficient, thermal diffusion, electrical and thermal conductivity of the multicomponent plasma

$$\left[S^{(0)}_{3/2}(W_1^2)\mathbf{W}_1, S^{(0)}_{3/2}(W_2^2)\mathbf{W}_2\right]_{12}\left(M_1^{1/2}M_2^{1/2}\right)^{-1} = -8\Omega_{12}^{(1,1)} \tag{E.1}$$

$$\left[S^{(0)}_{3/2}(W_1^2)\mathbf{W}_1, S^{(1)}_{3/2}(W_2^2)\mathbf{W}_2\right]_{12}\left(M_1^{3/2}M_2^{1/2}\right)^{-1} = -20\Omega_{12}^{(1,1)} + 8\Omega_{12}^{(2,2)} \tag{E.2}$$

$$\left[S^{(0)}_{3/2}(W_1^2)\mathbf{W}_1, S^{(2)}_{3/2}(W_2^2)\mathbf{W}_2\right]_{12}\left(M_1^{5/2}M_2^{1/2}\right)^{-1} = -35\Omega_{12}^{(1,1)} + 28\Omega_{12}^{(1,2)} - 4\Omega_{12}^{(1,3)} \tag{E.3}$$

$$\begin{aligned}&\left[S^{(0)}_{3/2}(W_1^2)\mathbf{W}_1, S^{(3)}_{3/2}(W_2^2)\mathbf{W}_2\right]_{12}\left(M_1^{7/2}M_2^{1/2}\right)^{-1}\\ &= -\frac{105}{2}\Omega_{12}^{(1,1)} + 63\Omega_{12}^{(1,2)} - 18\Omega_{12}^{(1,3)} + \frac{4}{3}\Omega_{12}^{(1,4)}\end{aligned} \tag{E.4}$$

$$\begin{aligned}&\left[S^{(0)}_{3/2}(W_1^2)\mathbf{W}_1, S^{(4)}_{3/2}(W_2^2)\mathbf{W}_2\right]_{12}\left(M_1^{9/2}M_2^{1/2}\right)^{-1}\\ &= -\frac{1155}{16}\Omega_{12}^{(1,1)} + \frac{231}{2}\Omega_{12}^{(1,2)} - \frac{99}{2}\Omega_{12}^{(1,3)} + \frac{22}{3}\Omega_{12}^{(1,4)} - \frac{1}{3}\Omega_{12}^{(1,5)}\end{aligned} \tag{E.5}$$

$$\begin{aligned}&\left[S^{(0)}_{3/2}(W_1^2)\mathbf{W}_1, S^{(5)}_{3/2}(W_2^2)\mathbf{W}_2\right]_{12}\left(M_1^{11/2}M_2^{1/2}\right)^{-1}\\ &= -\frac{3003}{32}\Omega_{12}^{(1,1)} + \frac{3003}{16}\Omega_{12}^{(1,2)} - \frac{429}{4}\Omega_{12}^{(1,3)} + \frac{143}{6}\Omega_{12}^{(1,4)} - \frac{13}{6}\Omega_{12}^{(1,5)} + \frac{1}{15}\Omega_{12}^{(1,6)}\end{aligned} \tag{E.6}$$

$$\begin{aligned}&\left[S^{(1)}_{3/2}(W_1^2)\mathbf{W}_1, S^{(1)}_{3/2}(W_2^2)\mathbf{W}_2\right]_{12}\left(M_1^{3/2}M_2^{3/2}\right)^{-1}\\ &= -110\Omega_{12}^{(1,1)} + 40\Omega_{12}^{(1,2)} - 8\Omega_{12}^{(1,3)} + 16\Omega_{12}^{(2,2)}\end{aligned} \tag{E.7}$$

S. Nguyen-Kuok, *Theory of Low-Temperature Plasma Physics*,
Springer Series on Atomic, Optical, and Plasma Physics 95,
DOI 10.1007/978-3-319-43721-7

$$\left[S_{3/2}^{(1)}(W_1^2)\mathbf{W}_1, S_{3/2}^{(2)}(W_2^2)\mathbf{W}_2\right]_{12}\left(M_1^{5/2}M_2^{3/2}\right)^{-1}$$
$$= -\frac{595}{2}\Omega_{12}^{(1,1)} + 189\Omega_{12}^{(1,2)} - 38\Omega_{12}^{(1,3)} + 4\Omega_{12}^{(1,4)} + 56\Omega_{12}^{(2,2)} - 16\Omega_{12}^{(2,3)} \tag{E.8}$$

$$\left[S_{3/2}^{(1)}(W_1^2)\mathbf{W}_1, S_{3/2}^{(3)}(W_2^2)\mathbf{W}_2\right]_{12}\left(M_1^{7/2}M_2^{3/2}\right)^{-1}$$
$$= -\frac{2415}{4}\Omega_{12}^{(1,1)} + 588\Omega_{12}^{(1,2)} - 162\Omega_{12}^{(1,3)} + \frac{64}{3}\Omega_{12}^{(1,4)} - \frac{4}{3}\Omega_{12}^{(1,5)} + 126\Omega_{12}^{(2,2)} - 72\Omega_{12}^{(2,3)} + 8\Omega_{12}^{(2,4)} \tag{E.9}$$

$$\left[S_{3/2}^{(1)}(W_1^2)\mathbf{W}_1, S_{3/2}^{(4)}(W_2^2)\mathbf{W}_2\right]_{12}\left(M_1^{9/2}M_2^{3/2}\right)^{-1} = -\frac{33{,}495}{32}\Omega_{12}^{(1,1)} + \frac{22{,}407}{16}\Omega_{12}^{(1,2)}$$
$$-\frac{2145}{4}\Omega_{12}^{(1,3)} + \frac{539}{6}\Omega_{12}^{(1,4)} - \frac{49}{6}\Omega_{12}^{(1,5)} + \frac{1}{3}\Omega_{12}^{(1,6)} + 231\Omega_{12}^{(2,2)} - 198\Omega_{12}^{(2,3)} + 44\Omega_{12}^{(2,4)} - \frac{8}{3}\Omega_{12}^{(2,5)} \tag{E.10}$$

$$\left[S_{3/2}^{(1)}(W_1^2)\mathbf{W}_1, S_{3/2}^{(5)}(W_2^2)\mathbf{W}_2\right]_{12}\left(M_1^{11/2}M_2^{3/2}\right)^{-1}$$
$$= -\frac{105{,}105}{64}\Omega_{12}^{(1,1)} + \frac{45{,}045}{16}\Omega_{12}^{(1,2)} - \frac{22{,}737}{16}\Omega_{12}^{(1,3)} + \frac{1859}{6}\Omega_{12}^{(1,4)} - \frac{143}{4}\Omega_{12}^{(1,5)}$$
$$+\frac{7}{3}\Omega_{12}^{(1,6)} - \frac{1}{15}\Omega_{12}^{(1,7)} + \frac{3003}{8}\Omega_{12}^{(2,2)} - 429\Omega_{12}^{(2,3)} + 143\Omega_{12}^{(2,4)} - \frac{52}{3}\Omega_{12}^{(2,5)} + \frac{2}{3}\Omega_{12}^{(2,6)} \tag{E.11}$$

$$\left[S_{3/2}^{(2)}(W_1^2)\mathbf{W}_1, S_{3/2}^{(2)}(W_2^2)\mathbf{W}_2\right]_{12}\left(M_1^{5/2}M_2^{5/2}\right)^{-1} = -\frac{8505}{8}\Omega_{12}^{(1,1)} + 833\Omega_{12}^{(1,2)}$$
$$-241\Omega_{12}^{(1,3)} + 28\Omega_{12}^{(1,4)} - 2\Omega_{12}^{(1,5)} + 308\Omega_{12}^{(2,2)} - 112\Omega_{12}^{(2,3)} + 16\Omega_{12}^{(2,4)} - 16\Omega_{12}^{(3,3)} \tag{E.12}$$

$$\left[S_{3/2}^{(2)}(W_1^2)\mathbf{W}_1, S_{3/2}^{(3)}(W_2^2)\mathbf{W}_2\right]_{12}\left(M_1^{7/2}M_2^{5/2}\right)^{-1}$$
$$= -\frac{42{,}735}{16}\Omega_{12}^{(1,1)} + \frac{22{,}071}{8}\Omega_{12}^{(1,2)} - \frac{2001}{2}\Omega_{12}^{(1,3)} + \frac{499}{3}\Omega_{12}^{(1,4)} - \frac{41}{3}\Omega_{12}^{(1,5)}$$
$$+\frac{2}{3}\Omega_{12}^{(1,6)} + 945\Omega_{12}^{(2,2)} - 522\Omega_{12}^{(2,3)} + 100\Omega_{12}^{(2,4)} - 8\Omega_{12}^{(2,5)} - 72\Omega_{12}^{(3,3)} + 16\Omega_{12}^{(3,4)} \tag{E.13}$$

$$\left[S_{3/2}^{(2)}(W_1^2)\mathbf{W}_1, S_{3/2}^{(4)}(W_2^2)\mathbf{W}_2\right]_{12}\left(M_1^{9/2}M_2^{5/2}\right)^{-1} = -\frac{705{,}705}{128}\Omega_{12}^{(1,1)} + \frac{234{,}927}{32}\Omega_{12}^{(1,2)}$$
$$-\frac{104{,}973}{32}\Omega_{12}^{(1,3)} + \frac{8437}{12}\Omega_{12}^{(1,4)} - \frac{623}{8}\Omega_{12}^{(1,5)} + \frac{29}{6}\Omega_{12}^{(1,6)} - \frac{1}{6}\Omega_{12}^{(1,7)} + \frac{4389}{2}\Omega_{12}^{(2,2)}$$
$$-1716\Omega_{12}^{(2,3)} + 440\Omega_{12}^{(2,4)} - \frac{160}{3}\Omega_{12}^{(2,5)} + \frac{8}{3}\Omega_{12}^{(2,6)} - 198\Omega_{12}^{(3,3)} + 88\Omega_{12}^{(3,4)} - 8\Omega_{12}^{(3,5)} \tag{E.14}$$

$$\left[S^{(2)}_{3/2}(W_1^2)\mathbf{W}_1, S^{(5)}_{3/2}(W_2^2)\mathbf{W}_2\right]_{12}\left(M_1^{11/2}M_2^{5/2}\right)^{-1} = -\frac{2,567,565}{256}\Omega_{12}^{(1,1)} + \frac{2,129,127}{128}\Omega_{12}^{(1,2)}$$
$$-\frac{579,579}{64}\Omega_{12}^{(1,3)} + \frac{75,933}{32}\Omega_{12}^{(1,4)} - \frac{16,237}{48}\Omega_{12}^{(1,5)} + \frac{655}{24}\Omega_{12}^{(1,6)} - \frac{79}{60}\Omega_{12}^{(1,7)} + \frac{1}{30}\Omega_{12}^{(1,8)}$$
$$+\frac{69,069}{16}\Omega_{12}^{(2,2)} - \frac{35,607}{8}\Omega_{12}^{(2,3)} + \frac{3003}{2}\Omega_{12}^{(2,4)} - \frac{715}{3}\Omega_{12}^{(2,5)} + \frac{59}{3}\Omega_{12}^{(2,6)} - \frac{2}{3}\Omega_{12}^{(2,7)}$$
$$-429\Omega_{12}^{(3,3)} + 286\Omega_{12}^{(3,4)} - 52\Omega_{12}^{(3,5)} + \frac{8}{3}\Omega_{12}^{(3,6)} \tag{E.15}$$

$$\left[S^{(3)}_{3/2}(W_1^2)\mathbf{W}_1, S^{(3)}_{3/2}(W_2^2)\mathbf{W}_2\right]_{12}\left(M_1^{7/2}M_2^{7/2}\right)^{-1} = -\frac{255,255}{32}\Omega_{12}^{(1,1)} + \frac{76,923}{8}\Omega_{12}^{(1,2)}$$
$$-\frac{34,119}{8}\Omega_{12}^{(1,3)} + 895\Omega_{12}^{(1,4)} - \frac{201}{2}\Omega_{12}^{(1,5)} + 6\Omega_{12}^{(1,6)} - \frac{2}{9}\Omega_{12}^{(1,7)} + \frac{14,553}{4}\Omega_{12}^{(2,2)}$$
$$-2430\Omega_{12}^{(2,3)} + 626\Omega_{12}^{(2,4)} - 72\Omega_{12}^{(2,5)} + 4\Omega_{12}^{(2,6)} - 444\Omega_{12}^{(3,3)} + 144\Omega_{12}^{(3,4)} - 16\Omega_{12}^{(3,5)} + \frac{32}{3}\Omega_{12}^{(4,4)} \tag{E.16}$$

$$\left[S^{(3)}_{3/2}(W_1^2)\mathbf{W}_1, S^{(4)}_{3/2}(W_2^2)\mathbf{W}_2\right]_{12}\left(M_1^{9/2}M_2^{7/2}\right)^{-1} = -\frac{4,879,875}{256}\Omega_{12}^{(1,1)} + \frac{3,516,513}{128}\Omega_{12}^{(1,2)}$$
$$-\frac{919,413}{64}\Omega_{12}^{(1,3)} + \frac{118,107}{32}\Omega_{12}^{(1,4)} - \frac{8353}{16}\Omega_{12}^{(1,5)} + \frac{339}{8}\Omega_{12}^{(1,6)} - \frac{71}{36}\Omega_{12}^{(1,7)} + \frac{1}{18}\Omega_{12}^{(1,8)}$$
$$+\frac{81,081}{8}\Omega_{12}^{(2,2)} - \frac{34,155}{4}\Omega_{12}^{(2,3)} + 2717\Omega_{12}^{(2,4)} - 422\Omega_{12}^{(2,5)} + 34\Omega_{12}^{(2,6)} - \frac{4}{3}\Omega_{12}^{(2,7)}$$
$$-1551\Omega_{12}^{(3,3)} + \frac{2222}{3}\Omega_{12}^{(3,4)} - 124\Omega_{12}^{(3,5)} + 8\Omega_{12}^{(3,6)} + \frac{176}{3}\Omega_{12}^{(4,4)} - \frac{32}{3}\Omega_{12}^{(4,5)} \tag{E.17}$$

$$\left[S^{(3)}_{3/2}(W_1^2)\mathbf{W}_1, S^{(5)}_{3/2}(W_2^2)\mathbf{W}_2\right]_{12}\left(M_1^{11/2}M_2^{7/2}\right)^{-1} = -\frac{20,165,145}{512}\Omega_{12}^{(1,1)} + \frac{4,327,323}{64}\Omega_{12}^{(1,2)}$$
$$-\frac{329,043}{8}\Omega_{12}^{(1,3)} + \frac{200,057}{16}\Omega_{12}^{(1,4)} - \frac{34,333}{16}\Omega_{12}^{(1,5)} + \frac{881}{4}\Omega_{12}^{(1,6)} - \frac{613}{45}\Omega_{12}^{(1,7)} + \frac{23}{45}\Omega_{12}^{(1,8)}$$
$$-\frac{1}{90}\Omega_{12}^{(1,9)} + \frac{1,486,485}{64}\Omega_{12}^{(2,2)} - \frac{389,961}{16}\Omega_{12}^{(2,3)} + \frac{150,293}{16}\Omega_{12}^{(2,4)} - \frac{3653}{2}\Omega_{12}^{(2,5)} + \frac{779}{4}\Omega_{12}^{(2,6)}$$
$$-\frac{35}{3}\Omega_{12}^{(2,7)} + \frac{1}{3}\Omega_{12}^{(2,8)} - \frac{8151}{2}\Omega_{12}^{(3,3)} + \frac{8008}{3}\Omega_{12}^{(3,4)} - \frac{1820}{3}\Omega_{12}^{(3,5)} + 64\Omega_{12}^{(3,6)} - \frac{8}{3}\Omega_{12}^{(3,7)}$$
$$+\frac{572}{3}\Omega_{12}^{(4,4)} - \frac{208}{3}\Omega_{12}^{(4,5)} + \frac{16}{3}\Omega_{12}^{(4,6)} \tag{E.18}$$

$$\left[S_{3/2}^{(4)}(W_1^2)\mathbf{W}_1, S_{3/2}^{(4)}(W_2^2)\mathbf{W}_2\right]_{12}\left(M_1^{9/2}M_2^{9/2}\right)^{-1} = -\frac{105{,}930{,}825}{2048}\Omega_{12}^{(1,1)} + \frac{10{,}735{,}725}{128}\Omega_{12}^{(1,2)}$$
$$-\frac{6{,}435{,}429}{128}\Omega_{12}^{(1,3)} + \frac{483{,}637}{32}\Omega_{12}^{(1,4)} - \frac{165{,}827}{64}\Omega_{12}^{(1,5)} + \frac{2123}{8}\Omega_{12}^{(1,6)} - \frac{1189}{72}\Omega_{12}^{(1,7)} + \frac{11}{18}\Omega_{12}^{(1,8)} -$$
$$-\frac{1}{72}\Omega_{12}^{(1,9)} + \frac{525{,}525}{16}\Omega_{12}^{(2,2)} - \frac{127{,}413}{4}\Omega_{12}^{(2,3)} + \frac{48{,}323}{4}\Omega_{12}^{(2,4)} - 2310\Omega_{12}^{(2,5)} + 247\Omega_{12}^{(2,6)} -$$
$$-\frac{44}{3}\Omega_{12}^{(2,7)} + \frac{4}{9}\Omega_{12}^{(2,8)} - \frac{26{,}169}{4}\Omega_{12}^{(3,3)} + \frac{11{,}374}{3}\Omega_{12}^{(3,4)} - \frac{2566}{3}\Omega_{12}^{(3,5)} + 88\Omega_{12}^{(3,6)} - 4\Omega_{12}^{(3,7)} +$$
$$+\frac{1232}{3}\Omega_{12}^{(4,4)} - \frac{352}{3}\Omega_{12}^{(4,5)} + \frac{32}{3}\Omega_{12}^{(4,6)} - \frac{16}{3}\Omega_{12}^{(5,5)} \tag{E.19}$$

$$\left[S_{3/2}^{(4)}(W_1^2)\mathbf{W}_1, S_{3/2}^{(5)}(W_2^2)\mathbf{W}_2\right]_{12}\left(M_1^{11/2}M_2^{9/2}\right)^{-1} = -\frac{489{,}834{,}345}{4096}\Omega_{12}^{(1,1)} + \frac{452{,}873{,}421}{2048}\Omega_{12}^{(1,2)}$$
$$-\frac{38{,}424{,}243}{256}\Omega_{12}^{(1,3)} + \frac{6{,}633{,}055}{128}\Omega_{12}^{(1,4)} - \frac{1{,}330{,}927}{128}\Omega_{12}^{(1,5)} + \frac{82{,}491}{64}\Omega_{12}^{(1,6)} - \frac{72{,}643}{720}\Omega_{12}^{(1,7)}$$
$$+\frac{199}{40}\Omega_{12}^{(1,8)} - \frac{109}{720}\Omega_{12}^{(1,9)} + \frac{1}{360}\Omega_{12}^{(1,10)} + \frac{10{,}975{,}965}{128}\Omega_{12}^{(2,2)} - \frac{6{,}239{,}805}{64}\Omega_{12}^{(2,3)} + \frac{1{,}378{,}663}{32}\Omega_{12}^{(2,4)}$$
$$-\frac{157{,}651}{16}\Omega_{12}^{(2,5)} + \frac{10{,}391}{8}\Omega_{12}^{(2,6)} - \frac{1229}{12}\Omega_{12}^{(2,7)} + \frac{85}{18}\Omega_{12}^{(2,8)} - \frac{1}{9}\Omega_{12}^{(2,9)} - \frac{160{,}875}{8}\Omega_{12}^{(3,3)}$$
$$+\frac{173{,}173}{12}\Omega_{12}^{(3,4)} - \frac{12{,}103}{3}\Omega_{12}^{(3,5)} + 562\Omega_{12}^{(3,6)} - \frac{122}{3}\Omega_{12}^{(3,7)} + \frac{4}{3}\Omega_{12}^{(3,8)} + \frac{4862}{3}\Omega_{12}^{(4,4)}$$
$$-676\Omega_{12}^{(4,5)} + \frac{296}{3}\Omega_{12}^{(4,6)} - \frac{16}{3}\Omega_{12}^{(4,7)} - \frac{104}{3}\Omega_{12}^{(5,5)} + \frac{16}{3}\Omega_{12}^{(5,6)} \tag{E.20}$$

$$\left[S_{3/2}^{(5)}(W_1^2)\mathbf{W}_1, S_{3/2}^{(5)}(W_2^2)\mathbf{W}_2\right]_{12}\left(M_1^{11/2}M_2^{11/2}\right)^{-1} = -\frac{2{,}505{,}429{,}927}{8192}\Omega_{12}^{(1,1)} + \frac{1{,}273{,}569{,}297}{2048}\Omega_{12}^{(1,2)}$$
$$-\frac{958{,}832{,}589}{2048}\Omega_{12}^{(1,3)} + \frac{23{,}205{,}897}{128}\Omega_{12}^{(1,4)} - \frac{10{,}626{,}655}{256}\Omega_{12}^{(1,5)} + \frac{1{,}923{,}623}{320}\Omega_{12}^{(1,6)} - \frac{181{,}901}{320}\Omega_{12}^{(1,7)}$$
$$+\frac{12{,}701}{360}\Omega_{12}^{(1,8)} - \frac{2047}{1440}\Omega_{12}^{(1,9)} + \frac{13}{360}\Omega_{12}^{(1,10)} - \frac{1}{1800}\Omega_{12}^{(1,11)} + \frac{257{,}041{,}785}{1024}\Omega_{12}^{(2,2)}$$
$$-\frac{20{,}383{,}935}{64}\Omega_{12}^{(2,3)} + \frac{10{,}212{,}345}{64}\Omega_{12}^{(2,4)} - \frac{673{,}803}{16}\Omega_{12}^{(2,5)} + \frac{210{,}763}{32}\Omega_{12}^{(2,6)} - \frac{2561}{4}\Omega_{12}^{(2,7)}$$
$$+\frac{1409}{36}\Omega_{12}^{(2,8)} - \frac{13}{9}\Omega_{12}^{(2,9)} + \frac{1}{36}\Omega_{12}^{(2,10)} - \frac{1{,}130{,}415}{16}\Omega_{12}^{(3,3)} + \frac{232{,}375}{4}\Omega_{12}^{(3,4)} - \frac{230{,}789}{12}\Omega_{12}^{(3,5)}$$
$$+3302\Omega_{12}^{(3,6)} - 321\Omega_{12}^{(3,7)} + \frac{52}{3}\Omega_{12}^{(3,8)} - \frac{4}{9}\Omega_{12}^{(3,9)} + \frac{15{,}015}{2}\Omega_{12}^{(4,4)} - \frac{11{,}492}{3}\Omega_{12}^{(4,5)} + \frac{2284}{3}\Omega_{12}^{(4,6)}$$
$$-\frac{208}{3}\Omega_{12}^{(4,7)} + \frac{8}{3}\Omega_{12}^{(4,8)} - \frac{4108}{15}\Omega_{12}^{(5,5)} + \frac{208}{3}\Omega_{12}^{(5,6)} - \frac{16}{3}\Omega_{12}^{(5,7)} + \frac{32}{15}\Omega_{12}^{(6,6)} \tag{E.21}$$

$$\left[S_{3/2}^{(0)}(W_1^2)\mathbf{W}_1, S_{3/2}^{(0)}(W_1^2)\mathbf{W}_1\right]_{12} = 8M_2\Omega_{12}^{(1,1)} \tag{E.22}$$

$$\left[S_{3/2}^{(0)}(W_1^2)\mathbf{W}_1, S_{3/2}^{(1)}(W_1^2)\mathbf{W}_1\right]_{12} = 20M_2^2\Omega_{12}^{(1,1)} - 8M_2^2\Omega_{12}^{(1,2)} \tag{E.23}$$

$$\left[S_{3/2}^{(0)}(W_1^2)\mathbf{W}_1, S_{3/2}^{(2)}(W_1^2)\mathbf{W}_1\right]_{12} = 35M_2^3\Omega_{12}^{(1,1)} - 28M_2^3\Omega_{12}^{(1,2)} + 4M_2^3\Omega_{12}^{(1,3)} \tag{E.24}$$

$$\left[S^{(0)}_{3/2}(W_1^2)\mathbf{W}_1, S^{(3)}_{3/2}(W_1^2)\mathbf{W}_1\right]_{12} = \frac{105}{2} M_2^4\Omega_{12}^{(1,1)} - 63M_2^4\Omega_{12}^{(1,2)} + 18M_2^4\Omega_{12}^{(1,3)} - \frac{4}{3} M_2^4\Omega_{12}^{(1,4)} \tag{E.25}$$

$$\left[S^{(0)}_{3/2}(W_1^2)\mathbf{W}_1, S^{(4)}_{3/2}(W_1^2)\mathbf{W}_1\right]_{12} = \frac{1155}{16} M_2^5\Omega_{12}^{(1,1)} - \frac{231}{2} M_2^5\Omega_{12}^{(1,2)} + \frac{99}{2} M_2^5\Omega_{12}^{(1,3)} - \frac{22}{3} M_2^5\Omega_{12}^{(1,4)} + \frac{1}{3} M_2^5\Omega_{12}^{(1,5)} \tag{E.26}$$

$$\left[S^{(0)}_{3/2}(W_1^2)\mathbf{W}_1, S^{(5)}_{3/2}(W_1^2)\mathbf{W}_1\right]_{12} = \frac{3003}{32} M_2^6\Omega_{12}^{(1,1)} - \frac{3003}{16} M_2^6\Omega_{12}^{(1,2)} + \frac{429}{4} M_2^6\Omega_{12}^{(1,3)} - \frac{143}{6} M_2^6\Omega_{12}^{(1,4)} + \frac{13}{6} M_2^6\Omega_{12}^{(1,5)} - \frac{1}{15} M_2^6\Omega_{12}^{(1,6)} \tag{E.27}$$

$$\left[S^{(1)}_{3/2}(W_1^2)\mathbf{W}_1, S^{(1)}_{3/2}(W_1^2)\mathbf{W}_1\right]_{12} = 110\left(\frac{6}{11} M_1^2 M_2 + \frac{5}{11} M_2^3\right)\Omega_{12}^{(1,1)} - 40M_2^3\Omega_{12}^{(1,2)} + 8M_2^3\Omega_{12}^{(1,3)} + 16M_1M_2^2\Omega_{12}^{(2,2)} \tag{E.28}$$

$$\left[S^{(1)}_{3/2}(W_1^2)\mathbf{W}_1, S^{(2)}_{3/2}(W_1^2)\mathbf{W}_1\right]_{12} = \frac{595}{2}\left(\frac{12}{17} M_1^2 M_2^2 + \frac{5}{17} M_2^4\right)\Omega_{12}^{(1,1)} - 189\left(\frac{4}{9} M_1^2 M_2^2 + \frac{5}{9} M_2^4\right)\Omega_{12}^{(1,2)} + 38M_2^4\Omega_{12}^{(1,3)} - 4M_2^4\Omega_{12}^{(1,4)} + 56M_1M_2^3\Omega_{12}^{(2,2)} - 16M_1M_2^3\Omega_{12}^{(2,3)} \tag{E.29}$$

$$\left[S^{(1)}_{3/2}(W_1^2)\mathbf{W}_1, S^{(3)}_{3/2}(W_1^2)\mathbf{W}_1\right]_{12} = \frac{2415}{4}\left(\frac{18}{23} M_1^2 M_2^3 + \frac{5}{23} M_2^5\right)\Omega_{12}^{(1,1)} - 588\left(\frac{9}{14} M_1^2 M_2^3 + \frac{5}{14} M_2^5\right)\Omega_{12}^{(1,2)} + 162\left(\frac{1}{3} M_1^2 M_2^3 + \frac{2}{3} M_2^5\right)\Omega_{12}^{(1,3)} - \frac{64}{3} M_2^5\Omega_{12}^{(1,4)} + \frac{4}{3} M_2^5\Omega_{12}^{(1,5)} + 126M_1M_2^4\Omega_{12}^{(2,2)} - 72M_1M_2^4\Omega_{12}^{(2,3)} + 8M_1M_2^4\Omega_{12}^{(2,4)} \tag{E.30}$$

$$\left[S^{(1)}_{3/2}(W_1^2)\mathbf{W}_1, S^{(4)}_{3/2}(W_1^2)\mathbf{W}_1\right]_{12} = \frac{33{,}495}{32}\left(\frac{24}{29} M_1^2 M_2^4 + \frac{5}{29} M_2^6\right)\Omega_{12}^{(1,1)} - \frac{22{,}407}{16}\left(\frac{72}{97} M_1^2 M_2^4 + \frac{25}{97} M_2^6\right)\Omega_{12}^{(1,2)} + \frac{2145}{4}\left(\frac{36}{65} M_1^2 M_2^4 + \frac{29}{65} M_2^6\right)\Omega_{12}^{(1,3)} - \frac{539}{6}\left(\frac{12}{49} M_1^2 M_2^4 + \frac{37}{49} M_2^6\right)\Omega_{12}^{(1,4)} + \frac{49}{6} M_2^6\Omega_{12}^{(1,5)} - \frac{1}{3} M_2^6\Omega_{12}^{(1,6)} + 231M_1M_2^5\Omega_{12}^{(2,2)} - 198M_1M_2^5\Omega_{12}^{(2,3)} + 44M_1M_2^5\Omega_{12}^{(2,4)} - \frac{8}{3} M_1M_2^5\Omega_{12}^{(2,5)} \tag{E.31}$$

$$\left[S^{(1)}_{3/2}(W_1^2)\mathbf{W}_1, S^{(5)}_{3/2}(W_1^2)\mathbf{W}_1\right]_{12} = \frac{105{,}105}{64}\left(\frac{6}{7}M_1^2M_2^5 + \frac{1}{7}M_2^7\right)\Omega_{12}^{(1,1)}$$
$$- \frac{45{,}045}{16}\left(\frac{4}{5}M_1^2M_2^5 + \frac{1}{5}M_2^7\right)\Omega_{12}^{(1,2)} + \frac{22{,}737}{16}\left(\frac{36}{53}M_1^2M_2^5 + \frac{17}{53}M_2^7\right)\Omega_{12}^{(1,3)}$$
$$- \frac{1859}{6}\left(\frac{6}{13}M_1^2M_2^5 + \frac{7}{13}M_2^7\right)\Omega_{12}^{(1,4)} + \frac{143}{4}\left(\frac{2}{11}M_1^2M_2^5 + \frac{9}{11}M_2^7\right)\Omega_{12}^{(1,5)}$$
$$- \frac{7}{3}M_2^7\Omega_{12}^{(1,6)} + \frac{1}{15}M_2^7\Omega_{12}^{(1,7)} + \frac{3003}{8}M_1M_2^6\Omega_{12}^{(2,2)} - 429M_1M_2^6\Omega_{12}^{(2,3)}$$
$$+ 143M_1M_2^6\Omega_{12}^{(2,4)} - \frac{52}{3}M_1M_2^6\Omega_{12}^{(2,5)} + \frac{2}{3}M_1M_2^6\Omega_{12}^{(2,6)} \tag{E.32}$$

$$\left[S^{(2)}_{3/2}(W_1^2)\mathbf{W}_1, S^{(2)}_{3/2}(W_1^2)\mathbf{W}_1\right]_{12} = \frac{8505}{8}\left(\frac{40}{243}M_1^4M_2 + \frac{168}{243}M_1^2M_2^3 + \frac{35}{243}M_2^5\right)\Omega_{12}^{(1,1)}$$
$$- 833\left(\frac{12}{17}M_1^2M_2^3 + \frac{5}{17}M_2^5\right)\Omega_{12}^{(1,2)} + 241\left(\frac{108}{241}M_1^2M_2^3 + \frac{133}{241}M_2^5\right)\Omega_{12}^{(1,3)}$$
$$- 28M_2^5\Omega_{12}^{(1,4)} + 2M_2^5\Omega_{12}^{(1,5)} + 308\left(\frac{4}{11}M_1^3M_2^2 + \frac{7}{11}M_1M_2^4\right)\Omega_{12}^{(2,2)} - 112M_1M_2^4\Omega_{12}^{(2,3)}$$
$$+ 16M_1M_2^4\Omega_{12}^{(2,4)} + 16M_1^2M_2^3\Omega_{12}^{(3,3)} \tag{E.33}$$

$$\left[S^{(2)}_{3/2}(W_1^2)\mathbf{W}_1, S^{(3)}_{3/2}(W_1^2)\mathbf{W}_1\right]_{12} = \frac{42{,}735}{16}\left(\frac{120}{407}M_1^4M_2^2 + \frac{252}{407}M_1^2M_2^4 + \frac{35}{407}M_2^6\right)\Omega_{12}^{(1,1)}$$
$$- \frac{22{,}071}{8}\left(\frac{120}{1051}M_1^4M_2^2 + \frac{756}{1051}M_1^2M_2^4 + \frac{175}{1051}M_2^6\right)\Omega_{12}^{(1,2)}$$
$$+ \frac{2001}{2}\left(\frac{450}{667}M_1^2M_2^4 + \frac{217}{667}M_2^6\right)\Omega_{12}^{(1,3)} - \frac{499}{3}\left(\frac{198}{499}M_1^2M_2^4 + \frac{301}{499}M_2^6\right)\Omega_{12}^{(1,4)}$$
$$+ \frac{41}{3}M_2^6\Omega_{12}^{(1,5)} - \frac{2}{3}M_2^6\Omega_{12}^{(1,6)} + 945\left(\frac{8}{15}M_1^3M_2^3 + \frac{7}{15}M_1M_2^5\right)\Omega_{12}^{(2,2)}$$
$$- 522\left(\frac{8}{29}M_1^3M_2^3 + \frac{21}{29}M_1M_2^5\right)\Omega_{12}^{(2,3)} + 100M_1M_2^5\Omega_{12}^{(2,4)} - 8M_1M_2^5\Omega_{12}^{(2,5)}$$
$$+ 72M_1^2M_2^4\Omega_{12}^{(3,3)} - 16M_1^2M_2^4\Omega_{12}^{(3,4)} \tag{E.34}$$

$$\left[S_{3/2}^{(2)}(W_1^2)\mathbf{W}_1, S_{3/2}^{(4)}(W_1^2)\mathbf{W}_1\right]_{12} = \frac{705{,}705}{128}\left(\frac{240}{611}M_1^4M_2^3 + \frac{336}{611}M_1^2M_2^5 + \frac{35}{611}M_2^7\right)\Omega_{12}^{(1,1)}$$
$$- \frac{234{,}927}{32}\left(\frac{80}{339}M_1^4M_2^3 + \frac{224}{339}M_1^2M_2^5 + \frac{35}{339}M_2^7\right)\Omega_{12}^{(1,2)}$$
$$+ \frac{104{,}973}{32}\left(\frac{240}{3181}M_1^4M_2^3 + \frac{2304}{3181}M_1^2M_2^5 + \frac{637}{3181}M_2^7\right)\Omega_{12}^{(1,3)}$$
$$- \frac{8437}{12}\left(\frac{480}{767}M_1^2M_2^5 + \frac{287}{767}M_2^7\right)\Omega_{12}^{(1,4)} + \frac{623}{8}\left(\frac{208}{623}M_1^2M_2^5 + \frac{415}{623}M_2^7\right)\Omega_{12}^{(1,5)}$$
$$- \frac{29}{6}M_2^7\Omega_{12}^{(1,6)} + \frac{1}{6}M_2^7\Omega_{12}^{(1,7)} + \frac{4389}{2}\left(\frac{12}{19}M_1^3M_2^4 + \frac{7}{19}M_1M_2^6\right)\Omega_{12}^{(2,2)}$$
$$- 1716\left(\frac{6}{13}M_1^3M_2^4 + \frac{7}{13}M_1M_2^6\right)\Omega_{12}^{(2,3)} + 440\left(\frac{1}{5}M_1^3M_2^4 + \frac{4}{5}M_1M_2^6\right)\Omega_{12}^{(2,4)}$$
$$- \frac{160}{3}M_1M_2^6\Omega_{12}^{(2,5)} + \frac{8}{3}M_1M_2^6\Omega_{12}^{(2,6)} + 198M_1^2M_2^5\Omega_{12}^{(3,3)} - 88M_1^2M_2^5\Omega_{12}^{(3,4)} + 8M_1^2M_2^5\Omega_{12}^{(3,5)} \tag{E.35}$$

$$\left[S_{3/2}^{(2)}(W_1^2)\mathbf{W}_1, S_{3/2}^{(5)}(W_1^2)\mathbf{W}_1\right]_{12} = \frac{2{,}567{,}565}{256}\left(\frac{80}{171}M_1^4M_2^4 + \frac{84}{171}M_1^2M_2^6 + \frac{7}{171}M_2^8\right)\Omega_{12}^{(1,1)}$$
$$- \frac{2{,}129{,}127}{128}\left(\frac{240}{709}M_1^4M_2^4 + \frac{420}{709}M_1^2M_2^6 + \frac{49}{709}M_2^8\right)\Omega_{12}^{(1,2)}$$
$$+ \frac{579{,}579}{64}\left(\frac{240}{1351}M_1^4M_2^4 + \frac{936}{1351}M_1^2M_2^6 + \frac{175}{1351}M_2^8\right)\Omega_{12}^{(1,3)}$$
$$- \frac{75{,}933}{32}\left(\frac{80}{1593}M_1^4M_2^4 + \frac{1128}{1593}M_1^2M_2^6 + \frac{385}{1593}M_2^8\right)\Omega_{12}^{(1,4)}$$
$$+ \frac{16{,}237}{48}\left(\frac{708}{1249}M_1^2M_2^6 + \frac{541}{1249}M_2^8\right)\Omega_{12}^{(1,5)} - \frac{655}{24}\left(\frac{36}{131}M_1^2M_2^6 + \frac{95}{131}M_2^8\right)\Omega_{12}^{(1,6)}$$
$$+ \frac{79}{60}M_2^8\Omega_{12}^{(1,7)} - \frac{1}{30}M_2^8\Omega_{12}^{(1,8)} + \frac{69{,}069}{16}\left(\frac{16}{23}M_1^3M_2^5 + \frac{7}{23}M_1M_2^7\right)\Omega_{12}^{(2,2)}$$
$$- \frac{35{,}607}{8}\left(\frac{48}{83}M_1^3M_2^5 + \frac{35}{83}M_1M_2^7\right)\Omega_{12}^{(2,3)} + \frac{3003}{2}\left(\frac{8}{21}M_1^3M_2^5 + \frac{13}{21}M_1M_2^7\right)\Omega_{12}^{(2,4)}$$
$$- \frac{715}{3}\left(\frac{8}{55}M_1^3M_2^5 + \frac{47}{55}M_1M_2^7\right)\Omega_{12}^{(2,5)} + \frac{59}{3}M_1M_2^7\Omega_{12}^{(2,6)} - \frac{2}{3}M_1M_2^7\Omega_{12}^{(2,7)}$$
$$+ 429M_1^2M_2^6\Omega_{12}^{(3,3)} - 286M_1^2M_2^6\Omega_{12}^{(3,4)} + 52M_1^2M_2^6\Omega_{12}^{(3,5)} - \frac{8}{3}M_1^2M_2^6\Omega_{12}^{(3,6)} \tag{E.36}$$

$$
\begin{aligned}
&\left[S^{(3)}_{3/2}(W_1^2)\mathbf{W}_1, S^{(3)}_{3/2}(W_1^2)\mathbf{W}_1\right]_{12}\\
&= \frac{255{,}255}{32}\left(\frac{112}{2431}M_1^6M_2 + \frac{1080}{2431}M_1^4M_2^3 + \frac{1134}{2431}M_1^2M_2^5 + \frac{105}{2431}M_2^7\right)\Omega_{12}^{(1,1)}\\
&- \frac{76{,}923}{8}\left(\frac{120}{407}M_1^4M_2^3 + \frac{252}{407}M_1^2M_2^5 + \frac{35}{407}M_2^7\right)\Omega_{12}^{(1,2)}\\
&+ \frac{34{,}119}{8}\left(\frac{440}{3791}M_1^4M_2^3 + \frac{2700}{3791}M_1^2M_2^5 + \frac{651}{3791}M_2^7\right)\Omega_{12}^{(1,3)}\\
&- 895\left(\frac{594}{895}M_1^2M_2^5 + \frac{301}{895}M_2^7\right)\Omega_{12}^{(1,4)} + \frac{201}{2}\left(\frac{26}{67}M_1^2M_2^5 + \frac{41}{67}M_2^7\right)\Omega_{12}^{(1,5)}\\
&- 6M_2^7\Omega_{12}^{(1,6)} + \frac{2}{9}M_2^7\Omega_{12}^{(1,7)} + \frac{14{,}553}{4}\left(\frac{8}{77}M_1^5M_2^2 + \frac{48}{77}M_1^3M_2^4 + \frac{21}{77}M_1M_2^6\right)\Omega_{12}^{(2,2)}\\
&- 2430\left(\frac{8}{15}M_1^3M_2^4 + \frac{7}{15}M_1M_2^6\right)\Omega_{12}^{(2,3)} + 626\left(\frac{176}{626}M_1^3M_2^4 + \frac{450}{626}M_1M_2^6\right)\Omega_{12}^{(2,4)}\\
&- 72M_1M_2^6\Omega_{12}^{(2,5)} + 4M_1M_2^6\Omega_{12}^{(2,6)} + 444\left(\frac{120}{444}M_1^4M_2^3 + \frac{324}{444}M_1^2M_2^5\right)\Omega_{12}^{(3,3)}\\
&- 144M_1^2M_2^5\Omega_{12}^{(3,4)} + 16M_1^2M_2^5\Omega_{12}^{(3,5)} + \frac{32}{3}M_1^3M_2^4\Omega_{12}^{(4,4)}
\end{aligned}
\tag{E.37}
$$

$$
\begin{aligned}
&\left[S^{(3)}_{3/2}(W_1^2)\mathbf{W}_1, S^{(4)}_{3/2}(W_1^2)\mathbf{W}_1\right]_{12}\\
&= \frac{4{,}879{,}875}{256}\left(\frac{448}{4225}M_1^6M_2^2 + \frac{2160}{4225}M_1^4M_2^4 + \frac{1512}{4225}M_1^2M_2^6 + \frac{105}{4225}M_2^8\right)\Omega_{12}^{(1,1)}\\
&- \frac{3{,}516{,}513}{128}\left(\frac{448}{15{,}223}M_1^6M_2^2 + \frac{6480}{15{,}223}M_1^4M_2^4 + \frac{7560}{15{,}223}M_1^2M_2^6 + \frac{735}{15{,}223}M_2^8\right)\Omega_{12}^{(1,2)}\\
&+ \frac{919{,}413}{64}\left(\frac{2480}{9287}M_1^4M_2^4 + \frac{5904}{9287}M_1^2M_2^6 + \frac{903}{9287}M_2^8\right)\Omega_{12}^{(1,3)}\\
&- \frac{118{,}107}{32}\left(\frac{1040}{10{,}737}M_1^4M_2^4 + \frac{7632}{10{,}737}M_1^2M_2^6 + \frac{2065}{10{,}737}M_2^8\right)\Omega_{12}^{(1,4)}\\
&+ \frac{8353}{16}\left(\frac{5304}{8353}M_1^2M_2^6 + \frac{3049}{8353}M_2^8\right)\Omega_{12}^{(1,5)} - \frac{339}{8}\left(\frac{40}{113}M_1^2M_2^6 + \frac{73}{113}M_2^8\right)\Omega_{12}^{(1,6)}\\
&+ \frac{71}{36}M_2^8\Omega_{12}^{(1,7)} - \frac{1}{18}M_2^8\Omega_{12}^{(1,8)} + \frac{81{,}081}{8}\left(\frac{8}{39}M_1^5M_2^3 + \frac{24}{39}M_1^3M_2^5 + \frac{7}{39}M_1M_2^7\right)\Omega_{12}^{(2,2)}\\
&- \frac{34{,}155}{4}\left(\frac{8}{115}M_1^5M_2^3 + \frac{72}{115}M_1^3M_2^5 + \frac{35}{115}M_1M_2^7\right)\Omega_{12}^{(2,3)}\\
&+ 2717\left(\frac{1364}{2717}M_1^3M_2^5 + \frac{1353}{2717}M_1M_2^7\right)\Omega_{12}^{(2,4)} - 422\left(\frac{52}{211}M_1^3M_2^5 + \frac{159}{211}M_1M_2^7\right)\Omega_{12}^{(2,5)}\\
&+ 34M_1M_2^7\Omega_{12}^{(2,6)} - \frac{4}{3}M_1M_2^7\Omega_{12}^{(2,7)} + 1551\left(\frac{660}{1551}M_1^4M_2^4 + \frac{891}{1551}M_1^2M_2^6\right)\Omega_{12}^{(3,3)}\\
&- \frac{2222}{3}\left(\frac{20}{101}M_1^4M_2^4 + \frac{81}{101}M_1^2M_2^6\right)\Omega_{12}^{(3,4)} + 124M_1^2M_2^6\Omega_{12}^{(3,5)} - 8M_1^2M_2^6\Omega_{12}^{(3,6)}\\
&+ \frac{176}{3}M_1^3M_2^5\Omega_{12}^{(4,4)} - \frac{32}{3}M_1^3M_2^5\Omega_{12}^{(4,5)}
\end{aligned}
\tag{E.38}
$$

$$
\begin{aligned}
&\left[S^{(3)}_{3/2}(W_1^2)\mathbf{W}_1, S^{(5)}_{3/2}(W_1^2)\mathbf{W}_1\right]_{12}\\
&= \frac{20{,}165{,}145}{512}\left(\frac{224}{1343}M_1^6M_2^3+\frac{720}{1343}M_1^4M_2^5+\frac{378}{1343}M_1^2M_2^7+\frac{21}{1343}M_2^9\right)\Omega_{12}^{(1,1)}\\
&-\frac{4{,}327{,}323}{64}\left(\frac{112}{1441}M_1^6M_2^3+\frac{720}{1441}M_1^4M_2^5+\frac{567}{1441}M_1^2M_2^7+\frac{42}{1441}M_2^9\right)\Omega_{12}^{(1,2)}\\
&+\frac{329{,}043}{8}\left(\frac{112}{6136}M_1^6M_2^3+\frac{2400}{6136}M_1^4M_2^5+\frac{3267}{6136}M_1^2M_2^7+\frac{357}{6136}M_2^9\right)\Omega_{12}^{(1,3)}\\
&-\frac{200{,}057}{16}\left(\frac{320}{1399}M_1^4M_2^5+\frac{918}{1399}M_1^2M_2^7+\frac{161}{1399}M_2^9\right)\Omega_{12}^{(1,4)}\\
&+\frac{34{,}333}{16}\left(\frac{200}{2641}M_1^4M_2^5+\frac{1857}{2641}M_1^2M_2^7+\frac{584}{2641}M_2^9\right)\Omega_{12}^{(1,5)}\\
&-\frac{881}{4}\left(\frac{525}{881}M_1^2M_2^7+\frac{356}{881}M_2^9\right)\Omega_{12}^{(1,6)}+\frac{613}{45}\left(\frac{765}{2452}M_1^2M_2^7+\frac{1687}{2452}M_2^9\right)\Omega_{12}^{(1,7)}\\
&-\frac{23}{45}M_2^9\Omega_{12}^{(1,8)}+\frac{1}{90}M_2^9\Omega_{12}^{(1,9)}+\frac{1{,}486{,}485}{64}\left(\frac{16}{55}M_1^5M_2^4+\frac{32}{55}M_1^3M_2^6+\frac{7}{55}M_1M_2^8\right)\Omega_{12}^{(2,2)}\\
&-\frac{389{,}961}{16}\left(\frac{16}{101}M_1^5M_2^4+\frac{64}{101}M_1^3M_2^6+\frac{21}{101}M_1M_2^8\right)\Omega_{12}^{(2,3)}\\
&+\frac{150{,}293}{16}\left(\frac{48}{1051}M_1^5M_2^4+\frac{640}{1051}M_1^3M_2^6+\frac{363}{1051}M_1M_2^8\right)\Omega_{12}^{(2,4)}\\
&-\frac{3653}{2}\left(\frac{128}{281}M_1^3M_2^6+\frac{153}{281}M_1M_2^8\right)\Omega_{12}^{(2,5)}+\frac{779}{4}\left(\frac{160}{779}M_1^3M_2^6+\frac{619}{779}M_1M_2^8\right)\Omega_{12}^{(2,6)}\\
&-\frac{35}{3}M_1M_2^8\Omega_{12}^{(2,7)}+\frac{1}{3}M_1M_2^8\Omega_{12}^{(2,8)}+\frac{8151}{2}\left(\frac{10}{19}M_1^4M_2^5+\frac{9}{19}M_1^2M_2^7\right)\Omega_{12}^{(3,3)}\\
&-\frac{8008}{3}\left(\frac{5}{14}M_1^4M_2^5+\frac{9}{14}M_1^2M_2^7\right)\Omega_{12}^{(3,4)}+\frac{1820}{3}\left(\frac{1}{7}M_1^4M_2^5+\frac{6}{7}M_1^2M_2^7\right)\Omega_{12}^{(3,5)}\\
&-64M_1^2M_2^7\Omega_{12}^{(3,6)}+\frac{8}{3}M_1^2M_2^7\Omega_{12}^{(3,7)}+\frac{572}{3}M_1^3M_2^6\Omega_{12}^{(4,4)}-\frac{208}{3}M_1^3M_2^6\Omega_{12}^{(4,5)}+\frac{16}{3}M_1^3M_2^6\Omega_{12}^{(4,6)}
\end{aligned}
\tag{E.39}
$$

$$
\begin{aligned}
&\left[S^{(4)}_{3/2}(W_1^2)\mathbf{W}_1, S^{(4)}_{3/2}(W_1^2)\mathbf{W}_1\right]_{12}=\frac{105{,}930{,}825}{2048}\\
&\times\left(\frac{1152}{91{,}715}M_1^8M_2+\frac{19{,}712}{91{,}715}M_1^6M_2^3+\frac{47{,}520}{91{,}715}M_1^4M_2^5+\frac{22{,}176}{91{,}715}M_1^2M_2^7+\frac{1155}{91{,}715}M_2^9\right)\Omega_{12}^{(1,1)}\\
&-\frac{10{,}735{,}725}{128}\left(\frac{448}{4225}M_1^6M_2^3+\frac{2160}{4225}M_1^4M_2^5+\frac{1512}{4225}M_1^2M_2^7+\frac{105}{4225}M_2^9\right)\Omega_{12}^{(1,2)}\\
&+\frac{6{,}435{,}429}{128}\left(\frac{5824}{195{,}013}M_1^6M_2^3+\frac{81{,}840}{195{,}013}M_1^4M_2^5+\frac{97{,}416}{195{,}013}M_1^2M_2^7+\frac{9933}{195{,}013}M_2^9\right)\Omega_{12}^{(1,3)}\\
&-\frac{483{,}637}{32}\left(\frac{1040}{3997}M_1^4M_2^5+\frac{2544}{3997}M_1^2M_2^7+\frac{413}{3997}M_2^9\right)\Omega_{12}^{(1,4)}+
\end{aligned}
\tag{E.40}
$$

$$
\begin{aligned}
&+\frac{165{,}827}{64}\left(\frac{15{,}600}{165{,}827}M_1^4M_2^5+\frac{116{,}688}{165{,}827}M_1^2M_2^7+\frac{33{,}539}{165{,}827}M_2^9\right)\Omega_{12}^{(1,5)}\\
&-\frac{2123}{8}\left(\frac{120}{193}M_1^2M_2^7+\frac{73}{193}M_2^9\right)\Omega_{12}^{(1,6)}+\frac{1189}{72}\left(\frac{408}{1189}M_1^2M_2^7+\frac{781}{1189}M_2^9\right)\Omega_{12}^{(1,7)}\\
&-\frac{11}{18}M_2^9\Omega_{12}^{(1,8)}+\frac{1}{72}M_2^9\Omega_{12}^{(1,9)}\\
&+\frac{525{,}525}{16}\left(\frac{64}{2275}M_1^7M_2^2+\frac{792}{2275}M_1^5M_2^4+\frac{1188}{2275}M_1^3M_2^6+\frac{231}{2275}M_1M_2^8\right)\Omega_{12}^{(2,2)}\\
&-\frac{127{,}413}{4}\left(\frac{8}{39}M_1^5M_2^4+\frac{24}{39}M_1^3M_2^6+\frac{7}{39}M_1M_2^8\right)\Omega_{12}^{(2,3)}\\
&+\frac{48{,}323}{4}\left(\frac{312}{4393}M_1^5M_2^4+\frac{2728}{4393}M_1^3M_2^6+\frac{1353}{4393}M_1M_2^8\right)\Omega_{12}^{(2,4)}\\
&-2310\left(\frac{52}{105}M_1^3M_2^6+\frac{53}{105}M_1M_2^8\right)\Omega_{12}^{(2,5)}+247\left(\frac{60}{247}M_1^3M_2^6+\frac{187}{247}M_1M_2^8\right)\Omega_{12}^{(2,6)}\\
&-\frac{44}{3}M_1M_2^8\Omega_{12}^{(2,7)}+\frac{4}{9}M_1M_2^8\Omega_{12}^{(2,8)}+\frac{26{,}169}{4}\left(\frac{56}{793}M_1^6M_2^3+\frac{440}{793}M_1^4M_2^5+\frac{297}{793}M_1^2M_2^7\right)\Omega_{12}^{(3,3)}\\
&-\frac{11{,}374}{3}\left(\frac{20}{47}M_1^4M_2^5+\frac{27}{47}M_1^2M_2^7\right)\Omega_{12}^{(3,4)}+\frac{2566}{3}\left(\frac{260}{1283}M_1^4M_2^5+\frac{1023}{1283}M_1^2M_2^7\right)\Omega_{12}^{(3,5)}\\
&-88M_1^2M_2^7\Omega_{12}^{(3,6)}+4M_1^2M_2^7\Omega_{12}^{(3,7)}+\frac{1232}{3}\left(\frac{3}{14}M_1^5M_2^4+\frac{11}{14}M_1^3M_2^6\right)\Omega_{12}^{(4,4)}\\
&-\frac{352}{3}M_1^3M_2^6\Omega_{12}^{(4,5)}+\frac{32}{3}M_1^3M_2^6\Omega_{12}^{(4,6)}+\frac{16}{3}M_1^4M_2^5\Omega_{12}^{(5,5)}
\end{aligned}
\tag{E.41}
$$

$$
\begin{aligned}
&\left[S^{(4)}_{3/2}(W_1^2)\mathbf{W}_1, S^{(5)}_{3/2}(W_1^2)\mathbf{W}_1\right]_{12} = \frac{489{,}834{,}345}{4096}\\
&\times\left(\frac{1152}{32{,}623}M_1^8M_2^2 + \frac{9856}{32{,}623}M_1^6M_2^4 + \frac{15{,}840}{32{,}623}M_1^4M_2^6 + \frac{5544}{32{,}623}M_1^2M_2^8 + \frac{231}{32{,}623}M_2^{10}\right)\Omega_{12}^{(1,1)}\\
&-\frac{452{,}873{,}421}{2048}\left(\frac{384}{50{,}269}M_1^8M_2^2 + \frac{9856}{50{,}269}M_1^6M_2^4 + \frac{26{,}400}{50{,}269}M_1^4M_2^6 + \frac{12{,}936}{50{,}269}M_1^2M_2^8 + \frac{693}{50{,}269}M_2^{10}\right)\\
&\times\Omega_{12}^{(1,2)} + \frac{38{,}424{,}243}{256}\left(\frac{8288}{89{,}567}M_1^6M_2^4 + \frac{44{,}880}{89{,}567}M_1^4M_2^6 + \frac{33{,}858}{89{,}567}M_1^2M_2^8 + \frac{2541}{89{,}567}M_2^{10}\right)\Omega_{12}^{(1,3)}\\
&-\frac{6{,}633{,}055}{128}\left(\frac{224}{9277}M_1^6M_2^4 + \frac{3696}{9277}M_1^4M_2^6 + \frac{4818}{9277}M_1^2M_2^8 + \frac{539}{9277}M_2^{10}\right)\Omega_{12}^{(1,4)}\\
&+\frac{1{,}330{,}927}{128}\left(\frac{24{,}400}{102{,}379}M_1^4M_2^6 + \frac{66{,}132}{102{,}379}M_1^2M_2^8 + \frac{11{,}847}{102{,}379}M_2^{10}\right)\Omega_{12}^{(1,5)}\\
&-\frac{82{,}491}{64}\left(\frac{6800}{82{,}491}M_1^4M_2^6 + \frac{57{,}420}{82{,}491}M_1^2M_2^8 + \frac{18{,}271}{82{,}491}M_2^{10}\right)\Omega_{12}^{(1,6)}\\
&+\frac{72{,}643}{720}\left(\frac{43{,}350}{72{,}643}M_1^2M_2^8 + \frac{29{,}293}{72{,}643}M_2^{10}\right)\Omega_{12}^{(1,7)} - \frac{199}{40}\left(\frac{190}{597}M_1^2M_2^8 + \frac{407}{597}M_2^{10}\right)\Omega_{12}^{(1,8)}\\
&+\frac{109}{720}M_2^{10}\Omega_{12}^{(1,9)} - \frac{1}{360}M_2^{10}\Omega_{12}^{(1,10)}\\
&+\frac{10{,}975{,}965}{128}\left(\frac{256}{3655}M_1^7M_2^3 + \frac{1584}{3655}M_1^5M_2^5 + \frac{1584}{3655}M_1^3M_2^7 + \frac{231}{3655}M_1M_2^9\right)\Omega_{12}^{(2,2)}\\
&-\frac{6{,}239{,}805}{64}\left(\frac{256}{14{,}545}M_1^7M_2^3 + \frac{4752}{14{,}545}M_1^5M_2^5 + \frac{7920}{14{,}545}M_1^3M_2^7 + \frac{1617}{14{,}545}M_1M_2^9\right)\Omega_{12}^{(2,3)}\\
&+\frac{1{,}378{,}663}{32}\left(\frac{1776}{9641}M_1^5M_2^5 + \frac{5984}{9641}M_1^3M_2^7 + \frac{1881}{9641}M_1M_2^9\right)\Omega_{12}^{(2,4)}\\
&-\frac{157{,}651}{16}\left(\frac{720}{12{,}127}M_1^5M_2^5 + \frac{7392}{12{,}127}M_1^3M_2^7 + \frac{4015}{12{,}127}M_1M_2^9\right)\Omega_{12}^{(2,5)}\\
&+\frac{10{,}391}{8}\left(\frac{4880}{10{,}391}M_1^3M_2^7 + \frac{5511}{10{,}391}M_1M_2^9\right)\Omega_{12}^{(2,6)} - \frac{1229}{12}\left(\frac{272}{1229}M_1^3M_2^7 + \frac{957}{1229}M_1M_2^9\right)\Omega_{12}^{(2,7)}\\
&+\frac{85}{18}M_1M_2^9\Omega_{12}^{(2,8)} - \frac{1}{9}M_1M_2^9\Omega_{12}^{(2,9)} + \frac{160{,}875}{8}\left(\frac{56}{375}M_1^6M_2^4 + \frac{220}{375}M_1^4M_2^6 + \frac{99}{375}M_1^2M_2^8\right)\Omega_{12}^{(3,3)}\\
&-\frac{173{,}173}{12}\left(\frac{56}{1211}M_1^6M_2^4 + \frac{660}{1211}M_1^4M_2^6 + \frac{495}{1211}M_1^2M_2^8\right)\Omega_{12}^{(3,4)}\\
&+\frac{12{,}103}{3}\left(\frac{370}{931}M_1^4M_2^6 + \frac{561}{931}M_1^2M_2^8\right)\Omega_{12}^{(3,5)} - 562\left(\frac{50}{281}M_1^4M_2^6 + \frac{231}{281}M_1^2M_2^8\right)\Omega_{12}^{(3,6)}\\
&+\frac{122}{3}M_1^2M_2^8\Omega_{12}^{(3,7)} - \frac{4}{3}M_1^2M_2^8\Omega_{12}^{(3,8)} + \frac{4862}{3}\left(\frac{6}{17}M_1^5M_2^5 + \frac{11}{17}M_1^3M_2^7\right)\Omega_{12}^{(4,4)}\\
&-676\left(\frac{2}{13}M_1^5M_2^5 + \frac{11}{13}M_1^3M_2^7\right)\Omega_{12}^{(4,5)} + \frac{296}{3}M_1^3M_2^7\Omega_{12}^{(4,6)} - \frac{16}{3}M_1^3M_2^7\Omega_{12}^{(4,7)}\\
&+\frac{104}{3}M_1^4M_2^6\Omega_{12}^{(5,5)} - \frac{16}{3}M_1^4M_2^6\Omega_{12}^{(5,6)}
\end{aligned}
\tag{E.42}
$$

$$
\begin{aligned}
&\left[S^{(5)}_{3/2}(W_1^2)\mathbf{W}_1, S^{(5)}_{3/2}(W_1^2)\mathbf{W}_1\right]_{12} \\
&= \frac{2{,}505{,}429{,}927}{8192}\left(\frac{2816}{834{,}309}M_1^{10}M_2 + \frac{74{,}880}{834{,}309}M_1^8M_2^3 + \frac{320{,}320}{834{,}309}M_1^6M_2^5 + \frac{343{,}200}{834{,}309}M_1^4M_2^7\right. \\
&\left. + \frac{90{,}090}{834{,}309}M_1^2M_2^9 + \frac{3003}{834{,}309}M_2^{11}\right)\Omega_{12}^{(1,1)} - \frac{1{,}273{,}569{,}297}{2048}\left(\frac{1152}{32{,}623}M_1^8M_2^3 + \frac{9856}{32{,}623}M_1^6M_2^5\right. \\
&\left. + \frac{15{,}840}{32{,}623}M_1^4M_2^7 + \frac{5544}{32{,}623}M_1^2M_2^9 + \frac{231}{32{,}623}M_2^{11}\right)\Omega_{12}^{(1,2)} + \frac{958{,}832{,}589}{2048}\left(\frac{17{,}280}{2{,}235{,}041}M_1^8M_2^3\right. \\
&\left. + \frac{430{,}976}{2{,}235{,}041}M_1^6M_2^5 + \frac{1{,}166{,}880}{2{,}235{,}041}M_1^4M_2^7 + \frac{586{,}872}{2{,}235{,}041}M_1^2M_2^9 + \frac{33{,}033}{2{,}235{,}041}M_2^{11}\right)\Omega_{12}^{(1,3)} \\
&- \frac{23{,}205{,}897}{128}\left(\frac{1120}{12{,}483}M_1^6M_2^5 + \frac{6160}{12{,}483}M_1^4M_2^7 + \frac{4818}{12{,}483}M_1^2M_2^9 + \frac{385}{12{,}483}M_2^{11}\right)\Omega_{12}^{(1,4)} \\
&+ \frac{10{,}626{,}655}{256}\left(\frac{19{,}040}{817{,}435}M_1^6M_2^5 + \frac{317{,}200}{817{,}435}M_1^4M_2^7 + \frac{429{,}858}{817{,}435}M_1^2M_2^9 + \frac{51{,}337}{817{,}435}M_2^{11}\right)\Omega_{12}^{(1,5)} \\
&- \frac{1{,}923{,}623}{320}\left(\frac{34{,}000}{147{,}971}M_1^4M_2^7 + \frac{95{,}700}{147{,}971}M_1^2M_2^9 + \frac{18{,}271}{147{,}971}M_2^{11}\right)\Omega_{12}^{(1,6)} \\
&+ \frac{181{,}901}{320}\left(\frac{129{,}200}{1{,}637{,}109}M_1^4M_2^7 + \frac{1{,}127{,}100}{1{,}637{,}109}M_1^2M_2^9 + \frac{380{,}809}{1{,}637{,}109}M_2^{11}\right)\Omega_{12}^{(1,7)} \\
&- \frac{12{,}701}{360}\left(\frac{570}{977}M_1^2M_2^9 + \frac{407}{977}M_2^{11}\right)\Omega_{12}^{(1,8)} + \frac{2047}{1440}\left(\frac{630}{2047}M_1^2M_2^9 + \frac{1417}{2047}M_2^{11}\right)\Omega_{12}^{(1,9)} \\
&- \frac{13}{360}M_2^{11}\Omega_{12}^{(1,10)} + \frac{1}{1800}M_2^{11}\Omega_{12}^{(1,11)} + \frac{257{,}041{,}785}{1024} \\
&\times\left(\frac{640}{85{,}595}M_1^9M_2^2 + \frac{13{,}312}{85{,}595}M_1^7M_2^4 + \frac{41{,}184}{85{,}595}M_1^5M_2^6 + \frac{27{,}456}{85{,}595}M_1^3M_2^8 + \frac{3003}{85{,}595}M_1M_2^{10}\right)\Omega_{12}^{(2,2)} \\
&- \frac{20{,}383{,}935}{64}\left(\frac{256}{3655}M_1^7M_2^4 + \frac{1584}{3655}M_1^5M_2^6 + \frac{1584}{3655}M_1^3M_2^8 + \frac{231}{3655}M_1M_2^{10}\right)\Omega_{12}^{(2,3)} \\
&+ \frac{10{,}212{,}345}{64}\left(\frac{1280}{71{,}415}M_1^7M_2^4 + \frac{23{,}088}{71{,}415}M_1^5M_2^6 + \frac{38{,}896}{71{,}415}M_1^3M_2^8 + \frac{8151}{71{,}415}M_1M_2^{10}\right)\Omega_{12}^{(2,4)} \\
&- \frac{673{,}803}{16}\left(\frac{720}{3987}M_1^5M_2^6 + \frac{2464}{3987}M_1^3M_2^8 + \frac{803}{3987}M_1M_2^{10}\right)\Omega_{12}^{(2,5)} \\
&+ \frac{210{,}763}{32}\left(\frac{12{,}240}{210{,}763}M_1^5M_2^6 + \frac{126{,}880}{210{,}763}M_1^3M_2^8 + \frac{71{,}643}{210{,}763}M_1M_2^{10}\right)\Omega_{12}^{(2,6)} \\
&- \frac{2561}{4}\left(\frac{272}{591}M_1^3M_2^8 + \frac{319}{591}M_1M_2^{10}\right)\Omega_{12}^{(2,7)} \\
&+ \frac{1409}{36}\left(\frac{304}{1409}M_1^3M_2^8 + \frac{1105}{1409}M_1M_2^{10}\right)\Omega_{12}^{(2,8)} - \frac{13}{9}M_1M_2^{10}\Omega_{12}^{(2,9)} + \frac{1}{36}M_1M_2^{10}\Omega_{12}^{(2,10)} \\
&+ \frac{1{,}130{,}415}{16}\left(\frac{48}{2635}M_1^8M_2^3 + \frac{728}{2635}M_1^6M_2^5 + \frac{1430}{2635}M_1^4M_2^7 + \frac{429}{2635}M_1^2M_2^9\right)\Omega_{12}^{(3,3)} \\
&- \frac{232{,}375}{4}\left(\frac{56}{375}M_1^6M_2^5 + \frac{220}{375}M_1^4M_2^7 + \frac{99}{375}M_1^2M_2^9\right)\Omega_{12}^{(3,4)}
\end{aligned}
$$

$$
\begin{aligned}
&+\frac{230{,}789}{12}\left(\frac{840}{17{,}753}M_1^6M_2^5+\frac{9620}{17{,}753}M_1^4M_2^7+\frac{7293}{17{,}753}M_1^2M_2^9\right)\Omega_{12}^{(3,5)}\\
&-3302\left(\frac{50}{127}M_1^4M_2^7+\frac{77}{127}M_1^2M_2^9\right)\Omega_{12}^{(3,6)}+321\left(\frac{170}{963}M_1^4M_2^7+\frac{793}{963}M_1^2M_2^9\right)\Omega_{12}^{(3,7)}\\
&-\frac{52}{3}M_1^2M_2^9\Omega_{12}^{(3,8)}+\frac{4}{9}M_1^2M_2^9\Omega_{12}^{(3,9)}+\frac{15{,}015}{2}\left(\frac{16}{315}M_1^7M_2^4+\frac{156}{315}M_1^5M_2^6+\frac{143}{315}M_1^3M_2^8\right)\Omega_{12}^{(4,4)}\\
&-\frac{11{,}492}{3}\left(\frac{6}{17}M_1^5M_2^6+\frac{11}{17}M_1^3M_2^8\right)\Omega_{12}^{(4,5)}+\frac{2284}{3}\left(\frac{90}{571}M_1^5M_2^6+\frac{481}{571}M_1^3M_2^8\right)\Omega_{12}^{(4,6)}\\
&-\frac{208}{3}M_1^3M_2^8\Omega_{12}^{(4,7)}+\frac{8}{3}M_1^3M_2^8\Omega_{12}^{(4,8)}+\frac{4108}{15}\left(\frac{14}{79}M_1^6M_2^5+\frac{65}{79}M_1^4M_2^7\right)\Omega_{12}^{(5,5)}\\
&-\frac{208}{3}M_1^4M_2^7\Omega_{12}^{(5,6)}+\frac{16}{3}M_1^4M_2^7\Omega_{12}^{(5,7)}+\frac{32}{15}M_1^5M_2^6\Omega_{12}^{(6,6)}
\end{aligned}
\tag{E.43}
$$

Index

S. Nguyen-Kuok, *Theory of Low-Temperature Plasma Physics*,
Springer Series on Atomic, Optical, and Plasma Physics 95,
DOI 10.1007/978-3-319-43721-7

MIX
Papier aus verantwortungsvollen Quellen
Paper from responsible sources
FSC® C105338

If you have any concerns about our products,
you can contact us on
ProductSafety@springernature.com

In case Publisher is established outside the EU,
the EU authorized representative is:
Springer Nature Customer Service Center GmbH
Europaplatz 3, 69115 Heidelberg, Germany

Printed by Libri Plureos GmbH
in Hamburg, Germany